Algebra 1

REVISED EDITION

Mary P. Dolciani
William Wooton
Edwin F. Beckenbach

EDITORIAL ADVISER
Andrew M. Gleason

HOUGHTON MIFFLIN COMPANY/BOSTON

ATLANTA   DALLAS   GENEVA, ILL.
LAWRENCEVILLE, N.J.   PALO ALTO   TORONTO

THE AUTHORS

Mary P. Dolciani, Professor of Mathematics, Hunter College of the City University of New York. Dr. Dolciani has been a director and teacher in numerous National Science Foundation and New York State Education Department institutes for mathematics teachers and Visiting Secondary School Lecturer for the Mathematical Association of America.

William Wooton, former Professor of Mathematics, Los Angeles Pierce College. Mr. Wooton has taught mathematics at both the junior and senior high school levels. He has also been a team member of the National Council of Teachers of Mathematics (NCTM) summer writing projects.

Edwin F. Beckenbach, Professor of Mathematics, Emeritus, University of California, Los Angeles. Dr. Beckenbach has been a team member and coordinator of the NCTM summer writing projects for elementary mathematics teachers. He has served as Chairman of the Committee on Publications of the Mathematical Association of America.

EDITORIAL ADVISER

Andrew M. Gleason, Hollis Professor of Mathematics and Natural Philosophy, Harvard University. Professor Gleason is a well-known research mathematician. His many affiliations include membership in the National Academy of Sciences.

ISBN: 0-395-32115-8

Contents

Chapter 6 *Functions, Relations, and Graphs* 187

Chapter 7 *Systems of Open Sentences* 237

Chapter 8 *Polynomials and Their Factors* 279

Chapter 9   *Polynomials and Rational Expressions*   *323*

Chapter 10   *Rational Expressions in Open Sentences*   *357*

Chapter 14 *Statistics and Probability* 473

Graphical Representation of Data

Arithmetical Description of Data

Probability

Symbols

Two separate athletic events may be held simultaneously in this prototype design for a double stadium.

1

Numbers and Variables

Naming Numbers

OBJECTIVES *for Sections 1-1 through 1-4:*
1. *Simplify numerical expressions which involve arithmetic operations, powers, or grouping symbols.*
2. *Evaluate variable expressions.*
3. *Graph numbers, and state the coordinates of points, on the number line.*
4. *Express measurements by means of directed numbers.*

1-1 Numbers and Variables in Expressions

One kind of symbol used in algebra is a *numeral,* or *numerical expression*. A **numerical expression** is simply a name for a number. The number is called the **value** of the expression. For example,

> 4×9 is a numerical expression for the number 36.
> 36 is the value of 4×9.

You can use the equality symbol, $=$, and write

$$4 \times 9 = 36.$$

Whenever you replace a numerical expression with the simplest, or most common, name of its value, you say that you have **simplified the expression**.

In simplifying a numerical expression, you use the following fact:

Substitution Principle

Changing the numeral by which a number is named in an expression does not change the value of the expression.

Below are several expressions and their values:

Numerical Expression	Value	You write:
$15 + 12$	27	$15 + 12 = 27$
$5 - 1$	4	$5 - 1 = 4$
7×8 (or $7 \cdot 8$)	56	$7 \times 8 = 56$
$42 \div 6$	7	$42 \div 6 = 7$

Another kind of symbol used in algebra is a *variable*, or a *variable expression*. A **variable** is a letter such as n, a, or x, or a symbol such as _?_, used to stand for a number or numbers. A **variable expression** is an expression containing at least one variable, but it may also contain other symbols, including numerals. The set of numbers that a variable can stand for is called the **replacement set**, or **domain**, of the variable. The members of the domain are the **values** of the variable. For example, if n can stand for 2 or 5, then you write $n = 2$ or $n = 5$. $3 \times n$ is a variable expression.

$$\text{If } n = 2, \text{ then } 3 \times n = 3 \times 2 = 6.$$
$$\text{If } n = 5, \text{ then } 3 \times n = 3 \times 5 = 15.$$

This is called **evaluating the expression.** In a variable expression like $3 \times n$, the multiplication sign is often omitted.

$$3 \times n \text{ may be written } 3n.$$

EXAMPLE Evaluate each expression for the given value of each variable.

a. $x - 3$; $x = 7$

b. $5y$; $y = 8$

c. $9 - \dfrac{12}{n}$; $n = 3$

d. $a + b - 2$; $a = 4$ and $b = 11$

SOLUTION

a. $x - 3 = 7 - 3 = 4$

b. $5y = 5 \times y = 5 \times 8 = 40$

c. $9 - \dfrac{12}{n} = 9 - \dfrac{12}{3} = 9 - 4 = 5$

d. $a + b - 2 = 4 + 11 - 2 = 13$

Oral Exercises

Give the simplest name of the value of each expression.

1. $12 + 31$
2. $57 - 31$
3. $120 \div 4$
4. 8×24

Evaluate each expression when $x = 4$.

5. $x - 2$
6. $7x$
7. $8 - \dfrac{12}{x}$
8. $10 + x$

Evaluate each expression when $a = 3$ and $b = 5$.

9. $a + b$
10. $b - a$
11. $\dfrac{a + b}{2}$
12. ab

13. If $y = 5$, write four variable expressions using y that have a value of 10.

14. If $m = 1$, how many variable expressions can be written using m that have a value of 1?

Written Exercises

Give the simplest name of the value of each expression.

A
1. $352 + 226$
2. $537 + 138$
3. $1642 - 853$
4. $1980 - 1974$
5. $1185 \div 5$
6. $9690 \div 19$
7. 215×13
8. 57×348
9. $3 + 12 + 413$
10. $8753 - 985$
11. $9 \times 12 \times 21$
12. $5656 \div 56$

Evaluate the expression.

13. $x + 5$ when $x = 7$
14. $7 - x$ when $x = 0$
15. $10 + x$ when $x = 3$
16. $3x$ when $x = 9$
17. $5x$ when $x = 1$
18. $\dfrac{15}{x}$ when $x = 3$
19. $\dfrac{x}{8}$ when $x = 0$
20. $x - 9$ when $x = 11$
21. $c + d$ when $c = 5$ and $d = 8$
22. $d - c$ when $c = 4$ and $d = 7$
23. $c - d$ when $c = 1$ and $d = 0$
24. cd when $c = 9$ and $d = 8$
25. $\dfrac{c}{d}$ when $c = 14$ and $d = 2$
26. $\dfrac{d}{c}$ when $c = 75$ and $d = 25$
27. $c + d - 4$ when $c = 9$ and $d = 3$
28. $c - d + 5$ when $c = 18$ and $d = 9$

B
29. $c + d$ when $c = \frac{1}{3}$ and $d = \frac{3}{4}$
30. $c + d - 1$ when $c = \frac{3}{4}$ and $d = \frac{5}{8}$
31. $c - d + 1$ when $c = 1.25$ and $d = 0.33$
32. $c - d + 1$ when $c = 0.65$ and $d = 0.27$
33. $cd + 1$ when $c = \frac{7}{8}$ and $d = \frac{4}{3}$
34. $cd - 1$ when $c = 2\frac{1}{4}$ and $d = 4$

Evaluate the expression.

35. $\dfrac{c}{d}$ when $c = \frac{1}{2}$ and $d = \frac{3}{4}$

36. $\dfrac{c}{d} - 1$ when $c = 1.25$ and $d = 0.5$

37. $\dfrac{d}{c} + 1$ when $c = 1.25$ and $d = 0.5$

38. $e + f + g$ when $e = \frac{1}{2}$, $f = \frac{1}{3}$, and $g = \frac{1}{4}$

39. efg when $e = \frac{2}{3}$, $f = \frac{3}{4}$, and $g = 16$

40. gfe when $e = \frac{2}{3}$, $f = \frac{3}{4}$ and $g = 16$

41. Show that the expressions $x + y$ and $y + x$ have the same value when $x = \frac{1}{3}$ and $y = \frac{1}{5}$.

42. Show that the expressions $\dfrac{x + y}{2}$ and $\dfrac{x}{2} + \dfrac{y}{2}$ have the same value when $x = \frac{1}{2}$ and $y = \frac{1}{4}$.

43. Find values for x and y that make the expressions $x - y$ and $y - x$ have the same value.

44. Find values for x and y that make the expressions $x + y$ and xy have the same value.

1-2 Grouping Symbols

When more than one operation must be done to find the value of an expression, **parentheses** ( ) are often used to make clear the order in which the operations are to be carried out.

$$2 \cdot 5 + 8 \text{ could mean}$$

Either	Or
$(2 \cdot 5) + 8 =$	$2 \cdot (5 + 8) =$
$10 + 8 =$	$2 \cdot 13 =$
$18 \longleftarrow$ different $\longrightarrow 26$	
values	

A pair of parentheses is a **grouping symbol**. Brackets and braces may also be used as grouping symbols:

Parentheses	Brackets	Braces
$2 \cdot (5 + 8)$	$2 \cdot [5 + 8]$	$2 \cdot \{5 + 8\}$

The product $2 \cdot (5 + 8)$ is usually written without the multiplication dot simply as $2(5 + 8)$. A product like $2 \cdot 13$ may be written

$$2(13) \quad \text{or} \quad (2)13 \quad \text{or} \quad (2)(13)$$

Variables are grouped in the same way as numbers. But:

$$2n + 8 \text{ always means } (2n) + 8$$
$$3n - 5 \text{ always means } (3n) - 5$$
$$19 \div 4n \text{ always means } 19 \div (4n)$$

Notice that in a fraction like $\frac{3 + 7}{11}$ or $\frac{n + 2}{9}$, the bar is both a division sign and a grouping symbol. Thus, $\frac{3 + 7}{11} = \frac{10}{11}$.

When $n = 5$, $\frac{n + 2}{9} = \frac{5 + 2}{9} = \frac{7}{9}$.

In the numerical expression in Example 1, below, you see a pair of parentheses inside a pair of brackets. Notice that to simplify such an expression, you first simplify the numeral in the innermost grouping symbol, then work toward the outermost grouping symbol.

EXAMPLE 1   Find the value of $5[24 - 3(7)]$.

SOLUTION
$$5[24 - 3(7)] = 5[24 - 21]$$
$$= 5(3)$$
$$= 15$$

EXAMPLE 2   Find the value of $3(x + 4) - 2x$ when $x = 5$.

SOLUTION
$$3(x + 4) - 2x = 3(5 + 4) - 2(5)$$
$$= 3(9) - 2(5)$$
$$= 27 - 10$$
$$= 17$$

EXAMPLE 3   If n has the replacement set $\{2, 3, 4\}$, find the corresponding values of $5n - 7$.

SOLUTION

When $n = 2$,
$$5n - 7 = 5(2) - 7$$
$$= 10 - 7$$
$$= 3$$

When $n = 3$,
$$5n - 7 = 5(3) - 7$$
$$= 15 - 7$$
$$= 8$$

When $n = 4$,
$$5n - 7 = 5(4) - 7$$
$$= 20 - 7$$
$$= 13$$

Oral Exercises

Find the value of each expression.

1. $6(3) + 5$

2. $56 - 3(8)$

3. $4(3 + 7)$

4. $9(8 - 4)$

5. $(4 \cdot 4) + (5 \cdot 5)$

6. $(7 \cdot 6) - (5 \cdot 4)$

7. $\frac{23 + 4}{3 + 6}$

8. $\frac{5(6)}{4(15)}$

9. $\frac{5 \times 9}{2 + 7}$

10. $2(5) + \frac{18}{6}$

11. $\frac{18 - 4}{1 \times 2}$

12. $(7 - 4)(3 + 5)$

13. $2x + (x + 1)$ when $x = 4$

14. $3(x - 1) - x$ when $x = 6$

15. $5x - 3$ when $x = 3$

16. $x + 2(x + 3)$ when $x = 5$

17. If y has the replacement set $\{1, 2, 3\}$, find the corresponding values of $3y - 2$.

18. Can you find the replacement set for $2x + 3$ if the corresponding values are 3, 13, and 19?

Written Exercises

Find the value of each expression.

A

1. $20(5)(3)$

2. $17 + (5 + 15)$

3. $(23 + 4) + 7$

4. $8(3) + 30(3)$

5. $5(25 + 15)$

6. $(17 + 8)9$

7. $\dfrac{80 - 10}{15 + 20}$

8. $\dfrac{(5 \cdot 7) + 5}{(9 \cdot 3) - 7}$

9. $\dfrac{(6)(6) + (8)(8)}{(4)(4) + (3)(3)}$

10. $[18 - (9)(2)]3$

11. $4[3(3) + 4(5)]$

12. $[35 - 3(4)] + 25$

13. $[(8)(7) - (2)(3)] \div 2$

14. $[5(5) + 4(6)] \div 7$

15. $[9(7) - 8(0)] \div 9$

16. $2(x + 2) - 2$ when $x = 2$

17. $16 - 3(y - 5)$ when $y = 7$

18. $5(a + 5) + 5$ when $a = 10$

19. $15 + 6(w + 1)$ when $w = 6$

20. $2u + (u + 3)$ when $u = 8$

21. $9v - 10$ when $v = 25$

22. $(7x - 5) + 3$ when $x = 12$

23. $15 + (5y - 5)$ when $y = 1$

24. If x has the replacement set $\{2, 4, 6\}$, find the corresponding values of $4x - 1$.

25. If y has the replacement set $\{1, 3, 5\}$, find the corresponding values of $5y + 3$.

26. If t has the replacement set $\{2, 3, 5\}$, find the corresponding values of $100t - 10$.

B

27. $[7(2) - 4] + [9 + 8(4)]$

28. $2[3(5) - 7] + [10 + 4(6)]$

29. $[5(9) + 8][8 + 9(5)]0$

30. $5[(8)(4) + 8] - (8)5$

31. $\{6 + 2[3(2) + 1] + 1\}3$

32. $2\{[3(2) + 5] + 3\}$

33. $\{(6 \cdot 250) \div [8(8) \div (4 \cdot 4)]\} + [25(12 \div 6)]$

34. $\{[(23 - 7)(6 + 4)] + 340\} - \{(48 \div 4)[24(7) \div (18 - 4)]\}$

35. $2(x - 1) + 3(x + 2) + 1$ when $x = 10$

36. $3(2y + 3)(y - 2)$ when $y = 3$

37. $5[(3w - 4) + (2w + 4)]$ when $w = 2$

38. $(3w - 1)(3w + 1) + (2w + 1)(2w - 1)$ when $w = 2$

39. $\dfrac{4u + 3}{2u + 3} + 2u + 1$ when $u = 6$

40. If x has the replacement set $\{10, 20, 30\}$, find the corresponding values of $10x + 2(x + 10)$.

41. If y has the replacement set $\{0, 2, 4, 6, 8\}$, find the corresponding values of $\dfrac{4y + 4}{2}$.

42. If t has the replacement set $\{10, 100, 1000, 10{,}000\}$, find the corresponding values of $4(t - 1) + 2(t - 1)$.

43. If u has the replacement set $\{1, 2, 3, 4, 5, 6, 7, 8, 9\}$, find the greatest corresponding value of $\dfrac{2u - 1}{2u}$.

44. If v has the replacement set $\{\frac{1}{2}, \frac{1}{3}, \frac{1}{4}, \frac{1}{5}\}$, find the corresponding values of $\dfrac{1}{2v}$. Which is the least corresponding value?

1-3 Powers in Numerical and Variable Expressions

To write products whose factors are the same, you can use *exponents*. For example,

instead of writing   you can write

$\underbrace{5 \times 5 \times 5 \times 5}_{4 \text{ factors}},$ 5^4

You read this as "five to the fourth power," or, "the fourth power of five."

In the expression 5^4, 5 is called the **base** and 4 is called the **exponent**.

$5 \times 5 = 5^2$. This is usually read as "five squared," or "the square of five," rather than "five to the second power," because 5^2 can represent the area of a square with sides each of length 5 units.

$5 \times 5 \times 5 = 5^3$. This is usually read as "five cubed" rather than "five to the third power," because 5^3 can represent the volume of a cube with edges of length 5 units.

This way of naming products can also be used in variable expressions.

$\underbrace{n \times n \times n \times n \times n}_{5 \text{ factors}}$ can be written n^5.

EXAMPLE 1   Simplify:

    a. 4^3

    b. 3×4

    c. $5(4^3)$

    d. $(5 \times 4)^3$

SOLUTION   **a.** $4^3 = 4 \times 4 \times 4 = 64$

    b. $3 \times 4 = 12$

    c. $5(4^3) = 5 \times 64 = 320$

    d. $(5 \times 4)^3 = 20^3 = 20 \times 20 \times 20 = 8000$

To avoid questions about the meaning of an expression in which grouping symbols have been omitted, mathematicians have agreed on the following steps to simplify such expressions. However, it is best to use enough grouping symbols to avoid possible differences in interpretation.

1. Simplify the names of powers.
2. Then simplify the names of products and quotients in order from left to right.
3. Then simplify the names of sums and differences in order from left to right.

EXAMPLE 2 Simplify $7(2^2) - 15 \div 5$

SOLUTION $\begin{aligned} 7(2^2) - 15 \div 5 &= 7(4) - 15 \div 5 \\ &= 28 - 3 \\ &= 25 \end{aligned}$

EXAMPLE 3 Find the value of $8n^2 - 5n$ when $n = 3$.

SOLUTION $\begin{aligned} 8n^2 - 5n &= 8(3^2) - 5(3) \\ &= 8 \cdot 9 - 5(3) \\ &= 72 - 15 \\ &= 57 \end{aligned}$

Oral Exercises

Simplify.

1. 5^3 2. 2^4 3. 3^3 4. $4(6^2)$ 5. $(3 \cdot 2)^2$ 6. $(3)^2(2)^2$

7. $6(4)^2$ 8. $[4(2)]^2$ 9. $5^2(8)$ 10. $3^2(2^2)$ 11. $10(9^2)$ 12. $2(4)(5)^2$

13. $3(3)^2 - 4(5)$ 14. $4^2(2) + 16 \div 2$ 15. $36 \div 3 - 2^2(3)$

Find the value of each expression.

16. $3x^2 - 2x$ when $x = 2$ 17. $5y^2 - 2y + 1$ when $y = 3$

18. $5p + p^2$ when $p = 5$ 19. $2w^2 + 6w - 2$ when $w = 1$

20. Insert grouping symbols in $2 \cdot 3^3 \cdot 5 - 2^2 \cdot 3 + 5$ so that it will have a value of 253.

Written Exercises

Simplify.

A   1. $11^2 - (2)(4)^2$ 2. $5^3 + 3(4^2)$ 3. $(4^3 - 3^2)^2(4)$

4. $3(4^3 - 2^3)$ 5. $(1^2 + 2^3 + 3^4) \div 9$ 6. $(4^3 \div 2^4) + 5$

7. $\dfrac{8^2 - 6^2}{8 - 6}$ 8. $\dfrac{5^3 - 3}{5 - 3}$ 9. $\dfrac{10^2 - 5^2}{10 + 5}$

10. $6(2^3) - 5(2^2) + 3$ 11. $4^3 - 2(4^2) + 3(4) - 5^2$ 12. $2^3(3) - 2(3^2) + 2(2) + 1$

Find the value of each expression.

13. $2x^2 - x$ when $x = 5$

14. $y^2 + y + 1$ when $y = 3$

15. $3 - 2t + 3t^2$ when $t = 1$

16. $w^2 - 5w + 6$ when $w = 6$

17. $7u^2 - 4u$ when $u = 4$

18. $9 + 9v + 9v^2$ when $v = 10$

19. $6m^2 - 5m + 4$ when $m = 6$

20. $56 - 3n - 2n^2$ when $n = 3$

21. $\dfrac{x^2 - x + 6}{x + 3}$ when $x = 3$

22. $\dfrac{2y^2 + 3y + 4}{3y^2 - y - 2}$ when $y = 2$

Simplify.

B 23. $2^5 \div 8 \div 2^2 \div 2$

24. $7(4^2 + 1) - 5(2^3) \div 10$

25. $8[6^2 - 3(11)] \div 8 + 3$

26. $12^2 - (5 \cdot 3)(7 - 5) + 9$

27. $(5 - 1)^3 + (11 - 2)^2 + (7 - 4)^3$

28. $\frac{1}{2}[12 \div 2(6)] \div (12 \cdot 2 \div 6)$

29. $(14 - 2 + 16 \div 4) \div 8(2^2)$

30. $(6^2 - 34)^2 - (5^2 - 21) + (4^3 - 2^3)$

31. $(1.4^2 - 0.2^2) - (1.4 + 0.2)(1.4 - 0.2)$

32. $[(4.2 + 2.5 + 1)(4.2 + 2.5 - 1)] \div [(11)(19)(0.01)]$

33. $[(5 \cdot 3)^2 + (4 \cdot 3)^2 - (2 \cdot 3)^2] \div [3^2(37)]$

34. $2 + \left(\dfrac{5 - 2}{5}\right)^2 + \left(\dfrac{5 + 3}{10}\right)^2$

35. $\left(\dfrac{9 - 4}{3}\right)^2 - 1 - \left(\dfrac{\frac{16}{3}}{4}\right)^2$

Find the value of each expression.

36. $3x - 5x^2$ when $x = 0.2$

37. $10y - y^2$ when $y = \frac{2}{5}$

38. $k^2 + 6k - 5$ when $k = 1.1$

39. $w^2 + w + 4$ when $w = \frac{2}{3}$

40. $0.25(1 - 2u + 3u^2)$ when $u = \frac{1}{2}$

41. $(v + 1)(v^2 - v + 1)$ when $v = 0.5$

programming in BASIC

These optional sections appear at appropriate places throughout the book. They may be used with any computer that accepts the language BASIC. In Section 1-3, you learned the steps for simplifying expressions. The BASIC language uses the same method! Moreover, it uses

$+$ for addition   and $-$ for subtraction,

but it uses

$*$ for multiplication   and $/$ for division.

Thus, you would write

$$\frac{2 \times 15}{6} \quad \text{as} \quad 2 * 15/6.$$

You could also have used parentheses and written (2 * 15)/6. Sometimes you need to use parentheses. Thus, you would write

$$\frac{3 \times 20}{4 \times 6} \quad \text{as} \quad 3 * 20/(4 * 6).$$

To the computer, 3 * 20/4 * 6 would mean $\frac{3 \times 20}{4} \times 6$.

One more symbol is ↑, which is used to write powers. Thus, you would write

$$3^4 \quad \text{as} \quad 3 \uparrow 4.$$

Putting these all together, you would write

$$3 \times 4^5 + \frac{17 - 2}{3 \times 5} + \frac{6^3}{2^2}$$

as

$$.3 * 4 \uparrow 5 + (17 - 2)/(3 * 5) + 6 \uparrow 3/2 \uparrow 2.$$

Notice that in BASIC all symbols are written in a line as on a typewriter. That is important because to write a statement for a computer, you put each line on a card or you type the line on a terminal that has a keyboard that looks like a typewriter.

Now you can write a program in BASIC! Each statement must have a line number, and each program must end with an END statement. Thus, you might write:

```
10   PRINT 3*4↑5+(17−2)/(3*5)+6↑3/2↑2
20   END
```

If you are working at a terminal, you type in the program, pressing the carriage RETURN key at the end of each line. Then you type the command RUN (a command has no line number) and press RETURN. The "PRINT" will cause the computer to find the value of the expression and then print it as shown at the right.

RUN

3127

END

To find the value of another expression, simply retype line 10 with the new expression following the word PRINT, and type RUN. (Line 20 is still there unchanged. To check this, type the command LIST. The computer will then print out all the statements it has in the program.)

Exercises

Write each of the following expressions in BASIC. Then put each into a two-line program like that shown above, and RUN it on your computer.

1. $3 \times 5 + 8$
2. $3 + 5 \times 8$
3. $(6 + 7) \times 10$

4. $(23 - 19) \times 3$
5. $4 \times 6 + 3^2$
6. $3^4 - 5 \times 8$

7. $\dfrac{8^2}{4^2}$
8. $\dfrac{15^2}{5^2}$
9. $\dfrac{16 + 2}{9}$

10. $\dfrac{8 \times 15}{3}$
11. $\dfrac{4 + 8 \times 5}{4 \times 5 + 2}$
12. $\dfrac{3 \times 6 + 5^2}{6 + 4 \times 16}$

13. $\dfrac{4 \times 3^2 - 3 \times 2^2}{6 \times 4 - 2 \times 9}$
14. $\dfrac{8 \times 7 + 3 \times 4^2}{5^3 - 3 \times 33}$
15. $\dfrac{4 \times 5^2 \times 3 - 6^3}{3^5 - 37 \times 6}$

16. 5^4 17. 5^5 18. 5^6 19. 5^7 20. 5^8 21. 5^9

If you find 5^{10} by multiplying 5^9 by 5, you will get 9765625. If you ask the computer to find $5 \uparrow 10$, it will give you $9.76563E + 06$. The "E + 06" means "multiplied by 10^6, or 1000000." That is,

$$9.76563E + 06 \text{ means } 9765630.$$

Thus, the computer has given you 9765625 rounded to six digits.

Use the computer to find each of the following values. Check your interpretation of the computer result by multiplying the preceding hand-computed result by 5.

22. 5^{11} 23. 5^{12} 24. 5^{13} 25. 5^{14} 26. 5^{15} 27. 5^{16}

1-4 Numbers and Their Graphs

Numbers can be placed in correspondence with points on a line. Here is a way to make this correspondence:

1. Choose any point on a line and label it 0. This point corresponds to the number 0 and is called the **origin.**

Figure 1

2. Choose any length to be one unit. Locate the point which is one unit to the right of the origin and label it 1.

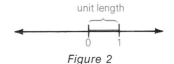

Figure 2

3. Using the same unit length, label points to the right of 1 as 2, 3, and so on.

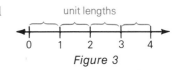

Figure 3

So far, you have labeled points on the line with the number 0 and with the numbers we call the **positive integers,** namely 1, 2, 3, and so on. Do you see that it makes sense to label the point halfway between 0 and 1 as $\frac{1}{2}$, and the point halfway between 1 and 2 as $1\frac{1}{2}$? Similarly, the point $\frac{3}{4}$ of the way from 0 to 1 is labeled $\frac{3}{4}$, and so on.

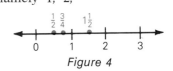

Figure 4

It is a surprising fact that even if you could label all the points on the line to the right of 0 that correspond to positive integers and fractions, there would still be many points that would not be labeled. The numbers that correspond to such points will be discussed in the next section.

So far, you have not labeled any points to the *left* of the origin. As you may already know, it is sometimes necessary to use numbers other than positive numbers. For example,

| owing $10: ⁻10 | being paid $10: 10 |
| 4 km west: ⁻4 | 4 km east: 4 |

The numbers suggested in the left-hand column above are usually called **negative** numbers. Because positive and negative numbers suggest opposite directions, they are sometimes called **directed** numbers.

To continue with the correspondence between numbers and points on a line:

4. Label the point one unit length to the left of 0 as ⁻1.

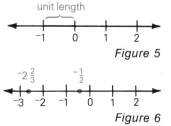

Figure 5

5. Using the unit length, label the points to the left of ⁻1 as ⁻2, ⁻3, and so on. These numbers are called the **negative integers.** All other points on the line to the left of 0 may be assigned numbers in the same way as was done for points to the right of 0.

Figure 6

A line with points labeled to correspond to numbers is called a **number line.** On a number line the point paired with a number is called the **graph** of the number. For example, in Figure 6 the point one-half unit to the left of the origin is the graph of ⁻$\frac{1}{2}$. The number paired with a point is called the **coordinate** of the point.

EXAMPLE   What is the coordinate of each of the points *A*, *B*, and *C* on the given number line?

SOLUTION   *A:* 3; *B:* ⁻1; *C:* ⁻$4\frac{1}{2}$

The set of numbers corresponding to all the points on the line is called the set of **real numbers.** By definition, then:

1. There is exactly one point on the number line corresponding to any given real number.

2. There is exactly one real number corresponding to any given point on the number line.

Oral Exercises

In Exercises 1–17 identify the points that are the graphs of the given numbers or state the coordinates of the given points. Use the number line below.

EXAMPLE   **a.** D **b.** 2   **c.** $4\frac{1}{2}$

SOLUTION   **a.** $^-2$ **b.** B **c.** The point halfway from R to A, or the midpoint of $\overline{RA}$ ($\overline{RA}$ is read "line segment R, A").

1. N	**2.** P	**3.** V	**4.** H	**5.** M
6. $^-8$	**7.** 6	**8.** 7	**9.** $3\frac{1}{2}$	**10.** $^-\frac{1}{2}$
11. $^-1.5$	**12.** 7.5	**13.** $\frac{20}{4}$	**14.** $\frac{-15}{6}$	**15.** $\frac{36}{8}$

16. Point halfway from P to E **17.** Point halfway from D to T

18. The chart below shows two examples of directed numbers. List as many more examples of directed numbers as you can.

Positive number	Negative number
A gain of $10	A loss of $10
Six degrees above zero	Six degrees below zero

Written Exercises

Draw a horizontal number line and on it show the graphs of the given numbers.

A

1. 5	**2.** $^-1$	**3.** $^-4$	**4.** 0	**5.** $3\frac{1}{2}$
6. $^-3\frac{1}{2}$	**7.** 2.5	**8.** $^-6.5$	**9.** $\frac{22}{4}$	**10.** $\frac{-30}{10}$

State the coordinate of the point on the number line at which you would arrive if you were to:

11. Start at the origin, move 4 units in the positive direction, and then move 3 more units in that direction.

12. Start at the origin, move 5 units in the positive direction, and then move 7 units in the negative direction.

13. Start at the origin, move 3 units in the negative direction, and then move 4 more units in that direction.

14. Start at the origin, move 2 units in the negative direction, and then move 3 units in the positive direction.

In Exercises 15–34 name the coordinate of the point described. Use the number line below.

15. The midpoint of $\overline{NC}$. 16. The midpoint of $\overline{PF}$.

17. The point one fourth of the distance from H to M.

18. The point one third of the distance from D to M.

EXAMPLE The point three fourths of the distance from C to K.

SOLUTION $\overline{CK}$ is 6 units long. $\frac{3}{4} \times 6 = \frac{9}{2}$, or $4\frac{1}{2}$, units. Therefore to find the required point, X, you move $4\frac{1}{2}$ units to the right from C. The coordinate of X is found from the diagram to be $3\frac{1}{2}$.

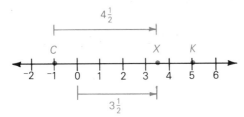

B 19. The point three eighths of the distance from L to R.

20. The point seven eighths of the distance from G to R.

21. The point one half of the distance from C to L.

22. The point one half of the distance from G to A.

23. The point one fourth of the distance from C to K.

24. The point three fourths of the distance from U to C.

25. The point one third of the distance from I to M.

26. The point one eighth of the distance from N to M.

27. The point two thirds of the distance from H to B.

28. The point five eighths of the distance from D to K.

C 29. The point between H and T that is twice as far from H as it is from T.

30. The point between G and R that is three times as far from R as it is from G.

31. The point to the left of S that is twice as far from S as it is from T.

32. The point to the right of H that is half as far from H as it is from U.

33. The point to the right of M that is one third as far from M as it is from P.

34. The point to the left of T that is one fourth as far from T as it is from F.

14 | *Chapter 1*

Self-Test 1

VOCABULARY   numerical expression (p. 1)   exponent (p. 7)
                simplify (p. 1)   origin (p. 11)
                variable (p. 2)   integer (p. 11)
                replacement set (p. 2)   number line (p. 12)
                domain (p. 2)   graph (p. 12)
                evaluate (p. 2)   coordinate (p. 12)
                grouping symbol (p. 4)   real numbers (p. 12)

Simplify.

1. $358 - 276$

2. $1218 \div 6$

Obj. 1, p. 1

Evaluate the expression.

3. $x + 16$ when $x = 3$

4. $5y$ when $y = 13$

Obj. 2, p. 1

Draw a horizontal number line and on it show the graphs of the given numbers.

5. $^-3$

6. $1\frac{1}{4}$

7. $^-2.5$

8. 0.4

Obj. 3, p. 1

Express each measurement by a directed number.

9. A gain of ten points

Obj. 4, p. 1

10. Five minutes before liftoff

Check your answers with those at the back of the book.

Sets

***OBJECTIVES** for Sections 1-5 through 1-7:*
1. Identify rational and irrational numbers.
2. Specify sets by roster or description.
3. Use the symbols $=$, $\in$, $\subset$, and $\emptyset$ to write statements about sets.
4. Draw Venn diagrams.
5. Graph sets of numbers on the number line.

1-5 Real Numbers

Every real number can be expressed as a *decimal numeral.* In earlier courses you learned how to find a decimal equivalent for a fraction, or *rational number.* A **rational number** is a number that can be expressed in the form $\frac{n}{d}$, where n and d are both integers and $d \neq 0$. For example, to find a decimal equivalent for $\frac{3}{8}$, you divide:

$$\begin{array}{r} 0.375 \\ 8\overline{)3.000} \end{array}$$

$\therefore$ (read "therefore") $\frac{3}{8} = 0.375$. 0.375 is a *terminating decimal*.

Some rational numbers are equivalent to *nonterminating decimals*. For example, to find a decimal equivalent for $\frac{6}{11}$, you divide:

$$\begin{array}{r} 0.5454 \\ 11\overline{)6.0000} \\ \underline{5\,5} \\ 5\,0 \\ \underline{4\,4} \\ 6\,0 \\ \underline{5\,5} \\ 5\,0 \\ \underline{4\,4} \\ 6 \end{array}$$

$\frac{6}{11} = 0.545454\ldots$ (The three dots mean that the pattern continues without end.)

Notice that as far as you carry out this division, the quotient will never end (or *terminate*) because 0 will never appear as a remainder. Notice also that no matter how far this division is carried out, the block of two digits, 54, will keep appearing and no other digits will appear. For this reason, the quotient is called a **repeating decimal**. Frequently a bar is used to show the block of digits which repeats, as in $0.\overline{54}$.

If the decimal equivalent for a rational number does not terminate, then it must repeat. If you examine the division above, you may see why. The only remainders that appear (printed in red) are 5 and 6. Actually, if any integer is divided by 11, the only remainders that could appear are the integers 0 through 10. (Of course, if the remainder is 0, the decimal terminates.) Therefore, after *at most* 10 digits beyond the decimal point, the next remainder would be the same as one of the remainders already obtained. At this point, the decimal quotient would begin to repeat. Similarly, the decimal equivalent of a rational number with a denominator of 7 starts repeating after at most 6 digits beyond the decimal point.

The decimal equivalent of a rational number with a denominator of d will start repeating after at most $d - 1$ digits beyond the decimal point. Of course, there may be fewer than $d - 1$ digits in the repeating block.

The preceding discussion suggests the following:

> Every rational number can be named by either a terminating or a repeating decimal.

The following can also be shown:

> Every repeating or terminating decimal names a unique rational number.

It is possible to construct a decimal numeral, however, that is non-terminating and nonrepeating. For example, the decimal numerals,

$$0 . 3 7 \underline{3 3} 7 \underline{3 3 3} 7 \underline{3 3 3 3} 7 \ldots$$

one 3   two 3's   three 3's   four 3's

and

$$0 . 2 4 6 8 1 0 1 2 \underline{1 4} \underline{1 6} \underline{1 8} \ldots$$

fourteen   sixteen   eighteen

both follow patterns and could be carried out to any number of decimal places, but no block of digits would keep repeating. According to the results stated above, these decimal numerals cannot be equivalent to rational numbers. These decimal numerals name what are called *irrational numbers*. An **irrational number** is any real number that is not a rational number. It should be clear that the two irrational numbers named above do correspond to points on the number line. For example, we can close in on the graph of 0.37337333733337 . . . on the number line by finding narrower and narrower gaps in which the graph must appear. First, we can locate it roughly between 0.3 and 0.4, then between 0.37 and 0.38, then between 0.373 and 0.374, and so on, as shown in Figure 7.

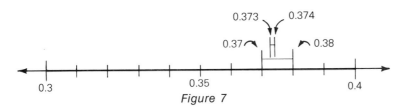

Figure 7

Oral Exercises

Complete Exercises 1–5 by using either the word "rational" or "irrational."

1. Every terminating decimal numeral represents a(n) _?_ number.

2. 0.334333443333444 . . . names a(n) _?_ number.

3. Nonterminating, nonrepeating decimal numerals represent _?_ numbers.

4. Every repeating or terminating decimal names a unique _?_ number.

5. 0.656565 . . . names a(n) _?_ number.

6. The decimal equivalent of a rational number with a denominator of 13 will start repeating after at most _?_ digits beyond the decimal point.

7. Construct a decimal numeral that names an irrational number.

Written Exercises

Find the decimal equivalent for each rational number.

A **1.** $\frac{1}{8}$ **2.** $\frac{2}{9}$ **3.** $\frac{2}{3}$ **4.** $\frac{4}{25}$ **5.** $\frac{5}{6}$

6. $-\frac{3}{20}$ **7.** $\frac{5}{12}$ **8.** $\frac{3}{16}$ **9.** $\frac{-7}{6}$ **10.** $\frac{21}{40}$

11. $\frac{22}{15}$ **12.** $\frac{-10}{11}$ **13.** $\frac{3}{13}$ **14.** $\frac{-19}{50}$ **15.** $\frac{37}{16}$

Name three rational numbers between each given pair of rational numbers.

EXAMPLE   0.6, 0.7

SOLUTION   0.61, 0.62, 0.63 (Many other answers are possible.)

16. 0.1, 0.2   **17.** 0.9, 1.0   **18.** 0.34, 0.35

19. ⁻0.0022, ⁻0.0021   **20.** 0.3333 . . . , 0.4444 . . .

Find a rational number that is halfway between the given numbers. Express your answer in decimal form.

EXAMPLE   6, 7

SOLUTION $\dfrac{6+7}{2} = \dfrac{13}{2} = 6.5.$ 6.5 is halfway between 6 and 7.

B **21.** 4, 5   **22.** ⁻6, ⁻5   **23.** ⁻1, 0   **24.** $\frac{1}{4}, \frac{1}{3}$

25. $\frac{1}{4}, \frac{1}{2}$ **26.** $\frac{1}{2}, \frac{5}{8}$ **27.** $\frac{-5}{6}, \frac{-2}{3}$ **28.** ⁻4.2, ⁻4.1

29. 4.08, 4.09   **30.** $3\frac{1}{2}, 3\frac{7}{8}$ **31.** $-2\frac{1}{3}, -2\frac{1}{5}$ **32.** 1.05, $1\frac{1}{10}$

Use All, Some, or No to complete each of the following statements.

33. _?_ negative rational numbers can be expressed as terminating decimals.

34. _?_ rational numbers can be expressed in the form $\dfrac{a}{b}$ where a is a positive integer and $b \neq 0$.

35. _?_ rational numbers can be expressed in the form $\dfrac{a}{b}$ where a and b are integers and $b \neq 0$.

36. _?_ rational numbers are irrational numbers.

ON THE CALCULATOR

If we find the decimal equivalent of $\frac{1}{7}$ on a calculator by dividing 1 by 7, the result is shown as 0.1428571. Since we know that the decimal equivalent for $\frac{1}{7}$ must repeat after at most 6 digits to the right of the decimal point, we can write $\frac{1}{7} = 0.\overline{142857}$.

If we find the decimal equivalent for $\frac{1}{17}$ on a calculator, the result is shown as 0.0588235. Can you tell whether or not the decimal equivalent has begun to repeat?

Exercises

Use a calculator to find the decimal equivalent for the given fractions.

1. $\frac{2}{7}, \frac{3}{7}, \frac{4}{7}$, and so on to $\frac{6}{7}$

2. $\frac{1}{9}, \frac{2}{9}, \frac{3}{9}$, and so on to $\frac{8}{9}$

3. $\frac{1}{11}, \frac{2}{11}, \frac{3}{11}$, and so on to $\frac{10}{11}$

4. $\frac{1}{27}, \frac{2}{27}, \frac{3}{27}$, and so on to $\frac{12}{27}$

1-6 Sets and Their Members

A **set** is any collection of objects. The objects are called the **members,** or **elements,** of the set. To specify a set you can either list the members of the set in a **roster,** or describe the set in words or symbols. In either case, you use braces { } to indicate that a set is being named. The braces are read, "the set whose members are . . . ." For example, to name the set whose members are the numbers 1, 3, 5, and 7, you can use either

the roster method: {1, 3, 5, 7} *or*
the description method: {the positive odd integers less than 8}.

Sets having exactly the same members are **equal** sets. You write:

{1, 2, 3} = {the first three positive integers}.

Here are some other terms and symbols used in dealing with sets:

Symbols	Words
$\in$	is a member (or element) of
$2 \in$ {the even integers}	2 is a member of the set of even integers.
$\{1, 3, 5\} = \{3, 1, 5\}$	Sets having exactly the same members are equal sets.
$S =$ {the even integers}	S equals the set of even integers. (Capital letters are often used to name sets.)
$2 \in S$	2 is a member of S.
$3 \notin S$	3 is *not* a member of S.
$\subset$	is a subset of (A **subset** of a set S is any set all of whose members are also in S.)
$\{1, 3\} \subset \{1, 2, 3\}$	The set whose members are 1 and 3 is a subset of the set whose members are 1, 2, and 3.
$S \subset S$	Every set is a subset of itself.
$\emptyset$	The **empty** set. This is the set with no members. Do not confuse $\emptyset$ with $\{0\}$, which has one member, namely 0.
$\emptyset \subset S$	The empty set is a subset of S. (The empty set is a subset of every set.)

Often it is useful to draw a diagram that shows how sets are related. Such diagrams are called Venn diagrams (named after the British logician John Venn, who first used them). The Venn diagram below shows that A is a subset of B:

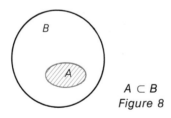

$A \subset B$
Figure 8

Oral Exercises

Specify each of the following sets by roster.

1. {the days of the week with names beginning with S}
2. {the name of the season that follows winter}
3. {the months having 31 days}

Specify each of the following sets by description.

4. $\{0, 2, 4, 6, 8\}$

5. $\{w, x, y, z\}$

6. $\{0, 1, 2, 3, 4, 5, 6, 7, 8, 9\}$

What symbol would you use in place of _?_ to make a true statement?

7. {the months of the year with names beginning with C} = _?_

8. $\{2, 4, 6\}$ _?_ $\{4, 6, 2\}$

9. 3 _?_ $\{1, 2, 3, 4, 5\}$

10. 6 _?_ $\{1, 2, 3, 4, 5\}$

11. A _?_ D

12. B _?_ C

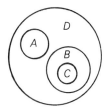

Exs. 11 and 12

13. Which two sets are subsets of any set with at least one member?

14. Draw a Venn diagram that shows that $R \not\subset T$.

Written Exercises

Replace _?_ by $\in$ or $\subset$ to make a true statement.

A   **1.** 2 _?_ $\{1, 2, 3\}$ **2.** $\{2, 3\}$ _?_ $\{1, 2, 3\}$ **3.** $\emptyset$ _?_ $\{0\}$

     4. $\emptyset$ _?_ $\emptyset$ **5.** 0 _?_ $\{0\}$ **6.** $\{1, 2, 3\}$ _?_ $\{1, 2, 3\}$

     7. $\frac{1}{4}$ _?_ $\{\frac{1}{2}, \frac{1}{3}, \frac{1}{4}\}$ **8.** $\{0, 2, 4\}$ _?_ $\{4, 0, 2\}$ **9.** 4 _?_ $\{0, 4, 8\}$

Classify each statement as true or false.

10. $\{2^6\} = \{4^3\}$ **11.** $13^2 \in \{168, 169, 170\}$ **12.** $\{1\} = \left\{\dfrac{5^2 - 6}{5^2 - 6}\right\}$

13. $\{24 - 2^3 \div 2^2\} = \{(24 - 2^3) \div 2^2\}$ **14.** $6^2 \in \{1^2 + 2^2 + 3^2\}$

Draw a Venn diagram to illustrate each statement.

15. $A \subset B$ **16.** $B \subset C$ **17.** $A \subset B \subset C$ **18.** $A \subset B \subset C \subset D$

B   **19.** $A \subset C$ and $B \subset C$, but $A \not\subset B$ and $B \not\subset A$.

     20. $A \subset B$ and $B \subset A$.

21. If $A \subset B$ and $B \subset A$, then what is true about sets A and B?

22. If $A \neq \emptyset$ and $A \subset \{a, b, c\}$, list all the sets which A might possibly be.

23. If $x \in C$ and $C \subset \{\text{Mike}\}$, does x stand for Mike?

C   **24.** Show that if $A \subset B$ and $B \subset C$, then $A \subset C$.

The set $\{p\}$ has the following subsets:

> 1   subset with no members—Ø
> 1   subset with one member—$\{p\}$
> 2 = total number of subsets

The set $\{p, q\}$ has the following subsets:

> 1   subset with no members—Ø
> 2   subsets with one member—$\{p\}$, $\{q\}$
> 1   subset with two members—$\{p, q\}$
> 4 = total number of subsets

Exercises

For each of the sets in Exercises 1–3 below, list the subsets containing no members, one member, two members, three members, and so on. How many subsets are there in each category? What is the total number of subsets for each of the given sets? Do you notice any number patterns in your data?

1. $\{p, q, r\}$ **2.** $\{p, q, r, s\}$ **3.** $\{p, q, r, s, t\}$

1-7 Sets of Numbers

Given a set of numbers, you can graph the points that correspond to the numbers in your set on the number line. The set of points is called the **graph** of the set of numbers.

In the examples below and on the next page, red is used to indicate the graph of the given set.

Set	Graph

$\{-2, -\frac{3}{2}, 3\}$

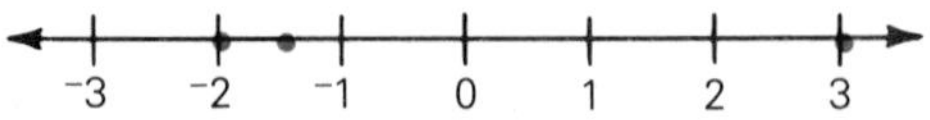

Figure 9

$\{$the whole numbers$\}$

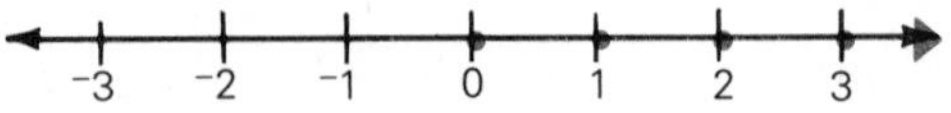

Figure 10

Note that the heavy red arrow in Figure 10 indicates that the set continues indefinitely to the right.

{the real numbers between ⁻1 and 2½}

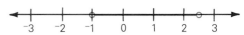

Figure 11

Note that neither ⁻1 nor 2½ is included in the set. Open circles on the number line in Figure 11 indicate the absence of the corresponding points from the graph.

{the real numbers greater than or equal to ³⁄₂}

Figure 12

Note two things in Figure 12: **(1)** that a solid dot at ³⁄₂ indicates that the point corresponding to ³⁄₂ is included in the set, and **(2)** that the arrow indicates that the set continues indefinitely to the right.

{the numbers greater than ⁻2 and less than or equal to 3}

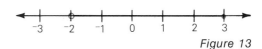

Figure 13

A set such as {⁻2, ⁻³⁄₂, 0, 3} is called a **finite** set because the number of members of the set is a whole number, in this case 4. (The word *finite* comes from a Latin word that means "end.") The empty set is finite because it has 0 members (zero is a whole number).

S = {the positive integers less than 10,000} is finite.
T = {the positive multiples of 3} is *not* finite, because there is no whole number for the number of elements of the set. A set that is not finite is called an **infinite** set.

You can use three dots in specifying either a finite set or an infinite set by the roster method when it is inconvenient or impossible to list all the elements of the set.
For example, you can write, for the set S above,

$$S = \{1, 2, 3, 4, 5, \ldots, 9999\},$$

and for the set T above

$$T = \{3, 6, 9, 12, \ldots\}.$$

If a set is finite, you should indicate that fact by writing the last number (or last few numbers) after the dots. If the set is infinite, the dots with no number following should be used. In either case you must list enough members of the set to show clearly the pattern of the set.
The set J, where

$$J = \{\ldots, ⁻3, ⁻2, ⁻1, 0, 1, 2, 3, \ldots\}$$

is called the set of **integers.** Here, the dots at the beginning and end of the roster indicate that the set continues indefinitely in both directions.

The set N, where

$$N = \{1, 2, 3, 4, 5, \ldots\}$$

is called the set of natural numbers (or counting numbers).

Also, recall the set R of rational numbers which was defined in Section 1-5:

$$R = \left\{\text{numbers of the form } \frac{n}{d}, \text{ where } n \text{ and } d \text{ are integers and } d \neq 0\right\}.$$

Note that all the integers (including 0) are in R, since, for example, $^-3 = \frac{^-3}{1}$ and $0 = \frac{0}{1}$. The Venn diagram in Figure 14 suggests the relationships among the sets N, J, R, and $\mathcal{R}$ (the set of all real numbers):

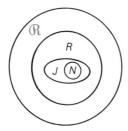

Figure 14

Remember that, as was shown in Section 1-5, there are numbers in $\mathcal{R}$ that are not in R, namely, the irrational numbers.

Oral Exercises

In Exercises 1–7, specify the graph of each set by referring to the number line below.

EXAMPLE   **a.** $\{^-2, 0, 6\}$

  b. {the real numbers between $^-3$ and 3}

SOLUTION   **a.** $\{N, S, R\}$

  b. $\overline{GA}$, without endpoints G and A

1. $\{^-8, ^-4, 1, 7\}$ **2.** $\{^-3, ^-1, 1, 3\}$

3. {the real numbers between $^-1$ and 5}

4. {the real numbers greater than or equal to $^-3$ and less than or equal to 3}

5. {the real numbers greater than ⁻2 and less than or equal to 4}
6. {the negative integers greater than ⁻5}
7. {the positive integers less than 8}
8. Which of the sets in Exercises 1–7 are finite and which are infinite?

Written Exercises

Draw the graph of each set on a number line.

A   1. $\{^-2, ^-4, ^-6, 3, 5\}$ 2. $\{^-\frac{9}{2}, ^-\frac{7}{2}, ^-\frac{5}{2}, 0, \frac{1}{2}\}$ 3. $\{^-4\frac{1}{2}, ^-\frac{1}{2}, 1\frac{1}{2}, 3\}$

4. {the real numbers between ⁻6 and ⁻3}
5. {the real numbers between ⁻4 and 2}
6. {the real numbers greater than ⁻1}
7. {the real numbers less than $2\frac{1}{2}$}
8. {the real numbers greater than 0 and less than or equal to $3\frac{1}{2}$}
9. {the real numbers less than 0 and greater than or equal to ⁻$3\frac{1}{2}$}
10. {the negative real numbers}
11. {the negative real numbers greater than ⁻3}
12. {the positive real numbers less than $5\frac{1}{2}$}
13. {the integers less than or equal to 4}
14. {the integers greater than or equal to 3}
15. {the integers greater than ⁻3 and less than 3}
16. {the real numbers less than or equal to $2\frac{1}{2}$ and greater than or equal to ⁻$2\frac{1}{2}$}
17. {the positive and negative real numbers}
18. {the real numbers that are neither positive nor negative}
19. {the integers less than 0 and greater than ⁻1}
20. Which of the sets in Exercises 1–19 are finite and which are infinite?

Let $A = \{^-1, \frac{3}{5}, 2, 7, 9\}$ and $B = \{^-2, ^-1, ^-\frac{3}{5}, \frac{2}{3}, 9\}$. Specify each of the following sets.

B   21. {the integers belonging to A and to B}
22. {the integers belonging to A or to B}
23. {the integers in A but not in B}
24. {the integers in B but not in A}
25. {the rational numbers belonging to A and to B}
26. {the rational numbers belonging to A or to B}
27. {the nonintegers belonging to A and to B}

Let $A = \{^-1, \frac{3}{5}, 2, 7, 9\}$ and $B = \{^-2, ^-1, ^-\frac{3}{5}, \frac{2}{3}, 9\}$. Specify each of the following sets.

28. {the nonintegers belonging to B and not to A}
29. {the positive integers that are neither in A nor in B}
30. {the positive integers not in both A and B}

Use All, Some, or No to complete each of the following statements.

31. _?_ rational numbers are real numbers.
32. _?_ real numbers are irrational numbers.
33. _?_ irrational numbers are integers.
34. _?_ natural numbers are integers.
35. _?_ real numbers are natural numbers.
36. _?_ rational numbers are integers.
37. _?_ integers are rational numbers.
38. _?_ rational numbers and irrational numbers are real numbers.
39. _?_ real numbers are rational numbers or irrational numbers.
40. _?_ natural numbers are irrational numbers.
41. _?_ rational numbers are negative integers.

DIVERSION

If you were asked "Which set of numbers has more members, the set of all the positive integers or the set of only the even positive integers?", what would your answer be?

Strangely enough, by pairing off the members of the two sets, we can show that neither set has more members than the other. The display below indicates how this is done.

$$\{1, 2, 3, 4, \ 5, \ldots\}$$
$$\updownarrow \updownarrow \updownarrow \updownarrow \ \updownarrow$$
$$\{2, 4, 6, 8, 10, \ldots\}$$

This kind of pairing is called a *one-to-one correspondence*. A **one-to-one correspondence** between two sets is a pairing which assigns to each member of each set one, and only one, member of the other set. We used this kind of correspondence in Section 1-4 when we established a pairing between the set of real numbers and the set of points on a line. In everyday life, we use one-to-one correspondence whenever we count. For example, to count the petals in the daisy at the right, we form a one-to-one correspondence between the petals and the counting numbers 1 through 8.

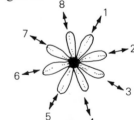

Exercises

1. Sally saves quarters in one piggy bank and dimes in another. She wishes to find out whether she has more dimes or more quarters. One way to find out would be to count the coins. How could she find out without counting?

2. You have a jar full of nickels and dimes. One way to find their total value is to form a correspondence between the nickels and dimes. Would you use a one-to-one correspondence?

3. In the diagram at the right, line segment *CD* is longer than line segment *AB*. Would you guess that segment *CD* has more points than segment *AB*? Look at the red line segments drawn from point *P*. They suggest a method of putting the points of segment *AB* into one-to-one correspondence with the points of segment *CD*. Can you describe this method? Would you believe that segment *AB* has just as many points as segment *CD*?

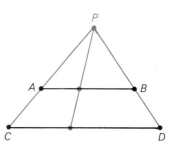

Self-Test 2

VOCABULARY   rational number (p. 15)   subset (p. 20)
terminating decimal (p. 16)   empty set (p. 20)
repeating decimal (p. 16)   graph (p. 22)
irrational number (p. 17)   finite set (p. 23)
set (p. 19)   infinite set (p. 23)
member of a set (p. 19)   integer (p. 23)

Tell whether the number is rational or irrational.

1. $\frac{1}{3}$ **2.** $6.\overline{103}$ **3.** $0.1020304\ldots$ **4.** $-\frac{1}{5}$ *Obj. 1, p. 15*

5. Specify the following set by roster: *Obj. 2, p. 15*
{the letters in the word NEPTUNE}.

6. Specify the following set by description:
{January, June, July}.

Replace ? by ∈ or ⊂ to make a true statement.

7. 8 ? {2, 4, 6, 8}   **8.** {1, 3} ? {1, 3, 5}   *Obj. 3, p. 15*

9. Draw a Venn diagram to illustrate $x \in A$, and $A \subset B$.   *Obj. 4, p. 15*

10. Graph the set of numbers, $\{-3, 0, \frac{1}{2}, 2\}$, on a number line.   *Obj. 5, p. 15*

Check your answers with those at the back of the book.

Number Sentences

1-8 Equations and Inequalities

If two real numbers are not equal, one is greater than the other. Instead of writing "7 is greater than 2," you can write

$$7 > 2.$$

The symbol $>$ means "is greater than." In order to determine which of two numbers is the greater, you can graph them on the number line:

Figure 15

The graphs of numbers show their *order*. Since 7 is to the right of 2, you know that $7 > 2$. On a horizontal number line, "greater than" always means "to the right of."

You can also write "2 is less than 7," or

$$2 < 7.$$

The symbol $<$ means "is less than." On a horizontal number line, "less than" always means "to the left of."

Notice that, although $^-3$ looks like a number greater than $^-1$, it is to the left of $^-1$ on the number line:

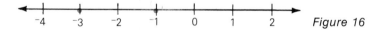

Figure 16

and therefore, $^-3 < {}^-1$ and $^-1 > {}^-3$.

Sometimes it is convenient to indicate that one number is greater than or equal to another. This can be written

$$x \geq 2 \ (x \text{ is greater than or equal to 2})$$

Similarly, you can write $x \leq 2$ (x is less than or equal to 2). Notice that the statements

$$4 \geq 1 \quad \text{and} \quad 4 \geq 2 + 2$$

are *both* true, since a "greater than or equal to" sentence is true if *either* the first number is greater than the second *or* if the first number equals the second.

On the other hand, the sentence

$$^-1 < x < 5$$

means "$^-1$ is less than x and x is less than 5."

The sentence

$$^-1 < 6 < 5$$

is false since 6 is not less than 5 (although $^-1$ is less than 6). In order for a statement like the one above to be true, both "less than" relationships must be true.

Sentences that involve the symbols $>$, $<$, $\geq$, or $\leq$, together with those involving $=$ or $\neq$, are called mathematical sentences, and the expressions that appear on either side of these symbols are called the members of the sentence. If a sentence contains a variable, it is called an open sentence. A mathematical sentence containing the symbol $=$ is called an equation. A sentence containing any of the other symbols is called an inequality.

Here are two mathematical sentences.

EXAMPLE 1   If $x \in$ {the positive integers}, find the values of x for which $x + 5 = 8$.

SOLUTION   Since $3 + 5 = 8$, the given equation is true if (and only if) $x = 3$.

EXAMPLE 2   If $y \in \{0, 1, 2\}$, for what values of y is $^-1 < y + 5 < 7$?

SOLUTION   Replace y in turn with 0, 1, and 2.

$^-1 < 0 + 5 < 7$ true
$^-1 < 1 + 5 < 7$, or $-1 < 6 < 7$ true
$^-1 < 2 + 5 < 7$, or $-1 < 7 < 7$ false

$\therefore$ the given sentence is true if $y = 0$ or $y = 1$.

The numbers in the replacement set (or domain) of a variable that make a given open sentence true when substituted for the variable are called solutions, or roots, of the sentence. The set of all such numbers is called the solution set of the open sentence. To solve an open sentence over a given domain means to determine its solution set. The graph of an open sentence is the graph of its solution set.

EXAMPLE 3   Solve and graph $x + 7 > 8$ over $\mathcal{R}$, the set of real numbers.

SOLUTION   In order for the sentence $x + 7 > 8$ to be true, x must be greater than 1. (It is incorrect to say x must be greater than or equal to 2, since x could be $\frac{3}{2}$, for example.)

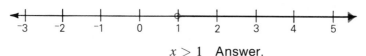

$x > 1$ Answer.

Oral Exercises

If $x \in \{2, 4, 6\}$, find the values of x for which the given statement is true.

1. $x - 2 = 4$
2. $9 - x = 5$
3. $x + 2 > 5$
4. $x - 2 < 7$
5. $x + 3 \geq 5$
6. $8 - x \leq 3$
7. $8 - x \geq 2$
8. $x + 4 \leq 6$

Find the solution set of the given open sentence if $x \in \{$the positive integers$\}$.

9. $x + 4 = 8$
10. $x + 5 \leq 7$
11. $7 > 2 + x$
12. $6 - x < 3$
13. $2x = 10$
14. $^-5 < x < 5$
15. $6 \geq x > 0$
16. $2 \leq x < 5$

17. What is the solution set of $x > 0$ if $x \in \{$the integers$\}$?
18. What is the solution set of $x \leq 0$ if $x \in \{$the integers$\}$?

Written Exercises

Find the solution set of the given open sentence in the given domain.

A
1. $3x + 2 = 2$; $x \in \{0, 1, 2\}$
2. $13 - 2x = 9$; $x \in \{0, 1, 2\}$
3. $3x + 2 > 2$; $x \in \{0, 1, 2\}$
4. $13 - 2x > 9$; $x \in \{0, 1, 2\}$
5. $2y \neq 8$; $y \in \{2, 4, 6\}$
6. $2y = y^2$; $y \in \{0, 2, 4, 6\}$
7. $2y < y^2$; $y \in \{0, 2, 4, 6\}$
8. $2y^2 - 1 \geq 0$; $y \in \{1, 3, 5\}$

Find and graph the solution set of the given open sentence if the domain of each variable is $\{$the positive real numbers$\}$.

9. $a \leq 5$
10. $b \geq 1$
11. $2 < c \leq 5$
12. $6 \geq d > 3$
13. $e + 5 = 11$
14. $f + 5 > 11$
15. $5 - x = 3$
16. $x - 5 = 3$
17. $4y + 3 = 19$
18. $4y + 3 > 19$
19. $w(w + 4) = 5$
20. $m(m + 4) > 5$

B
21. $4 < w + 3 < 8$
22. $7 > w - 2 > 3$
23. $3u + 4u = 7u$
24. $u \cdot u^2 = u^3$
25. $m \geq m - 1$
26. $n \geq n + 1$
27. $t - 1 \leq t$
28. $3 - x \leq x$

In Exercises 29–41, substitute the members of the given domain in the open sentence. List the resulting statements that are true.

EXAMPLE $2x + 3y = 13$; $x \in \{1, 2\}$, $y \in \{3, 4\}$

SOLUTION $2(1) + 3(3) = 13$; false $2(1) + 3(4) = 13$; false
                $2(2) + 3(3) = 13$; true $2(2) + 3(4) = 13$; false

29. $x + y = 8$; $x \in \{3, 4\}$, $y \in \{3, 4\}$ **30.** $h - w = 7$; $h \in \{9, 10\}$, $w \in \{0, 3\}$

31. $2r + s = 12$; $r \in \{5, 6\}$, $s \in \{3, 4\}$ **32.** $m < 2n$; $m \in \{3, 4\}$, $n \in \{1, 2\}$

33. $q + 2 > 3p$; $p \in \{2, 3\}$, $q \in \{4, 5\}$

34. $a + bc = a(b + c)$; $a \in \{1\}$, $b \in \{2\}$, $c \in \{0, 1\}$

35. $de + f = d + ef$; $d \in \{0\}$, $e \in \{1\}$, $f \in \{2, 3\}$

36. $3r = 2t + 9$; $r \in \{5, 6, 7\}$, $t \in \{3, 4, 5\}$

37. $x + 2y = 10$; $x \in \{3, 4, 5\}$, $y \in \{2, 3, 4\}$

38. $3w + 2k > 26$; $w \in \{4, 5, 6\}$, $k \in \{4, 5, 6\}$

39. $p < 2q + 1$; $p \in \{4, 5, 6\}$, $q \in \{0, 1, 2\}$

40. $2x - y \neq 3x - 2y$; $x \in \{2, 3, 4\}$, $y \in \{1, 2, 3\}$

41. $2w + x \neq w + 3x$; $w \in \{0, 1, 2\}$, $x \in \{0, 2, 4\}$

Write an open sentence for which the solution set over the set of real numbers is the given set.

C   **42.** $\{4\}$ **43.** $\{\frac{1}{2}\}$ **44.** $\{0.6\}$ **45.** $\emptyset$

46. {the positive real numbers}   **47.** {the real numbers}

1-9 Applying Open Sentences

When using algebra to solve problems, you often need to translate word phrases and sentences into mathematical phrases and sentences. You already know that the word "equals" is translated as "=" in a mathematical sentence, and that "is greater than" is translated as ">," and so on. Here is a short "dictionary" of other words and the symbols associated with them:

Word phrase	Mathematical phrase
the sum of 6 and 5	$6 + 5$
the product of 7 and 3	$(7)(3)$
the difference of 8 and 1	$8 - 1$
the quotient of 5 and 2	$5 \div 2$ or $\frac{5}{2}$
three more than x	$x + 3$
three less than x	$x - 3$

EXAMPLE   Represent each word sentence by an open mathematical sentence.

    a. The sum of 7 and twice a certain number n is greater than 31.

    b. The number of boys, b, in a class, plus half that number from another class is 27.

    c. The quotient of a certain number y and three is 7 less than 5 times the number.

SOLUTION   **a.** $7 + 2n > 31$

             b. $b + \frac{1}{2}b = 27$

             c. $\dfrac{y}{3} = 5y - 7$

Oral Exercises

Translate the given word sentence into a mathematical sentence.

1. The sum of three and nine is twelve.
2. The difference of thirteen and eight is five.
3. The product of six and twelve is seventy-two.
4. The quotient of twenty-four and four is six.
5. The difference of a number x and two is two.
6. The quotient of eleven and a number y is one.
7. The sum of seventeen and a number n is thirty.
8. The product of a number w and five is forty.

9. Write a word sentence for $x + 5 = 13$.
10. Write a word sentence for $9y = 36$.

Written Exercises

Translate the given word sentence into an open mathematical sentence and find the solution set of this sentence over the set of real numbers.

A   1. The sum of five and x is eleven.
    2. The difference of twelve and y is four.
    3. The product of two and g is ten.
    4. The quotient of twenty and w is five.

Translate the given open mathematical sentence into a word sentence.

5. $x - 4 = 5$ 6. $3y = 21$ 7. $w + 8 = 17$

8. $\dfrac{x}{7} = 6$ 9. $14 - y = 8$ 10. $11 + w = 20$

Represent the given word sentence by an open mathematical sentence and find the solution set of this sentence over the set of real numbers.

B 11. The sum of two times x and five is twenty-five.

12. The product of nine and the sum of two and y is thirty-six.

13. The sum of sixteen and the product of three and f is twenty-five.

14. The sum of twice w and seven is twenty-three.

15. The quotient of m and seven, decreased by five, is one.

16. Eight is one less than the product of three and n.

17. Thirty-one is three more than the product of seven and t.

18. Three times the sum of r and eight is thirty.

19. There are x oatmeal cookies and three times as many bran cookies in a box of two dozen.

20. There are g girls, and seven more boys than girls, on a tennis team of thirty-five members.

21. Nine more than three times x is twenty-four.

22. Of twelve birds at the bird feeder, there are c chickadees and four more blue jays than chickadees.

Self-Test 3

VOCABULARY   mathematical sentence (p. 29)
open sentence (p. 29)
equation (p. 29)
inequality (p. 29)
solution (p. 29)
root (p. 29)
solution set (p. 29)

1. Find and graph the solution set of $x \leq 4$. The domain of x is {the positive real numbers}.   *Obj. 1, p. 28*

Translate each word sentence into a mathematical sentence.

2. The product of two and the sum of n and three is twelve.   *Obj. 2, p. 28*

3. Six less than x is three times the quotient of x and 9.

Check your answers with those at the back of the book.

programming in BASIC

A computer can handle variables just as you can. For example, it can find the value of

$$X \uparrow 2 + 3$$

whenever you give it a value of X.

How do you give the computer a value of X?

If you are working at a terminal, you can use an INPUT statement. When the computer comes to that statement in the program, it types a question mark and waits for you to type in the value you want. Your program might look like this:

```
10   INPUT  X
20   PRINT  X↑2+3
30   END
```

You RUN it for each value of X. RUNS for 0, 1, 2, and 3 are shown at the right. Thus, the set of values of $x^2 + 3$ when the domain of x is $\{0, 1, 2, 3\}$ is $\{3, 4, 7, 12\}$.

If you are sending cards to a distant computer, it is not very practical to give only one value at a time. We will postpone this work until page 46.

RUN

?0
3

END
RUN

?1
4

END
RUN

?2
7

END
RUN

?3
12

END

Exercises

Using a program like that above, find the value of each expression if the domain of x is $\{0, 1, 2, 3, 4\}$. Write $3x$ as $3 * X$, $6x^2$ as $6 * X \uparrow 2$, and so on.

1. a. $2x$ b. $2x + 1$ c. $2x + 2$
2. a. $5x$ b. $5x + 1$ c. $5x + 2$
3. a. $2x^2$ b. $2x^2 + 1$ c. $2x^2 + x + 1$
4. a. $5x^2$ b. $5x^2 + 1$ c. $5x^2 + x + 1$
5. a. x^3 b. $x^3 + 1$ c. $x^3 + 2x + 1$
6. a. $3x^3$ b. $3x^3 + 1$ c. $3x^3 + x^2 + 1$

Write a program to find the value of each expression if the domain of x is $\{\frac{1}{2}, \frac{3}{4}, 1, \frac{5}{4}\}$.

7. a. $2x$ b. $2x^2$ c. $2x^3$
8. a. $16x^2$ b. $16x^2 + 1$ c. $16x^2 + x + 1$
9. a. $64x^3$ b. $64x^3 + 2$ c. $64x^3 + 16x^2 + 4x + 2$

Chapter Summary

1. A *numerical expression* is a name for a number. The number is called the *value* of the expression. A numerical expression is *simplified* when it is replaced by the simplest name of its value.

2. A letter or symbol used to stand for a number or numbers is called a *variable*. The set of numbers a variable can stand for is called the *replacement set*, or *domain*, of the variable.

3. In the expression 4^2, 4 is the *base* and 2 is the *exponent*.

4. On a number line the point paired with a number is called the *graph* of that number. The number paired with a point is called the *coordinate* of that point. The set of numbers corresponding to all the points on a number line is called the set of *real numbers*.

5. Every *rational number* can be named by a terminating or repeating decimal, and every terminating or repeating decimal names a unique rational number.

6. Decimal numerals that are nonterminating and nonrepeating name *irrational numbers*.

7. The set of rational numbers together with the set of irrational numbers make up the set of real numbers.

8. A *set* is any collection of objects called the *members*, or *elements*, of the set.

9. Equations or inequalities which contain one or more variables are called *open sentences*. The set consisting of the members of the domain of the variable(s) for which an open sentence is true is called the *solution set* of the sentence.

Chapter Review

1. Give the simplest name for the value of $623 + 326$.   *1-1*
   a. 949   b. 297   c. 303   d. 999

2. Evaluate the expression $p - q$ when $p = 25$ and $q = 8$.
   a. 33   b. 7   c. 17   d. 13

3. Simplify $5[24 - 6(3)]$.   *1-2*
   a. 102   b. 120   c. 90   d. 30

4. Find the value of $9n - 2(n + 11)$ when $n = 4$.
   a. 39   b. 6   c. 66   d. 50

5. Simplify $3^2 + 5(2)^2$.   *1-3*
   a. 29   b. 109   c. 19   d. 26

6. Evaluate $5y^2 - 2$ when $y = 3$.

    a. 13   **b.** 47   **c.** 173   **d.** 43

7. State the coordinate of the point on the number line at which you would arrive if you were to start at the origin, move 2 units in the negative direction, then 5 units in the positive direction.   *1-4*

    a. $^-7$ **b.** 3   **c.** $^-3$ **d.** 7

8. Find the decimal equivalent for $\frac{3}{8}$.   *1-5*

    a. 0.275   **b.** 0.33   **c.** 0.625   **d.** 0.375

9. Find a rational number halfway between 2.4 and 2.7.

    a. 2.55   **b.** 2.50   **c.** 2.45   **d.** 2.525

10. Replace $\underline{\ ?\ }$ in "3 $\underline{\ ?\ }$ $\{1, 3, 5\}$" by a symbol to make a true statement.   *1-6*

    a. $\in$ **b.** $\subset$ **c.** $=$ **d.** $<$

11. Which choice names the set of even positive integers?   *1-7*

    a. $\{0, 2, 4, 6, \ldots\}$ **b.** $\{1, 3, 5, \ldots\}$ **c.** $\{2, 4, 6, \ldots\}$

12. Find the solution set of $3y + 4 = 13$ if the domain of y is $\{^-3, ^-1, 1, 3\}$.   *1-8*

    a. $\{^-3\}$ **b.** $\{^-1\}$ **c.** $\{1\}$ **d.** $\{3\}$

13. Find the solution set of $x + 9 = 2$ if the domain of x is {the positive real numbers}.

    a. $\{^-7\}$ **b.** $\{7\}$ **c.** $\emptyset$ **d.** $\{0\}$

14. Translate the following word sentence into a mathematical sentence: "The product of three and the sum of y and five is fifteen."   *1-9*

    a. $3y + 5 = 15$ **b.** $3 + 5y = 15$ **c.** $3(y + 5) = 15$

15. Translate $9x = 72$ into a word sentence.

    a. The product of 9 and x is 72.

    b. The quotient of 9 and x is 72.

    c. 72 is the difference of 9 and x.

Chapter Test

1. Evaluate the expression $x + y$ when $x = 9$ and $y = 12$.   *1-1*
2. Find the value of $4(x + 3) - 2$ when $x = 6$.   *1-2*
3. Simplify $5(6^2 - 3^2)$.   *1-3*
4. State the coordinate of the point on the number line at which you will arrive if you start at the origin, move 4 units in the positive direction, then 7 units in the negative direction.   *1-4*

5. Find the decimal equivalent for $\frac{5}{12}$. 1-5

6. Replace _?_ by $\in$ or $\subset$ in "5 _?_ {5, 10, 15}" to make a true statement. 1-6

7. Draw the graph of {the real numbers between $^-1$ and 2} on the number line. 1-7

8. Find the solution set of $3y - 2 = 4$ if the domain of y is {0, 1, 2}. 1-8

9. Translate into an open mathematical sentence and find the solution set: The product of 3 and n is 21. 1-9

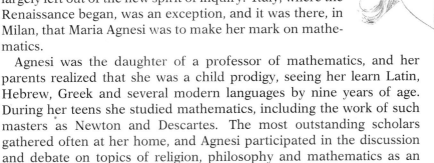

Maria Gaetana Agnesi
1718–1799

"The Witch of Agnesi"—what an ironic designation for a woman whose life was filled with devotion to science and philosophy and selfless service to others! Yet, through an accident of translation, this is the name by which Maria Agnesi became known over the centuries.

As the Renaissance dawned over Europe, dispelling the ignorance that had prevailed for so long, women were largely left out of the new spirit of inquiry. Italy, where the Renaissance began, was an exception, and it was there, in Milan, that Maria Agnesi was to make her mark on mathematics.

Agnesi was the daughter of a professor of mathematics, and her parents realized that she was a child prodigy, seeing her learn Latin, Hebrew, Greek and several modern languages by nine years of age. During her teens she studied mathematics, including the work of such masters as Newton and Descartes. The most outstanding scholars gathered often at her home, and Agnesi participated in the discussion and debate on topics of religion, philosophy and mathematics as an equal, earning their respect and admiration.

When she was twenty, Agnesi began a ten-year project that started as a guide to her younger brothers' studies but became her two-volume *Analytical Institutions,* one of the first comprehensive textbooks of calculus. Her distinction was that she synthesized the work of many mathematicians that had been found in bits and pieces in many different sources. Her knowledge of foreign languages helped her greatly in her task. The "witch" is a curve having the equation $xy^2 = a^2(a - x)$, which Agnesi included in her book. Her work was so highly regarded that she was presented with gifts of diamonds and other precious stones by Empress Maria Theresa and Pope Benedict XIV.

When her father died in 1752, Agnesi withdrew to a more secluded life and gave up her studies of mathematics completely. She spent the latter years of her life in service to the poor and sick.

This research submarine can carry four scientists and up to 3150 kg of payload to depths of 2400 m.

2

Basic Properties
of Real Numbers

Axioms and Theorems

OBJECTIVES *for Sections 2-1 through 2-3:*
1. *Determine whether a statement involving a quantifier is true or false.*
2. *Use the addition, multiplication, and equality axioms to simplify expressions and give reasons for statements in proofs.*

2-1 Quantifiers

Do any real numbers satisfy the equation

$$x + 3 = 3 + x?$$

If you replace x with 2, you obtain the true statement

$$2 + 3 = 3 + 2.$$

In fact, whatever numeral you use in place of x in the given equation, you obtain a true statement. Here are several ways to state this fact:

For all real numbers x, $x + 3 = 3 + x$.

For each real number x, $x + 3 = 3 + x$.

For every real number x, $x + 3 = 3 + x$.

For any real number x, $x + 3 = 3 + x$.

If x is any real number, then $x + 3 = 3 + x$.

Now consider the following statement:

$$\text{For each integer } y, \ y + 2 > 3.$$

This assertion is certainly false, because when you replace y with 1 you obtain a false statement:

$$1 + 2 > 3.$$

On the other hand, when you write 4 in place of y, you convert the open sentence "$y + 2 > 3$" into a true statement:

$$4 + 2 > 3.$$

Because there is an integer y for which "$y + 2 > 3$" is a true statement, you can make the following true assertion:

$$\text{There is an integer } y \text{ such that } y + 2 > 3.$$

Of course, there are many integers which satisfy "$y + 2 > 3$." But the existence of *at least one* is enough to guarantee the truth of the given assertion. Some other ways to state this assertion are:

There exists an integer y such that $y + 2 > 3$.

For at least one integer y, $y + 2 > 3$.

For some integer y, $y + 2 > 3$.

Such key words and phrases as *all, each, every, any, there is, there exists, there are, there exist, some,* and *at least one* involve the idea of "how many" or of "quantity." For this reason you call such an expression a **quantifier** when it is used in combination with a variable in an open sentence.

EXAMPLE   Which, if any, of the following statements are true?
a. There exists a real number m such that $3m + 7 = 13$.
b. For some real number r, $r < r + 1$.

SOLUTION   a. True, since $3 \cdot 2 + 7 = 13$.
b. True; in fact, the "universal" statement, "For *all* real numbers r, $r < r + 1$" is true.

Oral Exercises

Tell which of the following statements are true and which are false.

1. All integers are either positive or negative numbers.
2. There exists a real number c such that $(^-c) > 0$.

3. For some real number x, $x^4 = 16$.

4. For all real numbers y, y^3 is a positive number.

5. If n is any real number, then $n \le n + 1$.

6. Which of the integers greater than $^-5$ and less than 5 will, when substituted for t, make $2t - 1 > 0$ a true statement?

Written Exercises

Show that each of the following statements is true by finding a value of the variable for which the statement is true.

A   1. There is a real number k such that $2k + 4 = 0$.

2. At least one whole number satisfies $w^2 = 2w$.

3. There exists a whole number t such that $t^2 = t$.

4. For some integer y, $3 + 3y \le 4$.

Show that each of the following statements is false by finding a value of the variable for which the statement is false.

5. For every integer p, $p + 2 \ne 2$.

6. Any whole number w is less than $2w$.

7. For all real numbers n, $n^2 > n$.

8. For each integer x, $x > \dfrac{x}{2}$.

Write each statement as an expression for y in terms of x.

EXAMPLE   For each real number x, there is a real number y that is three times as great as x.

SOLUTION $y = 3x$

B   9. For each real number x, there is a real number y that is twice as great as x.

10. For every real number x, there is a real number y that is 5 less than x.

11. For any real number x, there is a real number y that is 6 more than twice x.

12. For each natural number x, there is a natural number y that exceeds x by 1.

13. There is a nonzero whole number y such that for every whole number x, the quotient of x divided by y equals y.

14. For any real number x, there is a real number y that is 4 greater than x.

2-2 Some Axioms of Addition and Multiplication

Addition assigns to any *two* real numbers, the **addends,** a *unique* (one and only one) real number, the **sum** of the addends (Figure 1). There-fore, we call addition a **binary operation** on real numbers (**bi** means "two").

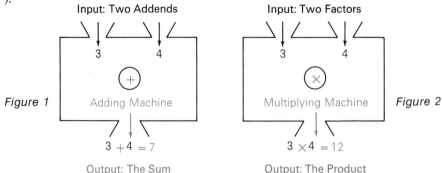

Figure 1   Adding Machine   Multiplying Machine   Figure 2

Output: The Sum   Output: The Product

Similarly, multiplication is a binary operation on real numbers be-cause it assigns to any *two* real numbers, the **factors,** a unique real number, the **product** of the factors (Figure 2).

All *properties* or facts concerning sums and products in the set $\Re$ of real numbers follow from a handful of basic statements, called **axioms, postulates,** or **assumptions,** that are accepted as true.

Some Axioms of Addition and Multiplication in $\Re$		
Name of Axiom	Statement of Axiom	Example
1. Closure	For all real numbers a and b, $a + b$ and ab are real numbers.	$4 + 3 = 7 \in \Re$ and $4 \cdot 3 = 12 \in \Re$.
2. Commutative	For all real numbers a and b, $a + b = b + a$ and $ab = ba$.	$4 + 3 = 3 + 4$ and $4 \cdot 3 = 3 \cdot 4$.
3. Associative	For all real numbers a, b, and c, $(a + b) + c = a + (b + c)$ and $(ab)c = a(bc)$.	$(4 + 3) + 7 = 4 + (3 + 7)$ and $(4 \cdot 3)7 = 4(3 \cdot 7)$.
4. Identity	There are real numbers 0 and 1, with $0 \neq 1$, such that if a is any real number, then $a + 0 = a$ and $0 + a = a$; $a \cdot 1 = a$ and $1 \cdot a = a$.	$4 + 0 = 4, 0 + 4 = 4$; $3 \cdot 1 = 3, 1 \cdot 3 = 3$.

Because of Axiom 4, we say that 0 is the **identity element for addition** and 1 is the **identity element for multiplication** in $\Re$.

The commutative and associative axioms allow you to add or multiply numbers in any order and in any grouping. When you are adding or multiplying three or more numbers, you can sometimes use this fact to simplify your work.

EXAMPLE 1 $27 + 3\frac{1}{2} + 43 + 1\frac{1}{2} = (27 + 43) + (3\frac{1}{2} + 1\frac{1}{2})$
$$= 70 + 5$$
$$= 75$$

EXAMPLE 2 $\frac{9}{7} \times 4 \times 21 \times 25 = (\frac{9}{7} \times 21)(4 \times 25)$
$$= 27 \cdot 100$$
$$= 2700$$

The following assumptions, in addition to the substitution principle (page 2), govern the use of the equality symbol $=$.

Axioms of Equality

For all real numbers a, b, and c:

Reflexive $a = a$.

Symmetric   If $a = b$, then $b = a$.

Transitive   If $a = b$ and $b = c$, then $a = c$.

EXAMPLE 3   Name the axiom of equality that is illustrated.
  a. If $4 + {}^-3 = 1$, then $1 = 4 + {}^-3$.
  b. If $1 + 5 = 6$ and $6 = 4 + 2$, then $1 + 5 = 4 + 2$.
  c. If $a = b$, $b = c$, and $c = d$, then $a = d$.

SOLUTION   **a.** The symmetric axiom.
  b. The transitive axiom.
  c. The transitive axiom in two steps:
    Step 1. If $a = b$ and $b = c$, then $a = c$.
    Step 2. If $a = c$ and $c = d$, then $a = d$.

Oral Exercises

Name the axiom that guarantees the truth of each statement. Assume that the replacement set of each variable is the set $\mathcal{R}$ of real numbers.

EXAMPLE   For every p, $3 + (p + 5) = 3 + (5 + p)$.

SOLUTION   Commutative axiom of addition.

1. $a = 1 \cdot a$
2. For each n, $3(2n) = (3 \cdot 2)n$.
3. ${}^-8({}^-6)$ is a real number.
4. $(x + y) + z = x + (y + z)$
5. If $d = 3$ and $3 = e$, then $d = e$.
6. If ${}^-4 + 4 = 0$, then $0 = {}^-4 + 4$.
7. For each x and each y, $[5 + 6(x + y)] + (x + y) = 5 + [6(x + y) + (x + y)]$.

Name the axioms that justify the lettered steps in these chains of equalities. In writing each chain, the transitive axiom of equality is used at least twice.

8. $(156 + 237) + 44 = (237 + 156) + 44$ (a)
$$= 237 + (156 + 44) \quad \text{(b)}$$
$$= 237 + 200 \quad \text{(c)}$$
$$= 437 \quad \text{(d)}$$

9. $(\frac{1}{2} \times 137) \times 4 = \frac{1}{2} \times (137 \times 4)$ (a)
$$= \frac{1}{2} \times (4 \times 137) \quad \text{(b)}$$
$$= (\frac{1}{2} \times 4) \times 137 \quad \text{(c)}$$
$$= 2 \times 137 \quad \text{(d)}$$
$$= 274 \quad \text{(e)}$$

10. $[5 \times (116 + 47)] \times \frac{1}{5} = (5 \times 163) \times \frac{1}{5}$ (a)
$$= (163 \times 5) \times \frac{1}{5} \quad \text{(b)}$$
$$= 163 \times (5 \times \frac{1}{5}) \quad \text{(c)}$$
$$= 163 \times 1 \quad \text{(d)}$$
$$= 163 \quad \text{(e)}$$

11. For all real numbers m: $4 \times (m \times 16) = 4(16m)$ (a)
$$= (4 \cdot 16)m \quad \text{(b)}$$
$$= 64m \quad \text{(c)}$$

In Exercises 12–14, state whether the relationship is reflexive, symmetric, and/or transitive.

12. is less than   **13.** is greater than or equal to   **14.** is twice as large as

Written Exercises

In Exercises 1–9 simplify the expression.

A

1. $48 + 902 + 9 + 5791$ **2.** $24(5)(12)$ **3.** $3\frac{1}{4} + 6 + 2\frac{1}{4}$

4. $5\frac{1}{3} + 2\frac{2}{5} + \frac{2}{3} + 3\frac{3}{5}$ **5.** $36(\frac{5}{6})(81)(\frac{4}{9})$ **6.** $\frac{5}{8} + 6\frac{1}{3} + \frac{3}{4} + \frac{5}{6}$

7. $2\frac{1}{8} + 10\frac{1}{5}(\frac{5}{6}) + 9\frac{3}{4}$ **8.** $97\frac{3}{7} + 3\frac{1}{4} + 2\frac{2}{7} + \frac{27}{28}$ **9.** $6\frac{3}{10} + \frac{2}{5} + 4(3\frac{1}{8}) + 7\frac{4}{5}$

Given that the replacement set of the variable in each of the following sentences is the set $\mathcal{R}$ of real numbers, specify the solution set of the sentence.

EXAMPLE 1 $(x + 5) + 6 = x + (5 + 6)$

SOLUTION   The associative axiom of addition guarantees that the equation is true for every real value of x.

∴ the solution set is $\mathcal{R}$.

10. $(6 + w) + 3 = 3 + (w + 6)$ **11.** $5 \cdot 9c < 9 \cdot 5c$

12. $4(p \cdot 7) \leq (4 \cdot p)7$

13. $3 + (4 \cdot 5)u = u(12 + 8) + 2$

B

14. $5 + (m + 3) < (m + 5) + 3$

15. $3 \cdot k(3 + 2) = 5 \cdot 3k$

16. $[(\frac{3}{5} + \frac{2}{5})g + 6] + 9 = 7 + [(\frac{4}{7} + \frac{3}{7})g + 8]$

17. $(5 + y)4 = 5(4) + y(4)$

18. $3x + [(2 + 0) + 6] = (3x + 20) + 6$

19. $4n \cdot 12 = 4 \cdot 6 \cdot 4n$

20. $(2 + 5)4 \cdot d = 4 \cdot 10d$

21. $(3h + 2h) + 6 = 6 + (5h + 3h)$

22. $(6l + 3) + 2l = 6l + (3 + 2l) + l$

23. $50 + 3(w + 2) = 6 + w(2 + 3) \cdot 0 + (1 + 2) \cdot w$

In each of Exercises 24–35 an operation ★ is defined over the set W of whole numbers, {0, 1, 2, 3, 4, . . .}. (Note that ★ is defined differently in each exercise.) In each case:

a. Find 3 ★ 4.

b. Determine whether the statement "For all whole numbers a and b, a ★ b is a whole number" is true or false.

c. State whether or not ★ is (1) commutative, (2) associative.

d. Does W contain an identity element for ★?

EXAMPLE 2 $a ★ b = ab$

SOLUTION   **a.** 3 ★ 4 = 12.   **b.** true   **c.** ★ is commutative and associative
d. yes, 1

EXAMPLE 3 $a ★ b = a^2b$

SOLUTION   **a.** $3 ★ 4 = 3^2(4) = 36$ **b.** true

c. Since $b ★ a = b^2a$, which is not, in general, equal to a^2b, ★ is not commutative. Since $(a ★ b) ★ c = (a^2b)^2c$ and $a ★ (b ★ c) = a^2(b^2c)$, ★ is not associative.   **d.** no

C

24. $a ★ b = a + b$

25. $a ★ b = a - b$

26. $a ★ b = a^2b^2$

27. $a ★ b = a + 2b$

28. $a ★ b = 2a + b$

29. $a ★ b = ab + 1$

30. $a ★ b = (a + b) + 1$

31. $a ★ b = a^2 + b^2$

32. $a ★ b = a(b + 1)$

33. $a ★ b = a^3b$

34. $a ★ b = \dfrac{a}{3} + b$

35. $a ★ b = a^2\left(\dfrac{b}{2}\right)$

DIVERSION

Arrange 16 toothpicks to form 5 squares as shown at the right. Can you move 3 of the toothpicks so that only 4 squares are formed?

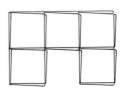

programming in BASIC

You can make the computer add by using two INPUTs:

```
10   INPUT A
20   INPUT B
30   PRINT A+B
40   END
```

By changing line 30 to

```
30   PRINT A*B
```

you can make the computer multiply.

You can make the computer repeat the program without waiting for you to type RUN each time by inserting the statement:

```
35   GOTO 10
```

This means that as soon as the computer has printed the value of $A + B$ (or $A * B$), it will *go to* line 10 and print another question mark. Such a program will NEVER END by itself. After you have found all the values you need, you can stop the computer by some special step. (On some systems, after it prints ?, you press the CONTROL and C keys on the terminal together—see your computer manual.)

You see now why we numbered the lines 10, 20, 30, 40 instead of 1, 2, 3, 4. By leaving intervals, we can insert a statement such as 35. *The computer will go through the statements in numerical order, no matter in what order you typed them in.*

If you know ahead of time the values whose sums (or products) you want to find, you can put them all into the program before you RUN it by using READ and DATA statements. You can use these statements also if you prepare cards to be sent to a computer. Look at this program:

```
10   READ A, B
20   DATA 10, 23, 14, 72, 125, 442, 1492, 1776
30   PRINT A+B
35   GOTO 10
40   END
```

The computer will take the first two values in the DATA statement for A and B, compute and PRINT A + B, GOTO 10, READ the next pair, and so on, until it has used all the values in the DATA statement of line 20:

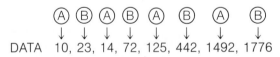

```
         Ⓐ  Ⓑ  Ⓐ  Ⓑ  Ⓐ   Ⓑ   Ⓐ    Ⓑ
         ↓   ↓   ↓   ↓   ↓    ↓    ↓     ↓
DATA   10, 23, 14, 72, 125, 442, 1492, 1776
```

Then it will print "OUT OF DATA" and stop.

When you use READ and DATA statements, you must arrange the DATA exactly in the order in which the READ statement will read them.

Exercises

1. RUN the program in the text several times using different INPUTs (or different DATA).

When you want to clear one program from the computer before starting a new one, type the command SCR (short for *scratch*).

2. Rewrite the program on page 34 using READ and DATA statements. RUN the resulting program.

3. Write a program that will INPUT A, B, C and PRINT (A + B) * C.

4. Add a line to the preceding program so that it will PRINT A * C + B * C also.

Write a program to find a value for each of these expressions.

5. L * W * H

6. 2 * L * W + 2 * W * H + 2 * H * L

7. 2 * 3.14159 * R

8. 3.14159 * R * R

2-3 The Distributive Axiom

The left-hand array in Figure 3 below contains 3×10 dots. On the right, the array has been separated into two smaller arrays, one containing 3×7 dots, the other 3×3 dots.

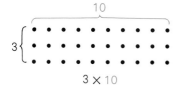

3 × 10

3 × 7   3 × 3   *Figure 3*

The separation does not change the number of dots in the array. Thus:

$$3 \times 10 = 3 \times 7 + 3 \times 3$$
$$\text{or} \quad 3(7 + 3) = 3 \times 7 + 3 \times 3$$

This last equation suggests another axiom: Multiplication is *distributive* with respect to addition.

Distributive Axiom of Multiplication with Respect to Addition

For all real numbers *a*, *b*, and *c*,

$$a(b + c) = ab + ac \quad \text{and} \quad (b + c)a = ba + ca.$$

By applying the symmetric axiom of equality, you can also state the distributive axiom in the following way:

For all real numbers a, b, and c,

$$ab + ac = a(b + c) \quad \text{and} \quad ba + ca = (b + c)a.$$

The following examples apply the distributive axiom.

EXAMPLE 1 **a.** $80(\frac{1}{4} + \frac{1}{5}) = 80 \cdot \frac{1}{4} + 80 \cdot \frac{1}{5} = 20 + 16 = 36$

b. $(1\frac{2}{3})3 = (1 + \frac{2}{3})3 = 1 \cdot 3 + \frac{2}{3} \cdot 3 = 3 + 2 = 5$

c. $\frac{1}{6} \cdot 13 + \frac{1}{6} \cdot 11 = \frac{1}{6}(13 + 11) = \frac{1}{6} \cdot 24 = 4$

EXAMPLE 2 Show that for every real number c,

$$5c + 2c = 7c.$$

SOLUTION

Statement	Reason
1. c is a real number.	Given
2. $5c + 2c = (5 + 2)c$	Distributive axiom
3. $5 + 2 = 7$	Addition fact
4. $(5 + 2)c = 7c$	Substitution principle
5. $\therefore 5c + 2c = 7c$	Transitive axiom of equality

The reasoning from the given statement, or **hypothesis**, "c is a real number," to the final statement, or **conclusion**, "$5c + 2c = 7c$," is called a **direct proof**. Each statement in a proof must be guaranteed by the hypothesis, or by known facts, definitions, or axioms. Assertions that are proved are called **theorems**.

Because properties of real numbers guarantee that for *all* values of the variable, the expressions

$$5c + 2c \quad \text{and} \quad 7c$$

denote the same number, we call them **equivalent expressions**. The expression $5c + 2c$ has two *terms*, $5c$ and $2c$, and the expression $7c$ has only one. When you replace a given variable expression with an equivalent expression having as few terms as possible, you say that you have **simplified** the expression.

Although you would not ordinarily write all the steps given in Example 2, you should know how the simplification depends on the properties of real numbers. The next example shows a brief way of presenting a simplification. Notice that the commutative and associative axioms are used in the first step.

EXAMPLE 3   **a.** Simplify: $4x^2 + 3 + 9x^2 + 8$

b. Use the results of part **(a)** to state a theorem.

SOLUTION   **a.** $4x^2 + 3 + 9x^2 + 8 = (4x^2 + 9x^2) + (3 + 8)$
$$= (4 + 9)x^2 + 11$$
$$= 13x^2 + 11$$

$\therefore 4x^2 + 3 + 9x^2 + 8 = 13x^2 + 11.$

b. For all real numbers x, $4x^2 + 3 + 9x^2 + 8 = 13x^2 + 11.$

Oral Exercises

Name the axioms that justify the steps in each simplification.

1. $28(2\frac{1}{7}) = 28(2 + \frac{1}{7})$
$$= 28(2) + 28(\frac{1}{7})$$
$$= 56 + 4$$
$$= 60$$

2. $(\frac{1}{6} + \frac{3}{4})48 = \frac{1}{6}(48) + \frac{3}{4}(48)$
$$= 8 + 36$$
$$= 44$$

3. $72(2\frac{1}{3} + \frac{1}{4}) = 72(2\frac{1}{3}) + 72(\frac{1}{4})$
$$= 72(2 + \frac{1}{3}) + 18$$
$$= 72(2) + 72(\frac{1}{3}) + 18$$
$$= (144 + 24) + 18$$
$$= 168 + 18$$
$$= 186$$

4. $\frac{2}{5}(67) + 9\frac{3}{5}(67) = (\frac{2}{5} + 9\frac{3}{5})67$
$$= [\frac{2}{5} + (9 + \frac{3}{5})]67$$
$$= [(9 + \frac{3}{5}) + \frac{2}{5}]67$$
$$= [9 + (\frac{3}{5} + \frac{2}{5})]67$$
$$= (9 + 1)67$$
$$= 10(67)$$
$$= 670$$

In Exercises 5–7 state the reason for each statement in the proof.

5. Prove: For each real number n, $n + (1 + 2n) = 3n + 1$.

PROOF

1. n is a real number.
2. $n + (1 + 2n) = n + (2n + 1)$
3. $= (n + 2n) + 1$
4. $= (1 \cdot n + 2 \cdot n) + 1$
5. $= (1 + 2)n + 1$
6. $= 3n + 1$
7. $\therefore n + (1 + 2n) = 3n + 1$

6. Prove: For all real numbers a, b, c, and d, $[(a + b) + c]d = ad + (bd + cd)$.

PROOF

1. a, b, c, and d are real numbers.
2. $[(a + b) + c]d = (a + b)d + cd$
3. $= (ad + bd) + cd$
4. $= ad + (bd + cd)$
5. $\therefore [(a + b) + c]d = ad + (bd + cd)$

State the reason for each statement in the proof.

7. Prove: If a and b are any real numbers, then $ab(a + b) = a^2b + ab^2$.

PROOF

1. a and b are real numbers.
2. ab is a real number.
3. $\quad ab(a + b) = (ab)a + (ab)b$
4. $\qquad\qquad = a(ab) + (ab)b$
5. $\qquad\qquad = (a \cdot a)b + a(b \cdot b)$
6. $\qquad\qquad = a^2b + ab^2$
7. $\therefore\ ab(a + b) = a^2b + ab^2$

8. Can you show a proof for each of the following?

a. Prove: For every real number s, $2(4s + 1) = 8s + 2$.

b. Prove: For all real numbers x, y, z, and w,
$$x[(y + z) + w] = x(y + z) + xw.$$

In Exercises 9–20 simplify the expression.

EXAMPLE $2x + 5y + 4x + y$

SOLUTION $6x + 6y$

9. $5y + 16y$
10. $5p + p + 9p$
11. $3 + 4(k + 2)$
12. $2m(4 + 6)$
13. $\frac{1}{4}x + 2x + 2\frac{3}{4}x$
14. $2(3 + 2r) + 6r$
15. $3c + 5(2 + c)$
16. $\frac{1}{3}(6m + 2)$
17. $3q + 6m + 5q + m + q$
18. $(2u + 5u^2) + (2u^2 + 5u)$
19. $2x + 4y + z + 8x + 3y$
20. $(4w^4 + 2w^2) + (w + 6w^4)$

Written Exercises

Simplify each expression.

EXAMPLE 1 $16p + 27q + 9p + 13r + 3q$

SOLUTION $16p + 9p + 27q + 3q + 13r = (16 + 9)p + (27 + 3)q + 13r$
$\qquad\qquad\qquad\qquad\qquad\qquad = 25p + 30q + 13r$

A
1. $9x^2 + 3x^2 + 11x^2$
2. $4y^4 + 3y^3 + y^4$
3. $16a^2 + 16 + 16a^2$
4. $10n + 3n^2 + 9n^2$
5. $4c + 12 + 9c + 15c + 9$
6. $5h + 8 + 19 + 18h + h$
7. $2l^2 + 3m + 5l + l^2 + m$
8. $9x + y + 4x + 7x + 9y + 2$
9. $p^3 + 2p + 3p^3 + p^2 + p$
10. $3a + 4c + a + 3b + c + 9a$
11. $(4 + 6x + 2x^2) + (5x + x^2)$
12. $(5 + t^2 + 4t) + (2t^2 + 3 + 6t)$
13. $(5y + y^3 + 2y^2) + (6y^3 + y)$
14. $(w^4 + 5w^2) + (2w^2 + 5w + w^4 + 1)$

15. $(3 + s^2 + 2s^4 + 9s) + (3s^3 + 7 + 6s^4 + s)$

16. $(5w^3 + 3w^2 + 2w + 6) + (8 + 5w + 2w^2 + w^3)$

17. $(9h + 2k^2 + h^2 + 5k) + (2k + h + 3h^2 + 5k^2)$

18. $(12r^3 + 5r^2 + 2r + 8) + (3r^4 + 9 + 3r^2 + r^3)$

EXAMPLE 2 $5(3x + 4) + 3(6x + 2)$

SOLUTION   1. Apply the distributive axiom.

$$5(3x + 4) + 3(6x + 2) = (15x + 20) + (18x + 6)$$

2. Simplify the resulting expression.

$$(15x + 20) + (18x + 6) = (15 + 18)x + (20 + 6)$$
$$= 33x + 26$$
$$\therefore\ 5(3x + 4) + 3(6x + 2) = 33x + 26$$

B

19. $3(x + y) + 2(x + y) + 4x$

20. $5(p + q) + 6q + 2(p + q)$

21. $5(a^2 + 3a) + 3(4a^2 + 5) + a$

22. $6(3b + 2) + 4(b^2 + 2b) + b^2$

23. $9(z + 3z^2) + 7(z^2 + 4z)$

24. $(7d^2 + 3d)2 + (d + 2d^2)3$

25. $5 + 3(k + 1) + 6(2k + 1)$

26. $4(l + 2) + 3l + 5(3l + 3)$

27. $4(h + 3i + j) + 2(h + 3j)$

28. $3(2p + q + 3r) + 8(2q + r)$

29. $6(3y^2 + 2y + 4) + 4(2y^2 + 5y)$

30. $5(5 + 3u + 2u^2) + 3(u^2 + 5)$

31. $2(2d^2 + 3d + 4) + 7(1 + 2d + d^2)$

32. $4(2x + y + z) + 5(2z + x + 3z)$

33. Four times the sum of r and s, increased by twice the sum of r and $2s$.

34. Twice the sum of $2x$ and y, increased by three times the sum of x and $2y$.

35. Twice the sum of two and the square of t, increased by the sum of t-squared and 4.

36. Triple the sum of three and the cube of w, increased by twice the sum of w-cubed and 3.

37. The product of six and the square of z, increased by the sum of seven z-squared and 6.

38. Quadruple the product of two and the cube of c, increased by the sum of three and two times c-cubed.

39. Eleven more than the sum of two, k, and the fifth power of k.

C

40. $5[3x + 4(3 + 2x)] + 6$

41. $3[2y + 3(y + 1) + 5] + 3(5 + 3y)$

42. $2[3(2x + 5) + 2(3x + 5)] + [3 + 5(4x + 1) + 2x]3$

43. $16 + 4[2m^2 + 11(3m^2 + 7)] + 5(2 + m^2)$

44. $3[4(r + 2) + 3(2s + r + 1)] + 6(2r + s + 3)$

45. $5[2(k + 2l + 7)] + 3[2(3k + l) + 3(2l + k)]$

46. $2[p + 6(p + 3q + 1) + 3(2p + q + 1)] + 4(3p + q)$

47. $4[x + 2(y + z + 2)] + 2[x + 2(y + z + 2)] + x + 2(y + z + 2)$

DIVERSION

A company has three robots, named KT, KC, and KG, which are rented out to do certain jobs. One of the robots washes windows, another mows lawns, and the other rakes leaves.

1. The window washer, which has its own power source, is the laziest.
2. KG, who shares KT's power source, works harder than the lawn mower.

Which robot does which job?

Self-Test 1

VOCABULARY quantifier (p. 40) hypothesis (p. 48)
binary operation (p. 42) conclusion (p. 48)
axioms (p. 42) direct proof (p. 48)
postulates (p. 42) theorem (p. 48)
assumptions (p. 42) equivalent expression (p. 48)
identity element (p. 42) simplified (p. 48)

Show that each of the following statements is true by finding a value of the variable for which the statement is true.

1. For some natural number n, $n + 2n = 15$. *Obj. 1, p. 39*
2. There exists an integer y such that $3y + 4 < 5$.

Complete.

3. The sentence "$m + m + 1 = 1 + 2m$" is true
   a. for all real numbers m.
   b. for some, but not all, real numbers m.
   c. for no real numbers m.

Simplify.

4. $5 + 2m^2 + 7 + 5m^2$ 5. $2(3x + 5) + 3(4 + 2x)$ *Obj. 2, p. 39*

6. Name the axioms that justify the lettered steps in the following chain of equalities. For each real number n:

$$2n + n = 2(n) + 1(n) \quad \text{(a)}$$
$$= (2 + 1)n \quad \text{(b)}$$
$$= 3n \quad \text{(c)}$$
$$\therefore 2n + n = 3n \quad \text{(d)}$$

Check your answers with those at the back of the book.

Addition of Real Numbers

OBJECTIVES for Sections 2-4 through 2-6:
1. *Picture the addition of real numbers on the number line.*
2. *Add two or more real numbers, using the number line if necessary.*
3. *Use addition of real numbers to solve word problems.*
4. *Simplify expressions, and specify and graph solution sets of equations, involving additive inverses or absolute values.*

2-4 Addition on the Number Line

Figure 4 shows how to picture the addition of real numbers on a horizontal number line. The short *black* arrow represents the *positive* number 3 by a move, or *displacement,* of 3 units to the **right** from the origin.

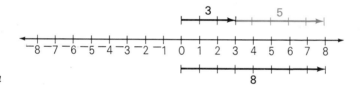

Figure 4

The short *colored* arrow pictures 5 by a move of 5 units to the **right,** starting at the graph of 3. Together, the two moves amount to a move of 8 units to the **right** from the origin. Thus, the diagram pictures the fact that

$$3 + 5 = 8.$$

Moves to the *left* represent *negative* numbers. Thus, to add ⁻5 to 3, start at the head of the arrow representing 3 and move 5 units to the **left.**

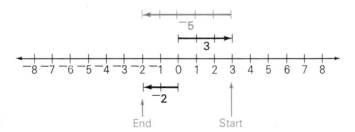

Figure 5

This displacement brings you 2 units to the left of the origin, that is, to the graph of ⁻2.

Figure 5 suggests that

$$3 + {}^{-}5 = {}^{-}2.$$

EXAMPLE State the addition fact suggested by each diagram.

SOLUTION

a.

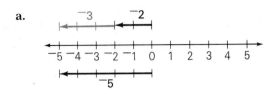

a. $^-2 + {}^-3 = {}^-5$

b.

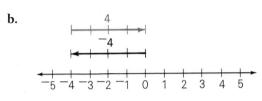

b. $^-4 + 4 = 0$

c.

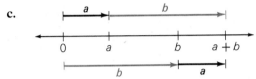

c. $a + b = b + a$

You can use the following rule to find sums on the number line:

To add two real numbers on the number line:

Start at the origin and draw an arrow representing the first number. Then, from the head of that arrow, draw an arrow representing the second number. The arrow from the origin to the head of the second arrow represents the sum of the two numbers.

Oral Exercises

State an addition fact pictured by each diagram.

1.

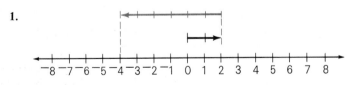

2.

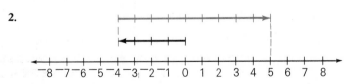

Simplify each expression. Think of displacements along a number line.

3. $^-3 + 5$ **4.** $0 + {}^-8$ **5.** $2 + {}^-2$ **6.** $^-3 + {}^-5$

7. $^-9 + 5$ **8.** $^-2\frac{1}{2} + \frac{1}{2}$ **9.** $^-\frac{9}{5} + {}^-\frac{2}{5}$ **10.** $^-\frac{3}{7} + 0$

11. $200 + {}^-220$ **12.** $^-3.7 + 1.9$ **13.** $^-4.15 + 2.15$ **14.** $90.45 + 9.55$

Written Exercises

In Exercises 1–14:
a. Using the number line if necessary, simplify each given expression.
b. Use the associative axiom of addition to regroup the terms in the given expression, and then simplify the resulting expression.

EXAMPLE 1 $(8 + {}^-10) + 1$

SOLUTION   **a.** $(8 + {}^-10) + 1 = {}^-2 + 1 = {}^-1.$
                b. $(8 + {}^-10) + 1 = 8 + ({}^-10 + 1) = 8 + {}^-9 = {}^-1.$

A   **1.** $(6 + {}^-10) + 3$ **2.** $({}^-8 + 5) + 3$ **3.** $({}^-12 + {}^-9) + 7$

     4. $({}^-7 + 10) + {}^-3$ **5.** $({}^-1.3 + {}^-5.8) + {}^-2.1$ **6.** $({}^-35 + 0) + 35$

     7. $^-6 + (9 + {}^-11)$ **8.** $10 + ({}^-11 + 12)$ **9.** $2\frac{1}{4} + (\frac{3}{4} + {}^-\frac{1}{4})$

     10. $^-3.6 + (2.1 + {}^-3.5)$ **11.** $^-\frac{5}{8} + ({}^-\frac{1}{8} + {}^-\frac{1}{4})$ **12.** $10 + ({}^-110 + {}^-1110)$

     13. $(15 + {}^-7) + ({}^-13 + 9)$ **14.** $({}^-8 + 16) + ({}^-9 + {}^-12)$

In Exercises 15–47 solve each equation over the set $\Re$ of real numbers.

EXAMPLE 2 $4 + a = {}^-8$

SOLUTION   To move from the graph of 4 to the graph of $^-8$, go 12 units to the left.

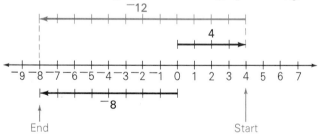

       Check: $4 + a = {}^-8$
                   $4 + {}^-12 \overset{?}{=} {}^-8$ ($\overset{?}{=}$ means "Is this statement true?")
                      $^-8 = {}^-8 \; \checkmark$ ($\checkmark$ means "Yes, it is.")

15. $2 + y = {}^-5$ **16.** $^-3 + p = {}^-5$ **17.** $x + 6 = 4$ **18.** $c + 9 = 2$

19. $^-5 + h = {}^-5$ **20.** $^-3 + l = 3$ **21.** $^-10 = v + {}^-13$ **22.** $9 = {}^-6 + n$

23. $0 = 13 + q$ **24.** $^-8 = r + 8$ **25.** $x + {}^-2\frac{3}{4} = {}^-5$ **26.** $^-2\frac{3}{8} + s = {}^-3$

Solve each equation over the set $\mathcal{R}$.

B 27. $z + {}^-3 = {}^-5 + 6$ 28. $x + x = 0$ 29. $p + p + p = {}^-9$

30. $^-5 + u = {}^-3 + {}^-6$ 31. $^-8 + 6 = y + {}^-4$ 32. $^-6 + {}^-6 = {}^-6 + h + 3$

33. $s + {}^-3 = 6 + {}^-10$ 34. $^-12 = l + l$ 35. $^-19 = 2 + w + {}^-5$

36. $^-11 + m = {}^-10 + 7$ 37. $2\frac{1}{2} + {}^-1\frac{1}{4} = x + x$ 38. $^-5\frac{7}{8} + 1\frac{5}{8} = k + {}^-\frac{7}{8}$

39. $n + {}^-2.75$
$= 2 + {}^-3.25$ 40. $c + 25$
$= {}^-101.1 + 2.25$ 41. $^-68.3 = k + 13.5$

C 42. $^-2 + {}^-x = {}^-3 + 8$ 43. $^-y + 5 = {}^-7 + 3$ 44. $^-5 + {}^-15 = {}^-n + {}^-3$

45. $12 + {}^-7 = 5 + {}^-z$ 46. $^-\frac{1}{2} + {}^-k = {}^-\frac{1}{4} + {}^-\frac{1}{4}$ 47. $^-h + \frac{3}{4} = {}^-\frac{1}{8} + {}^-\frac{5}{8}$

Problems

a. Write an expression for a sum of real numbers that answers the question in each problem.
b. Simplify the expression.
c. Answer the question.

EXAMPLE   An elevator starts at the ground floor and goes up to the 34th floor. It then goes down 15 floors, up 6 floors, and down 10 floors. At what floor is the elevator now?

SOLUTION   **a.** $34 + {}^-15 + 6 + {}^-10$ **b.** $34 + {}^-15 + 6 + {}^-10 = 15$

       c. The elevator is at the 15th floor.

A 1. An elevator starts at the ground floor and goes up to the 36th floor. It then goes down 13 floors, up 10 floors, down 6 floors, and up 5 floors. At what floor is the elevator now?

2. A weather balloon rose to a height of 28,050 m above sea level. It dropped 3182 m and then rose 5796 m. How far above sea level was it at that point?

3. A helicopter travels directly north from its base at Eagle Point for 48 km and then flies directly south for 53 km. Where is the helicopter then located with reference to Eagle Point?

4. If the temperature at midnight is 8°C and if it falls a total of 13° in the next six hours, what is the temperature at 6:00 A.M.?

B 5. Ethel bought four calculators for $12.15, $13.05, $13.40, and $12.85. She later sold these for $12.25, $12.85, $13.15, and $12.75, respectively. What was the net financial result of these transactions?

6. Hank Sander's normal pulse rate is 72 beats per min. After jogging for 10 min, his rate rises by 33 beats. It drops by 28 beats after a 5 min rest period, and then rises by 37 beats after another 10 min of jogging. What is his pulse rate now?

7. On Monday, Sue Brooks bought 550 shares of stock in the Callen Corporation for $8.00 a share. On Tuesday she sold 45 shares for $9.50 a share, on Wednesday she sold 325 shares for $9.00 a share, and on Thursday she sold 180 shares for $7.75 a share. How much money did she make as a result of these transactions?

8. A substance at "absolute zero" ($^-273°C$) is heated until its temperature rises 150°. Then it is cooled until its temperature drops 15°. What is its new temperature?

9. On a certain bank statement a credit is shown by a positive number such as $8.55, while a debit is shown by a negative number like $^-$5.15. What is the net change in the bank balance as a result of the following credits and debits: $19.25, $37.50, $^-$23.20, $74.40, $^-$54.70?

10. On a revolving charge account, Juan Perez purchased $30.50 worth of clothing, and $220.75 worth of electronics equipment. He then made two monthly payments of $54.00 each. If the interest charges for the period of two months were $5.05, what did Juan then owe on the account?

11. A stock selling for $35 per share rose $2.50 per share each of two days and then fell $1.85 per share for each of three days. What was the selling price per share of the stock after these events?

12. The pilot of a jet traveling at 9550 m above sea level is ordered by ground control to descend 1900 m. Later, the pilot is ordered to climb 4500 m to a new altitude. At what altitude will the jet then be flying?

C   13. It takes 4.184 newton-meters of work to raise 1 g of water 1°C. How many newton-meters of work must be used to raise the temperature of a 25 g sample of H_2O from $^-112°C$ to 35°C?

2-5 The Additive Inverse of a Number

Figure 6 pictures an important pairing of real numbers. On the number line, the graphs of paired nonzero numbers, like $^-1$ and 1, or $^-3$ and 3,

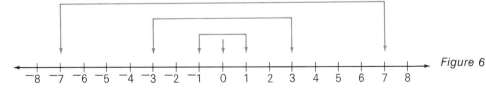

Figure 6

are at the same distance from the origin, but on *opposite* sides of the origin. Zero is paired with itself.

You can check that adding two such paired numbers on a number line gives 0. For example, as Figure 7 suggests,

$$7 + {}^-7 = 0.$$

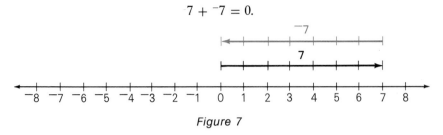

Figure 7

Each number in such a pair is called the **opposite**, or the **additive inverse,** or the **negative** of the other number. The symbol

$$-a \text{ (note the lowered position of the minus sign)}$$

denotes **the opposite of** a, or **the additive inverse of** a, or **the negative of** a.

For example:

$-7 = {}^-7$, read "**the additive inverse of** seven equals negative seven";
$-({}^-3) = 3$, read "**the additive inverse of** negative three equals three";
$-0 = 0$, read "**the additive inverse of** zero equals zero."

The following axiom is a formal way of saying that every real number has a unique opposite, and that the sum of a number and its opposite is always zero.

Axiom of Additive Inverses

For every real number a, there is a unique real number $-a$ such that

$$a + (-a) = 0 \quad \text{and} \quad (-a) + a = 0.$$

The equation $-7 = {}^-7$ indicates that the numerals -7 and ${}^-7$ name the same number. This means that you can always use the numeral -7 (lowered minus sign) in place of the numeral ${}^-7$ (raised minus sign). It also means that you can read -7 either as "negative 7" or as "the opposite of 7." *Throughout the rest of this book, lowered minus signs will be used in the numerals for negative numbers.*

Caution! A *variable* expression like $-a$ should always be read "the opposite of a" or "the additive inverse of a" or "the negative of a." Never call it "negative a," because $-a$ may denote either a negative number, a positive number, or zero.

By looking at the number line in Figure 6 on page 57, you can see that the following statements are true:

1. If a is a positive number, then $-a$ is a negative number; if a is a negative number, then $-a$ is a positive number; if a is 0, then $-a$ is 0.

2. The opposite of $-a$ is a; that is, $-(-a) = a$.

EXAMPLE 1 Simplify: **a.** $-(-4)$ **b.** -0 **c.** $-(1 + 2)$

SOLUTION   **a.** 4   **b.** 0   **c.** -3.

EXAMPLE 2 $-(-2) + (-5) = 2 + (-5) = -3$

Oral Exercises

Name the additive inverse (opposite) of each number.

EXAMPLE $2 + (-3)$

SOLUTION $2 + (-3) = -1$
        The additive inverse of -1 is $-(-1) = 1$.

1. 6   2. -4 3. $\frac{3}{4}$ 4. -2.6 5. $-(-5)$ 6. 7.5
7. $-(-\frac{1}{2})$ 8. 0   9. $2 + (-7)$ 10. $-6 + (-8)$ 11. $-3 + 1$ 12. $-9 + 12$

Simplify each expression.

13. $7 + [4 + (-4)]$ 14. $(-9 + 9) + 15$ 15. $-6 + (-9) + 6$
16. $[6 + (-3)] + 3$ 17. $-y + 7 + y$ 18. $p + (-p + 10)$
19. $-(-105)$ 20. $-[-(-6)]$ 21. $-(-2) + 8$
22. $-(-10) + (-10)$ 23. $-(-3 + 2) + 7$ 24. $-3[x - (6 - x)]$
25. If x is a positive number, then $-x$ is a __?__ number.
26. If x is a negative number, then $-x$ is a __?__ number.

Written Exercises

Simplify each expression.

EXAMPLE 1 $-[-(-5) + (-3)]$

SOLUTION $-[-(-5) + (-3)] = -[5 + (-3)] = -(2) = -2$.

A   1. $-(-8) + 15$ 2. $-10 + [-(-6)]$ 3. $-(-4 + 8)$
    4. $-(-2 + 6) + (-10)$ 5. $[2 + (-5)] + [-(-5)]$ 6. $-[-3.5 + (-2.9)]$
    7. $-\frac{5}{6} + [-(-\frac{1}{6})]$ 8. $-(-2\frac{3}{5}) + \frac{7}{5}$ 9. $-[-(-0.9) + (-1)]$

For the values of x and y given in Exercises 10–15, compute:
a. $-(x + y)$; **b.** $-[-(-x) + (-y)]$.

10. $x = 1$, $y = -2$ 11. $x = -3$, $y = 0$ 12. $x = -4$, $y = -2$

13. $x = -9$, $y = 1$ 14. $x = 0$, $y = 3$ 15. $x = 24$, $y = -30$

In Exercises 16–43 replace the variable, in turn, by the name of each member of $\{-2, -1, 0, 1, 2\}$, and then specify the solution set of each open sentence over the given replacement set. (Recall that $\emptyset$ is the solution set of an equation with no solution.)

EXAMPLE 2 $-x \leq 0$

SOLUTION $-(-2) \leq 0$; that is, $2 \leq 0$; false
$-(-1) \leq 0$; that is, $1 \leq 0$; false
$-0 \leq 0$; that is, $0 \leq 0$; true
$-1 \leq 0$; true
$-2 \leq 0$; true
$\therefore$ over $\{-2, -1, 0, 1, 2\}$, the solution set is $\{0, 1, 2\}$.

B   16. $-x = 1$ 17. $-y = -1$ 18. $0 \leq a$

19. $3 = -x$ 20. $-l = -2$ 21. $-h = -1$

22. $p + (-1) = 1$ 23. $q + 1 = -1$ 24. $-z > 0$

25. $-z < 0$ 26. $-k \geq 0$ 27. $-m > -2$

28. $-e < -1$ 29. $-(-f) + 1 = 0$ 30. $1 + [-(-w)] = -1$

31. $n + 1 \geq -1$ 32. $n + 1 \leq 1$ 33. $x + [-(-2)] = 1$

34. $0 < -r < 2$ 35. $-1 < -p < 2$ 36. $-2 < y < 2$

37. $-2 \leq y \leq 2$ 38. $0 \geq -z > 2$ 39. $0 \geq -z > 1$

40. $-1 < -q < 1$ 41. $-2 < q < 2$

42. $-2 < q \leq 2$ 43. $-1 < t - 1 < 1$

Evaluate each expression if the value of $a = -1$, $b = \frac{1}{2}$, $c = -\frac{3}{4}$, and $d = 1$.

C   44. $-[a - b + (-c + d)]$ 45. $-[a + (-b) + c - d]$

46. $a + [-(b + c + d)]$ 47. $(d + b) + [-(a + c)]$

48. $[-(a + b)] + [-(c + d)]$ 49. $-[-(-b) + c + (-b) + 1]$

50. $-[-(a + b + c)] + (a + b + c)$ 51. $-(a + b) + [-(c + d - 1)]$

52. $-[-a + b + (-c + d)]$ 53. $-[(-a + b) + c + d + 2]$

54. If $x = \frac{1}{3}$, $y = -\frac{1}{2}$, and $z = -\frac{1}{4}$, find an expression in terms of x, y, and z that has a value of $-\frac{2}{3}$.

programming in BASIC

You can make the computer decide between two choices by using a BASIC statement as follows:

IF (statement) THEN (line number)

IF the statement is true for the current value(s) of the variable(s), THEN the computer goes to the given line for its next instruction. Study the following:

```
10  INPUT X
20  INPUT Y
30  IF X>=Y THEN 60
40  PRINT X+Y+1
50  GOTO 70
60  PRINT X+Y+−1
70  END
```

Notice that $\geq$ is written as $>=$. Similarly, $\leq$ is written as $<=$. Usually $\neq$ is written as $<>$.

Notice also line 50 where one branch skips over the other to avoid printing both outputs.

Exercises

1. Write a program that will INPUT A, B, C and PRINT A + B if A + B is greater than C or PRINT C if A + B is not greater than C.

2. Write a program that will input three numbers and print the product of the second and third numbers if this product is less than the product of the first and second numbers. Otherwise, the product of the first and second numbers is to be printed.

2-6 Absolute Value

In any pair of nonzero opposites, like 4 and -4, one number is a positive number and the other is a negative number. You call the positive member of the pair the absolute value of each of the numbers. Thus, 4 is the absolute value of 4; 4 is also the absolute value of -4.

The absolute value of a number is denoted by writing the name of the number between a pair of vertical bars | |. For example,

$$|4| = 4 \quad \text{and} \quad |-4| = 4.$$

The absolute value of 0 is defined to be 0:

$$|0| = 0.$$

On a number line, the absolute value of a number is the length of the arrow representing the number, without regard to the direction of the arrow (Figure 8). Thus, the absolute value of a number is the *distance* between the origin and the graph of the number.

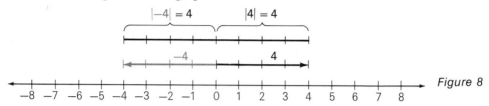

Figure 8

EXAMPLE 1   Draw the graph of the equation $|a| = 1$ if $a \in \Re$.

SOLUTION   The graph consists of the two points that are one unit from the origin, that is, the graphs of 1 and -1.

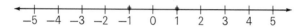

EXAMPLE 2   Draw the graph of the inequality $|x| < 3$ if $x \in \Re$.

SOLUTION   The graph consists of all the points which are *less than* 3 units from the origin in either direction.
Another inequality with this graph is $-3 < x < 3$.

EXAMPLE 3   Draw the graph of the set of all real numbers y such that $|y| \geq 2$.

SOLUTION   The graph consists of all the points which are *at least* 2 units from the origin on either side of the origin.

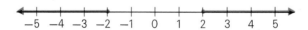

Oral Exercises

State the absolute value of each number.

1. 2	**2.** -2	**3.** $-\frac{3}{8}$	**4.** $\frac{5}{3}$
5. -6.9	**6.** 6.9	**7.** $-\frac{10}{9}$	**8.** 0

Simplify each expression.

9. $4 + |-4|$	**10.** $|4| + |-4|$	**11.** $-2 + |-2|$	**12.** $-2 + |2|$
13. $|0| + |-4|$	**14.** $-(8 + |-9|)$	**15.** $-(|-7| + 10)$	**16.** $-(|-2| + |-5|)$

Tell whether the statement is true or false. Give a reason for your answer.

EXAMPLE $-10 < |-10|$

SOLUTION   True, because $|-10| = 10$, and $-10 < 10$ is a true statement.

17. $|-12| > |10|$ **18.** $|36| > |-36|$ **19.** $|-9| < |-8|$ **20.** $8 \geq |-8|$

21. $|-0| = |0|$ **22.** $-2 < |-7|$ **23.** $|-1| > 0$ **24.** $|-3| \leq 3$

25. $-|13| = |-13|$ **26.** $-|-5| = -5$ **27.** $-|-9| < |-9|$ **28.** $|-6| \geq |-5|$

29. If x is any real number, then $|x| = x$ if $x \geq 0$;
$$|x| = -x \text{ if } x < 0.$$

Explain why the statement is true or why it is false.

Written Exercises

Determine whether the given statement is true or false.

A   **1.** $|-2 + (-5)| < |-2| + |-5|$ **2.** $|7 + (-4)| > |7| + |-4|$

       3. $|-9 + 6| \leq |-9| + |6|$ **4.** $|-3| + |-10| \geq |-3 + (-10)|$

       5. The absolute value of every real number is greater than 0.

       6. Some real numbers do not have absolute values.

       7. The absolute value of any real number equals the absolute value of the opposite of the real number.

       8. The additive inverse of the opposite of any real number is the number itself.

Find the value of each expression.

       9. $6|-4|$ **10.** $2|-5| + |-2|$ **11.** $7|-7| + |-7|$

12. $|4 + (-5)| + 3$ **13.** $|-1 + 9| + (-7)$ **14.** $-|-5 + 18| + |-2|$

15. $-|-(5 + 3) + (-1)|$ **16.** $-(3|3|) + (-|5|)$ **17.** $-4|-7| + |-9|$

Solve each equation over the set $\Re$ of real numbers.

EXAMPLE $|p| + 3 = 5$

SOLUTION   Since 5 is the sum of 2 and 3, $|p|$ must equal 2; that is, $|p| = 2$. The only real numbers whose absolute value is 2 are -2 and 2.

      $\therefore$ the solution set is $\{-2, 2\}$.

18. $|p| = 2$ **19.** $|y| = -2$ **20.** $|l| = 5$

21. $|z| + 1 = 3$ **22.** $|u| + (-3) = 5$ **23.** $-4 + |h| = 6$

24. $-8 + |n| = 0$ **25.** $|a| + |-2| = |5|$ **26.** $|-7| + |w| = |7|$

27. $|-d| = 2$ **28.** $-|-q| = 3$ **29.** $6 = |-x|$

Basic Properties of Real Numbers | **63**

Graph the solution set of each open sentence.

B
30. $|x| < 1$ **31.** $|y| > 1$ **32.** $|z| \leq 1$

33. $|m| < -1$ **34.** $|n| > -1$ **35.** $|p| \geq 1$

36. $-|t| < 2$ **37.** $|-q| > 2$ **38.** $|-n| \leq 2$

39. $|k| > 0$ **40.** $|c| < 0$ **41.** $|f| \geq 0$

42. $-2 \leq |-y|$ **43.** $3 < |-x|$ **44.** $-1 > |-z|$

45. $-2 < -|p|$ **46.** $-3 \leq -|-l|$ **47.** $|-y| \geq 0$

C
48. For what real values of n is it true that $|n| > -n$?

49. For what real values of m is it true that $|m| < -m$?

Find a value of x and a value of y that give a false statement.

50. $|x + y| > |x| + |y|$ **51.** $|x + y| = |x| + |y|$ **52.** $|x + y| < |x| + |y|$

53. Replace $\underline{\ ?\ }$ by $<, >, =, \leq,$ or $\geq$ so that the resulting statement is true: For all real numbers x and y, $|x + y| \underline{\ ?\ } |x| + |y|$.

Self-Test 2

1. Simplify the expression "$-3 + 4$" by picturing the addition of the numbers -3 and 4 on the number line.   *Obj. 1, p. 53*

Simplify.

2. $(12 + {}^-33) + {}^-3$ **3.** $(14 + {}^-17) + ({}^-9 + 5)$ *Obj. 2, p. 53*

4. Midge walked $10\frac{1}{2}$ blocks south, 5 blocks east, and then 15 blocks north and 3 blocks west. At the end of the walk, where was Midge relative to her starting position?   *Obj. 3, p. 53*

Simplify.

5. $-(-5 + 8) + (-3)$ **6.** $-[|-4| + (-|3|)]$ *Obj. 4, p. 53*

Solve each equation over $\mathcal{R}$ and graph the solution set on the number line.

7. $|-2| + |y| = |3|$ **8.** $|z| \leq 1$

Check your answers with those at the back of the book.

---

programming in BASIC

If you need to use the same expression several times in a program, you can give it a special name—a variable—by using a LET statement.
Study the following:

```
10   INPUT A
20   INPUT B
30   LET S=A+B
40   IF S>=0 THEN 70
50   PRINT −S
60   GOTO 80
70   PRINT S
80   END
```

Exercises

1. Write a program that will INPUT three numbers and PRINT the additive inverse of each, the sum of the three numbers, and the additive inverse of the sum.

2. Write a program that will INPUT three numbers and PRINT their sum and their average. (The *average* of three numbers is their sum divided by the number of numbers, that is, 3.)

Chapter Summary

The following statements are true for all real values of each variable.

1. Axioms of equality: *Reflexive property* $a = a$
Symmetric property If $a = b$, then
$b = a$.

Transitive property If $a = b$ and $b = c$,
then $a = c$.

2. Axioms of closure: The sum $a + b$ is a unique real number.
The product ab is a unique real number.

3. Commutative axioms: $a + b = b + a$ $ab = ba$

4. Associative axioms: $(a + b) + c = a + (b + c)$
$(ab)c = a(bc)$

5. $a + b + c = (a + b) + c,\ a + b + c + d = (a + b + c) + d$, and so on.
$abc = (ab)c,\ abcd = (abc)d$, and so on.

6. **Distributive axiom:** $a(b + c) = ab + ac$ and $(b + c)a = ba + ca$. Also, $ab + ac = a(b + c)$ and $ba + ca = (b + c)a$.

7. **Identity axioms:** The set $\mathfrak{R}$ contains a unique element 0 and a unique element 1 $(0 \neq 1)$ such that
$$a + 0 = a \quad \text{and} \quad 0 + a = a, \text{ and}$$
$$a \cdot 1 = a \quad \text{and} \quad 1 \cdot a = a.$$

8. **Axiom of additive inverses:** For every a, there is a unique real number $-a$ such that $a + (-a) = 0$ and $(-a) + a = 0$.

Chapter Review

1. For which value of x is the following statement true? There is a real number x such that $3x - 2 = 4$.　　　　　*2-1*
   a. 0　　　　b. 2　　　　c. 1　　　　d. 3

Simplify each expression in Review Items 2–9.

2. $12 + 374 + 88 + 226$　　　　　*2-2*
   a. 600　　　　b. 750　　　　c. 700　　　　d. 590

3. $5y + 3(y + 2) - 6$　　　　　*2-3*
   a. $8y$　　　　b. $8y - 6$　　　　c. $2y$　　　　d. $8y - 4$

4. $5b^2 + a^3 - b^2 + 3a$
   a. $4b^2 + 4a$　　b. $6b^2 + 4a^3$　　c. $4b^2 + 2a$　　d. $4b^2 + a^3 + 3a$

5. $(^-12 + 13) + 6$　　　　　*2-4*
   a. 31　　　　b. $^-7$　　　　c. 7　　　　d. 19

6. $24 + (18 + {}^-24)$
   a. 30　　　　b. 66　　　　c. $^-18$　　　　d. 18

7. $-6 + -(-9)$　　　　　*2-5*
   a. 3　　　　b. 15　　　　c. -15　　　　d. -3

8. $-[-5 + (-1)]$
   a. 6　　　　b. 4　　　　c. -4　　　　d. -6

9. $|6 + (-8)| - 3$　　　　　*2-6*
   a. -5　　　　b. 5　　　　c. -1　　　　d. 1

Chapter Test

1. Find a value for n to show that "For all n, $n + 3 = 3n$" is false.　　　　　*2-1*

Name the axiom that guarantees the truth of each of the following. Assume that the domain of each variable is $\mathcal{R}$.

2-2

2. $4(3 + 2) = 4(3) + 4(2)$

3. If $x = y$, then $y = x$.

Simplify.

2-3

4. $2x^2 + 5(y + 3x^2) + 2y$

2-4

5. $^-7 + (9 + {}^-12)$

2-5

6. $-[13 + (-23)]$

7. Specify the solution set of $-n > -4$ if the replacement set for n is $\{-4, 0, 4\}$

2-6

8. Simplify $3|-12| - (3|-4|)$

programming in BASIC

The computer has a special built-in function for finding absolute values. It is written ABS(X). To see how it works, try this program:

```
10   PRINT ABS(-12)
20   END
```

You can now find the value of an expression such as $2|-4| + (-|5|)$. Change line 10 to:

```
10   PRINT 2*ABS(-4)+(-ABS(5))
```

A computer can print other things besides numerical results. If you put an expression in quotation marks in a PRINT statement, the computer will copy that expression and print back all the symbols and spaces exactly as you wrote them (except that you must not use quotation marks within quotation marks). For example, try:

```
10   PRINT "ABS(-12)"
```

You can use a computer to determine if a statement is true. Try this:

```
10   IF ABS(-20)<20 THEN 40
20   PRINT "FALSE"
30   GOTO 50
40   PRINT "TRUE"
50   END
```

Exercise

1. Write a program that will find which of the values $-2, -1, 0, 1, 2$ of the variable make the statement true:

  a. $|x| < 2$ **b.** $|x| > 1$

Extinct in the wild, this Pere David's deer is a member of a herd maintained in captivity. Zoologists use modern techniques to preserve endangered species such as this deer.

3
Using Number Properties

Addition Properties

OBJECTIVES *for Sections 3-1 and 3-2:*
1. *Use addition properties to find the sum of two or more real numbers.*
2. *Use the properties of real numbers to justify steps in proving theorems about addition.*

3-1 Proving Theorems about Addition

With a five-gram mass in each pan, the pans of the beam balance pictured in Figure 1 balance each other. **Add** a **ten-gram** mass to each pan and the pans remain in balance (Figure 2). These drawings suggest the **additive property of equality:** *If the same number is added to equal numbers, the sums are equal.*

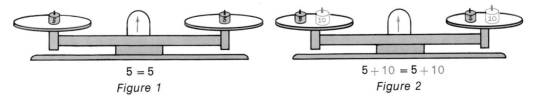

$$5 = 5$$
Figure 1

$$5 + 10 = 5 + 10$$
Figure 2

Additive Property of Equality

If a, b, and c are real numbers and $a = b$, then

$$a + c = b + c \quad \text{and} \quad c + a = c + b.$$

PROOF

Statement	Reason
1. a, b, and c are real numbers; $a = b$.	Given
2. $a + c$ and $c + a$ are real numbers.	Closure axiom of addition
3. $\therefore a + c = b + c$ and $c + a = c + b$	Substitution principle (substituting b for a)

Similar reasoning will prove the **multiplicative property of equality** (see Oral Exercise 11, page 72).

Multiplicative Property of Equality

If a, b, and c are real numbers and $a = b$, then

$$ac = bc \quad \text{and} \quad ca = cb.$$

Numerical work often suggests theorems about real numbers. For instance, look at the following computation.

$$(5 + 2) + (-2) = 5 + [2 + (-2)]$$
$$= 5 + 0$$
$$= 5$$
$$\therefore (5 + 2) + (-2) = 5$$

This computation suggests a useful idea: *To "undo" the addition of a given number to another number, add the opposite of the given number to the sum.* This idea is stated in the following assertion:

Theorem. For all real numbers b and c,

$$(b + c) + (-c) = b.$$

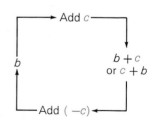

PROOF

Statement	Reason
1. b and c are real numbers.	Given
2. $-c$ is a real number.	Axiom of additive inverses

3. $(b + c) + (-c) = b + [c + (-c)]$ Associative axiom of addition
4. $(b + c) + (-c) = b + 0$ Axiom of additive inverses
5. $b + 0 = b$ Identity axiom of addition
6. $\therefore (b + c) + (-c) = b$ Transitive axiom of equality

Do you agree that the following equations are also true for all real numbers b and c?

$$[b + (-c)] + c = b,$$
$$(-c + b) + c = b,$$
and
$$-c + (b + c) = b.$$

For example,
$$(-38 + 15) + 38 = 15$$
$$[6 + (-8)] + (-6) = -8.$$

Oral Exercises

Simplify each expression. Assume that each variable denotes a real number.

1. $(8 + 6) + (-8)$
2. $[-6 + (-4)] + 4$
3. $8 + (-8 + p)$
4. $[7x^2 + (-5)] + 5$
5. $-60 + (r^2s^2 + 60)$
6. $-8 + (5x + 8) + 2x$

State the additive property of equality when:

7. b is replaced with $-b$.
8. c is replaced with $-(r + s)$.
9. a and b are replaced with h and g, respectively.
10. a is replaced with $(a + d)$.

The theorems stated in Exercises 11 and 12 are true for all real numbers a, b, and c. Justify each step in the given proofs.

EXAMPLE   Prove: If $a = b$, then $-a = -b$.

PROOF

Statement	*Reason*
1. a and b are real numbers, and $a = b$.	Given
2. $-a$ and $-b$ are unique real numbers.	Axiom of additive inverses
3. $a + (-b) = b + (-b)$	Additive property of equality
4. $b + (-b) = 0$	Axiom of additive inverses
5. $a + (-b) = 0$	Transitive axiom of equality
6. $-a + [a + (-b)] = -a + 0$	Additive property of equality
7. $(-a + a) + (-b) = -a + 0$	Associative axiom of addition
8. $\qquad 0 + (-b) = -a + 0$	Axiom of additive inverses
9. $\qquad\qquad -b = -a$	Identity axiom of addition
10. $\qquad\qquad -a = -b$	Symmetric axiom of equality

Justify each step in the given proofs.

11. Prove: If $a = b$, then $ac = bc$ and $ca = cb$.

PROOF

1. a, b, and c are real numbers, and $a = b$.
2. ac and ca are real numbers.
3. $\therefore ac = bc$ and $ca = cb$.

12. Prove: If $a + c = b + c$, then $a = b$.

PROOF

1. a, b, and c are real numbers, and $a + c = b + c$.
2. $-c$ is a real number.
3. $(a + c) + (-c) = (b + c) + (-c)$
4. $a + [c + (-c)] = b + [c + (-c)]$
5. $a + 0 = b + 0$
6. $\therefore a = b$

Written Exercises

Use the theorem stated on page 70 to simplify each expression.

EXAMPLE $42 + (-8)$

SOLUTION $42 + (-8) = (34 + 8) + (-8) = 34$

A
1. $17 + (-5)$
2. $31 + (-7)$
3. $49 + (-13)$
4. $95 + (-25)$
5. $-53 + 175$
6. $-99 + 100$
7. $-43 + 87$
8. $-24 + 61$

The theorems stated in Exercises 9–16 are true for all real numbers. State reasons for the steps in the given proofs.

9. Prove: $(a + c) + (-a) = c$.

PROOF

1. a, c are real numbers.
2. $-c$ is a real number.
3. $(a + c) + (-a) = (c + a) + (-a)$
4. $= c + [a + (-a)]$
5. $= c + 0$
6. $= c$
7. $\therefore (a + c) + (-a) = c$

10. Prove: If $a = b$, $b = d$, and $a = c$, then $d = c$.

PROOF

1. a, b, c, and d are real numbers.
2. $a = b$ and $b = d$
3. $a = d$
4. $d = a$
5. $a = c$
6. $\therefore d = c$

B   **11.** Prove: $(-c) + (c + b) = b$.

PROOF

1. b and c are real numbers.
2. $-c$ is a real number.
3. $-c + (c + b) = (-c + c) + b$
4. $= 0 + b$
5. $= b$
6. $\therefore -c + (c + b) = b$

12. Prove: If $1 + a = 1$, then $a = 0$.

PROOF

1. a is a real number and $1 + a = 1$.
2. $-1 + (1 + a) = -1 + 1$
3. $(-1 + 1) + a = -1 + 1$
4. $0 + a = 0$
5. $\therefore a = 0$

13. Prove: If $a = c$ and $a + b = c + d$, then $b = d$.

PROOF

1. a, b, c, and d are real numbers; $a = c$ and $a + b = c + d$.
2. $a + b = a + d$
3. $b + a = d + a$
4. $\therefore b = d$ (Oral Ex. 12, p. 72)

14. Prove: If $a + b = 0$, then $a = -b$.

PROOF

1. a and b are real numbers, and $a + b = 0$.
2. $(a + b) + (-b) = 0 + (-b)$
3. $a + [b + (-b)] = 0 + (-b)$
4. $\qquad\qquad a + 0 = 0 + (-b)$
5. $\qquad\qquad \therefore a = -b$

15. Prove: If $r + b = a$ and $s + b = a$, then $r = s$.

PROOF

1. a, b, r, and s are real numbers; $r + b = a$ and $s + b = a$.
2. $\qquad\qquad a = s + b$
3. $\therefore r + b = s + b$
4. $\qquad \therefore r = s$ (Oral Ex. 12, p. 72)

16. Prove: If $b + c = a$, then $b = a + (-c)$.

PROOF

1. a, b, and c are real numbers, and $b + c = a$.
2. $(b + c) + (-c) = b$
3. $\qquad \therefore a + (-c) = b$
4. $\qquad\qquad \therefore b = a + (-c)$

C **17.** Prove: For all real numbers a and b, $a + b + [(-a) + (-b)] = 0$.

18. Prove: For all real numbers a, b, and c, if $a + c = c + b$, then $a = b$.

3-2 Adding Real Numbers

Try to state a theorem that the following example suggests.

EXAMPLE 1   On a number line, show that:

a. $-(4 + 3) = -7$

b. $-4 + (-3) = -7$, and, therefore, that

c. $-(4 + 3) = -4 + (-3)$

SOLUTION   **a.** Add 4 and 3, obtaining 7. Then find the opposite of this sum, -7.

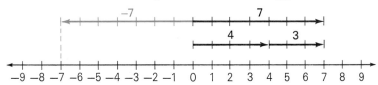

b. Add -4 and -3. The sum is -7.

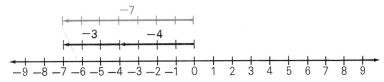

(Solution continued on next page.)

c. By part **(a)**, $-(4 + 3) = -7$. By part **(b)**, $-4 + (-3) = -7$.
∴ by the symmetric and transitive axioms of equality,
$$-7 = -4 + (-3) \qquad \text{and} \qquad -(4 + 3) = -4 + (-3).$$

This example suggests the following theorem, whose proof is outlined in Example 1 on page 76.

Property of the Opposite of a Sum

The opposite of a sum of real numbers is the sum of the opposites of the numbers; that is, for all real numbers a and b,

$$-(a + b) = -a + (-b).$$

By using the property of the opposite of a sum and the other properties that you have learned, you can compute sums of any real numbers without having to think of a number line.

EXAMPLE 2 Simplify: $-9 + (-2)$

SOLUTION

$-9 + (-2) = -(9 + 2)$	Property of the opposite of a sum
$\quad\quad\quad\quad = -11$	Substitution principle
$\therefore -9 + (-2) = -11$	Transitive axiom of equality

EXAMPLE 3 Simplify: $12 + (-8)$

SOLUTION

$12 + (-8) = (4 + 8) + (-8)$	Substitution principle
$\quad\quad\quad\quad = 4$	Theorem on page 70
$\therefore 12 + (-8) = 4$	Transitive axiom of equality

After computing many sums using either the number line or the methods of Examples 2 and 3, you may discover the short-cut methods given in the following rules.

Rules for Addition

1. If $a \geq 0$ and $b \geq 0$, then $a + b = |a| + |b|$.
Example. $6 + 3 = 9$

2. If $a < 0$ and $b < 0$, then $a + b = -(|a| + |b|)$.
Example. $-1 + (-5) = -(1 + 5) = -6$

3. If $a > 0$ and $b < 0$ and $|a| \geq |b|$, then $a + b = |a| - |b|$.
Example. $15 + (-7) = 15 - 7 = 8$

4. If $a \geq 0$ and $b < 0$ and $|b| > |a|$, then $a + b = -(|b| - |a|)$.
Example. $2 + (-7) = -(7 - 2) = -5$

EXAMPLE 4   Simplify: $8 + (-13) + 7 + (-4)$

SOLUTION 1

Step 1	*Step 2*	*Step 3*
$8 + (-13) = -5;$	$-5 + 7 = 2;$	$2 + (-4) = -2.$

SOLUTION 2

Step 1	*Step 2*	*Step 3*
8	-13	15
$\underline{7}$	$\underline{-4}$	$\underline{-17}$
15	-17	-2

Oral Exercises

Add.

1. 6 $\underline{7}$	**2.** -6 $\underline{-9}$	**3.** 17 $\underline{-8}$	**4.** -5 $\underline{5}$	**5.** -7 $\underline{-7}$
6. -2.1 $\underline{-4.9}$	**7.** -4.9 $\underline{-2.1}$	**8.** $\frac{7}{4}$ $\underline{-\frac{3}{4}}$	**9.** $-\frac{7}{8}$ $\underline{\frac{1}{8}}$	**10.** 5.6 $\underline{-7.2}$

State whether each expression names a positive number, a negative number, or zero. Then simplify the expression.

11. $-5 + (-7)$

12. $-40 + 20$

13. $-6 + 8 + 6$

14. $-9 + (-3) + 9$

15. $-8 + (-14) + 22$

16. $13 + [-(-4 + 2)]$

17. $4 + x + (-6) + (-x)$

18. $-y + 5 + y + (-5)$

19. $-(x + 2) + x + 2$

Replace each ? with one of the words *always, sometimes, never* to convert the given sentence into a true statement having the widest application.

EXAMPLE   The sum of a negative number and zero is ? a negative number.

SOLUTION   Replacing the ? with either *sometimes* or *always* yields a true statement; however, the statement with *always* has the widest application. Thus: The sum of a negative number and zero is **always** a negative number.

20. The sum of two negative numbers is ? a negative number.

21. The sum of two positive numbers is ? a positive number.

22. The sum of a positive number and a negative number is ? a negative number.

23. The sum of two negative numbers is ? zero.

24. The sum of a positive number and a negative number is ? zero.

Written Exercises

Simplify each expression for a sum.

A   **1.** $17 + (-6) + (-14) + 9$

     2. $-36 + (-4) + 21 + (-29)$

  3. $290 + (-495) + 95 + (-128)$

     4. $496 + (-621) + 49 + (-208)$

  5. $9 + 3\frac{1}{12} + (-1\frac{1}{12}) + (-8)$

     6. $7.3 + (-2.7) + (-3.3) + (-6.5)$

  7. $38 + (-33) + (-49) + (-21)$

     8. $-65 + (-29) + 77 + (-19)$

  9. $-18 + 61 + 90 + (-87)$

    10. $82 + (-88) + 105 + (-73)$

 11. $-[46 + (-8)] + [-(-7) + 43]$

    12. $[-19 + (-8)] + [-(4 + 13)]$

 13. $-(-36 + 18) + [-(-9 + 60)]$

    14. $-[83 + (-55)] + [-(64 + 46)]$

 15. $5.4 + (-4.8) + (-9.6) + 0 + (-7.5) + 19.2$

 16. $-0.9 + 4.25 + (-3.69) + (-11.8) + 12$

 17. $-\frac{7}{4} + \frac{3}{2} + (-3\frac{1}{2}) + \frac{3}{4} + (-2\frac{3}{4}) + (-3)$

    18. $-\frac{5}{9} + 3\frac{2}{9} + (-4\frac{4}{9}) + 1\frac{1}{9}$

Simplify each expression.

 19. $a^2 + b + (-4)a^2 + 6b + 4a^2$

    20. $3x^3 + 2x + (-2) + (-3)x^3 + (-5)x + 3$

 21. $6n^4 + (-3)n + (-6)n + n^4 + (-5n^4)$

 22. $(-9p^2) + 4pt + (-t^2) + 9p^2 + (-6pt) + (-10)t^2$

 23. $(-7)y^2 + (-8)xy + 3x^2 + (-y^2) + 3xy + (-1)x^2$

 24. $k^2 + 7k + (-12) + (-6)k + (-3k^2) + 14$

The statements in Exercises 25–27 are true for all real numbers a, b, and c. Give reasons for the statements in each proof. You may use the property of the opposite of a sum (proved below), and the fact that $-(-a) = a$ for every real number a.

EXAMPLE 1   Prove: $-(a + b) = -a + (-b)$

SOLUTION   By definition, $-(a + b)$ denotes the *one and only* number whose sum with $a + b$ is zero; that is, $(a + b) + [-(a + b)] = 0$. Therefore, you can prove that $-(a + b) = -a + (-b)$ by showing that

$$(a + b) + [-a + (-b)] = 0.$$

The reasoning goes like this:

PROOF

1. a and b are real numbers.	Given
2. $-a$ and $-b$ are real numbers.	Axiom of additive inverses
3. $(a + b) + [-a + (-b)]$ $\quad = [a + (-a)] + [b + (-b)]$	Commutative and associative axioms of addition
4. $\quad = 0 + 0$	Axiom of additive inverses
5. $\quad = 0$	Identity axiom of addition
6. $\therefore (a + b) + [-a + (-b)] = 0$	Transitive axiom of equality

B **25.** Prove: $-(a + b) + b = -a$

PROOF

1. a and b are real numbers.
2. $-a$ and $-b$ are real numbers.
3. $\qquad -(a + b) = -a + (-b)$
4. $\quad -(a + b) + b = [-a + (-b)] + b$
5. $\qquad\qquad\qquad = -a + [(-b) + b]$
6. $\qquad\qquad\qquad = -a + 0$
7. $\qquad\qquad\qquad = -a$
8. $\therefore \ -(a + b) + b = -a$

26. Prove: $-(-a + b) + (-a) = -b$

PROOF

1. a and b are real numbers.
2. $-a$ and $-b$ are real numbers.
3. $\quad -(-a + b) + (-a) = [-(-a) + (-b)] + (-a)$
4. $\qquad\qquad\qquad\qquad = a + [(-b) + (-a)]$
5. $\qquad\qquad\qquad\qquad = a + [(-a) + (-b)]$
6. $\qquad\qquad\qquad\qquad = [a + (-a)] + (-b)$
7. $\qquad\qquad\qquad\qquad = 0 + (-b)$
8. $\therefore \ -(-a + b) + (-a) = -b$

27. Prove: $-[(a + b) + c] = [-a + (-b)] + (-c)$

PROOF

1. a, b, and c are real numbers.
2. $-a$, $-b$, and $-c$ are real numbers.
3. $\quad -[(a + b) + c] = -(a + b) + (-c)$
4. $\qquad\qquad\qquad\ = [-a + (-b)] + (-c)$
5. $\therefore \ -[(a + b) + c] = [-a + (-b)] + (-c)$

In each of Exercises 28–35 write a chain of equations leading to the stated equation. Justify each step.

EXAMPLE 2 $(a + b) + [(-a) + (-b)] = 0$

SOLUTION
$(a + b) + [(-a) + (-b)]$ Commutative and associative
$= [a + (-a)] + [b + (-b)]$ axioms of addition
$= 0 + 0$ Axiom of additive inverses
$= 0$ Identity axiom of addition
$\therefore \ (a + b) + [(-a) + (-b)] = 0$ Transitive axiom of equality

C **28.** $(-2) + [(-a) + (a + 2)] = 0$

29. $[(-a) + (a + 1)] + (-1) = 0$

30. $-[(-a) + b] + b = a$

31. $-[(-a) + (-b)] + (-a) = b$

32. $(a + b) + [-(a + b + c)] = -c$

33. $-(a + b) + [(a + b) + (-c)] = -c$

34. $a + [-(a + b)] = -b$

35. $(-a) + [a + (-b)] + b = 0$

Problems

A 1. Jane Toohey had a balance of $221.50 in her checking account. During the week, she wrote checks for $67.80, $29.35, $14.75, and $40.04. On Thursday she made a deposit of $75. What was the balance at the end of the week?

2. Sam was on a diet. Over a 4-week period he lost 5.4 kg, gained 3.375 kg, lost 3.6 kg, and gained 1.575 kg. How much did he gain or lose over all?

3. Marci Cummings made the following entries in her checking account:

Deposits:　　　$63.78, $2.07, and $1.56
Withdrawals:　$36.12, $0.75

What was the balance of her checking account if there was $150 in it initially?

4. The Barford Book Company had 18,500 copies of a nonfiction book in its warehouse. During a 2-week period, it shipped out 9145 books and received 3700 more copies; then shipped out 8750 books and received 6075 more copies. What was the number of copies of this book in the warehouse at the end of the 2-week period?

B 5. Lee Ann has 18 more paintings than Jody. Together they have more than 31 paintings. How many paintings does Jody have?

6. Plant X grows 5 cm each week, and it is now 50 cm tall. Plant Y grows 7 cm each week, and it is now 28 cm tall. How many weeks from now will plant Y be as tall as plant X?

Self-Test 1

Simplify.

1. $-6 + 15 + 26 + (-5)$　　　　　　　**2.** $13 + (-56) + 37 + 6$　　　*Obj. 1, p. 69*

State the reasons justifying the steps in the given chain of equalities.

3.　　$64 + (-28) = (36 + 28) + (-28)$　　　　　　　*Obj. 2, p. 69*

4.　　　　　　　$= 36 + [28 + (-28)]$

5.　　　　　　　$= 36 + 0$

6.　　　　　　　$= 36$

7. $\therefore 64 + (-28) = 36$

Check your answers with those at the back of the book.

Multiplication Properties

OBJECTIVES for Sections 3-3 and 3-4:
1. *Use multiplication properties to simplify expressions involving products of real numbers.*
2. *Use the properties of real numbers to justify steps in proving theorems about multiplication.*

3-3 Multiplying Real Numbers

The equations

$$3 \times 0 = 0 \quad \text{and} \quad 0 \times 3 = 0$$

illustrate the theorem called the **multiplicative property of zero**: *When one of the factors of a product is 0, the product itself is 0.* (For a proof of this property, see Example 3, page 82.)

Multiplicative Property of 0

For each real number *a*,

$$a \cdot 0 = 0 \quad \text{and} \quad 0 \cdot a = 0.$$

Would you guess that $a(-1) = -a$? You might if you noticed that

$$2 \times (-1) = -1 + (-1) = -2,$$
$$3 \times (-1) = -1 + (-1) + (-1) = -3,$$

and so on.

These examples suggest the **multiplicative property of -1**: *Multiplying any real number by -1 produces the additive inverse of the number.*

Multiplicative Property of -1

For each real number *a*,

$$a(-1) = -a \quad \text{and} \quad (-1)a = -a.$$

A special case of this property occurs when $a = -1$; you have

$$(-1)(-1) = 1.$$

The following examples show how to use the multiplicative property of -1, along with the multiplication facts for positive numbers and the associative and commutative axioms, to multiply any two real numbers.

(1) $7 \cdot 3 = 21$

(2) $(-7)3 = (-1 \cdot 7)3 = -1(7 \cdot 3) = -1(21) = -21$

(3) $7(-3) = 7[3(-1)] = (7 \cdot 3)(-1) = 21(-1) = -21$

(4) $(-7)(-3) = (-1 \cdot 7)(-1 \cdot 3) = [-1(-1)](7 \cdot 3) = 1 \cdot 21 = 21$

These examples suggest the following theorem.

Property of Opposites in Products

For all real numbers a and b,

$$(-a)b = -ab, \quad a(-b) = -ab, \quad (-a)(-b) = ab.$$

Practice in simplifying products should lead you to discover the following rules for multiplication.

Rules for Multiplication

1. The absolute value of the product of two (or more) real numbers is the product of the absolute values of the numbers:

$$|ab| = |a| \times |b|.$$

2. A product of nonzero real numbers of which an even number are negative is a positive number. A product of nonzero real numbers of which an odd number are negative is a negative number.

EXAMPLE 1 State whether the given expression names a positive number, a negative number, or zero. Then simplify the expression.

    a. $-8(-5)(-9)$ **b.** $7^2(-2)^3(0)$ **c.** $(-1)^5(-3)^3(4)$

SOLUTION   **a.** A negative number; -360 **b.** 0   **c.** A positive number; 108.

EXAMPLE 2 Simplify each expression.

    a. $(-3y)(-12y)$ **b.** $6x + (-7x)$

SOLUTION   **a.** $(-3y)(-12y) = [-3(-12)]y \cdot y = 36y^2$.

               b. $6x + (-7x) = 6x + (-7)x = [6 + (-7)]x = (-1)x = -x$.

Oral Exercises

Simplify each expression.

  1. $-9(4)$ **2.** $-6(-5)$ **3.** $(-3)(-7)(-5)$

4. $(-2)(-y)$ **5.** $-19 \cdot 0 \cdot (-4)$ **6.** $(-3p)(8p)$

7. $(-1)^3$ **8.** $(-2)^5$ **9.** $2(-3)^2$

10. $|(-4)^3|$ **11.** $3k(-1)^2(-k)$ **12.** $(-t)(-5)^2(-2t)$

EXAMPLE $-3xy + (-6xy)$

SOLUTION $-3xy + (-6xy) = -3xy + (-6)xy = [-3 + (-6)]xy = (-9)xy = -9xy.$

13. $8w + (-10w)$ **14.** $-5t^3 + t^3$ **15.** $-6lh + (-5lh)$

Written Exercises

In Exercises 1–45 simplify each expression.

EXAMPLE 1 $-84(14) + 16(-14)$

SOLUTION $-84(14) + 16(-14) = -84(14) + 16(-1)(14) = [-84 + (-16)]14$
$$= -100 \cdot 14 = -1400$$

A

1. $(-52 + 49)13$ **2.** $25[13 + (-19)]$

3. $-4[-6 + (-9)]$ **4.** $[-9 + (-8)]0$

5. $-75 + 75(-9)$ **6.** $-58 + (-98)58$

7. $-42 \cdot 26 + (-42)(-26)$ **8.** $-19(51) + (-51)$

9. $53 + 53(-23)$ **10.** $(-63)(-14) + 14(-63)$

11. $(-1)^3(79) + 79(-210)$ **12.** $(-123)(199) + (-1)^5(123)$

EXAMPLE 2 $-5[3x + (-2y)]$

SOLUTION $-5[3x + (-2y)] = -5(3x) + (-5)(-2y)$
$$= -5(3x) + (-5)[(-2)y]$$
$$= (-5 \cdot 3)x + (-5 \cdot -2)y$$
$$= -15x + 10y$$

13. $-3(5a + 2b)$ **14.** $-4[2b + (-4)d]$

15. $8[-p + (-4)q]$ **16.** $-7[-k + (-1)m]$

17. $-5[-3y + (-4y^2)]$ **18.** $6[2c + (-4c^3)]$

19. $7l + 8h + (-3l)$ **20.** $5w + 6u + (-7w)$

21. $9d + (-2e) + (-6d) + e$ **22.** $-8x + (-2y) + (-3x) + 6y$

23. $-2ef + 5ef + (-5ef) + ef$ **24.** $-7lk + 12lk + (-lk) + (-6lk)$

25. $r + 8s + [-(2r + 5s)]$ **26.** $3x + y + [-(2x + 4y)]$

27. $26n^2 + (-3n^2) + 11n^3 + (-19n^2) + (-10n^3)$ **28.** $-t^4 + 4t^6 + (-3t^6) + (-2t^6)$

29. $x^2 + (-4x) + (-10) + (-x^2) + (-13x) + 8$

30. $-6i^3 + 3 + (-2i) + 2i^3 + 15i + (-12)$

Simplify each expression.

B 31. $5(-y + 3t) + (-2)(4y + t)$ 32. $6[k + (-3m)] + (-4)(2k + m)$

33. $-7(-2a + b) + 11[a + (-2b)]$ 34. $-3[2x + (-y)] + (-4)(-3x + 4y)$

35. $6[-3y^2 + (-3w)] + [-(-4y^2 + 13w)]$ 36. $[-(-7x^3 + p^3)] + 12(-x^3 + 6p^3)$

37. $3[4(-u^2 + 4v) + (-5)] + (-6v) + (-6u^2)$

38. $-5r + (-3r^2) + 4[-8 + 3(-2r^2 + r)]$

39. $-6[4(-3 + 2bc) + (-2)] + (-4)(-2bc + 1)$

40. $-2[-3 + 5(-3rx + 2)] + 3[11 + (-5rx)]$

41. $5[-x + 2(3x + y)] + 2[-y + 3(2y + x)]$

42. $-3[2(p + 2q) + (-3)(q + 2p) + (-3p) + (-2)(p + q)]$

43. $4[-r + 2(3r + s) + s + (-2)(2s + r) + (-3r)(2)]$

44. $-1[2(-x) + 3(x + y) + (-1)(x + y)] + (2x + y)(-2)$

45. $-3[-2(a + 3b) + (-1)(a + 3b) + 3(a + 3b)] + 3(3a + b)$

In Exercises 46–55, evaluate the expression if the value of $x = -1$, $y = 3$, $m = -2$, and $t = 0$.

46. $-3(x + 2m)^2$ 47. $x^4(-3y + tx)$

48. $(2x + 3y + 4m)^3$ 49. $m^3(2y + 3m) + m(2y + 3m)$

50. $[2x^2 + (-3y)](-3m^2)$ 51. $m(x^2 + t^2) + m^2 + (-y)(2x + t^3)$

52. $[(m^2 + y^2)x + 10] + (-3)(x + y)^2$ 53. $x^3 + x^2(2x + y) + x(2x + y)$

54. $-y^2(-x + y + t) + (-x^2)(x + y + t)$ 55. $-m^4(-y^2 + x^3 + x^4 + t)$

Give reasons for the statements in the following proofs.

EXAMPLE 3   Prove: For each real number a, $a \cdot 0 = 0$ and $0 \cdot a = 0$.

PROOF

Statement	*Reason*
1. a is a real number.	Given
2. $0 + 0 = 0$	Identity axiom of addition
3. $a(0 + 0) = a \cdot 0$	Multiplicative prop. of equality
4. $a(0 + 0) = a \cdot 0 + a \cdot 0$	Distributive axiom
5. $a \cdot 0 + a \cdot 0 = a(0 + 0)$	Symmetric axiom of equality
6. $\therefore a \cdot 0 + a \cdot 0 = a \cdot 0$	Transitive axiom of equality
7. $(a \cdot 0 + a \cdot 0) + [-(a \cdot 0)]$ $= a \cdot 0 + [-(a \cdot 0)]$	Additive property of equality
8. $a \cdot 0 + [a \cdot 0 + [-(a \cdot 0)]]$ $= a \cdot 0 + [-(a \cdot 0)]$	Associative axiom of addition
9. $a \cdot 0 + 0 = 0$	Axiom of additive inverses
10. $a \cdot 0 = 0$	Identity axiom of addition
11. $0 \cdot a = a \cdot 0$	Commutative axiom of addition
12. $\therefore 0 \cdot a = 0$	Transitive axiom of equality

56. Prove: For all real numbers a, $a(-1) = -a$ and $(-1)a = -a$.

1. a is a real number.
2. $a(-1)$ is a real number.
3. $\quad a + a(-1) = a \cdot 1 + a(-1)$
4. $\quad\quad\quad\quad\quad = a[1 + (-1)]$
5. $\quad\quad\quad\quad\quad = a \cdot 0$
6. $\quad\quad\quad\quad\quad = 0$
7. But $\quad\quad 0 = a + (-a)$
8. $\therefore\ a + a(-1) = a + (-a)$
9. $-a + [a + a(-1)] = -a + [a + (-a)]$
10. $(-a + a) + a(-1) = (-a + a) + (-a)$
11. $0 + a(-1) = 0 + (-a)$
12. $\therefore\ a(-1) = -a$
13. $\quad (-1)a = a(-1)$
14. $\therefore\ (-1)a = -a$

57. Prove: For all real numbers a and b, $(-a)(-b) = ab$.

1. a and b are real numbers.
2. $-a = a(-1)$
3. $-b = (-1)b$
4. $\quad (-a)(-b) = [a(-1)][(-1)b]$
5. $\quad\quad\quad\quad\quad = a[(-1)(-1)]b$
6. $\quad\quad\quad\quad\quad = (a \cdot 1)b$
7. $\quad\quad\quad\quad\quad = ab$
8. $\therefore\ (-a)(-b) = ab$

Show that the given expressions are equivalent by writing a chain of equations and justifying the steps with reasons.

EXAMPLE 4 $\quad -a(b + c)$ and $-ab + (-ac)$

PROOF

Step	Reason
$-a(b + c) = (-a)b + (-a)c$	Distributive axiom
$\quad\quad\quad = [(-1)a]b + [(-1)a]c$	Multiplicative property of -1
$\quad\quad\quad = (-1)(ab) + (-1)(ac)$	Associative axiom of multiplication
$\quad\quad\quad = -ab + (-ac)$	Multiplicative property of -1
$\therefore\ -a(b + c) = -ab + (-ac)$	Transitive axiom of equality

C **58.** $-(-a)$ and a

59. $(-a)b$ and $-ab$

60. $a(-b)$ and $-ab$

Evaluate each expression using the following values for the variables.

$$x = -0.5 \qquad y = 3.06 \qquad t = -12.4 \qquad w = 0.33$$

1. $-3(x + 2t)^2 + 2w$ **2.** $(x + 3t)[x + (-3t)]$

3. $x + (-2y) + (-3t) + 4w$ **4.** $-2w + (-2)(x + 3t) + (-3x)$

5. $x^2 + y^2 + t^2 + w^2(x + 1)$ **6.** $(x + y + t + w)^2 + (-w)$

7. $(x + y)(t + w)(x + w)$ **8.** $(x + y)^2 + (-2)[(t + w) + (-1)(x + y)]$

3-4 The Multiplicative Inverse of a Number

Numbers like $\frac{3}{2}$ and $\frac{2}{3}$, whose product is 1, are called **reciprocals**, or **multiplicative inverses**, of each other. For example:

1. -8 and $-\frac{1}{8}$ are reciprocals of each other because $-8(-\frac{1}{8}) = 1$.
2. 0.2 is the reciprocal of 5, and 5 is the reciprocal of 0.2, because $0.2 \times 5 = 1$.
3. -1 is its own reciprocal because $(-1)(-1) = 1$.
4. 0 has no reciprocal because the product of 0 and any real number is 0, *not* 1.

The symbol $\dfrac{1}{a}$ denotes the *reciprocal*, or *multiplicative inverse*, of a. For example:

$$\frac{1}{\frac{3}{4}} = \frac{4}{3}$$

may be read "the reciprocal of three-fourths equals four-thirds."

$$\frac{1}{-2} = -\frac{1}{2}$$

may be read "the reciprocal of negative two equals negative one-half."
   That *every real number except 0 has a reciprocal* is an axiom.

Axiom of Multiplicative Inverses

For every real number a except zero, there is a unique real number $\dfrac{1}{a}$ such that

$$a \cdot \frac{1}{a} = 1 \quad \text{and} \quad \frac{1}{a} \cdot a = 1.$$

EXAMPLE 1   Simplify $(-a)\left(-\dfrac{1}{a}\right)$ for $a \neq 0$.

SOLUTION   By the multiplicative property of -1 (page 79),

$$-a = (-1)a \quad \text{and} \quad -\frac{1}{a} = (-1)\frac{1}{a}.$$

$$\therefore (-a)\left(-\frac{1}{a}\right) = (-1)a \cdot (-1)\frac{1}{a}$$

$$= (-1)(-1) \cdot a \cdot \frac{1}{a}$$

$$= \underbrace{\quad}_{1} \quad \cdot \quad \underbrace{\quad}_{1} \quad = 1$$

$$\therefore (-a)\left(-\frac{1}{a}\right) = 1$$

Do you see that the result of Example 1 means that $-a$ and $-\dfrac{1}{a}$ are reciprocals of each other? Thus,

$$\frac{1}{-a} = -\frac{1}{a}$$

for every *nonzero* real number a.

EXAMPLE 2   Simplify each expression:

   a. $(8 \cdot 4)\frac{1}{4}$ **b.** $(-\frac{1}{3})(-3x^4)$ **c.** $-24 \cdot \frac{1}{6}$

SOLUTION   **a.** $(8 \cdot 4)\frac{1}{4} = 8(4 \cdot \frac{1}{4}) = 8 \cdot 1 = 8.$
   b. $(-\frac{1}{3})(-3x^4) = (-\frac{1}{3})[(-3)x^4] = [-\frac{1}{3}(-3)]x^4 = 1 \cdot x^4 = x^4.$
   c. $-24 \cdot \frac{1}{6} = (-4 \cdot 6)\frac{1}{6} = -4(6 \cdot \frac{1}{6}) = -4 \cdot 1 = -4.$

Example 2 uses the following theorem, whose proof is left for you to complete in Exercises 23 and 24, page 87.

Theorem.   For all real numbers b and all nonzero real numbers c,

$$(bc)\frac{1}{c} = b \quad \text{and} \quad \frac{1}{c}(cb) = b.$$

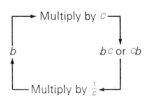

Here is an informal way to state this theorem: *To "undo" the multiplication of one number by a given nonzero number, multiply the product by the reciprocal of the given nonzero number.*

Study the example below.

EXAMPLE 3 Simplify each expression:

$$\textbf{a. } (5 \cdot 9)(\tfrac{1}{5} \cdot \tfrac{1}{9}) \qquad \textbf{b. } (a \cdot b)\left(\frac{1}{a} \cdot \frac{1}{b}\right), \; a \neq 0, \; b \neq 0$$

SOLUTION **a.** $(5 \cdot 9)(\tfrac{1}{5} \cdot \tfrac{1}{9}) = (5 \cdot \tfrac{1}{5})(9 \cdot \tfrac{1}{9}) = 1 \cdot 1 = 1.$

b. $(a \cdot b)\left(\dfrac{1}{a} \cdot \dfrac{1}{b}\right) = \left(a \cdot \dfrac{1}{a}\right)\left(b \cdot \dfrac{1}{b}\right) = 1 \cdot 1 = 1.$

This example suggests the following Property of the Reciprocal of a Product.

Property of the Reciprocal of a Product

The reciprocal of a product of real numbers, each different from zero, is the product of the reciprocals of the numbers; that is, for all real numbers a and b such that $a \neq 0$ and $b \neq 0$,

$$\frac{1}{ab} = \frac{1}{a} \cdot \frac{1}{b}.$$

Oral Exercises

State the reciprocal, or multiplicative inverse, of each number.

1. 3

2. -3

3. $\dfrac{1}{5}$

4. $-\dfrac{3}{7}$

5. $\dfrac{7}{3}$

6. $\dfrac{x}{2}, \; x \neq 0$

7. $-\dfrac{1}{t}, \; t \neq 0$

8. $\dfrac{r}{s}, \; r \neq 0, \; s \neq 0$

Find the value of n.

9. $\dfrac{1}{n} = -2$

10. $\dfrac{1}{n} = -\dfrac{1}{9}$

11. $\dfrac{1}{n} = \dfrac{5}{7}$

12. $\dfrac{1}{n} = -1$

Simplify each product.

EXAMPLE 1 $\dfrac{-1}{3}\left(\dfrac{1}{-10}\right)$

SOLUTION $\dfrac{-1}{3}\left(\dfrac{1}{-10}\right) = \dfrac{-1(1)}{3(-10)} = \dfrac{-1}{-30} = \dfrac{1}{30}$

13. $\dfrac{1}{5}\left(\dfrac{1}{6}\right)$

14. $-\dfrac{1}{7}\left(\dfrac{1}{3}\right)$

15. $-\dfrac{1}{9}\left(\dfrac{1}{-2}\right)$

16. $\dfrac{1}{-7}\left(\dfrac{-1}{8}\right)$

17. $\dfrac{1}{-11}\left(\dfrac{1}{-1}\right)$

18. $\dfrac{-1}{8}\left(\dfrac{1}{-3}\right)$

19. $\dfrac{-1}{20}\left(\dfrac{-1}{4}\right)$

20. $-\dfrac{1}{3}\left(\dfrac{1}{33}\right)$

In Exercises 21–24 show that the given expressions are equivalent by stating the reasons justifying the steps in the given chain of equalities.

EXAMPLE 2 $78 \cdot \frac{1}{3}$ and 26

SOLUTION

Step	Reason
$78 \cdot \frac{1}{3} = (26 \cdot 3)\frac{1}{3}$	Substitution principle
$= 26\left(3 \cdot \frac{1}{3}\right)$	Associative axiom of multiplication
$= 26 \cdot 1$	Axiom of multiplicative inverses
$= 26$	Identity axiom of multiplication
$\therefore 78 \cdot \frac{1}{3} = 26$	Transitive axiom of equality

21. $\frac{1}{13}(-1001)$ and -77

$$\frac{1}{13}(-1001) = \frac{1}{13}[1001(-1)]$$
$$= \left(\frac{1}{13} \cdot 1001\right)(-1)$$
$$= \left[\frac{1}{13}(13 \cdot 77)\right](-1)$$
$$= \left[\left(\frac{1}{13} \cdot 13\right)77\right](-1)$$
$$= (1 \cdot 77)(-1)$$
$$= 77(-1)$$
$$= -77$$
$$\therefore \frac{1}{13}(-1001) = -77$$

22. $\left(-\frac{1}{6}\right)(-198)$ and 33

$$\left(-\frac{1}{6}\right)(-198) = \left[(-1)\frac{1}{6}\right][(-1)198]$$
$$= [(-1)(-1)]\left[\frac{1}{6}(198)\right]$$
$$= 1\left[\frac{1}{6}(198)\right]$$
$$= \frac{1}{6}(198)$$
$$= \frac{1}{6}[6(33)]$$
$$= \left[\frac{1}{6}(6)\right]33$$
$$= 1(33)$$
$$= 33$$
$$\therefore \left(-\frac{1}{6}\right)(-198) = 33$$

23. $(bc)\frac{1}{c}$ and b, if $c \neq 0$

$$(bc)\frac{1}{c} = b\left(c \cdot \frac{1}{c}\right)$$
$$= b \cdot 1$$
$$= b$$
$$\therefore (bc)\frac{1}{c} = b$$

24. $\frac{1}{c}(cb)$ and b, if $c \neq 0$

$$\frac{1}{c}(cb) = \left(\frac{1}{c} \cdot c\right)b$$
$$= 1 \cdot b$$
$$= b$$
$$\therefore \frac{1}{c}(cb) = b$$

Show that the given expressions are equivalent. State the reasons that justify the steps.

25. $42\left(-\frac{1}{7}\right)$ and -6

26. $\frac{1}{12}(8)(-3)$ and -2

Written Exercises

Simplify each expression.

EXAMPLE 1 $-52mn\left(-\frac{1}{4}\right)$

SOLUTION $-52mn\left(-\frac{1}{4}\right) = \left[-52\left(-\frac{1}{4}\right)\right]mn = [(-1)(-1)]\left(52 \cdot \frac{1}{4}\right)mn$

$$= 1 \cdot \left(13 \cdot 4 \cdot \frac{1}{4}\right)mn = 13mn$$

A   **1.** $-\frac{1}{4}(-252)$ **2.** $-\frac{1}{5}(275)$ **3.** $-72\left(-\frac{1}{8}\right)$

     4. $-92\left(-\frac{1}{2}\right)$ **5.** $-\frac{1}{3}(60)\frac{1}{2}$ **6.** $-\frac{1}{6}(-318)(-2)$

     7. $\frac{-1}{1}(300)\left(-\frac{1}{20}\right)$ **8.** $\frac{1}{-8}(-288)\left(-\frac{1}{4}\right)$ **9.** $3xy\left(-\frac{1}{3}\right)$

     10. $-9pq\left(-\frac{1}{9}\right)$ **11.** $(-72)\left(-\frac{1}{8}\right)\left(\frac{1}{9}\right)$ **12.** $\frac{1}{11}(-220)\left(-\frac{1}{10}\right)$

     13. $\left(-\frac{1}{7}\right)49x^2$ **14.** $-\frac{1}{29}(-348s^2)$ **15.** $\frac{1}{k}(2kh),\ k \neq 0$

     16. $(8mn)\frac{1}{m},\ m \neq 0$ **17.** $\frac{1}{p^2}(4p^2)\left(-\frac{1}{4}\right),\ p \neq 0$ **18.** $-\frac{1}{5}(-35x^4)\frac{1}{x^4},\ x \neq 0$

EXAMPLE 2 $\frac{1}{4}[28x + (-16y)]$

SOLUTION $\frac{1}{4}[28x + (-16y)] = \frac{1}{4}(28x) + \frac{1}{4}(-16y) = \left[\frac{1}{4}(28)\right]x + \left[\frac{1}{4}(-16)\right]y$

$$= 7x + (-4y)$$

     19. $\frac{1}{5}(15p + 30)$ **20.** $\frac{1}{3}(36y + 21)$

     21. $-\frac{1}{8}(-24k + 40h)$ **22.** $[-63w + 99u]\left(-\frac{1}{9}\right)$

     23. $[-84y + (28)xu]\left(-\frac{1}{7}\right)$ **24.** $-\frac{1}{12}(48mn + 108n)$

     25. $\frac{1}{4}(-4r^2 + 4s^2)$ **26.** $-\frac{1}{13}[13x^4 + (-13y^4)]$

B   **27.** $6\left[\frac{1}{3}x + \left(-\frac{1}{2}\right)y\right] + \left(-\frac{1}{6}\right)[18y + (-48x)]$

     28. $-25\left[\left(-\frac{1}{5}\right)a + \frac{1}{5}b\right] + \frac{1}{7}(-56a + 21b)$

     29. $-\frac{1}{4}[8p^2 + 16p + (-20)] + \frac{1}{2}[-24 + 14p + (-2p^2)]$

     30. $\frac{1}{9}[18x^2 + (-9xy) + 27y^2] + \frac{1}{5}(-10y^2 + 15xy + 20x^2)$

31. $8\left[m + \left(-\frac{1}{4}\right)w\right] + (-4)\left(\frac{1}{2}w + 2m\right) + [-(-w + 2)]$

32. $-4(k + 2h) + \frac{1}{2}[6k + (-4h)] + \left(-\frac{1}{6}\right)[-(-18k + 12h)]$

33. $-\frac{1}{3}(42a^2 + 3b^2) + \frac{1}{2}(-8a^2 + 12b^2) + \left(-\frac{1}{4}\right)[12b^2 + (-32a^2)]$

34. $\frac{1}{2}(-10A^2 + 16B^2) + (-6)\left(\frac{1}{2}A^2 + \frac{1}{3}B^2\right) + (-2)[B^2 + (-2A^2)]$

35. $\frac{1}{5}(-5k + 10h) + \frac{1}{2}[2k + (-8h)] + 6\left[\frac{1}{2}k + \left(-\frac{1}{3}h\right)\right]$

36. $-6\left[\frac{1}{2}(4x + 1) + \left(-\frac{1}{2}\right)\right] + (-4)\left[\frac{1}{2}(4x + 1) + \left(-\frac{1}{2}\right)\right] + \left(-\frac{1}{4}\right)(-80x)$

37. $-8\left[\frac{1}{8} + \left(-\frac{1}{2}\right)(x + 2y) + \left(-\frac{1}{4}\right)(x + 2y) + \left(-\frac{1}{4}\right)\right]$

38. $-12\left[\left(\frac{1}{4}x + \left(-\frac{1}{2}y\right) + \frac{1}{3}z\right] + 18\left[-\frac{1}{6}x + \frac{1}{9}y + \left(-\frac{1}{3}z\right)\right]$

39. $-1[x^2 + (-2y^2) + (-z^2)] + 16\left[-\frac{1}{8}z^2 + \frac{1}{2}y^2 + \left(-\frac{1}{16}x^2\right)\right]$

40. $-\frac{1}{9}\left[-18h + 9i + \frac{1}{7}(-7i + 21h) + \left(-\frac{1}{12}\right)(-24h + 12i) + 9i\right]$

C   **41.** Prove: For all real numbers x and all nonzero real numbers y,
$\left(-\frac{1}{y}\right)(-yx) = x$.

42. Prove: For all real numbers t and all nonzero real numbers s,
$\frac{1}{s}(ts) = t$.

programming in BASIC

There is a special shortcut for repeating operations. You use a pair of statements, one beginning with FOR and the other beginning with NEXT. To see how this combination works, RUN this program:

```
10   FOR I=1 to 5
20   PRINT I, "*"
30   NEXT I
40   END
```

Notice that the comma in line 20 makes the value of I and the asterisk print on the same line with a wide space between them.

The combination FOR . . . NEXT makes what is called a *loop*. The value of I is set at 1 for the first time through the loop. "NEXT I" in this program adds 1 to the value of I, making it 2 for the second time through the loop, and so on. After the loop has been executed with the value of I at 5, the computer goes to the step following NEXT I, which in this case is END.

You can also use the FOR . . . NEXT combination to find successive values of an expression. For example, change line 20 to

20   PRINT I, 3*I

and RUN the program again.

You have seen that a comma spaces out the items printed on the same line. A semicolon will cause symbols to be printed closer together, right beside each other on some systems. Putting a comma or a semicolon at the *end* of a PRINT statement keeps the terminal carriage on the same line. For example, change line 20 to

20   PRINT "*";

and RUN the program.

Any letter may be used in place of I, but you must use the same letter in the FOR and NEXT statements that go together.

Exercises

1. Change line 10 in the program in the text to 10 FOR I = 5 TO 10 and compare the RUNS for:

   20   PRINT I, "*"
   20   PRINT I, 3*I
   20   PRINT "*";

2. Write a program that will find values of X + 2 for values of X from 0 to 10.

3. Write a program that will PRINT:

   + − + − + − + − +

   (*Hint:* PRINT "+−" four times and then PRINT "+" outside the loop with no punctuation after it.)

4. RUN this program:

   ```
    5   FOR J=1 TO 7
   10   FOR I=1 TO 5
   20   PRINT "*";
   30   NEXT I
   32   PRINT ⟵──── This line is necessary to bring the terminal
   35   NEXT J           carriage back to begin a new line.
   40   END
   ```

5. Make these changes in the preceding program:

   ```
    1   INPUT M
    2   INPUT N
    5   FOR J=1 TO N
   10   FOR I=1 TO M
   ```

   Try various values of M and N.

Self-Test 2

VOCABULARY   reciprocal (p. 84)
                 multiplicative inverse (p. 84)

Simplify.

1. $5[16 + (-17)]$ 2. $-8(11) + 4(-22)$ *Obj. 1, p. 79*

State the reasons justifying the steps in the given chain of equalities.

3. $\frac{1}{6}(96) = \frac{1}{6}(6 \cdot 16)$ *Obj. 2, p. 79*

4. $= \left(\frac{1}{6} \cdot 6\right)16$

5. $= (1)16$

6. $= 16$

7. $\therefore \frac{1}{6}(96) = 16$

Check your answers with those at the back of the book.

EXTRA FOR EXPERTS

Indirect Proof

The proofs you have studied in this book are each made up of a sequence of true statements leading *directly* from the hypothesis to the conclusion of a theorem (called a *direct proof* on page 48). Another kind of reasoning is illustrated in the following situation.

   Suppose that a detective investigating a theft narrowed the list of suspects to three people, Mr. Abel, Mr. Bolton, and Mr. Carter. Further investigation revealed that at the time of the theft, Mr. Abel was out of town, and Mr. Bolton was in the hospital.

   On the basis of these facts, the detective knew that Abel and Bolton were innocent because neither of them could have been in two places at the same time. So he arrested Carter. "You see," said the detective, "I eliminated all possible suspects except one. That one had to be the prime suspect!"

In mathematics this kind of reasoning appears in an **indirect proof.** To write an indirect proof of a theorem, you start with the hypothesis and with two possibilities:

1. the conclusion is false; or
2. the conclusion is true.

Next, you show that supposing that the first possibility is the case ("the conclusion is false") leads you to contradict an accepted fact, such as the hypothesis, an axiom, or a previously proved theorem. As a result, you can eliminate the first possibility and know that the second possibility must be the case ("the conclusion is true").

EXAMPLE Prove that for all real numbers a and b, if $a \neq 0$ and $ab = 0$, then $b = 0$.

SOLUTION

Statement	*Reason*
1. $ab = 0$	1. Given
Assume that $b \neq 0$. Then:	
2. $(ab) \cdot \dfrac{1}{b} = 0 \cdot \dfrac{1}{b}$	2. Multiplicative axiom of equality
3. $a\left(b \cdot \dfrac{1}{b}\right) = 0 \cdot \dfrac{1}{b}$	3. Associative axiom of multiplication
4. $a \cdot 1 = 0 \cdot \dfrac{1}{b}$	4. Axiom of multiplicative inverses
5. $a = 0 \cdot \dfrac{1}{b}$	5. Identity axiom for multiplication
6. $a = 0$	6. Multiplicative property of 0

But this contradicts the given fact that $a \neq 0$. Therefore the assumption "$b \neq 0$" is false, and "$b = 0$" is true.

Exercises

Prove each theorem indirectly.

1. For all real numbers a, b, and c, if $a + c \neq b + c$, then $a \neq b$.
2. For all real numbers a, b, and c, if $a + (-c) \neq b + (-c)$, then $a \neq b$.
3. For all real numbers a, b, and c, if $ac \neq bc$, then $a \neq b$.
4. For all real numbers a, b, and c, if $\dfrac{a}{c} \neq \dfrac{b}{c}$ and $c \neq 0$, then $a \neq b$.
5. For all nonzero numbers a and b, if $a \neq b$, then $\dfrac{1}{a} \neq \dfrac{1}{b}$.
6. For all nonzero numbers a and b, if $a \neq b$, then $-a \neq -b$.
7. $1 \neq 2$ 8. $-1 \neq 0$

Chapter Summary

The following statements are true for all real values of each variable unless noted otherwise.

1. **Additive property of equality:** If $a = b$, then $a + c = b + c$ and $c + a = c + b$.

2. **Multiplicative property of equality:** If $a = b$, then $ac = bc$ and $ca = cb$.

3. $(b + c) + (-c) = -c + (b + c) = b$.

   If $c \neq 0$, $(bc)\frac{1}{c} = \frac{1}{c}(cb) = b$.

4. **Property of the opposite of a sum:** $-(a + b) = (-a) + (-b)$

5. **Multiplicative property of 0:** $a \cdot 0 = 0$ and $0 \cdot a = 0$

6. **Multiplicative property of -1:** $a(-1) = -a$ and $(-1)a = -a$

7. **Property of opposites in products:** $(-a)b = -ab$, $a(-b) = -ab$, $(-a)(-b) = ab$

8. **Axiom of multiplicative inverses:** For every $a \neq 0$, there is a unique real number $\frac{1}{a}$ such that
   $$a \cdot \frac{1}{a} = 1 \text{ and } \frac{1}{a} \cdot a = 1.$$

9. **Property of the reciprocal of a product:** For every $a \neq 0$ and $b \neq 0$,
   $$\frac{1}{ab} = \frac{1}{a} \cdot \frac{1}{b}.$$

Chapter Review

1. Which expression equals $29 + (-2)$? *3-1*

   a. $(31 + 2) + (-2)$ **b.** $(29 + 2) + (-2)$ **c.** $(31 - 2) + (-2)$

Simplify each expression in Review Items 2–7.

2. $74 + (-47) + 103 + (-4)$ *3-2*

   a. 126   **b.** 130   **c.** 79   **d.** 261

3. $-235 + (-35) + 54 + (-173)$

   a. -319 **b.** 81   **c.** -389 **d.** 27

Simplify each expression.

4. $23(-8) + (-5)(-23)$ 3-3

a. 69 b. 299 c. -299 d. -69

5. $12x + (-8y) + 38x + 9y^2$

a. $40x + y$ b. $26x + y^2$ c. $50x + (-8y) + 9y^2$

6. $-9k\left(\frac{1}{9}\right)$ 3-4

a. $9k$ b. $-k$ c. $-9k$ d. k

7. $\frac{1}{5}(-25m + 15n)$

a. $-5m + 3n$ b. $-5m + (-3n)$ c. $5m + (-3n)$

Chapter Test

1. $-27 + 47 = -27 + (\underline{?} + 20) = \underline{?}$ 3-1
2. $-6 + (-23) + 27 = -6 + (-23) + (23 + \underline{?}) = -6 + \underline{?} = \underline{?}$
3. $86 + 59 + (-186) = 59 + \underline{?} + (-186) = 59 + 86 + (-\underline{?}) - 100 = \underline{?}$
4. $-45 + 12 + 90 = -45 + \underline{?} + 12 = -45 + (45 + \underline{?}) + 12 = \underline{?}$

Simplify.

5. $-(63 + 4) + 4 + (-7)$ 3-2
6. $-(13 + 87) + 24 + (-76)$
7. $(-15)x^2 + y + 3x^2 + (-8)y$
8. $16m + 7n^2 + (-4n^2) + (-16)m$

Solve.

9. High tide one morning at Anchorage, Alaska, had a height of 8.7 m. Subsequent low and high tides fluctuated from that height as follows: down 7.3 m, up 7.8 m, down 9 m, up 8.3 m, down 6.5 m, up 6.7 m. What was the height of the last tide for which data was given?

Simplify.

10. $77(-9) + (-77)(51)$ 3-3
11. $-8[-12 + (-8)]$
12. $-3[2x + (-6y)]$
13. $13a + 5b - [7a + (-2b)]$
14. $\frac{1}{8}[-24y + (-8t)]$ 3-4
15. $-\frac{1}{3}(-42c + 27d)$
16. $18[\frac{5}{6}x^2 + (-\frac{1}{9}y^2)]$
17. $-27(\frac{2}{3}a) + (-\frac{1}{3})45b$

Elijah McCoy
1844–1929

The inventions of Elijah McCoy were to be the origin of the expression "the real McCoy" because they were valued so highly that users wanted to be sure they had a genuine McCoy device.

The son of Kentucky slaves who escaped via the Underground Railroad, McCoy was born in Canada and later settled in Michigan. An early interest in machines led him to an apprenticeship in Scotland, from which he returned as a mechanical engineer. Unable to find employment as an engineer, he went to work as a fireman for a railroad. His experience there stimulated McCoy to devise his first invention and set the direction for his future work.

In those days, locomotives and heavy machinery had to be shut down periodically for lubrication, and even then it was largely a guess as to when it was time to do so. McCoy was disturbed by the damage to machinery from poor lubrication, and especially by the waste of time and money in shutting down machines. To correct this problem, McCoy patented his first invention in 1872, an "automobile lubricator" that provided a continuous flow of lubricant directly to a machine's moving parts. McCoy improved his system and developed other lubricating devices, eventually holding over 50 patents. His inventions were widely applied to locomotives and industrial machinery throughout the world.

Careers
in Forestry

Forests are one of our most important natural resources. Foresters protect, manage, and develop forest land. They plan reforestation and cutting projects, direct fire prevention programs, and assist on projects to control floods, insects, soil erosion, and tree diseases. Managed properly, our forests can be used over and over again. Foresters must have a college education, including courses in mathematics and the biological and physical sciences.

An artist's impression of a Satellite Power System which would convert sunlight in space to electrical energy and then to microwave energy for transmission to Earth. As designed, it would provide as much electricity as five nuclear power plants.

Solving Equations and Problems

Addition and Subtraction Transformations

OBJECTIVES *for Sections 4-1 and 4-2:*
1. Use addition and subtraction transformations to solve equations.
2. Simplify expressions involving sums and differences.

4-1 Transforming Equations by Addition

You can use the additive property of equality to solve certain equations. Study the following sequence of equations.

(1) $\qquad y + 7 = -3$
(2) $(y + 7) + (-7) = -3 + (-7)$ Add -7 to each member of the equation.
(3) $\qquad \therefore y = -10$ Simplify each member.
$\qquad\qquad\qquad\qquad\qquad$ [*Recall:* $(b + c) + (-c) = b$.]

On the other hand, if you add 7 to each member of the third equation, you obtain the first equation:

$$y = -10$$
$$y + 7 = -10 + 7$$
$$\therefore y + 7 = -3$$

Thus, any root of either of the equations

$$y + 7 = -3 \quad \text{and} \quad y = -10$$

must also satisfy the other equation. Therefore the equations have the same solution set, namely, $\{-10\}$.

Equations having the same solution set over a given set are called **equivalent equations** over that set. To solve an equation you usually try to change, or **transform**, it into an equivalent equation whose solution set can be found by inspection. The properties of real numbers guarantee that each of the following transformations always produces an equivalent equation.

Transformation by Substitution: Substituting for either member of a given equation an expression equivalent to that member.

Transformation by Addition: Adding the same real number to each member of a given equation.

In the following example, and, in fact, from here on in this book, you can assume unless otherwise stated that:

Open sentences in one variable are to be solved over the set $\mathcal{R}$ of real numbers.

EXAMPLE Solve $-5 + x = -13$.

SOLUTION
$$-5 + x = -13$$
$$5 + (-5 + x) = 5 + (-13)$$
$$x = -8$$

$\left\{\begin{array}{l}\text{Since "}-5 + x\text{" shows } -5 \text{ added} \\ \text{to } x\text{, you add 5 (the opposite of } -5) \\ \text{to the sum to obtain } x.\end{array}\right.$

Because errors may occur in transforming equations, you should always check each root of the transformed equation in the given equation.

Check: $-5 + x = -13 \leftarrow$ given equation
$$-5 + (-8) \overset{?}{=} -13$$
$$-13 = -13 \ \checkmark$$

Oral Exercises

Name the number that must be added to each member of the first equation to obtain the second equation. Then name the root of the first equation.

1. $x + 9 = 17$; $x = 8$

2. $y + (-7) = 5$; $y = 12$

3. $-3 + d = 0$; $d = 3$

4. $6 + n = -13$; $n = -19$

5. $14 = m + (-4)$; $18 = m$

6. $-8 = -3 + k$; $-5 = k$

7. $-15 = 15 + t$; $-30 = t$

8. $3x + (-2) = 4$; $3x = 6$

9. $2c + 3 = 5$; $2c = 2$

10. $-4 + 5e = 6$; $5e = 10$

Written Exercises

Solve each of the following equations, either by inspection or by transforming the equation into an equivalent equation which you can solve by inspection.

A
1. $x + (-39) = 18$
2. $y + (-27) = 51$
3. $-18 + h = 47$

4. $-46 + w = 70$
5. $51 + s = 51$
6. $36 = r + 36$

7. $u + 40 = 84$
8. $c + 19 = 41$
9. $d + 23 = 0$

10. $59 + p = 0$
11. $216 = -104 + x$
12. $48 = -175 + y$

13. $p + (-4) = \frac{3}{4}$
14. $w + (-\frac{3}{5}) = 5$
15. $s + (-2.5) = 5.93$

16. $r + (-0.95) = 0.15$
17. $\frac{7}{8} = 0.875 + u$
18. $3\frac{1}{7} = -\frac{6}{7} + c$

B
19. $-x + 5 = 21$
20. $-y + 5 = 25$
21. $-3 + (-w) = -5$

22. $-2 + (-a) = -11$
23. $(b + 2) + (-4) = 7$
24. $(c + 5) + (-8) = 4$

25. $7 + (3 + p) = -5$
26. $-3 + [r + (-5)] = -8$
27. $-4 + [s + (-3)] = 5$

28. $9 + (t + 4) = -7$
29. $-5 + (-x) + 6 = -4$
30. $-y + 7 + (-4) = -5$

31. $(k + 5) + (-8) = -11$
32. $-t + 10 + (-21) = 11$
33. $-7 + (-x) + 7 = 0$

C
34. $|x| + (-1) = 6$
35. $|y| + (-4) = 0$
36. $-6 + |m| = -2$

37. $7 + |r| = 3$
38. $|s| + 3 = -4$
39. $-10 + |t| = -4$

4-2 Subtracting Real Numbers

If you buy a 15-cent pencil and give the clerk a quarter, he may hand you a dime and say:

"15 and 10 is 25."

The clerk uses addition, $15 + 10 = 25$, to do subtraction, $25 - 15 = 10$. In the language of algebra, you say, "The equations

$$15 + x = 25 \quad \text{and} \quad 25 - 15 = x$$

have the same solution."

In general, for any real numbers a and b, the equations

$$b + x = a \quad \text{and} \quad a - b = x$$

have the same solution, called the *difference* between a and b. Thus the **difference** $a - b$ between any two real numbers a and b is the number whose sum with b is a.

EXAMPLE 1 State an equation involving x to fit the diagram below, and solve the equation.

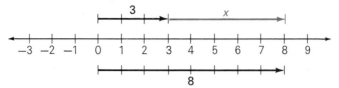

SOLUTION Each of the following equations fits the diagram.

(1) $3 + x = 8$ (2) $8 - 3 = x$

The solution of each equation is 5.

EXAMPLE 2 Simplify: $-7 - (-4)$

SOLUTION The difference $-7 - (-4)$ is the number whose sum with -4 is -7. On the number line, to move from the graph of -4 to the graph of -7 requires a displacement of 3 units to the left.

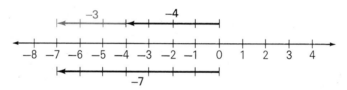

Thus $-4 + (-3) = -7$, and therefore

$$-7 - (-4) = -3.$$

You can find a useful expression for $a - b$ by transforming the equation $b + x = a$ by addition:

$$b + x = a$$
$$x + b = a$$
$$x + b + (-b) = a + (-b)$$
$$x + 0 = a + (-b)$$
$$\therefore x = a + (-b)$$

The last equation clearly has just one root, $a + (-b)$. Checking this root in the original equation, you have:

$$b + x = a$$
$$b + [a + (-b)] \overset{?}{=} a$$
$$b + (-b) + a \overset{?}{=} a$$
$$0 + a \overset{?}{=} a$$
$$a = a \ \checkmark$$

Since the *one and only* root of $b + x = a$ is $a + (-b)$, the following theorem has been proved.

Relationship between Addition and Subtraction

For all real numbers a and b,

$$a - b = a + (-b).$$

To subtract b from a, add the opposite of b to a.

Using this relationship, you may replace any difference with a sum.

Difference	Sum	Value	Check
$5 - 2$	$5 + (-2)$	3	$2 + 3 = 5$
$5 - (-2)$	$5 + 2$	7	$-2 + 7 = 5$
$-5 - 2$	$-5 + (-2)$	-7	$2 + (-7) = -5$
$-5 - (-2)$	$-5 + 2$	-3	$-2 + (-3) = -5$

EXAMPLE 3   Replace the symbol $\underline{\ ?\ }$ by $=$ or $\neq$ to produce a true statement.

     a. $5 - 2 \underline{\ ?\ } 2 - 5$

     b. $(10 - 7) - 1 \underline{\ ?\ } 10 - (7 - 1)$

SOLUTION   **a.** $5 - 2 = 3$; $2 - 5 = -3$

       $\therefore 5 - 2 \neq 2 - 5$.

     b. $(10 - 7) - 1 = 3 - 1 = 2$; $10 - (7 - 1) = 10 - 6 = 4$

       $\therefore (10 - 7) - 1 \neq 10 - (7 - 1)$.

Example 3 shows that *subtraction of real numbers is neither commutative nor associative*. On the other hand, you can prove (Example 3, page 104) that *multiplication is distributive with respect to subtraction.* Thus, for example, for each real number x,

$$5(x - 4) = 5 \cdot x - 5 \cdot 4 = 5x - 20.$$

Expressions for sums such as

$$-4 + (-10), \quad 7 + (-y), \quad \text{and} \quad 6y^2 + (-8y)$$

are usually written as differences:

$$-4 - 10, \quad 7 - y, \quad \text{and} \quad 6y^2 - 8y.$$

EXAMPLE 4   Simplify: $8 - 17 + 4 - 3 - 10$

SOLUTION 1 $8 - 17 = -9$; $-9 + 4 = -5$; $-5 - 3 = -8$; $-8 - 10 = -18$.

SOLUTION 2 $8 + 4 = 12$; $-17 + (-3) + (-10) = -30$; $12 + (-30) = -18$.

EXAMPLE 5   Simplify: $7 - y - 4 + 6y^2 - 8y - 10$

SOLUTION $7 - y - 4 + 6y^2 - 8y - 10 = 7 - 4 - 10 + (-y - 8y) + 6y^2$
$$= -7 - 9y + 6y^2.$$

EXAMPLE 6   Solve: **a.** $r - 2 = -23$ **b.** $x + 8 = -24$

SOLUTION   **a.** $r - 2 = -23$
$\qquad\quad r - 2 + 2 = -23 + 2$
$\qquad\qquad\qquad r = -21$

$\qquad$ *Check:* $-21 - 2 \overset{?}{=} -23$
$\qquad\qquad\qquad -23 = -23$ ✓
$\qquad \therefore$ the solution set is $\{-21\}$.
$\qquad\qquad\qquad\qquad$ Answer.

b. $x + 8 = -24$
$\qquad x + 8 - 8 = -24 - 8$
$\qquad\qquad\quad x = -32$

$\qquad$ *Check:* $-32 + 8 \overset{?}{=} -24$
$\qquad\qquad\qquad -24 = -24$ ✓
$\qquad \therefore$ the solution set is $\{-32\}$.
$\qquad\qquad\qquad\qquad$ Answer.

Because you can describe the method used in Example 6(b) as *subtracting 8* from each member of the equation (rather than adding -8 to each member), you may call the method **transformation by subtraction.** Of course, transformation by subtraction is just another form of transformation by addition.

Oral Exercises

a. Express each difference as a sum.
b. Simplify the resulting expression.

EXAMPLE 1 $8 - 13$

SOLUTION   **a.** $8 + (-13)$ **b.** -5

1. $3 - 16$

2. $5 - (-7)$

3. $-5 - 0$

4. $6 - (-6)$

5. $-8 - (-8)$

6. $11 - 6$

7. $6 - 11$

8. $-3.24 - 3.25$

9. $-\frac{5}{8} - (-\frac{3}{8})$

10. $3(y - 2) - y$

11. $x - (x + 1)$

12. $(h + 3) - (h - 3)$

Name the number that must be subtracted from each member of the first equation to obtain the second equation.

13. $x + 7 = 13$; $x = 6$

14. $y + 8 = 5$; $y = -3$

15. $18 - w = 5$; $-w = -13$

In Exercises 16 and 17, give the reason for each lettered statement in the proof.

EXAMPLE 2 Prove: If x is a real number and $x + 5 = 12$, then $x = 7$.

PROOF

Statement	*Reason*
x is a real number and $x + 5 = 12$.	(a) Given
$(x + 5) + (-5) = 12 + (-5)$	(b) Additive property of equality
$x + [5 + (-5)] = 12 + (-5)$	(c) Associative axiom of addition
$x + 0 = 12 + (-5)$	(d) Axiom of additive inverses
$x = 12 + (-5)$	(e) Identity axiom of addition
$x = 12 - 5$	(f) Relationship between addition and subtraction
$x = 7$	(g) Substitution principle

16. Prove: If y is a real number and $y + 4 = 13$, then $y = 9$.

y is a real number and $y + 4 = 13$. (a)
$(y + 4) + (-4) = 13 + (-4)$ (b)
$(y + 4) + (-4) = 13 - 4$ (c)
$(y + 4) + (-4) = 9$ (d)
$y + [4 + (-4)] = 9$ (e)
$y + 0 = 9$ (f)
$y = 9$ (g)

17. Prove: If x is a real number and $x = 5$, then $x - 2 = 3$.

x is a real number and $x = 5$. (a)
$x + (-2) = 5 + (-2)$ (b)
$x - 2 = 5 - 2$ (c)
$x - 2 = 3$ (d)

Written Exercises

Simplify and check. Assume that each variable denotes a real number.

EXAMPLE 1 $29 - 37$

SOLUTION $29 - 37 = 29 + (-37) = -8$

Check: $37 + (-8) = 29$ ✓

EXAMPLE 2 $(x + 5) - (x - 2)$

SOLUTION $(x + 5) - (x - 2) = x + 5 + \{-[x + (-2)]\}$
$= x + 5 + (-x) + 2 = 7$

Check: $x - 2 + 7 = x + 5$ ✓

A 1. $171 - (-275)$ 2. $-481 - 769$ 3. $-1362 - 1812$ 4. $1979 - (-1499)$

5. $4.92 - 5.93$ 6. $14.47 - 38.09$ 7. $0 - (-\frac{7}{9})$ 8. $0 - 3\frac{2}{3}$

9. $33 - 70 + 64 - 25 - 22$ 10. $90 - 96 - 14 + 48$ 11. $-375 + 225 - 435$

12. $-598 - 102 + 78 - 298$ 13. $11\frac{2}{7} - 4\frac{2}{7} - 5\frac{3}{7}$ 14. $-24\frac{1}{3} + 8\frac{2}{3} - 20\frac{1}{3}$

Simplify and check.

15. $-(1833 - 2002)$ **16.** $-(-744 - 81)$ **17.** $(x + 8) - (x + 11)$

18. $(y - 3) - (y + 5)$ **19.** $(t + 7) - (t - 5)$ **20.** $(4 - r) - (-3 - r)$

21. $4(\frac{3}{4} - \frac{1}{2}) - 10 - 34$ **22.** $18 + 36(\frac{2}{9} - \frac{5}{6}) - 38$

23. $5x^2 - 2x + 5 - x^2 - 3x - 8$ **24.** $8y^2 + 3y - 10 - 6y^2 + 7y + 5$

25. $-m^3 + 4m - 5 - 3m^3 - 6m + 5$ **26.** $3w^2 + 7w - 11 - 6w^2 + 7 - 6w$

Solve each equation.

B **27.** $(6x - 1) - (5x - 5) = 0$ **28.** $(5y - 4) - (4y + 8) = 4$

29. $17 - m = -8$ **30.** $25 - w = 10$

31. $-39 - r = -56$ **32.** $-18 - s = 16$

33. $5 - (t - 7) = 3$ **34.** $-8 - (p - 9) = -4$

35. $-k - 12 = -12$ **36.** $(3n - 2) - (7 + 2n) = 3$

37. $(1 - 2u) - (3 - 3u) = 0$ **38.** $(3 - 5v) - (-8 - 6v) = -5$

Write each phrase in algebraic symbols, and then simplify.

39. -8 decreased by 3 **40.** From -7 subtract -2

41. x less $(x - 7)$ **42.** y less $(4 + y)$

43. $(3x + 1)$ less $(3x - 1)$ **44.** $(1 - k)$ less $(k - 1)$

45. $(6 - 2k)$ subtracted from $(1 - k)$ **46.** $(7 - 3h)$ subtracted from $(h - 7)$

47. Four times $(1 - e)$ less $(5e + 1)$ **48.** Six times $(5f + 3)$ less $(5 - 3f)$

Give the reasons for the statements in the following proofs. Assume that a, b, and c are real numbers.

EXAMPLE 3 Distributive property of multiplication with respect to subtraction:

$$a(b - c) = ab - ac.$$

PROOF

Statement	Reason
1. a, b, and c are real numbers.	Given
2. $b - c = b + (-c)$	Relationship between addition and subtraction
3. $a(b - c) = a[b + (-c)]$	Substitution principle
4. $a[b + (-c)] = ab + a(-c)$	Distributive axiom of mult. with respect to add.
5. $a(-c) = -ac$	Property of opposites in products
6. $ab + a(-c) = ab + (-ac)$	Substitution principle
7. $ab + (-ac) = ab - ac$	Relationship between addition and subtraction
8. $\therefore a(b - c) = ab - ac$	Transitive axiom of equality

49. Prove: $-(a - b) = -a + b$

1. a and b are real numbers.
2. $-(a - b) = (-1)(a - b)$
3. $\quad = (-1)a - (-1)b$
4. $\quad = -a - (-b)$
5. $\quad = -a + [-(-b)]$
6. $\quad = -a + b$

50. Prove: $-(-a - b) = a + b$

1. a and b are real numbers.
2. $-(-a - b) = (-1)(-a - b)$
3. $\quad = (-1)(-a) - (-1)b$
4. $\quad = -(-a) - (-b)$
5. $\quad = -(-a) + [-(-b)]$
6. $\quad = a + b$

Prove the stated theorem.

C **51.** For all real numbers a, b, and c, if $a = b$, then $a - c = b - c$.

52. For all real numbers a, b, and c, if $a - c = b - c$, then $a = b$.

53. For all real numbers a, b, and c, $a - (b - c) = (a - b) + c$.

54. For all real numbers a, b, and c, $-(a + b + c) = -a - b - c$.

A set of numbers is closed under subtraction if all differences between members of the set also belong to the set. Which of the following sets are closed under subtraction? Explain.

55. $\{0, 1\}$

56. {the natural numbers}

57. {the rational numbers}

58. {the even integers}

59. {the integers}

60. {the real numbers}

61. {the odd integers}

62. {the positive real numbers and 0}

Problems

Solve each problem.

EXAMPLE How much less is the altitude of the Dead Sea, 389 m below sea level, than the altitude of Death Valley, California, 85 m below sea level?

SOLUTION $-85 - (-389) = -85 + 389 = 304.$ 304 m. Answer.

A **1.** The Kent family drove from Los Angeles, 83 m above sea level, to Death Valley, California, 85 m below sea level. What was the difference between their elevation in Los Angeles and in Death Valley?

2. If the Greek mathematician Pythagoras was born in 572 B.C. and died in 497 B.C., what was his age when he died?

3. Mercury melts at $-38.87°C$ and boils at $356.6°C$. What is the difference between these temperatures?

4. What is the difference between the melting and boiling points of fluorine, if it melts at $-219.6°C$ and boils at $-188.1°C$?

5. Rita rode the subway from a point 48 blocks south of Central Avenue to a point 36 blocks north of Central Avenue. How many blocks did she travel?

6. A corporation's expenses total $1,375,500. What must its sales be if it makes a net profit of $225,750?

7. Find the difference in latitude between Nome, Alaska, at 65°N and Buenos Aires, Argentina, at 35°S.

B 8. A satellite was 640 km above sea level when it passed over a point on the ocean floor 9000 m deep. At that moment, what was the distance between the floor of the ocean and the satellite?

9. A deep-sea diver rose from -50 m to -15 m, while another dived from -35 m to -65 m. Which deep-sea diver made the greater change in depth?

10. The Thibodeau family opened a revolving credit charge account with a deposit of $75. They charged five items costing $25.34, $6.96, $19.95, $5.50, and $33.49. What is the present status of their account?

11. A balloon at an altitude of 270 m climbs at 1 m/s for 1 minute, then drops for 2 min at 2 m/s, and levels off. What is the balloon's resulting altitude?

Self-Test 1

VOCABULARY   equivalent equations (p. 98)

Solve each equation over ℛ.

1. $x + (-17) = -13$

2. $10 + (-2y) + 3y = 14$ *Obj. 1, p. 97*

Simplify.

3. $-(400 - 136) + (97 - 159)$

4. $(4p + 3) - (2p - 2)$ *Obj. 2, p. 97*

Check your answers with those at the back of the book.

DIVERSION

Which of the figures at the right could be folded into a cube without a top?

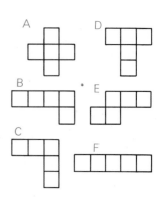

Multiplication and Division Transformations

OBJECTIVES *for Sections 4-3 through 4-5:*
1. Use multiplication and division transformations to solve equations.
2. Simplify expressions involving products and quotients.
3. Use inverse operations to solve equations.

4-3 Transforming Equations by Multiplication

The multiplicative property of equality states that *if equal numbers are multiplied by the same number, then the products are equal.* For example, since

$$8 + 4 = 12,$$

you have

$$(8 + 4) \cdot 5 = 12 \cdot 5 \quad \text{and} \quad 5 \cdot (8 + 4) = 5 \cdot 12.$$

This property guarantees that the following transformation always produces an equivalent equation.

> **Transformation by Multiplication:** Multiplying each member of a given equation by the same *nonzero* real number.

For example, if you multiply each member of the first equation below by $\frac{2}{3}$, you obtain the third equation, $x = 6$.

$$
\begin{array}{rl}
(1) & \frac{3}{2}x = 9 \\
(2) & \frac{2}{3} \cdot \frac{3}{2}x = \frac{2}{3} \cdot 9 \\
(3) & x = 6
\end{array}
$$

On the other hand, if you multiply each member of the third equation by $\frac{3}{2}$, you obtain the first equation.

$$
\begin{array}{c}
x = 6 \\
\frac{3}{2} \cdot x = \frac{3}{2} \cdot 6 \\
\frac{3}{2}x = 9
\end{array}
$$

Thus, any root of either of the equations

$$\frac{3}{2}x = 9 \quad \text{and} \quad x = 6$$

is a root of the other equation, and the two equations are equivalent.

EXAMPLE 1 Solve: $\frac{1}{3}n = -14$

SOLUTION

$$\frac{1}{3}n = -14$$
$$3 \cdot \frac{1}{3}n = 3 \cdot (-14)$$
$$n = -42$$

$\left\{\begin{array}{l} \text{Since } \frac{1}{3}n \text{ shows } n \text{ multiplied by } \frac{1}{3}, \\ \text{you multiply the product by 3 (the} \\ \text{reciprocal of } \frac{1}{3}) \text{ to obtain } n. \end{array}\right.$

Check:
$$\frac{1}{3}n = -14$$
$$\frac{1}{3}(-42) \overset{?}{=} -14$$
$$-14 = -14 \ \checkmark$$
$\therefore$ the solution set is $\{-42\}$.

EXAMPLE 2 Solve: $-4d = 160$

SOLUTION

$$-4d = 160$$
$$-\frac{1}{4}(-4d) = -\frac{1}{4} \cdot 160$$
$$d = -40$$

Check:
$$-4d = 160$$
$$-4(-40) \overset{?}{=} 160$$
$$160 = 160 \ \checkmark$$

$\therefore$ the solution set is $\{-40\}$.   **Answer.**

Will multiplication by **0** transform an equation into an equivalent equation? Generally, it will not! Consider the following sequence of equations:

$$\begin{array}{ll} (1) & 6x = 24 \\ (2) & 0 \cdot 6x = 0 \cdot 24 \\ (3) & (0 \cdot 6)x = 0 \cdot 24 \\ (4) & 0 \cdot x = 0 \end{array}$$

Equation (1) has just one root, namely 4. But every real number satisfies equation (4), since zero times any real number equals zero. Therefore, equations (1) and (4) are *not* equivalent equations.

Remember: *In transforming equations, never multiply by zero.*

Oral Exercises

Name the number by which each member of the first equation must be multiplied to obtain the second equation.

1. $\frac{1}{4}x = 24$; $x = 96$

2. $\frac{1}{3}y = -4$; $y = -12$

3. $\frac{1}{2}m = 0$; $m = 0$

4. $4w = 12$; $w = 3$

5. $6k = -18$; $k = -3$

6. $28 = -4h$; $-7 = h$

7. $\frac{3}{4}s = -\frac{6}{5}$; $s = -\frac{8}{5}$

8. $-6 = 0.5p$; $-12 = p$

9. $2.5 = -5x$; $-0.5 = x$

Fill in the ? to make each sentence true.

10. If $-3a = 5$, then $3a = \underline{?}$.

11. If $4m = 7$, then $8m = \underline{?}$.

12. If $-6x = 10$, then $3x = \underline{?}$.

13. If $-2c = 5$, then $5c = \underline{\ ?\ }$.

14. If $\frac{1}{2}d = 8$, then $\frac{1}{4}d = \underline{\ ?\ }$.

15. If $5x = y + z$, then $10x = \underline{\ ?\ }$.

16. If $2s = 3$ and $t \neq 0$, then $6st = \underline{\ ?\ }$.

Written Exercises

Solve each equation.

A
1. $\frac{1}{12}x = 3$
2. $\frac{1}{30}y = 6$
3. $\frac{1}{8}b = -12$
4. $\frac{1}{6}w = -13$

5. $14p = 182$
6. $15s = 165$
7. $-12r = 96$
8. $15a = -225$

9. $5 = -\frac{1}{16}c$
10. $-72 = -\frac{1}{7}m$
11. $-\frac{1}{13}n = -2.3$
12. $\frac{1}{17}t = -0.7$

13. $0.6 = 0.6h$
14. $-2.4 = 2.4g$
15. $\frac{1}{107}t = 0$
16. $-\frac{1}{0.25}d = 2$

B
17. $\frac{1}{4}x = 2\frac{1}{4}$
18. $\frac{1}{8}y = 4\frac{3}{8}$
19. $-\frac{1}{3}m = 1\frac{2}{3}$

20. $\frac{1}{7}w = -4\frac{6}{7}$
21. $8 = 6a + 20$
22. $19 = -5b + 19$

23. $\frac{1}{5}k + 3 = 9$
24. $\frac{1}{4}m - 6 = -7$
25. $\frac{1}{3}x + 5 = -8$

26. $6 = \frac{1}{6}y + 5$
27. $-10 = \frac{1}{10}h - 9$
28. $7 = \frac{1}{9}w + 7$

29. $2.5r + 5 = 7$
30. $0.75s - 8 = -4.25$
31. $18 = -5.5t - 37$

C
32. $\frac{1}{11}|x| = 13$
33. $\frac{1}{13}|y| = -9$
34. $\frac{1}{5}|-m| - 2 = 0$

35. $\frac{1}{9}|w| + 3 = 5$
36. $5 + \frac{1}{2}|u| = 9$
37. $16 = \frac{1}{20}|v|$

38. $23 = \frac{1}{10}|-p| + 3$
39. $16 = \frac{1}{7}|j| - 4$
40. $29 = 20 + \frac{1}{2}|-k|$

4-4 Dividing Real Numbers

Because the equation

$$7 \times 3 = 21 \quad \text{is true, the equation} \quad 21 \div 7 = 3$$

is also true. In general, for any real number a and any *nonzero* real number b, the equations

$$bx = a \quad \text{and} \quad a \div b = x$$

have the same solution, called the *quotient* of a and b or the *quotient* of a by b. Thus, the **quotient** $a \div b$, where $b \neq 0$, is the number whose product with b is a.

The quotient $a \div b$ can also be represented by a fraction:

$$a \div b = \frac{a}{b}$$

When you studied fractions, you learned facts like these:

$$\tfrac{9}{5} = 9 \times \tfrac{1}{5}; \quad \tfrac{2}{3} = 2 \times \tfrac{1}{3}; \quad \tfrac{40}{8} = 40 \times \tfrac{1}{8}.$$

These equations suggest how to use reciprocals to express quotients as products.

Relationship between Multiplication and Division

For all real numbers a and all nonzero real numbers b,

$$a \div b = a \times \frac{1}{b}.$$

To divide a by b, multiply a by the reciprocal of b.

You can show that this relationship is true by transforming the equation $bx = a$ by multiplication:

$$bx = a$$

$$\frac{1}{b}(bx) = \frac{1}{b}(a) \qquad (b \neq 0)$$

$$\left(\frac{1}{b} \cdot b\right)x = a\left(\frac{1}{b}\right)$$

$$1(x) = a\left(\frac{1}{b}\right)$$

$$\therefore x = a\left(\frac{1}{b}\right)$$

The last equation has just one root, $a\left(\frac{1}{b}\right)$. Checking this root in the original equation, you have:

$$bx = a$$

$$b\left(a \cdot \frac{1}{b}\right) \stackrel{?}{=} a$$

$$a\left(b \cdot \frac{1}{b}\right) \stackrel{?}{=} a$$

$$a(1) \stackrel{?}{=} a$$

$$a = a \; \checkmark$$

Since the *one and only* root of "$bx = a$" is $a\left(\frac{1}{b}\right)$, and since "$bx = a$" and "$a \div b = x$" have the same root, it follows that:

$$a \div b = a\left(\frac{1}{b}\right) \quad \text{or} \quad \frac{a}{b} = a\left(\frac{1}{b}\right).$$

Using this result, you can replace any quotient with a product.

Quotient	Product	Value	Check
$63 \div 7$ or $\dfrac{63}{7}$	$63 \times \dfrac{1}{7}$	9	$7 \times 9 = 63$
$63 \div -7$ or $\dfrac{63}{-7}$	$63 \times \left(-\dfrac{1}{7}\right)$	-9	$-7 \times (-9) = 63$
$-63 \div 7$ or $\dfrac{-63}{7}$	$-63 \times \dfrac{1}{7}$	-9	$7 \times (-9) = -63$
$-63 \div (-7)$ or $\dfrac{-63}{-7}$	$-63 \times \left(-\dfrac{1}{7}\right)$	9	$-7 \times 9 = -63$

Notice that $63 \div 7 = 9$, but $7 \div 63 = \frac{1}{9}$.

This example shows that *division of real numbers is not commutative. It is not associative*, either. For example,

$$(96 \div 8) \div 2 = 12 \div 2 = 6,$$

but

$$96 \div (8 \div 2) = 96 \div 4 = 24.$$

On the other hand, *division is distributive with respect to both addition and subtraction* (Exercises 35 and 36, page 114). For example:

$$\frac{35 + 49}{-7}$$

$$\frac{84}{-7} = -12 \qquad \frac{35}{-7} + \frac{49}{-7} = -5 + (-7) = -12$$

Why must division by zero be avoided in the set of real numbers? For any number a, $\dfrac{a}{0} = c$ would mean that $a = 0 \times c$. If $a \neq 0$, *no* value of c can make this last equation a true statement, since $0 \times c = 0$ for each value of c. But if $a = 0$, then every value of c makes the equation a true statement. Therefore, a "quotient" $\dfrac{a}{0}$ either would have *no* value or would have *an infinite set* of values. Therefore, *division by zero has no meaning in the set of real numbers.*

You cannot divide zero by zero. But can you divide zero by any other number? Consider these examples:

$$\tfrac{0}{9} = 0 \cdot \tfrac{1}{9} = 0$$
$$0 \div (-7) = 0 \times (-\tfrac{1}{7}) = 0$$

In general, for any real number $a \neq 0$:

$$0 \div a = 0 \cdot \frac{1}{a} = 0.$$

Thus, *the quotient of zero and any nonzero number is zero.*

EXAMPLE Solve each equation: **a.** $-61 = -\dfrac{t}{2}$ **b.** $3y = -105$

SOLUTION   **a.** $-61 = -\dfrac{t}{2}$

$-61 = (-\tfrac{1}{2})(t)$

$-2(-61) = -2[(-\tfrac{1}{2})(t)]$

$122 = [-2(-\tfrac{1}{2})]t$

$122 = t$

Check: $-61 \overset{?}{=} -\dfrac{122}{2}$

$-61 = -61 \ \checkmark$

$\therefore$ the solution set is $\{122\}$.

b. $3y = -105$

$\dfrac{3y}{3} = \dfrac{-105}{3}$

$y = -35$

$3(-35) \overset{?}{=} -105$

$-105 = -105 \ \checkmark$

$\therefore$ the solution set is $\{-35\}$.

Because you can describe the method used in part (b) of the preceding example as *dividing* each member of the equation *by 3* (rather than multiplying each member by $\tfrac{1}{3}$), you may call the method **transformation by division**. Of course, transformation by division is just another form of transformation by multiplication.

Oral Exercises

Simplify each expression.

EXAMPLE 1 $\dfrac{24m^2}{-8}$ **SOLUTION** $\dfrac{24m^2}{-8} = -3m^2$

Check: $-8(-3m^2) = 24m^2 \ \checkmark$

1. $\tfrac{36}{9}$

2. $\dfrac{50}{-5}$

3. $\dfrac{-48}{8}$

4. $\dfrac{-56}{-7}$

5. $0 \div 8$

6. $\dfrac{0}{-4}$

7. $\dfrac{-11}{-11}$

8. $-6 \div 6$

9. $\dfrac{-1}{1}$

10. $\dfrac{8x}{-4}$

11. $\dfrac{-16t^2}{8}$

12. $\dfrac{-63a^2}{-9}$

13. $p \div (-1)$

14. $-p \div (-1)$

15. $x \div x, \ x \neq 0$

16. $\dfrac{x}{-x}, \ x \neq 0$

Read each quotient as a product. Then state the value of the quotient.

EXAMPLE 2 $9t \div (-\tfrac{1}{5})$ **SOLUTION** $9t \div (-\tfrac{1}{5}) = 9t \times (-5) = -45t$

17. $8 \div \tfrac{1}{3}$

18. $-9 \div \tfrac{1}{8}$

19. $-8 \div (-\tfrac{1}{9})$

20. $6x \div \tfrac{1}{7}$

21. $-8y \div (-\tfrac{1}{5})$

22. $-12 \div \dfrac{1}{m}, \ m \neq 0$

Name the number by which each member of the first equation must be divided to obtain the second equation.

23. $14p = 42; \ p = 3$

24. $-6e = 6; \ e = -1$

25. $-9q = -81; \ q = 9$

26. $-5x = 0; \; x = 0$ **27.** $12x = -72; \; x = -6$ **28.** $-2.5h = 50; \; h = -20$

Solve each equation.

29. $7d = -63$ **30.** $-6f = 48$ **31.** $-5 = \dfrac{c}{6}$

32. $\dfrac{n}{-5} = 13$ **33.** $2.6 = x - 6.2$ **34.** $7 = h - 19$

Replace the ? by one of the words *positive, negative,* or *zero* to make a true statement.

35. The quotient of two negative numbers is always ?.

36. In a quotient the divisor can never be ?.

37. The quotient of zero by any negative number is always ?.

38. The quotient of a negative number by a positive number is always ?.

Written Exercises

Simplify each expression.

EXAMPLE 1 $-\frac{3}{5}x^2y \div \left(-\frac{1}{10}\right)$

SOLUTION $-\frac{3}{5}x^2y \div \left(-\frac{1}{10}\right) = -\frac{3}{5}x^2y \cdot (-10) = -3 \cdot \frac{1}{5} \cdot (-10) \cdot x^2y$
$$= -3 \cdot (-2) \cdot x^2y = 6x^2y$$

A **1.** $80 \div (-16)$ **2.** $-116 \div 29$ **3.** $-224 \div (-14)$ **4.** $0 \div (-95)$

5. $-18 \div \frac{1}{3}$ **6.** $175 \div \left(-\frac{1}{4}\right)$ **7.** $-\frac{11}{24} \div \left(-\frac{1}{8}\right)$ **8.** $-\frac{13}{22} \div \left(-\frac{1}{11}\right)$

9. $-2.72 \div 17$ **10.** $2.52 \div (-1.4)$ **11.** $\dfrac{8}{-\frac{1}{7}}$ **12.** $\dfrac{-12}{-\frac{1}{12}}$

13. $16ab \div (-4)$ **14.** $-78c^2 \div (-6)$ **15.** $-\frac{5}{6}d \div \left(-\frac{1}{12}\right)$ **16.** $-\frac{4}{3}xy^3 \div \left(-\frac{1}{6}\right)$

In Exercises 17–20 find the average of the numbers in the given set. (The average is the quotient of the sum of the numbers and the number of numbers in the set.)

EXAMPLE 2 $\{5, 6, 7, 8, 4\}$ **SOLUTION** $\dfrac{5 + 6 + 7 + 8 + 4}{5} = 6$

17. $\{-10, 12, 46, -32\}$ **18.** $\{-37, 28, -49, 16, 12\}$

19. $\{25, -33, -48, 9, 23, -9, 5\}$ **20.** $\{-22, 56, 64, 18, 19, -45\}$

Solve each equation.

21. $\dfrac{a}{18} = -13$ **22.** $476 = -68c$ **23.** $-14.4 = -2.4e$ **24.** $-1.7g = 13.6$

Simplify each expression.

B 25. $(-63 \div 7) \div [-6(-\frac{1}{4})]$

26. $[84 \div (-4)] \div [7 \div (-\frac{1}{5})]$

27. $\dfrac{-121 \div [11(6 - 7)]}{-3 - 5}$

28. $\dfrac{(-225 \div 25) \div (-81 \div 9)}{-36 \div (-2)}$

29. $[(-1)^3 - 3(-4)] \div [-(40 + 4)]$

30. $7(-4) \div [-4(-1)^2] - 4^2$

31. $\dfrac{-1275 \div (2^2 + 1^2)}{3^2 - 4}$

32. $\dfrac{(5^2 - 3^2) \div (6^2 - 20)}{8^2 - 7^2}$

33. $\dfrac{[(-2)^2 + (-3)^2] \div [(-3)^2 + 2^2]}{(-2)^2 + (-3)^2}$

34. $\dfrac{[(-256) \div (-16)] \div [2^2 + 2^2]}{[(-169) \div (-13)] \div [4^2 - 3]}$

Give the reasons for the statements in the following proofs. Assume that a, b, and c denote real numbers and that $c \neq 0$.

35. Prove: $\dfrac{a + b}{c} = \dfrac{a}{c} + \dfrac{b}{c}$

    1. a, b, and c are real numbers and $c \neq 0$.

    2. $\dfrac{a + b}{c} = (a + b)\dfrac{1}{c}$

    3. $\qquad = a \cdot \dfrac{1}{c} + b \cdot \dfrac{1}{c}$

    4. $\qquad = \dfrac{a}{c} + \dfrac{b}{c}$

    5. $\therefore \dfrac{a + b}{c} = \dfrac{a}{c} + \dfrac{b}{c}$

36. Prove: $\dfrac{a - b}{c} = \dfrac{a}{c} - \dfrac{b}{c}$

    1. a, b, and c are real numbers and $c \neq 0$.

    2. $\dfrac{a - b}{c} = (a - b)\dfrac{1}{c}$

    3. $\qquad = a \cdot \dfrac{1}{c} - b \cdot \dfrac{1}{c}$

    4. $\qquad = \dfrac{a}{c} - \dfrac{b}{c}$

    5. $\therefore \dfrac{a - b}{c} = \dfrac{a}{c} - \dfrac{b}{c}$

A set S of numbers is *closed under division*, excluding division by zero, if all quotients of numbers in S by nonzero numbers in S also belong to S. Which of the following sets are closed under division, excluding division by zero? Explain.

37. $\{1\}$

38. $\{-1, 0, 1\}$

39. {the positive real numbers}

40. $\{\frac{1}{3}, 1, 3\}$

41. $\{-1\}$

42. {the negative real numbers}

43. $\{\frac{1}{2}, 1, 2\}$

44. {the integers}

45. $\mathcal{R}$

Prove each of the following theorems.

C 46. For all nonzero real numbers a, $\dfrac{a}{a} = 1$.

47. For all nonzero real numbers a, $\dfrac{-a}{a} = -1$.

48. For all real numbers a and b, and all nonzero real numbers c, if $a = b$, then $\dfrac{a}{c} = \dfrac{b}{c}$.

49. For all real numbers a and b, and all nonzero real numbers c, if $\dfrac{a}{c} = \dfrac{b}{c}$, then $a = b$.

50. For all real numbers a and all nonzero real numbers c, $(a \div c) \times c = a$.

51. For all real numbers a and all nonzero real numbers c, $(a \times c) \div c = a$.

4-5 Using Inverse Operations

Using the relationship between addition and subtraction and the fact that

$$(b + c) + (-c) = b$$

for all real numbers b and c, you can prove that

$$(b + c) - c = b \quad \text{and} \quad (b - c) + c = b.$$

(See the Example and Exercise 7, page 117.) This suggests that to "undo" an addition, you subtract; to "undo" a subtraction, you add. We call addition of a given number and subtraction of the same number **opposite, or inverse, operations.** Can you tell why multiplication by a nonzero number and division by the same number are inverse operations?

The following display shows that in solving an equation, you undo the operations used in building the equation, but in *reverse order*.

Building an Equation

$x = -17$	Given.
$2 \cdot x = 2 \cdot (-17)$	Multiply by 2.
$2x = -34$	
$2x + 47 = -34 + 47$	Add 47.
$2x + 47 = 13$	
$5x - 3x + 47 = 13$	Substitute $5x - 3x$ for $2x$.

Solving an Equation

$5x - 3x + 47 = 13$	Given.
$2x + 47 = 13$	Substitute $2x$ for $5x - 3x$.
$2x + 47 - 47 = 13 - 47$	Subtract 47.
$2x = -34$	
$\dfrac{2x}{2} = \dfrac{-34}{2}$	Divide by 2.
$x = -17$	

The following steps are usually helpful in transforming an equation into an equivalent equation that can be solved by inspection.

Solving an Equation

1. Simplify each member of the equation.

2. If there are indicated additions or subtractions, use the inverse operations to undo them.

3. If there are indicated multiplications and divisions involving the variable, use the inverse operations to undo them.

4. If you can solve the transformed equation by inspection, check its root in the given equation.

EXAMPLE Solve: $7 - 2(3x - 1) = 33$

SOLUTION
1. Copy the equation; use the relationship between addition and subtraction and the distributive axiom; then simplify the left member.

$$7 - 2(3x - 1) = 33$$
$$7 + (-2)(3x - 1) = 33$$
$$7 - 6x + 2 = 33$$
$$9 - 6x = 33$$

2. Subtract 9 from each member.

$$9 - 6x - 9 = 33 - 9$$
$$-6x = 24$$

3. Divide each member by -6.

$$\frac{-6x}{-6} = \frac{24}{-6}$$
$$x = -4$$

Check: $7 - 2[3(-4) - 1] \overset{?}{=} 33$
$$7 - 2(-13) \overset{?}{=} 33$$
$$7 + 26 \overset{?}{=} 33$$
$$33 = 33 \; \checkmark$$

∴ the solution set is $\{-4\}$. **Answer.**

Oral Exercises

In Exercises 1–6 give the transformations used to produce:
a. the second equation from the first;
b. the third equation from the second.

EXAMPLE $\dfrac{x + 3}{5} = -2; \; x + 3 = -10; \; x = -13$

SOLUTION a. Multiply each member by 5.

b. Subtract 3 from each member.

1. $3x + 9 = 15$; $3x = 6$; $x = 2$

2. $19 = -3y - 5$; $24 = -3y$; $-8 = y$

3. $3(d - 1) + 4 = 13$; $3(d - 1) = 9$;
   $d - 1 = 3$

4. $w - \frac{1}{2}w = 8$; $2w - w = 16$;
   $w = 16$

5. $\dfrac{4t + 1}{3} = -5$; $4t + 1 = -15$;
   $4t = -16$

6. $\frac{2}{3}(9x) = 60$; $9x = \frac{3}{2}(60)$;
   $9x = 90$

Give the reasons for the statements in the following proofs.

EXAMPLE   Prove: For all real numbers b and c, $(b + c) - c = b$.

PROOF

Statement	*Reason*
1. b and c are real numbers.	Given
2. $-c$ is a real number.	Axiom of additive inverses
3. $b + c$ is a real number.	Closure axiom of addition
4. $(b + c) - c = (b + c) + (-c)$	Relationship between addition and subtraction
5. $(b + c) + (-c) = b$	Theorem on page 70
6. $\therefore (b + c) - c = b$	Transitive axiom of equality

7. Prove: For all real numbers b and c, $(b - c) + c = b$.

   1. b and c are real numbers.
   2. $b - c = b + (-c)$
   3. $(b - c) + c = [b + (-c)] + c$
   4. $= (b + c) + (-c)$
   5. $= b$
   6. $\therefore (b - c) + c = b$

8. Prove: For all real numbers b and all nonzero real numbers c, $(bc) \div c = b$.

   1. b and c are real numbers; $c \neq 0$.
   2. bc is a real number.
   3. $(bc) \div c = (bc)\dfrac{1}{c}$
   4. $(bc)\dfrac{1}{c} = b$
   5. $\therefore (bc) \div c = b$

Written Exercises

Solve each equation.

A

1. $3x + 7 = 31$

2. $5y - 12 = 33$

3. $-4 + 8m = 44$

4. $-63 = 3 + 6w$

5. $\dfrac{r}{4} - 2 = 6$

6. $\dfrac{s}{5} + 7 = -3$

7. $4t - t = -15$

8. $6p - 3p = -18$

9. $-72 = 6m + 4m - 2m$

10. $9n - 4n + 2n = -91$

11. $-51 = 5 - 7k$

12. $64 = 9 - 5h$

13. $3x + 1 + 2x - 1 = -20$

14. $3y + 4y + 7 = 49$

15. $6f + 4 - 5f = 2$

16. $8w - 3 - 7w = 0$

17. $70 = 9r + 4r + 18$

18. $5s + 216 + 7s = 228$

19. $t - \frac{1}{2}t - 6 = 0$

20. $\frac{3}{2}p - p - 7 = -4$

21. $0 = -f - 6f - 84$

Solve each equation.

22. $0 = 125 - 17k - 8k$ **23.** $21 + 5h - 7h = 56$ **24.** $41 - 9e + 5e - 5 = 0$

25. $12 - 3b - 2b = -3$ **26.** $7c - 5c + 8 - c = 15$ **27.** $(d - 3) + (d - 1) = 26$

28. $2(h + 3) - 6 = 0$ **29.** $3(y - 1) + 4 = -2$ **30.** $(x + 1) + (x + 2) = 41$

B **31.** $5(p - 2) - 2p + 21 = -1$ **32.** $4(k + 3) - 3 - 4k = 5$ **33.** $3q - (2q - 3) = -8$

34. $9t - (6t - 4) = -8$ **35.** $2u + 3(u + 4) = 7$ **36.** $-2(3 - j) + 5j = 1$

37. $-9(1 + e) + 8e = 10$ **38.** $-5r + 2(r + 4) - 6 = -19$

39. $5(v + 2) + 4(v - 4) + 6 = 45$ **40.** $3(i + 3) + 5(i + 3) + 4 = -4$

41. $1 = \frac{3}{4}(8 - x) + 1$ **42.** $1 - \frac{2}{5}(10 - p) = 1$

43. $3 = \frac{2}{3}(t + 9) - 1$ **44.** $\frac{3}{5}(5 - k) - 6 = 0$

45. $6(h + 2) - 3(h + 3) = 3$ **46.** $4(m - 3) + 6(m + 2) - 10m = 1$

47. $8 = 2(n + 5) + 3(n + 3) + 4$ **48.** $3(d - 4) + (4 - d)2 = 16$

49. $5x - 4(x + 3) + 3x = -24$ **50.** $9(y - 6) - (y + 3) + 5y = 8$

C **51.** $50 - 3[(25 + 2e) - 1] = 38$ **52.** $4w - [(2w - 4) + 6] - 3w = -117$

53. $-[(3t + 1) - (t + 2)] + 5t - (3 - 3t) = -2$

54. $3[u - 3(1 - u) - 2] + 2(2u - 2) - 8u = 5$

55. $2[2(3 - 2k) - (k + 3)] - 3[2(1 - 3k) + 8k] = 144$

56. $3(|x| - 2) - 2|x| - 12 = 0$ **57.** $|y| - (4 - 2|y|) = 2$

58. $(2|f| - 4) - (2 - 3|f|) = 9$ **59.** $|w| - (3|w| + 5) - 2|w| - 4 = 7$

Self-Test 2

VOCABULARY   transformation by multiplication (p. 107)
transformation by division (p. 112)
inverse operations (p. 115)

Solve.

1. $\frac{1}{13}p = 24$ **2.** $-72k = -24$ *Obj. 1, p. 107*

Simplify.

3. $84 \div (-6)$ **4.** $\dfrac{-12}{-\frac{1}{6}}$ **5.** $-\frac{1}{8}rs^2 \div (-\frac{1}{4})$ *Obj. 2, p. 107*

Solve.

6. $\frac{2}{7}y + 4 = 16$ **7.** $-3(m + 4) - 6 = 3$ *Obj. 3, p. 107*

8. $(2 - k) - 2(k + 3) = -22$

Check your answers with those at the back of the book.

Problem Solving

OBJECTIVES for Sections 4-6 through 4-9:
1. *Represent numerical relationships stated in words by mathematical expressions or equations.*
2. *Solve equations having the variable in both members.*
3. *Use equations and formulas to solve word problems.*

4-6   From Words to Symbols

A basic skill in solving problems is the use of mathematical expressions or sentences in place of words. Here are some simple examples.

Words	Mathematical Expressions or Sentences
1. a. Mrs. Hart received 153 more votes than Mr. Tully. b. In the Hart–Tully election 1289 votes were cast.	a. Let x = number of Tully votes. Then $x + 153$ = number of Hart votes. b. $x + (x + 153) = 1289$ Tully Hart Total
2. a. In a certain rectangle, the width is 6 cm less than the length.	a. Let y cm = length Then $(y - 6)$ cm = width
b. The perimeter of the rectangle is 80 cm.	b. $y + y + (y - 6) + (y - 6) = 80$ length width perimeter
3. a. On a trip by plane and bus, Mr. Salins traveled six times farther by plane than by bus. b. Mr. Salins' trip covered 2100 km.	a. Let t km = distance traveled by bus. Then $6t$ km = distance traveled by plane. b. $t + 6t = 2100$ bus plane total distance distance distance

Oral Exercises

In Exercises 1–10 let x represent a certain number. Express the number described in each exercise in terms of x.

1. It is 6 more than x.
2. It is 9 less than x.
3. It equals x reduced by 3.
4. It equals x increased by 2.
5. It is 3 less than twice x.
6. It is 4 more than one-half x.
7. The quotient of it and x is -2.
8. The product of x and it is 0.
9. Its sum with x is -3.
10. The difference between it and x is 10.

11. If n is an integer, the next greater integer is _?_ .
12. If n is an even integer, $n + 2$, $n + 4$, and _?_ are the next 3 greater even integers.
13. How many cents are d dimes worth? $d + 3$ dimes? $3d$ dimes?
14. How many cents are q quarters worth? $2q$ quarters? $q - 2$ quarters?
15. The sum of two integers is 14. If one integer is y, the other is _?_ .
16. The product of two numbers is -4. If one number is n, the other number is _?_ .
17. If an ancient vase is t years old now, how old was it last year? 2 years ago? 100 years ago?
18. If an antique spinning wheel is p years old now, how old will it be next year? 5 years from now? 100 years from now?

Written Exercises

Represent the English sentence by an equation.

EXAMPLE 1 The sum of negative three and one-half a certain number is two.

SOLUTION $-3 + \dfrac{n}{2} = 2$

EXAMPLE 2 The area of a certain square, s meters on a side, is 144 m².

SOLUTION $s^2 = 144$.

A

1. One number is two times another number, and their sum is -15.
2. One number is 8 more than another number, and their sum is -12.
3. Nine less than half a number is 10.
4. Ten more than twice a number is -26.
5. The perimeter of a square, s centimeters on a side, is 64 cm.

6. In a rectangle whose length is four times the width, the area is 150 m².

7. The degree measure of one angle is three times that of another angle, and the sum of the angles is 90°.

8. Louise has played the piano four years longer than her sister, who has played y years, and between them they have played the piano for a total of 18 years.

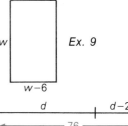

Ex. 9

9. The perimeter of a rectangle is 88 cm, and its length is 6 cm less than its width of w centimeters.

10. A 76 cm pipe is cut into two pieces so that one piece is 22 cm shorter than the other, which is d cm long.

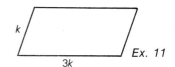

Ex. 10

11. One side of a parallelogram is three times as long as an adjacent side, which is k meters long, and the perimeter is 64 m.

Ex. 11

12. The area of a triangle whose base is 18 cm is 108 cm².

Ex. 12

13. The area of a rectangle 15 m long and w meters wide is 210 m².

14. In k slices of bread, each containing 72 calories, there are 576 calories.

Problems

In each exercise choose a variable and use it to represent the numbers described in part (a). Then write an equation based on the additional information given in part (b).

EXAMPLE a. The length of a rectangle is 2 m less than three times its width.

b. The perimeter of the rectangle is 60 m.

SOLUTION a. Let $x =$ the width and $3x - 2 =$ the length.

b. $2(3x - 2) + 2(x) = 60$

A 1. a. One number is 3 greater than another.
   b. The sum of the numbers is 42.

2. a. The width of a rectangle is 3 m greater than two times its length.
   b. The perimeter of the rectangle is 96 m.

In each exercise choose a variable and use it to represent the numbers described in part (a). Then write an equation based on the additional information given in part (b).

3. a. A molecule of sugar contains twice as many atoms of hydrogen as oxygen, and one more atom of carbon than oxygen.
   b. There are 45 atoms in a sugar molecule.

4. a. Mark has 15 more tape cassettes than Eric.
   b. Together, Mark and Eric have 95 tape cassettes.

5. a. It took Claire 10 min more to drive from her home to school than it did for her to return home.
   b. Her round trip took 50 minutes.

6. a. One side of a rectangle is 5 cm shorter than an adjacent side.
   b. The perimeter of the rectangle is 62 cm.

7. a. One secretary can type 68 more words over a given period of time than another secretary.
   b. Together, over the given period, these secretaries type 256 words.

8. a. One key punch operator can punch twice as many cards as another operator over a given period of time.
   b. Together, over the given period, these operators punch 123 cards.

9. a. In a certain school there are $\frac{2}{3}$ as many boys as girls.
   b. Together, there are 500 boys and girls.

B **10. a.** At the Randall Coliseum, each box seat costs $2 more than a reserved grandstand seat, and each bleacher seat costs $2 less than a grandstand seat.
   b. Two box seats and 3 bleacher seats cost $19.

11. a. Cal has 6 more quarters than he has dimes.
   b. Cal's quarters and dimes are worth $30.

12. a. Brand X costs $3 more than brand Y, and brand Z costs $2 more than brand X.
   b. Two brand Y's and a brand X cost $35.

13. a. Lynn has $250 more invested at 9% interest than she does at 8% interest.
   b. Lynn receives $75 in interest each year.

14. a. In a triangle ABC, the degree measure of $\angle A$ is 15 more than that of $\angle B$. The degree measure of $\angle C$ is twice that of $\angle B$.
   b. The sum of the degree measures of the three angles is 180.

15. a. During the summer Maria's sister earned $35 less than twice as much as Maria did.
   b. The girls' earnings totaled $850.

16. a. The length of a rectangular-shaped playground is 23 m less than twice its width.
   b. The perimeter of the playground is 230 m.

4-7 Using Equations to Solve Problems

In a word problem, you are told how certain numbers relate to one another. If you can represent the relationships by an open sentence, then you can solve the problem by solving the open sentence.

EXAMPLE 1 After studying the Glenoaks section of the city, an urban planner recommended that a development be built there to provide 150 condominiums, each having one, two, or three bedrooms.

  She proposed that there be twice as many two-bedroom as one-bedroom units, and 10 fewer three-bedroom than one-bedroom units. How many of each type of condominium did she propose?

SOLUTION 1. **Read the problem carefully and decide what numbers are asked for.** This problem asks for three numbers: the number of one-bedroom, the number of two-bedroom, and the number of three-bedroom condominiums, in the proposed development.

  2. **Choose a variable, and, with the given facts, use it to represent the numbers asked for.**

  a. Given any number of 1-bedroom units:
  b. there are twice as many 2-bedroom units;
  c. there are 10 fewer 3-bedroom units.

  a. Let x = the number of 1-bedroom units.
  b. Then $2x$ = the number of 2-bedroom units;
  c. $x - 10$ = the number of 3-bedroom units.

  3. **Write an open sentence based on the given facts.**

number of one-bedroom units	added to	number of two-bedroom units	added to	number of three-bedroom units	equals	total number of units
x	$+$	$2x$	$+$	$x - 10$	$=$	150

  4. **Solve the open sentence and find the required numbers.**

$$x + 2x + x - 10 = 150$$
$$4x - 10 = 150$$
$$4x - 10 + 10 = 150 + 10$$
$$4x = 160$$
$$x = 40 \qquad \text{(one-bedroom units)}$$

Then:
$$2x = 2(40) = 80 \qquad \text{(two-bedroom units)}$$
$$x - 10 = 40 - 10 = 30 \qquad \text{(three-bedroom units)}$$

(Solution continued on p. 124)

5. Check your results with the words of the problem.

Are there twice as many two-bedroom as one-bedroom units?

$$80 = 2(40)$$
$$80 = 80 \checkmark$$

Are there 10 fewer three-bedroom than one-bedroom units?

$$30 = 40 - 10$$
$$30 = 30 \checkmark$$

Is there a total of 150 condominiums?

$$40 + 80 + 30 = 150$$
$$150 = 150 \checkmark$$

∴ the planner proposed 40 one-, 80 two-, and 30 three-bedroom condominiums. **Answer.**

In solving certain problems, sketches picturing the facts of the problem may help you to see relationships.

EXAMPLE 2 A copper wire 110 cm long and 0.1 cm in diameter was bent to form a rectangle. One side of the rectangle was 10 cm more than twice as long as an adjacent side. Find the length of the longer side.

SOLUTION 1. The problem asks for the length of the longer side of the rectangle (see diagram below).

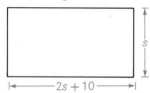

2.   Let s = the number of cm in the shorter side.

Then $2s + 10$ = the number of cm in the longer side.

3. The sum of the lengths of the four sides of the rectangle   equals   the length of the wire

$s + s + 2s + 10 + 2s + 10$	$=$	110

4.

$s + s + 2s + 10 + 2s + 10$	$=$	110
$6s + 20$	$=$	110
$6s + 20 - 20$	$=$	$110 - 20$
$6s$	$=$	90
s	$=$	15 (shorter side)

Then: $2s + 10 = 2(15) + 10 = 40$ (longer side)

5. Checking the results of Step 4 is left to you.

The length of the longer side of the rectangle was 40 cm. **Answer.**

Notice that in Example 2 you did not use the fact that the diameter of the wire was 0.1 cm. Sometimes word problems contain unnecessary information. It is your job to select the needed facts.

The five steps used to solve the problems in Examples 1 and 2 form a plan that usually helps in solving any word problem.

Plan for Solving a Word Problem

1. Read the problem carefully and decide what numbers are asked for. A sketch may be helpful.
2. Choose a variable and, with given facts, use it to represent the numbers asked for.
3. Write an open sentence based on the given facts.
4. Solve the open sentence and find the required numbers.
5. Check your results with the words of the problem.

Problems

Solve each problem.

A 1. The perimeter of a square is 152 m. How long is each side of the square?

2. The area of a rectangle 12 cm wide is 336 cm². How long is it?

3. The perimeter of a rectangle is 72 m. The length of the rectangle is 3 times its width. How long is the rectangle?

4. One record album sold for one fourth as much as another. If the less expensive one cost $2, how much did the other cost?

5. Together, a house and lot cost $60,000. The house cost 5 times as much as the lot. How much did the lot cost?

6. The sum of 3 times a number and 15 is −63. Find the number.

7. Together, a hat and coat cost $112. The coat cost 3 times as much as the hat. How much did the coat cost?

8. In the senior class, one-fifth of the students are enrolled in mathematics. Find the number of students in the class if 87 are enrolled in mathematics.

9. The area of a rectangle 7 m wide is 133 m². Find the length of the rectangle.

10. Helen rides the bus part of the way to school and runs the rest of the way. She runs 3 min longer than she rides. If it takes Helen 15 min to arrive at school, how long does she spend on the bus?

Solve each problem.

11. Janice will need 64 m of fence to enclose her rectangular garden. If the length of the garden is 4 m more than the width, what are the dimensions of the garden?

12. The difference between 4 times a number and the number is 163. What is the number?

13. Thirty-six more than 5 times a number is −34. What is the number?

14. Water is a compound made up of 8 parts by mass of oxygen and 1 part by mass of hydrogen. How many grams of oxygen are there in 234 g of water?

15. In a student election, 395 students voted for one or the other of the two candidates. The candidate elected had a majority of 103 votes. How many voted for each candidate?

16. The measure in degrees of one angle of a triangle exceeds the measure in degrees of another by 24. The third angle of the triangle measures 9° less than the lesser of the other two angles. Find the measure in degrees of the three angles.

17. Seventy-eight more than 3 times a number is 144. Find the number.

18. If each of the base angles of an isosceles triangle measures 20° less than twice the measure in degrees of the vertex angle, find the measure of each angle.

base angles

B 19. One number is 25 greater than a second number. If the lesser number is subtracted from 3 times the greater number, the difference is 195. Find the numbers.

20. The difference of two numbers is 15. If twice the lesser number is subtracted from 4 times the greater number, the difference is 36. Find the greater number.

21. The entertainment part of a 60-minute TV program lasted 10 min longer than 4 times the advertising part. How many minutes of the program were devoted to advertising?

22. Sal had 15 coins in his pocket. If he had 3 fewer dimes than nickels and 3 more pennies than nickels, how much money did Sal have in his pocket?

23. Carol has twice as many 8¢ stamps as 10¢ stamps and 4 more 15¢ stamps than 10¢ stamps. In all, she has 28 stamps. Does she have enough stamps to mail a parcel for which the postage is $3? Explain.

24. Mel is 6 years older than Hank, and Hank is 4 years older than Sid. The sum of their ages is 77. What is Sid's age?

25. When the length of a rectangle was decreased 2 m and the width was increased 2 m, the resulting figure was a square with a perimeter of 36 m. What were the dimensions of the original rectangle?

26. Eighteen hundred ninety-five dollars is to be shared among three people so that the second will have $125 more than the first, and the third $175 more than the second. How much will each person receive?

27. A collection of 30 nickels and quarters has a total value of $5.70. How many of each kind of coins are there in the collection?

28. In a theater, orchestra tickets cost $2 more apiece than balcony tickets, and each loge ticket costs $3 more than an orchestra ticket. Ellen Kenny paid $66 for 2 loge and 5 balcony tickets. How much do orchestra tickets cost?

29. One side of a triangular lot is 10 m less than twice the length of the second side. The third side is 9 m shorter than the first side. The perimeter of the lot is 61 m. How long is the shortest side?

30. In solving a puzzle problem, Marlo took 8 fewer moves than twice the number of moves Gino took. If together they made 13 moves, who made more moves and by how many moves did they differ?

C 31. Althea is 5 years older than Leah, and Leah is 4 years older than Jessica. The sum of their ages is 34. What is Althea's age?

4-8 Equations Having the Variable in Both Members

In the equation

$$7x = 18 - 2x$$

the variable x appears in both members. Are you permitted to use the addition transformation to add $2x$ to each member?

The answer is yes. For every real number x, $2x$ denotes a product of real numbers and, therefore, represents a real number. Thus, since you can add the same real number to each member of the equation without changing the solution set, you can also add $2x$ to each member without changing the solution set. Hence, you may solve the equation as follows:

$$7x = 18 - 2x$$
$$7x + 2x = 18 - 2x + 2x$$
$$9x = 18$$
$$\frac{9x}{9} = \frac{18}{9}$$
$$x = 2$$

Check:
$$7x = 18 - 2x$$
$$7 \cdot 2 \overset{?}{=} 18 - 2 \cdot 2$$
$$14 \overset{?}{=} 18 - 4$$
$$14 = 14 \ \checkmark$$

∴ the solution set is {2}. **Answer.**

The next two examples illustrate the fact that the solution set of an equation may be the empty set, or even the set of all real numbers.

EXAMPLE 1   Solve: $5 - (8 - 3t) = (4t + 7) - t$

SOLUTION

$$5 - (8 - 3t) = (4t + 7) - t$$
$$5 - 8 + 3t = 3t + 7$$
$$-3 + 3t = 3t + 7$$
$$-3 + 3t - 3t = 3t + 7 - 3t$$
$$-3 = 7$$

Since the given equation is equivalent to the false statement $-3 = 7$, the *equation has no root.*

$\therefore$ the solution set is $\emptyset$.   **Answer.**

EXAMPLE 2   Solve: $5(z + 1) + 6 = 3(8 + z) + (2z - 13)$

SOLUTION

$$5(z + 1) + 6 = 3(8 + z) + (2z - 13)$$
$$5z + 5 + 6 = 24 + 3z + 2z - 13$$
$$5z + 11 = 11 + 5z$$
$$5z + 11 - 5z = 11 + 5z - 5z$$
$$11 = 11$$

Since the given equation is equivalent to the true statement $11 = 11$, the equation is satisfied by every real number.

$\therefore$ the solution set is $\mathcal{R}$.   **Answer.**

Any equation which is a true statement for every numerical replacement of the variable in the equation is called an **identity.** Thus, in Example 2, $5(z + 1) + 6 = 3(8 + z) + (2z - 13)$ is an identity.

Oral Exercises

Tell how to transform the equation into an equivalent one in which one member is a variable and the other member is a constant. Then state the transformed equation.

EXAMPLE $4y + 14 = 11y$

SOLUTION

1. Subtract $4y$ from each member of the given equation.

$$4y + 14 = 11y$$
$$4y + 14 - 4y = 11y - 4y$$
$$14 = 7y$$

2. Divide each member of the resulting equation by 7.

$$\frac{14}{7} = \frac{7y}{7}$$

$\therefore$ The equation is $2 = y$.   **Answer.**

$$2 = y$$

1. $8x = 3 + 7x$
2. $y = 6 - y$
3. $9t - 13 = 10t$
4. $6w = 3w - 12$
5. $-5s = 5s + 5$
6. $8r - 18 = -r$
7. $1.5e = 0.7e - 24$
8. $-0.25 = 2.5k - 0.75$
9. $3y^2 - y^2 = 4 - y^2$

Written Exercises

Solve each equation. If the equation is an identity, state this fact.

A
1. $8x = 40 + 3x$
2. $15y = 12y + 30$
3. $9g = 80 - g$
4. $12w = 68 - 5w$
5. $25k = 17k - 56$
6. $30r = 19r - 44$
7. $64 - 8s = 8s$
8. $256 - 16t = 0$
9. $7a - 72 = a$
10. $0 = 152 - 8b$
11. $192 + 15c = 11c$
12. $2d + 5 = 2d - 5$
13. $9p = 6p - 27$
14. $9e = e - 88$
15. $4f - 63 = -3f$
16. $-m = 112 - 5m$
17. $6n + 6 = 4n$
18. $-16q - 18 = -10q$
19. $18 - 2x = 8x - 6$
20. $14y - 7 = 8 - y$
21. $8 + 4d = 3d + 13$
22. $41 + 8w = 16 + 3w$
23. $5g - 2 = -g + 22$
24. $-5r + 21 = 4r - 15$
25. $3x + x + 1 = 4x + 1$
26. $44 - s = 18 - 3s$
27. $6k + 12 - 2k = 2k - 4$

B
28. $3(x - 4) = -2(x + 6)$
29. $5(y + 2) = -(8 - 3y)$
30. $-2m + 3(2m + 1) = 8m - 9$
31. $-3w + 4(3w - 1) = 16 - w$
32. $4(5k - 5) + 19 = -9 + 2(3k - 8)$
33. $7(t - 3) + 4 = 33 + 3(t - 2)$
34. $-3(2 - 4r) = 2(3r - 4) + 8$
35. $4 - 6m = (3m + 14) - 2(3m + 5)$
36. $10n + (6n - 5) = 8n + (7n + 20)$
37. $4(p + 3) - 2 = -4 + 4(p - 1)$
38. $9h - (h^2 - 4h + 1) = 38 - h^2$
39. $28x - 2(1 + 2x - x^2) = 262 + 2x^2$
40. $2(3y - 4) - 3(2y - 1) = 6(2y + 1)$
41. $5(2m + 3) - 3m = -2(4m + 1) - 13$
42. $-1(1 - 4w) = 2(2w - 1) + 1$
43. $4(2e + 3) - 2 = -6 - 8(1 - e)$

C
44. $2[2x - 13 - 2(3x + 1)] = 6(5x - 2) - 3(x - 1) + 20$
45. $2(15 - 7y) = 3[y + 2 - 3(2y - 1)]$
46. $3[2(m - 5) - 3(2m + 1)] = 4(m - 2) - 2(3m + 1) + 1$
47. $-2[3(w - 2) - (w - 1)] = 3[2(w + 3) + 5w - 2(3w - 1)] + w - 2$
48. $c^2 + 12(c - 2) = 3[2(c - 4) + 2c] + c^2$
49. $-[4 + 5(2 - 3t)] + t + t^2 = -7[2t + 2(3 - 2t) + 1] + t^2$
50. $h^2 - [2(h + 4) - 3] = h^2 - (5 + 4h) - 2[2(2 - h) - 3]$
51. $5i - [i - 2(i + 4) + 3 - i^2] = -(5 + i - i^2)$
52. $2[p - 3(1 - p) + 1] - [3(2p - 1) - 3] = 2(3p + 2) - 4(2 - p) - 2$
53. $-1[1(m - 1) - m] - 1[1(1 - m) - (m - 1) - 1] = -1[1(m - 1) - m] + 3$

Problems

Solve each problem.

EXAMPLE Beth is 8 years older than her brother. Next year she will be twice as old as her brother. How old is Beth now?

SOLUTION
1. The problem asks for Beth's age now.
2. Let x years be her brother's age now. The facts of the problem are arranged in the chart below.

	Brother	Beth
Age now	x	$x + 8$
Age next year (1 year hence)	$x + 1$	$x + 8 + 1$

3. Next year, Beth will be **twice** as old as her brother.

$$x + 8 + 1 = 2(x + 1)$$

4. Solve the equation.   5. Check your results.

You should discover that Beth's age now is 15 years.

A
1. Find a number which is 10 less than its additive inverse.

2. Find a number which is 14 greater than its additive inverse.

3. Two more than a certain number is 15 less than twice the number. Find the number.

4. Six times a certain number is 12 more than 4 times the number. Find the number.

5. Paula is 3 times as old as Lola, while Sherry is 6 years older than Lola. If Paula and Sherry are twins, what is Paula's age?

6. Tim is twice as old as Art. Tim is also exactly 15 years older than Art was last year. What are their ages?

7. Jeff is 12 years old and his father is 63. In how many years will Jeff be one-fourth as old as his father?

8. The Green River is about twice as long as the Trinity River, and the Fraser River is about 380 km less than three times the length of the Trinity. If the total length of the three rivers is 3100 km, how long is each river?

9. At the Civic Center Arena, on an average, the attendance at a hockey game is 2000 more than twice the attendance at the circus. The attendance at a basketball game is 2000 less than twice the attendance at the circus. If the combined attendance at all three events is 40,000, what is the attendance at each event?

10. The sum of 5 more than a certain number and 10 more than twice the number is equal to the product of 2 and the number increased by 8. Find the number.

11. Four times a number subtracted from 5 times the number equals the number. Find the number.

12. In a baseball game the Dolphins had 3 more hits than the Sharks. If 4 times the number of hits by the Sharks was 18 greater than twice the number of hits by the Dolphins, how many hits did each team make?

13. The sum of two numbers is 25. Three times the lesser number is 5 more than twice the greater. Find the lesser number.

14. Juan's father is 4 times as old as Juan. Twenty-two years from now he will be twice as old as Juan will be. What is his father's age now?

15. Juanita has 10 coins of the same kind. One dollar and 6 of the coins have the same total value as all 10 coins. What kind of coin does Juanita have?

16. The perimeter of a rectangle is 25 cm greater than the sum of the length and width of the rectangle. If the length is 5 cm greater than the width, what are the dimensions of the rectangle?

17. If each of the base angles of an isosceles triangle is 10° less than twice the vertex angle, find the measure of each angle.

18. Each leg of an isosceles triangle exceeds twice the base by 5 cm. Find each side of the triangle if the perimeter is 60 cm.

19. Peggy has 4 times as many dimes as quarters. If her dimes and quarters total $3.90, how many dimes does she have?

B 20. The length of one rectangle is 4 m greater than the width of a second rectangle. The width of the first rectangle is 6 m and the length of the second rectangle is 9 m. The area of the second rectangle is 3 m² less than the area of the first rectangle. Find the length of the first rectangle.

21. The sum of 5 more than a certain number and 10 more than 4 times the number is equal to the product of 6 and the number increased by 3. Find the number.

22. The base of a triangle has the same length as the side of a square. A second side of the triangle is 1 cm longer than the base, and the third side is 5 cm shorter than 3 times the base. If the perimeter of the triangle equals that of the square, find the length of the longest side of the triangle.

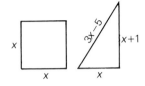

23. If one side of a square is increased by 8 m and an adjacent side is decreased by 2 m, a rectangle is formed whose perimeter is 40 m. Find the length of a side of the square.

Solve each problem.

24. The design for a space colony specifies that, per person, the amount of surface area planned for business use is 1.5 times the area planned for schools, while the amount of area for residential use is 8 times that for business use. Plant-growing area, 44 m² per person, is 0.5 m² more than 3 times the area for schools, business, and residential use combined. What is the area per person specified for residential use?

25. If the Washington Monument were 25.5 m taller, it would be just as tall as the Gateway Arch in St. Louis. If the height of the Washington Monument is 72.0 m more than one-half the height of the Gateway Arch, how tall is each of these structures?

26. Cheryl's age is 4 years less than 3 times that of her sister Monique. Half of Cheryl's age increased by Monique's age is 2 years more than twice Monique's age. Find their ages.

27. At one time, 4 kg of onions cost the same as 2 kg of green beans. At the same time, 1 kg of green beans cost 3 times as much as 1 kg of potatoes, while 1 kg of onions cost 4¢ less than 2 kg of potatoes. What was the cost of 1 kg of each vegetable?

28. Colleen has 80 coins which are pennies, nickels, and dimes. She has 4 more dimes than pennies, and 6 more nickels than dimes. How much money does she have?

29. In 1752, the year in which Benjamin Franklin conducted his famous kite experiment, he was 14 years less than twice as old as Samuel Adams. If Franklin was 16 years older than Adams, in what year was Franklin born?

30. In 1512, when Michelangelo finished painting the ceiling of the Sistine Chapel, he was 7 years more than half as old as Leonardo da Vinci. If da Vinci was 23 years old when Michelangelo was born, when was da Vinci born?

4-9 Working with Formulas

Formulas are equations stating numerical relationships between physical or other measurements. You have probably learned such formulas as:

$$\text{distance traveled} = \text{constant rate} \times \text{time traveled}$$
$$d = rt$$

$$\text{simple interest} = \text{principal} \times \text{rate of interest} \times \text{time}$$
$$I = Prt$$

$$\text{area of a rectangle} = \text{length of rectangle} \times \text{width of rectangle}$$
$$A = lw$$

Other formulas are given in Table 1 at the back of the book.

In using formulas, you may wish to find equivalent equations in which a particular variable is expressed in terms of other variables.

EXAMPLE The formula

$$s = \tfrac{1}{2}at^2$$

expresses the distance s in meters that a freely falling object will fall in t seconds near the surface of a planet where the acceleration due to gravity is a m/s^2.

a. Find an expression for a in terms of s and t.

b. If an object falls 82 m to the surface of the moon in 10 s, what is the acceleration due to gravity on the moon?

SOLUTION **a.** $s = \tfrac{1}{2}at^2$

$$2 \cdot s = 2 \cdot \tfrac{1}{2}at^2$$

$$\frac{2s}{t^2} = \frac{at^2}{t^2}, \text{ provided } t \neq 0$$

$$\frac{2s}{t^2} = a; \text{ or } a = \frac{2s}{t^2}, \text{ provided } t \neq 0. \quad \text{Answer.}$$

b. $a = \frac{2s}{t^2}$, $s = 82$, $t = 10$.

$$a = \frac{2 \cdot 82}{10^2} = \frac{164}{100} = 1.64$$

$\therefore$ the acceleration is 1.64 m/s^2. Answer.

Notice that in the solution to part (a) of the example, the formula for the acceleration a is not valid when $t = 0$, since division by $0^2 = 0$ is not defined.

Oral Exercises

Solve each equation for x, y, or t.

1. $x - a = b$ **2.** $y + c = d$ **3.** $t - e = 2e$ **4.** $4f - x = f$

5. $\dfrac{y + g}{2} = h$ **6.** $P = 1 + 3t$ **7.** $K = \pi + 4x$ **8.** $\dfrac{a + y - b}{4} = c$

9. The formula for the area A of a trapezoid with height h and bases b and b' is $A = \tfrac{1}{2}h(b + b')$. State a formula that expresses the height h of the trapezoid in terms of the bases b and b' and the area A.

Written Exercises

Solve for the variable shown in color.

A   **1.** $p = a + b + c$ **2.** $P = 2l + 2w$ **3.** $A = \frac{1}{2}bh$ **4.** $V = \dfrac{Bh}{3}$

     5. $A = 2\pi r h$ **6.** $C = 2\pi r$ **7.** $E = mc^2$ **8.** $V = \frac{1}{3}\pi r^2 h$

     9. $A = P + Prt$ **10.** $A = P + Prt$ **11.** $2x + 3y = 6$ **12.** $2x - 4y = 8$

     13. $d = \dfrac{a}{2}(2t - 1)$ **14.** $d = \dfrac{a}{2}(2t - 1)$ **15.** $E = am(T - t)$ **16.** $V = \pi h(R^2 - r^2)$

In Exercises 17–22 use the given formula or a suitable formula from Table 1 at the back of the book to obtain your solution.

B   **17.** A rectangular box 9 m long and 7 m wide has a volume of 315 m³. How deep is the box?

     18. O'Kennedy Pears are packed in a cylindrical can 8 cm in diameter. If the volume of the can is 552.64 cm³, how tall is the can? (Use 3.14 as an approximate value of π.)

     19. When a steel ball 30 cm in diameter was dropped into a full tub of water, it displaced about 14,130 cm³ of water. Use this fact to find an approximate value of π.

     20. The formula $s = c + O + p$ relates the selling price s of an item to the wholesale cost c, overhead expenses O, and profit p for the item. What is the profit on a microcomputer selling at \$599 if it cost the dealer \$449 and the overhead is \$25?

     21. The formula $R = \dfrac{S + F + P}{S + P}$ relates the mass ratio R of a rocket to the masses S, F, and P of the structure of the vehicle, the fuel, and the payload. How much fuel must be loaded in a rocket whose basic structure and payload each have a mass of 900 kg, if the mass ratio is to be 6?

     22. The formula $V = \frac{1}{3}\pi r^2 h$ is used to find the volume of a cone. How many cubic centimeters of ice cream can be packed in a "super" cone that has a diameter of 9 cm and a height of 10 cm?

Evaluate each expression.

C   **23.** Heat radiation: kT^4; let $k = 0.000000822$, $T = 6000°K$.

     24. Kinetic energy: $\dfrac{mv^2}{2}$; let $m = 25$ kg, $v = 100$ m/s.

     25. Centripetal force: $\dfrac{mv^2}{r}$; let $m = 6.75$ kg, $v = 6$ m/s, $r = 3$ m.

     26. Heat energy from electricity: $0.238I^2Rt$; let $I = 20$ amp, $R = 10$ ohms, $t = 300$ s.

Self-Test 3

VOCABULARY   identity (p. 128)   formula (p. 132)

1. Represent the following English sentence by an equation: April is 4 years older than Celeste; three years ago the sum of their ages was 24.

Obj. 1, p. 119

2. Solve: $3(k - 4) = 9 - 4k$

Obj. 2, p. 119

3. Thirty-six more than 3 times a number is -48. What is the number?

Obj. 3, p. 119

4. Find the length and width of a rectangle whose perimeter is 48 cm if the length is 6 cm greater than the width.

5. In a 30-minute TV program, 5 times as many minutes were devoted to entertainment as were devoted to commercials. How many minutes were devoted to entertainment?

6. Jody has 10 coins of the same kind. The value of all 10 of these coins is 75 cents more than the value of 7 of the coins. What kind of coins does she have?

7. Using the formula $V = \pi r^2 h$, with $\frac{22}{7}$ as an approximate value of π, find the height h of a cylindrical tank if its volume V is 176 m³ and its radius r is 2 m.

Check your answers with those at the back of the book.

ON THE CALCULATOR

A lunar gas gun has been designed to launch construction material for a space station from the surface of the moon. The mass of material, m, it can launch is given by the following formula:

$$m = \frac{Mw}{5.22 \, pv^2}$$ where

m = mass of material to be launched in kg
M = mass of barrel of gun in kg
w = working stress of barrel in Pa (pascals)
p = density of barrel in kg/m³
v = muzzle velocity in m/s

1. Find m, if $M = 1{,}400{,}000$ kg, $w = 1{,}650{,}000{,}000$ Pa, $p = 1380$ kg/m³, and $v = 2370$ m/s.

2. How many metric tons, t, of material can be launched? (1 t = 1000 kg)

programming in BASIC

The program which follows will let you INPUT a set of numbers, count them as you type them in, add each to the accumulated sum, and finally print out the number of numbers, their sum, and their average.

In order to tell the computer when you have put in your last number, you select a number different from those that you want added. For example, if all the numbers you are adding are positive, you can use a negative number as a "clue" number.

Study the program and the print-out shown to the right of the program. The variable K is used as the "counter," and the "clue" number is -1. Look carefully at lines 50 and 60 of the program. Line 50 means that the computer takes the value that is stored in K, adds 1 to it, and puts the sum back into K. Similarly, line 60 means that the computer takes the value that is stored in S, adds to it the number you have just put in, and puts the sum back into S.

The LET statements (in lines 10 and 20) give K and S their initial values. Line 10 starts the value of K at 0 and line 20 starts the value of S at 0.

```
10   LET K=0                         RUN
20   LET S=0
30   INPUT N                         ?24
40   IF N<0 THEN 80                  ?53
50   LET K=K+1                       ?58
60   LET S=S+N                       ?92
70   GOTO 30                         ?76
80   PRINT "NUMBER =";K              ?37
90   PRINT "SUM =";S                 ?−1
100  PRINT "AVERAGE =";S/K           NUMBER = 6
110  END                            SUM = 340
                                     AVERAGE = 56.6667
```

Notice that the value of N changes each time you INPUT a number. Notice also that the "clue" number is not counted.   END

Exercise

Write a program that lets you INPUT a series of orders giving the number of articles and the price of each. The program should count the number of orders and compute the total cost.

Hint: Begin the program with:

```
10   LET K=0
20   LET S=0
30   INPUT N
40   IF N<0 THEN
50   INPUT P
```

(In each case N is the number of articles at price P.)
Use these values for a test RUN:

>2 articles at $2.50 apiece
>3 articles at $1.75 apiece
>4 articles at $3.25 apiece

Chapter Summary

1. Equations having the same solution set over a given set are called *equivalent equations*. Each of the following transformations always produces an equivalent equation.

   a. Substituting for either member of a given equation an expression equivalent to that member.

   b. Adding the same real number to, or subtracting the same real number from, each member of a given equation.

   c. Multiplying or dividing each member of a given equation by the same nonzero real number.

   d. Adding to, or subtracting from, each member of a given equation the same variable expression in any variable(s) that appear in the equation.

2. For all real numbers a and b, $a - b = x$ if $b + x = a$; furthermore, $a - b = a + (-b)$ (*Relationship between Addition and Subtraction*).

3. The set of real numbers is closed under subtraction. Subtraction of real numbers is neither commutative nor associative; however, multiplication is distributive with respect to subtraction.

4. For all real numbers a and b, $b \neq 0$, $a \div b = x$ if $bx = a$; furthermore, $a \div b = a \cdot \dfrac{1}{b}$ (*Relationship between Multiplication and Division*).

5. Division of real numbers is neither commutative nor associative; however, division is distributive both with respect to addition and with respect to subtraction.

Chapter Review

Solve each equation.

1. $x + (-32) = 56$ *4-1*

   a. 24   b. -88 c. 88   d. -24

2. $n + 16 = 0$

   a. 0   b. -16 c. 16   d. 8

Simplify.

3. $2001 - (-1066)$
4-2

    a. 3067   **b.** 935   **c.** -935 **d.** -3067

4. $3t^2 - t + 18 + t^2 - 8t + 6$

    a. $4t^2 - 7t + 24$ **b.** $2t^2 - 9t + 24$ **c.** $4t^2 - 9t + 24$

Solve each equation.

5. $\frac{1}{3}x = 12$
4-3

    a. 4   **b.** 15   **c.** 36   **d.** 9

6. $-14y = \dfrac{28}{3}$

    a. $-130\frac{2}{3}$ **b.** $-1\frac{1}{3}$ **c.** $-\frac{3}{2}$ **d.** $-\frac{2}{3}$

7. Simplify $-119 \div 7$.
4-4

    a. -17 **b.** -19 **c.** -40 **d.** -13

8. Solve $\dfrac{x}{-16} = -4$.

    a. -4 **b.** 64   **c.** -64 **d.** -12

Solve each equation.

9. $4k - 15 = 33$
4-5

    a. $4\frac{1}{2}$ **b.** -12 **c.** 12   **d.** $-4\frac{1}{2}$

10. $-8y + 32 + 6y - 24 = 0$

    a. 6   **b.** -4 **c.** 4   **d.** -6

Represent the English sentence by an equation. Use n for the variable.

11. One number is 8 times another, and their sum is 108.
4-6

    a. $2n + 8 = 108$ **b.** $8n = 108$ **c.** $n + 8n = 108$

12. One number is 3 less than another, and 5 times their sum is 35.

    a. $5(n - 3) = 35$ **b.** $5(n + n - 3) = 35$ **c.** $n - 3 = n + 35$

13. The length of a rectangle is twice its width. It has a perimeter of 312 cm. What is its length?
4-7

    a. 104 cm   **b.** 52 cm   **c.** 156 cm   **d.** 140 cm

Solve each equation.

14. $6t - 38 = t + 7$
4-8

    a. 5   **b.** 9   **c.** -9 **d.** $6\frac{1}{5}$

15. $7 - 4d = -12d - 9$

    a. -1 **b.** 1 **c.** -2 **d.** 2

Solve for y.

4-9

16. $6x + 3y = 9$

    a. $y = 3x - 2$ **b.** $y = 3 - 2x$ **c.** $y = 2x - 3$

17. $x - 4y = 13x + 16$

    a. $y = -3x - 4$ **b.** $y = 3x + 4$ **c.** $y = 3x - 4$

Chapter Test

Solve.

1. $y + 34 = -160$ **2.** $x - 73 = 4$ 4-1

Simplify.

3. $73 - (-137)$ **4.** $-12 + 47$ 4-2

Solve.

5. $\frac{1}{8}x = 2664$ **6.** $-23y = 368$ 4-3

Simplify.

7. $87 \div (-29)$ **8.** $-1001 \div \frac{1}{13}$ 4-4

Solve.

9. $4x - 39 = 5$ **10.** $26 - 3m = 79$ 4-5

11. Write an equation for the following English sentence: One number, n, is 3 less than another number, and their sum is 45.   4-6

12. A jogger runs for 75 minutes each day. She spends one and a half times as much time jogging in the afternoon as she does in the morning. How much time does she spend jogging each morning and each afternoon?   4-7

Solve.

13. $6t = t + 115$ **14.** $7d - 36 = 9 + 4d$ 4-8

Solve for y.

15. $3y - 10 = 6x + 2$ **16.** $4y + 9x = x - 32$ 4-9

Cumulative Review   Chapters 1–4

1. Find the value of $9 - x(x - 2)$ when $x = 4$.

   a. 18   b. 10   c. 1   d. 23

2. Find the value of $3a^2 - 2a$ when $a = 4$.

   a. 136   b. 40   c. 184   d. 568

3. Which of these is an irrational number?

   a. $\frac{3}{17}$ b. 1.232323 . . .   c. 6.14144   d. 0.313113111 . . .

4. Which statement is false?

   a. $0 \in \emptyset$ b. $3 \notin$ {the even integers}   c. $\{2, 5\} \subset \{2, 3, 5\}$ d. $\emptyset \subset \emptyset$

5. Which of these is the graph of the solution set of $b \le -2$ over $\mathcal{R}$?

   a.

   b.

   c.

   d.

6. Translate into a mathematical sentence: The quotient of 12 and n is 2.

   a. $\frac{n}{12} = 2$ b. $\frac{n}{2} = 12$ c. $\frac{12}{n} = 2$ d. $12n = 2$

7. Which axiom guarantees that the statement $2a + (b + 5) = 2a + (5 + b)$ is true?

   a. Associative   b. Distributive   c. Transitive   d. Commutative

8. Simplify: $3(x + 5) + 2(3x + 1)$

   a. $9x + 17$ b. $9x + 6$ c. $18x + 17$ d. $6x + 6$

9. Simplify: $-(-27 + 16) + (-9)$

   a. -20 b. 34   c. 20   d. 2

10. Evaluate $-(b - a) + (-(-c))$ when $a = 3$, $b = 5$, and $c = 2$.

    a. -4 b. 0   c. 4   d. -6

11. Find the solution set: $8 - |m| = 2$.

    a. $\{6\}$ b. $\{-6\}$ c. $\{6, -6\}$ d. $\{10\}$

12. Replace the _?_ with a word to make the statement true: The sum of a positive number and a negative number is _?_ zero.

    a. sometimes   b. always   c. never

13. Simplify: $(-2x^2)(-5x^2)$

    a. $10x^2$ b. $-7x^2$ c. $10x^4$ d. $-20x^6$

14. Evaluate $-5x(a - 2m)$ when $a = -3$, $m = 1$ and $x = -2$.

    a. -32 b. -10 c. 50   d. -50

15. Which of these is *not* the reciprocal of -1.25?

    a. -0.8 **b.** 1.25 **c.** $\frac{-4}{5}$ **d.** $\dfrac{1}{-1.25}$

16. Simplify: $-\dfrac{1}{3a}[12ab + (-9ac^2)]$

    a. $-4b + 3c^2$ **b.** $-4b + (-9ac^2)$ **c.** $-12bc^2$ **d.** $-bc^2$

17. Simplify: $(3 - 2y) - 3(4 - 3y)$

    a. $-11y - 9$ **b.** $y - 9$ **c.** $7y + 15$ **d.** $7y - 9$

18. Solve: $\frac{1}{5}n + 8 = 5$

    a. $\{65\}$ **b.** $\{-\frac{3}{5}\}$ **c.** $\{-15\}$ **d.** $\{17\}$

19. Solve: $-2(3 - 5w) = -21$

    a. $\{3\}$ **b.** $\{-1.5\}$ **c.** $\{-2.7\}$ **d.** $\{-25\}$

20. Solve: $-2 - 6m = 3m + 16$

    a. $\{-2\}$ **b.** $\{2\}$ **c.** $\{6\}$ **d.** $\{-6\}$

21. Solve for a: $s = vt + \frac{1}{2}at^2$

    a. $a = s - vt - \dfrac{1}{2}t^2$ **b.** $a = \dfrac{t^2}{2}(s - vt)$

    c. $a = \dfrac{2(s - vt)}{t^2}$ **d.** $a = \dfrac{s}{vt} - 2t^2$

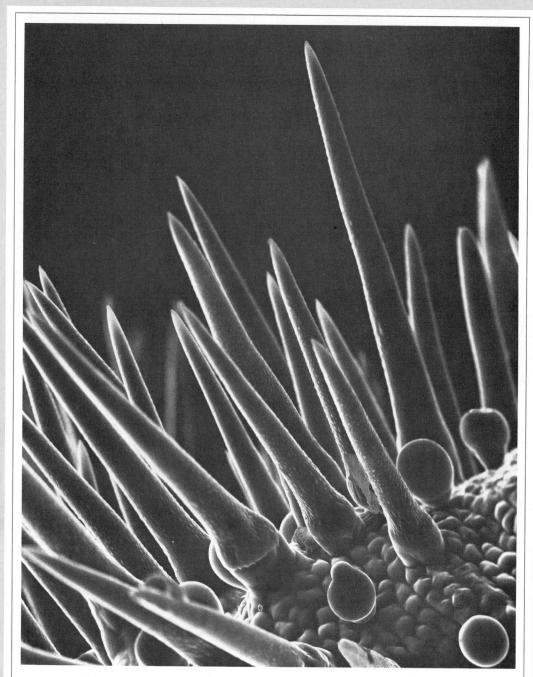

This photograph of a lantana leaf, magnified 800 times, was taken with the aid of a scanning electron microscope. The growths protruding from the surface of the leaf are trichomes, part of the epidermal hair structure of the plant.

5
Solving Inequalities and Problems

Solving Inequalities

OBJECTIVES for Sections 5-1 through 5-5:
1. Use the inequality axioms to justify steps in proofs.
2. Solve inequalities and graph their solution sets.
3. Find the intersection and the union of two sets.
4. Solve conjunctions and disjunctions of inequalities and graph their solution sets.
5. Solve open sentences involving absolute value.

5-1 Axioms of Inequality

How many of the following statements are true?

$$1 < -3 \qquad 1 = -3 \qquad -3 < 1$$

Of course, only the last one is true. In comparing real numbers, you use the following property:

Axiom of Comparison

For all real numbers a and b, one and only one of the following statements is true:

$$a < b \qquad a = b \qquad b < a.$$

EXAMPLE Let a, b, and c be real numbers such that $a < b$ and $b < c$. Show a, b, and c on a number line, and then compare a and c.

SOLUTION

$a < c$

The example suggests the following axiom.

> ## Transitive Axiom of Inequality
>
> For all real numbers a, b, and c:
> 1. If $a < b$ and $b < c$, then $a < c$.
> 2. If $a > b$ and $b > c$, then $a > c$.

What happens when the same number is added to each member of a true inequality such as "$-3 < 1$"?

Case 1. Add 5.

$$-3 < 1$$
Question: $-3 + 5 \underset{?}{\_} 1 + 5$
That is: $2 \underset{?}{\_} 6$
Answer: $2 < 6$
$\therefore -3 + 5 < 1 + 5$

Case 2. Add -5.

$$-3 < 1$$
Question: $-3 + (-5) \underset{?}{\_} 1 + (-5)$
That is: $-8 \underset{?}{\_} -4$
Answer: $-8 < -4$
$\therefore -3 + (-5) < 1 + (-5)$

This kind of reasoning suggests the next property.

> ## Additive Axiom of Inequality
>
> For all real numbers a, b, and c:
> 1. If $a < b$, then $a + c < b + c$ and $a - c < b - c$.
> 2. If $a > b$, then $a + c > b + c$ and $a - c > b - c$.

Now let us see what happens when each member of "$-3 < 1$" is multiplied by a nonzero real number.

Case 1. Multiply by 5.

$$-3 < 1$$
Question: $-3 \cdot 5 \underset{?}{\_} 1 \cdot 5$
That is: $-15 \underset{?}{\_} 5$
Answer: $-15 < 5$
$\therefore -3 \cdot 5 < 1 \cdot 5$

Case 2. Multiply by -5.

$$-3 < 1$$
Question: $-3(-5) \underset{?}{\_} 1(-5)$
That is: $15 \underset{?}{\_} -5$
Answer: $15 > -5$
$\therefore -3(-5) > 1(-5)$

The two cases suggest that multiplying each member of an inequality by:

1. a positive number preserves the *direction* (or *order*, or *sense*) of the inequality.
2. a negative number reverses the direction of the inequality.

These results illustrate the next axiom.

Multiplicative Axiom of Inequality

For all real numbers a, b, and c:

1. If $a < b$ and $c > 0$, then $ac < bc$ and $\dfrac{a}{c} < \dfrac{b}{c}$.

If $a > b$ and $c > 0$, then $ac > bc$ and $\dfrac{a}{c} > \dfrac{b}{c}$.

2. If $a < b$ and $c < 0$, then $ac > bc$ and $\dfrac{a}{c} > \dfrac{b}{c}$.

If $a > b$ and $c < 0$, then $ac < bc$ and $\dfrac{a}{c} < \dfrac{b}{c}$.

When you multiply an inequality by a *nonzero* real number, *you must take into account the direction associated with the multiplier.* But what if the multiplier is zero? Because of the multiplicative property of 0, multiplying each member of an inequality by 0 simply produces the uninteresting result $0 = 0$.

Knowing the axioms of inequality, you can prove theorems about order in the set $\mathfrak{R}$ of real numbers. Here is an example.

Theorem. If a is a real number and $a \neq 0$, then $a^2 > 0$.

PROOF

Since $a \neq 0$, the axiom of comparison tells you that there are two cases to consider: *Case 1: $a < 0$* and *Case 2: $a > 0$*.

The reasoning in Case 1 goes like this:

1. $a < 0$ Given
2. $\therefore a \cdot a > 0 \cdot a$ Multiplicative axiom of inequality (part 2)
3. But $a \cdot a = a^2$ Definition of a^2
4. and $0 \cdot a = 0$.   Multiplicative property of 0
5. $\therefore a^2 > 0$ Substitution principle

The reasoning in Case 2 is similar. (See Exercise 11, page 147.)

Oral Exercises

Name the property of real numbers that guarantees that each statement is true.

1. If $a \in \mathcal{R}$ and $a > 4$, then $a + 1 > 4 + 1$.
2. If $b \in \mathcal{R}$ and $b < 0$, then $b + (-3) < 0 + (-3)$.
3. Of two different real numbers, one must be less than the other.
4. If x and y are real numbers and $x < y < 5$, then it is true that $x < 5$.
5. If $a \in \mathcal{R}$ and $2a < 8$, then $\frac{2a}{2} < \frac{8}{2}$.
6. If $m \in \mathcal{R}$ and $n \in \mathcal{R}$, then at least one of the statements "$m > n$," "$n > m$" is false.
7. Any real number that is neither zero nor a negative number must be a positive number.
8. If $k \in \mathcal{R}$ and $k > -3$, then $(-1)k < (-1)(-3)$.
9. Explain the axioms of inequality in your own words.

Written Exercises

Give reasons for the statements in the proofs of the given theorems.
Assume that each variable denotes a real number.

A   **1.** If $a > 0$, then $-a < 0$.

PROOF

1. a is a real number and $a > 0$.
2. $a + (-a) > 0 + (-a)$
3. But $a + (-a) = 0$
4. and $0 + (-a) = -a$.
5. $\therefore 0 > -a$
6. $\therefore -a < 0$

2. If a and b are positive numbers and $a < b$, then $a^2 < b^2$.

PROOF

1. a and b are positive numbers and $a < b$.
2. $a \cdot a < b \cdot a$ and $b \cdot a < b \cdot b$
3. $\therefore a^2 < ba$ and $ba < b^2$
4. $\therefore a^2 < b^2$

3. If $a < b$, then $a - b < 0$.

PROOF

1. a and b are real numbers and $a < b$.
2. $a - b < b - b$
3. $b - b = b + (-b)$
4. $b + (-b) = 0$
5. $\therefore a - b < 0$

4. If $a - b < 0$, then $a < b$.

PROOF

1. a and b are real numbers and $a - b < 0$.
2. $a - b = a + (-b)$
3. $\therefore a + (-b) < 0$
4. $[a + (-b)] + b < 0 + b$
5. $[a + (-b)] + b = a$
6. $\qquad 0 + b = b$
7. $\qquad \therefore a < b$

5. For all positive numbers a, b, c, and d, if $a > b$ and $c > d$, then $ac > bd$.

PROOF

1. a, b, c, and d are positive real numbers; $a > b$; $c > d$.
2. $ac > bc$
3. $bc > bd$
4. $\therefore ac > bd$

6. If $a > 0$ and $b > 0$, then $ab > 0$.

PROOF

1. a and b are real numbers; $a > 0$; $b > 0$.
2. $ab > 0 \cdot b$
3. $0 \cdot b = 0$
4. $\therefore ab > 0$

7. If $a > b$ and $c > d$, then $a + c > b + d$.

PROOF

1. a, b, c, and d are real numbers; $a > b$; $c > d$.
2. $a + c > b + c$
3. $b + c > b + d$
4. $\therefore a + c > b + d$

8. If $a > 0$ and $b < 0$, then $ab < 0$.

PROOF

1. a and b are real numbers: $a > 0$; $b < 0$.
2. $a \cdot b < 0 \cdot b$
3. $0 \cdot b = 0$
4. $\therefore ab < 0$

B **9.** Show by an example that even though $x > y$ and $t > w$, it need not be true that $x - t > y - w$.

10. Show by an example that if $x > y$, it need not be true that $x^2 > y^2$.

Prove the given theorem.

11. If a is a positive real number, then $a^2 > 0$. (Case 2 of the proof begun on page 145.)

12. If x is a positive real number, then $x^3 > 0$.

13. If x is a real number and $x < 0$, then $-x > 0$.

14. For all real numbers x and y, if $x > y$, then $x - y > 0$.

15. For all real numbers x and y, if $x - y > 0$, then $x > y$.

16. For all real numbers x and y, if $x < 0$ and $y < 0$, then $xy > 0$.

C **17.** For all negative real numbers x and y, if $x < y$, then $x^2 > y^2$.

18. For all real numbers x and y, if $x < y$, then $x < \dfrac{x + y}{2} < y$. (*Hint:* If $x < y$, then $x + x < x + y$ and $x + y < y + y$.)

19. For all real numbers x,

$$|x| = x \text{ if } x \geq 0,$$
$$|x| = -x \text{ if } x < 0.$$

20. For all real numbers x and y, if $x > 0$ and $y \leq 0$, then $x + y = |x| - |y|$. (*Hint:* Use the theorem stated in Exercise 19.)

21. For all real numbers x and y, if $x < 0$ and $y < 0$, then $x + y = -(|x| + |y|)$. (*Hint:* Use the theorem stated in Exercise 19.)

5-2 Equivalent Inequalities

The axioms that have been stated guarantee that the following transformations of a given inequality always produce an **equivalent inequality,** that is, one with the same solution set.

Transformations that Produce an Equivalent Inequality

1. Substituting for either member of the inequality an expression equivalent to that member.

2. Adding to (or subtracting from) each member the same real number.

3. Multiplying (or dividing) each member by the same positive number.

4. Multiplying (or dividing) each member by the same negative number and *reversing* the direction of the inequality.

EXAMPLE   Solve $4(y - 2) + 5 \geq 2y + 17$ and graph its solution set.

SOLUTION
1. Copy the inequality. $4(y - 2) + 5 \geq 2y + 17$
2. Use the distributive axiom and $4y - 8 + 5 \geq 2y + 17$
   simplify the left member. $4y - 3 \geq 2y + 17$
3. Add 3 to each member. $4y - 3 + 3 \geq 2y + 17 + 3$
                                                 $4y \geq 2y + 20$
4. Subtract $2y$ from each member. $4y - 2y \geq 2y + 20 - 2y$
                                                 $2y \geq 20$
5. Divide each member by 2. $\dfrac{2y}{2} \geq \dfrac{20}{2}$
                                                 $y \geq 10$

$\therefore$ the solution set is {all real numbers greater than or equal to 10}.

6. Graph the solution set.

$-15 \ -10 \ -5 \quad 0 \quad 5 \quad 10 \ 15 \ 20 \ 25$

Oral Exercises

Tell how to transform the given inequality into an equivalent inequality in which one member is a variable and the other member is a numeral. State the transformed inequality.

EXAMPLE $-6x \le 54$

SOLUTION   Divide each member by -6 and reverse the direction of the inequality; $x \ge -9$.

1. $4x \le -12$ 2. $5y > -35$ 3. $-k < -5$ 4. $-\frac{w}{4} < 8$

5. $t + 5 \ge 7$ 6. $2 + m \le 1$ 7. $-2 \ge n - 4$ 8. $-9 < r - 6$

9. $7k \ge 0$ 10. $-12s \le -48$ 11. $-13 > -\frac{h}{2}$ 12. $0 \le -11p$

Written Exercises

Solve each inequality. In Exercises 1–20 also graph the solution set.

A
1. $x - 26 \ge -33$ 2. $y + 13 < 21$ 3. $14m < -112$ 4. $19w \ge 361$

5. $3r - 6 > 18$ 6. $6s + 4 \le 40$ 7. $9 - p > 12$ 8. $-5 - k < 1$

9. $-4h + 7 > 15$ 10. $-12t + 2 \ge 50$ 11. $-1 \le 1 + \frac{A}{5}$ 12. $-2 > 2 + \frac{B}{5}$

13. $8 - \frac{C}{2} < 8$ 14. $5 \le 5 - \frac{D}{4}$ 15. $6 > \frac{x-2}{4}$ 16. $\frac{y+3}{-4} > -2$

17. $2k - 4 \le 4 - 2k$ 18. $-4w + 9 > w - 21$

19. $14 - 7s \le 6s - 13 - 4s$ 20. $29 - 10x - 6x < -6 - 9x$

B
21. $4m - 5(16 - m) - 1 \ge 0$ 22. $-6k - 5(7 - k) \le 0$

23. $6(p - 2) < 7(p - 3)$ 24. $5(h + 12) \ge -(6 - 4h)$

25. $3(x - 6) \le 15 + 4(4 - x)$ 26. $-3(y - 5) + 9 < 6(y - 2) - 9$

27. $5(2h - 6) - 7(h + 7) > 4h$ 28. $6w + 2(w - 3) > 6(w - 2) - 12$

29. $5(k + 4) - 2(k + 6) \ge 5(k + 1) - 1$

30. $-4(h + 5) + 8(h + 3) \ge 4(h + 1)$

31. $6(A - 3) + (A - 3) \le 5(A - 3) - 6$

32. $3(B - 2) + 25 \ge 2(B - 3) + B$

C
33. $2[3(2 - x) - 2(2x - 1)] - 2 > -[2(3 - x) + 2(2x - 1)]$

34. $3[(2y - 1) - (1 - 2y) - 1] \le 2(y - 1) + 3(y + 1) + 4$

35. $5 - [2(m - 3) - (2m - 3)2] > 2[-3(2 - m) + (m - 2)] + 1$

Transform each inequality into an equivalent inequality in which one member is the variable shown in color.

36. $2(5x - 3a) > 7x + 3a$ 37. $3(y + 4b) < 3(6b - y)$

38. $-2(d + 3c) + 3c < 5c + 2d$ 39. $6(2w - d) - 4d \ge (3w - 2d)(-2)$

40. $6e - 4(r + 3e) < 10e - 2(3r + e)$

41. $3w - 18(3s - w) \ge 6(2w - 3s)$

42. $-5[m - (m + t)] + 16m \ge m + 7(2t - 3m)$

5-3 Intersection and Union of Sets

Suppose that $S = \{0, 2, 4, 6\}$ and $T = \{0, 4, 8, 12\}$. As shown in the Venn diagram in Figure 1, S and T are subsets of the set W of whole numbers.

Can you describe the set shown by red shading in Figure 1? It consists of the elements that belong to *both* S and T. It is $\{0, 4\}$ and is called the *intersection* of S and T. In symbols, you write

$$\{0, 2, 4, 6\} \cap \{0, 4, 8, 12\} = \{0, 4\},$$

and you say "the intersection of $\{0, 2, 4, 6\}$ and $\{0, 4, 8, 12\}$ equals $\{0, 4\}$."

In general, if S and T are any sets, then the set whose members are the elements belonging to both S *and* T is called the **intersection** of S and T and is denoted by $S \cap T$.

As another example, notice that

$$\{0, 2, 4, 6\} \cap \{1, 3, 5, 7, 9\} = \emptyset.$$

Sets like $\{0, 2, 4, 6\}$ and $\{1, 3, 5, 7, 9\}$ which have no members in common are called **disjoint sets**.

Figure 1

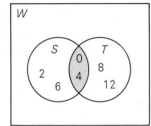

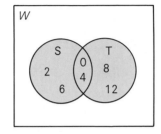

Figure 2

The shaded region in the Venn diagram in Figure 2 represents the set consisting of all the elements which belong to *at least one* of the sets $S = \{0, 2, 4, 6\}$ and $T = \{0, 4, 8, 12\}$. This set contains all the elements of S, together with all the elements of T; it is called the *union* of the two sets. You write

$$\{0, 2, 4, 6\} \cup \{0, 4, 8, 12\} = \{0, 2, 4, 6, 8, 12\},$$

and say "the union of $\{0, 2, 4, 6\}$ and $\{0, 4, 8, 12\}$ is $\{0, 2, 4, 6, 8, 12\}$."

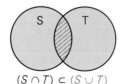

$(S \cap T) \subset (S \cup T)$ Figure 3

In general, if S and T are any sets, then the set whose members are the elements belonging to S or T (or to both S and T) is called the **union** of S and T and is denoted by $S \cup T$.

Notice, as shown in Figure 3, that the intersection of two sets is a subset of their union.

Oral Exercises

Refer to the adjoining diagram and specify each of the following sets by roster.

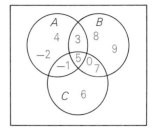

1. $A \cap B$
2. $B \cap C$
3. $A \cup B$
4. $B \cup C$
5. $C \cap A$
6. $C \cup A$
7. $(A \cap B) \cap C$
8. $A \cup (B \cup C)$
9. $(A \cap B) \cup C$
10. $A \cap (B \cup C)$

Replace the $\underline{?}$ with one of the symbols R, S, T, or $\emptyset$ to obtain a true statement.

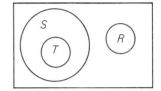

11. If R and S are disjoint sets, then $R \cap S = \underline{?}$.
12. If $T \subset S$, then $S \cup T = \underline{?}$ and $S \cap T = \underline{?}$.

Written Exercises

In Exercises 1–10 specify the intersection and union of the given sets.

A
1. $\{-1, 0, 6, 8\}, \quad \{-1, 5, 6\}$
2. $\{-8, -6, -4, -2\}, \quad \{-8, -4, -1\}$
3. $\{-2, 0, 2, 7\}, \quad \{-2, 7\}$
4. $\{0, 9\}, \quad \{-12, -9, 0, 9, 13\}$
5. $\{3, 5, 7, 9\}, \quad \{4, 6, 8, 10\}$
6. $\{-11, -5, 0\}, \quad \{5, 11\}$
7. {the odd whole numbers}, $\quad \{0, 1, 2, 3\}$
8. {the odd whole numbers}, $\quad$ {the even whole numbers}
9. $\mathfrak{R}, \emptyset$
10. {the nonnegative numbers}, $\quad$ {the negative numbers}

In Exercises 11–16 graph the sets

a. R b. S c. $R \cup S$ d. $R \cap S$

if R and S are as given. In case a required set is the empty set, state that fact and omit the graph.

EXAMPLE 1 $R = \{$the natural numbers less than or equal to 6$\}$,
$S = \{$the integers between -2 and 6$\}$

SOLUTION

a.
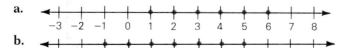

b.
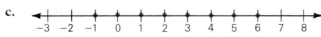

c.
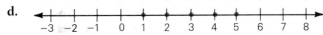

d.

Solving Inequalities and Problems | 151

Graph   a. R b. S c. $R \cup S$ d. $R \cap S$

11. $R = \{$the natural numbers less than 5$\}$,
    $S = \{$the natural numbers between 1 and 5$\}$

12. $R = \{$the whole numbers less than or equal to 7$\}$,
    $S = \{$5 and the natural numbers less than 5$\}$

13. $R = \{$the even whole numbers less than 8$\}$,
    $S = \{$the whole numbers less than or equal to 6$\}$

14. $R = \{$the negative integers greater than $-5\}$,
    $S = \{$the integers between -4 and 4$\}$

15. $R = \{-1, 4,$ and the real numbers between -1 and 4$\}$,
    $S = \{$the nonnegative real numbers less than or equal to 5$\}$

16. $R = \{$the real numbers less than or equal to 7$\}$,
    $S = \{$the real numbers greater than 5$\}$

In Exercises 17–31, A is the solution set of the first inequality and B is the solution set of the second inequality. Graph:

         a. A b. B c. $A \cup B$ d. $A \cap B$

EXAMPLE 2 $x < 2,\ x \geq -1$

SOLUTION

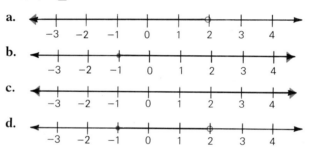

   a.
   b.
   c.
   d.

B

17. $-2 < x < 2,\ x > 0$	**18.** $-1 < x < 1,\ x < 0$	**19.** $y \geq 1,\ y \leq 3$
20. $y \geq -2,\ y \leq 4$	**21.** $m \leq -3,\ m < -4$	**22.** $d > 0,\ d > 2$
23. $w \geq 1,\ w \leq 1$	**24.** $w > 0,\ w \leq 0$	**25.** $r > -3,\ r \leq -3$
26. $r \leq 0,\ -1 < r < 1$	**27.** $0 < s < 3,\ 3 < s < 5$	**28.** $s < -4,\ s > -4$
29. $t \geq 2,\ t \geq -2$	**30.** $t < 5,\ t \geq -5$	**31.** $3 < x \leq 5,\ x \geq 0$

In Exercises 32–39 copy the Venn diagram shown. On your copy, shade the region representing the set named.

32. $A \cup C$

33. $A \cap C$

34. $(A \cap B) \cap C$

35. $A \cap (B \cap C)$

C   **36.** $(A \cup B) \cup C$

37. $A \cup (B \cup C)$

38. $(A \cup B) \cap (A \cup C)$

39. $(A \cap B) \cup (A \cap C)$

Let *A*, *B*, and *C* be subsets of *U* = {0, 1, 2, 3, 4, 5, 6}. Specify *B* by roster.

EXAMPLE 3 *A* = {0, 1, 2, 3, 4}, *A* ∩ *B* = {1, 4}, and *A* ∪ *B* = {0, 1, 2, 3, 4, 5, 6}.

SOLUTION   1. Draw a diagram. Mark it to show the elements of *A* ∩ *B*.
2. Mark it to show the remaining members of *A*.
3. Indicate, as members of *B* (but not *A*), the elements of *A* ∪ *B* not shown in Steps 1 and 2.

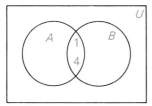 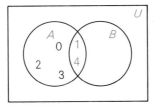 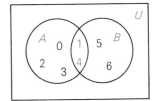

∴ *B* = {1, 4, 5, 6}.   Answer.

40. *A* = {1, 3, 5}, *A* ∩ *B* = {3}, and *A* ∪ *B* = {1, 3, 5, 6}

41. *A* = {2, 4, 6}, *A* ∩ *B* = {4, 6}, and *A* ∪ *B* = {0, 2, 3, 4, 6}

42. *A* = {3, 5}, *B* ∩ *C* = {3}, *B* ∪ *C* = {1, 2, 3, 4, 6}, *A* ∪ *C* = {1, 3, 4, 5}

43. *A* ∩ *A* = ∅, *C* ∪ *A* = {6}, *B* ∪ *C* = *U*, *B* ∩ *C* = ∅

5-4 Combining Inequalities

A sentence such as

$$-3 < x \quad \text{and} \quad x \le 4,$$

which is formed by joining two sentences by the word *and*, is called a
conjunction of sentences. Another way to write the given conjunction is

$$-3 < x \le 4.$$

For a conjunction to be true, *both* of the joined sentences must be
true. As the diagram below suggests, the solution set of

$$-3 < x \le 4$$

is the *intersection* of the solution sets of $-3 < x$ and $x \le 4$.

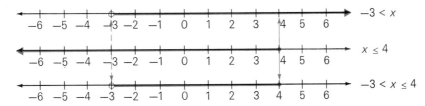

EXAMPLE 1   Solve the conjunction $-2 \le z - 1 < 5$, and graph its solution set.

SOLUTION   1. Copy the given sentence.

$$-2 \le z - 1 < 5$$

2. Add 1 to each member

$$-2 + 1 \le z - 1 + 1 < 5 + 1$$
$$-1 \le z < 6$$

∴ the solution set is $\{-1$ and all real numbers between -1 and $6\}$.

A sentence formed by joining two sentences by the word *or*, such as

$$y < -2 \quad \text{or} \quad y > 3,$$

is called a **disjunction** of sentences. For a disjunction to be true, *at least one* of the joined sentences must be true. Notice that the solution set of $y < -2$ or $y > 3$ is the *union* of the solution sets of $y < -2$ and $y > 3$.

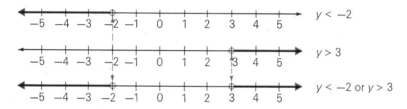

EXAMPLE 2   Solve the disjunction $3r + 2 \le -5$ or $2r - 1 \ge 0$ and graph its solution set.

SOLUTION   1. Copy the given sentence.

$$3r + 2 \le -5 \qquad \text{or} \qquad 2r - 1 \ge 0$$

2. *In the first inequality,* subtract 2 from each member; then divide each member by 3.

3. *In the second inequality,* add 1 to each member; then divide each member by 2.

$$\begin{array}{c|c}
3r + 2 - 2 \le -5 - 2 & 2r - 1 + 1 \ge 0 + 1 \\
3r \le -7 & 2r \ge 1 \\
\dfrac{3r}{3} \le \dfrac{-7}{3} & \dfrac{2r}{2} \ge \dfrac{1}{2} \\
r \le -\frac{7}{3} \quad \text{or} & r \ge \frac{1}{2}
\end{array}$$

∴ the solution set is $\{$all real numbers that are less than or equal to $-\frac{7}{3}$ or greater than or equal to $\frac{1}{2}\}$.

Oral Exercises

1. In a conjunction, sentences are joined by the word __?__.
2. In a disjunction, sentences are joined by the word __?__.
3. A conjunction of two sentences is true provided __?__ of the joined sentences are true.
4. A disjunction of sentences is true provided __?__ or more of the joined sentences are true.

Tell whether the given statement is true or false. Give a reason for your answer.

EXAMPLE $-6 < 0$ and $3 > 6$.

SOLUTION False. $-6 < 0$ is true, but $3 > 6$ is false, and a conjunction is false unless both of the joined statements are true.

5. $6 < 8$ or $0 > -2$
6. $4 < 7$ and $5 < 9$
7. $-3 \leq -2$ or $0 \leq -7$
8. $1 > 2$ or $3 = -3$
9. $7 > \frac{1}{7}$ and $-4 \geq 4$
10. $0 < -3$ and $4 < 0.4$

In Exercises 11–16 match each graph with one of the open sentences given in **a** through **g**.

11.
12.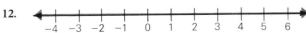
13.
14.
15.
16.

a. $-3 < x$ and $x \leq 2$
b. $x^2 \leq 4$
c. $x \leq -1$ or $x \geq 1$
d. $x \leq 2$ and $x > 0$
e. $x > -1$ or $x > 3$
f. $x \geq -2$ or $x \leq 1$
g. $x < -2$ or $x \geq 1$

Written Exercises

Solve each open sentence and graph each solution set that is not the empty set.

A
1. $-1 \leq x + 3 < 5$
2. $2 \leq y + 6 < 8$
3. $-4 \leq -1 + z \leq 0$
4. $-6 < -2 + w < 4$
5. $3 < 2t + 1 < 13$
6. $-14 \leq 4k - 2 \leq -6$

Solving Inequalities and Problems | 155

Solve each sentence and graph the solution set.

7. $p + 2 > 6$ or $p + 2 < -6$

8. $4 + h \leq -3$ or $4 + h \geq 5$

9. $2s - 1 \geq 1$ or $2s - 1 \leq -1$

10. $4m - 5 > 7$ or $4m - 5 < -9$

11. $4 \geq -4 - 2n \geq -18$

12. $8 > 5 - 3q > -13$

13. $1 - 2a < -5$ or $1 - 2a \geq 5$

14. $8 - 2b > -6$ or $8 - 2b \leq 6$

B 15. $x - 3 \geq -4$ or $x + 1 < 0$

16. $6 + y < -2$ or $4 - y \geq 4$

17. $-m + 5 > 5$ or $m - 8 \leq -m$

18. $3w + 8 < 2$ or $w + 12 > 1$

19. $10 - 2p > 12$ and $7p < 4p + 9$

20. $6t - 12 \geq 12t$ and $13 + 4t < 1$

21. $-3 < -2a - 1$ and $0 \geq -1 - a$

22. $0 < 2 - b$ and $3b - 2 \geq 2b + 1$

23. $6 - c > c$ or $3c - 1 < c + 13$

24. $d + 8 \leq 3d + 2$ or $2d - 8 < 3d - 2$

25. $13x \leq 7x - 12$ and $1 - 4x > 13$

26. $3y + 4 \geq y + 10$ or $3y - 3 \geq 2y - 9$

27. $8 - 2w > 2w$ or $5w - 1 < 2w + 14$

28. $-3 < p - 3$ or $3p - 8 \leq p$

C 29. $2x - 1 \leq 3(x - 1) \leq 2(x - 2)$

30. $3y - 2 \leq 3y + 7 \leq 3y + 3$

31. $2 + 5t < 5t + 9 < 1 + 5t$

32. $2 - 3p < 2(1 - p) < 5 - 3p$

33. $w - 1 < 2w + 3 < w + 4$

34. $2j - 1 < 3(j + 1) < 2(j + 2)$

35. $7 - 3k \geq 6 - 4k \geq 4 - 3k$

36. $7 - 4s \geq 6 - 5s \geq 4(1 - s)$

37. $3 + 6r < 6r + 10 < 2 + 6r$

38. $2 + 3t < 3t + 6 < 3t + 10$

5-5 Absolute Values in Open Sentences (Optional)

Equations and inequalities involving absolute value (Section 2-6) appear in many areas of mathematics.

EXAMPLE 1 Solve the equation $|2x + 3| = 11$.

SOLUTION $|2x + 3| = 11$ is equivalent to the *disjunction*

$$2x + 3 = -11 \quad \text{or} \quad 2x + 3 = 11$$
$$2x + 3 - 3 = -11 - 3 \quad \mid \quad 2x + 3 - 3 = 11 - 3$$
$$2x = -14 \quad \mid \quad 2x = 8$$
$$\frac{2x}{2} = \frac{-14}{2} \quad \mid \quad \frac{2x}{2} = \frac{8}{2}$$
$$x = -7 \quad \text{or} \quad x = 4$$

$\therefore$ the solution set is $\{-7, 4\}$.

EXAMPLE 2 Solve the inequality $|2x + 3| < 11$ and graph the solution set.

SOLUTION $|2x + 3| < 11$ is equivalent to the *conjunction*

$$-11 < 2x + 3 < 11$$

$$-11 - 3 < 2x + 3 - 3 < 11 - 3$$
$$-14 < 2x < 8$$
$$\frac{-14}{2} < \frac{2x}{2} < \frac{8}{2}$$
$$-7 < x < 4$$

∴ the solution set is {all real numbers between -7 and 4}.

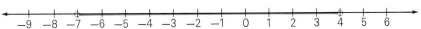

EXAMPLE 3   Solve the inequality $|2x + 3| > 11$ and graph the solution set.

SOLUTION $|2x + 3| > 11$ is equivalent to the *disjunction*

$$2x + 3 < -11 \quad \text{or} \quad 2x + 3 > 11.$$

Completing the solution is left to you. You should find that the solution set is {all real numbers less than -7 or greater than 4}.

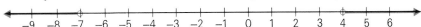

Oral Exercises

State a conjunction or a disjunction of open sentences that is equivalent to the given open sentence.

1. $|x - 4| = 6$ 2. $|y + 3| = 5$ 3. $|t| \geq 1$ 4. $|w| < 1$

5. $|r - 3| < 6$ 6. $|3 - s| \geq 5$ 7. $|8 - t| \geq 3$ 8. $|6 + k| > 0$

9. $|1 - 4x| = 3$ 10. $|1 - 3y| = 10$ 11. $|2h - 9| \leq 1$ 12. $|3e - 7| < 2$

13. $|3 - 3f| \geq 7$ 14. $|5 - 2g| > 3$ 15. $|k| + 4 = 3$ 16. $3 + |m| > -3$

Written Exercises

A   1–16. Solve each of the open sentences in the preceding set of Oral Exercises and graph its solution set.

Solve each open sentence.

B   17. $|1 - (3 - x)| = 4$ 18. $|6 - (y - 1)| = 3$ 19. $|1 - (2 - z)| < 4$

20. $|8 - (w - 1)| \leq 9$ 21. $|1 + 2(x - 1)| = 7$ 22. $|4 - 3(2 - y)| = 0$

23. $5 - 3(1 - |z|) \geq 8$ 24. $-8 + 4(1 - 3|w|) \leq 18$

25. $-3 + 4(2 - |x|) \geq 7$ 26. $5(1 + |y - 3|) - 12 \leq -7$

27. $4(2 + |1 - z|) + 3 \geq 11$ 28. $2(2 + |w - 5|) - 11 \leq -1$

29. $5 + 2(2 + |1 - k|) + 5 \geq 9$ 30. $4(2 + |1 - t|) + 8 \geq 16$

31. $-2|3 + m| \leq -6$ 32. $\frac{1}{2}|4 - 2n| > 4$

Solve each open sentence.

C   **33.** $|3x + 5| = x + 7$ **34.** $|3 + 2y| = y - 3$ **35.** $|2z + 1| = z - 4$

36. $|3w + 2| = w + 5$ **37.** $|3 - 5t| \geq 1 + 3t$ **38.** $|2p + 6| < 4p - 1$

39. $|5k - 3| \geq 1 + 3k$ **40.** $4l - 1 > 2|l + 3|$ **41.** $2 < |a| < 6$

42. $2 < |b| < 4$ **43.** $-1 < |c| < 1$ **44.** $-3 \leq |d| < 3$

45. $4 \leq |e - 1| < 5$ **46.** $4 \leq |2 - f| \leq 5$ **47.** $-3 \geq |x + 5| \geq -6$

Self-Test 1

VOCABULARY   axiom of comparison (p. 143)
transitive axiom of inequality (p. 144)
equivalent inequalities (p. 148)
disjoint sets (p. 150)
intersection of sets (p. 150)
union of sets (p. 150)
conjunction of sentences (p. 153)
disjunction of sentences (p. 154)

1. Supply the missing reasons in the following proof.   *Obj. 1, p. 143*

Prove: If $x \in \Re$ and $-2x + 6 < 14$, then $x > -4$.

a. $-2x + 6 < 14$ a. Given
b. $-2x + 6 + (-6) < 14 + (-6)$ b. ?
c. $-2x + [6 + (-6)] < 8$ c. Associative axiom of
                                            addition, substitution
                                            principle
d. $-2x < 8$ d. Axiom of additive
                                            inverses, identity
                                            axiom of addition
e. $x > -4$ e. ?

2. Solve over $\Re$ and graph the solution set of $3(x - 4) \leq -6$.   *Obj. 2, p. 143*

3. Find $A \cup B$ and $A \cap B$ if $A = \{$the integers between -4 and $4\}$ *Obj. 3, p. 143*
and $B = \{$the whole numbers less than or equal to $4\}$.

Solve each sentence over $\Re$ and graph its solution set.

4. $-2 \leq y - 2 < 4$ **5.** $z + 2 > 4$ or $z - 4 < -6$ *Obj. 4, p. 143*

6. (Optional) $|k - 4| > 3$ *Obj. 5, p. 143*

Check your answers with those at the back of the book.

Solving Problems

OBJECTIVE for Sections 5-6 through 5-9:
1. Solve word problems concerning consecutive integers and consecutive multiples, angle relationships, uniform motion, and mixtures.

5-6 Problems about Integers

If you count by ones from any given integer, you obtain **consecutive integers**.

EXAMPLE 1   **a.** 4, 5, 6, 7, and 8 are five consecutive integers.

   b. -3, -2, -1, and 0 are four consecutive integers.

   c. If x is an integer, then $x - 1$, x, and $x + 1$ are three consecutive integers.

EXAMPLE 2   The atomic numbers of carbon, nitrogen, oxygen, and fluorine are consecutive integers in increasing order. The sum of the atomic numbers of these four elements is five times the atomic number of carbon. What are the atomic numbers of the four elements?

SOLUTION   1. The problem asks for four consecutive integers whose sum is 5 times the least of the integers.

   2.   Let x = the atomic number of carbon (the least of the integers).
   Then $x + 1$ = the atomic number of nitrogen;
   $x + 2$ = the atomic number of oxygen;
   $x + 3$ = the atomic number of fluorine.

   3. The sum of the four atomic numbers is 5 times the atomic number of carbon

$$x + (x + 1) + (x + 2) + (x + 3) \quad = \quad 5x$$

   4.
$$4x + 6 = 5x$$
$$4x + 6 - 4x = 5x - 4x$$
$$6 = x$$

   ∴ the four consecutive integers are 6, 7, 8, and 9.

   5. Does the sum of the four integers equal 5 times the least one?
$$6 + 7 + 8 + 9 \overset{?}{=} 5 \cdot 6$$
$$30 = 30 \ \checkmark$$

Element	Carbon	Nitrogen	Oxygen	Fluorine
Atomic Number	6	7	8	9

Answer.

Some problems following this section deal with *consecutive multiples* of a given number n. When you multiply consecutive integers by n, you obtain **consecutive multiples of n:**

$$\ldots, -3n, -2n, -n, 0, n, 2n, 3n, \ldots$$

For example, the consecutive multiples of 2 are the *even integers:*

$$\ldots, -6, -4, -2, 0, 2, 4, 6, \ldots$$

Notice that the integers that are not even are the *odd integers,* which, in *consecutive* order are:

$$\ldots, -5, -3, -1, 1, 3, 5, \ldots$$

The key to solving problems about consecutive even or odd integers is this fact:

If a is an even integer,

then a, $a + 2$, $a + 4$, $a + 6$, . . . are consecutive even integers;

but if a is odd,

then a, $a + 2$, $a + 4$, $a + 6$, . . . are consecutive odd integers.

EXAMPLE 3 Find all sets of three consecutive positive odd integers such that the greatest integer in the set is no less than twice the least integer in the set.

SOLUTION 1. The problem asks for three consecutive odd integers that are positive and such that the greatest is *no less than* (that is, *greater than or equal to*) twice the least.

2. Let x = the least of the three odd integers. Then the next two odd integers are

$$x + 2 \quad \text{and} \quad x + 4.$$

3. The greatest integer   is no less than   twice the least integer.

$x + 4$	$\geq$	$2x$
4. $x + 4 - x$	$\geq$	$2x - x$
4	$\geq$	x

Since the least odd integer must be less than or equal to 4 and must be positive, the only choices for least integer are 1 and 3.

∴ the only possible sets are $A = \{1, 3, 5\}$ and $B = \{3, 5, 7\}$.

5. *Check:* In each set, is the greatest integer no less than twice the least?

In A: $5 \overset{?}{\geq} 2 \cdot 1$; $5 \geq 2$. ✓

In B: $7 \overset{?}{\geq} 2 \cdot 3$; $7 \geq 6$. ✓

∴ the required sets are $\{1, 3, 5\}$ and $\{3, 5, 7\}$.   **Answer.**

As the preceding example shows, some word problems involve inequalities. The following chart may help you to translate word sentences into inequalities.

Words	Inequality
a is greater than b a is more than b	$a > b$
a is less than b	$a < b$
a is at least b a is no less than b	$a \geq b$
a is at most b a is no greater than b a is no more than b	$a \leq b$

Oral Exercises

1. If $x = 12$, represent 13, 14, and 15 in terms of x.
2. If $y = 21$, represent 18, 19, and 20 in terms of y.
3. Let e be an even integer. What is the next greater even integer? the preceding even integer?
4. Let w be the last integer in the list 3, 5, 7, and 9. Express the other integers in terms of w.
5. Let t represent any integer. Is $2t$ an odd or an even integer? $2t + 1$? $2t + 2$? $2t + 3$?
6. If $s = 16$, represent 8, 16, 24, and 32 in terms of s.
7. Let r be a multiple of 6. What are the next two greater multiples of 6? the preceding multiple of 6?
8. If k is a multiple of 5, what are the next two multiples of 5? the preceding multiple of 5?

State an inequality for the given situation.

EXAMPLE   One number is 7 more than another number and their sum is at least -23.

SOLUTION $(n + 7) + n \geq -23$.

9. In a rectangle whose length is 4 times its width, the perimeter is more than 40 cm.

State an inequality for the given situation.

10. In a triangle in which the lengths of the sides are given by three consecutive integers, the perimeter is less than 50 cm.

11. Hank is x years old. Ed is twice as old as Hank. Also, Ed is at least 8 years older than Hank.

12. During a certain week, a bag of onions cost y cents. A bag of potatoes cost half as much as a bag of onions. To buy one bag of each of these vegetables that week, you had to spend no less than 89¢.

13. The sum of two consecutive odd integers is greater than 18.

14. The sum of two consecutive even integers is at most 70.

15. Three consecutive integers are such that the sum of the first and third is more than twice the second.

16. Construct a consecutive integer problem that can be represented by $3n < 2(n + 2)$.

Problems

Solve the problem.

A 1. Find two consecutive integers whose sum is 375.

2. Find two consecutive even integers whose sum is 446.

3. Find three consecutive integers whose sum is -36.

4. The sum of three consecutive integers is 213. Find the numbers.

5. The sum of three consecutive integers is -153. Find the numbers.

6. The sum of four consecutive even integers is 260. Find the numbers.

7. Find three consecutive integers, if the sum of the first and third is 246.

8. Find four consecutive integers, if the sum of the second and fourth is 352.

9. The lengths of adjacent sides of a parallelogram are consecutive even integers. The perimeter is 108 m. What are the dimensions of the parallelogram?

10. The atomic numbers of radium, thorium, and uranium are consecutive even integers in increasing order. Three times the atomic number of radium is 82 more than the sum of the atomic numbers of thorium and uranium. What are the atomic numbers of these three elements?

11. Find the least two consecutive odd integers whose sum is greater than 28.

12. Find the greatest two consecutive even integers whose sum is less than 100.

B 13. The greater of two consecutive even integers is 6 less than twice the lesser. Find the numbers.

14. The lesser of two consecutive integers is 1 more than twice the greater. Find the numbers.

15. Find four consecutive integers such that 4 times the second decreased by twice the fourth is 90.

16. Find four consecutive even integers such that twice the third is the sum of the fourth and 3 times the second.

17. Four times the lesser of two consecutive even integers is less than 3 times the greater. What are the greatest possible values for the integers?

18. Three consecutive integers are such that their sum is more than 30 decreased by twice the second integer. What are the least possible values for the integers?

19. Three brothers were born in consecutive years. The sum of their ages is more than 42 diminished by the age of the youngest. What are the least possible ages of the brothers?

20. The lesser of two consecutive integers is less than 6 more than three fourths the greater. Find the greatest possible values for the integers.

21. Three consecutive integers are such that the sum of the second and third is greater than the difference when 4 is subtracted from one fourth of the first. Find their least possible values.

C 22. The ages in years of three sisters are consecutive multiples of 5. Six years ago the sum of their ages was 72. Find their present ages.

23. Find four consecutive multiples of 4 such that twice the sum of the first and fourth integers exceeds 3 times the least by 32.

24. Find the three least consecutive multiples of 6 such that the middle one is less than their sum diminished by 24.

25. Find all sets of three consecutive multiples of 4 whose sum is between −64 and −16.

26. Find all sets of three consecutive multiples of 5 whose sum is between −75 and −15.

27. Find four consecutive multiples of 8 such that 3 times the sum of the two lesser integers exceeds twice the sum of the two greater integers by 40.

28. Find the three greatest consecutive multiples of 3 such that the middle one is greater than their sum decreased by 18.

5-7 Problems about Angles

The facts shown in the chart below will help you to solve certain problems about angles.

Right angle	Two adjacent right angles
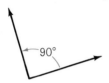 Degree measure is 90.	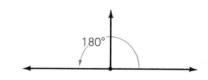 The sum of the degree measures is 180.

Complementary angles	Supplementary angles
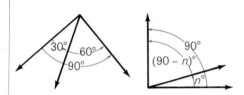 The sum of the degree measures is 90.	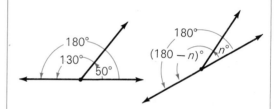 The sum of the degree measures is 180.

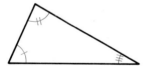

The sum of the degree measures of the angles of a triangle is 180.

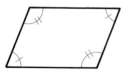   

The opposite angles of a parallelogram are equal in measure.

Two adjacent angles of a parallelogram are supplementary.

EXAMPLE $\angle A$ (read "angle A") measures $15°$ more than twice its complement. What is the degree measure of $\angle A$?

SOLUTION   1. The problem asks for the degree measure of an angle that measures $15°$ more than twice its complement.

2.   Let $x° =$ the measure of $\angle A$.
Then $(90 - x)° =$ the measure of its complement, and
$2(90 - x)° =$ twice the measure of its complement.

3. Measure of $\angle A$ is $15°$ more than twice the measure of complement

$$x = 15 + 2(90 - x)$$

4.
$$x = 15 + 180 - 2x$$
$$x = 195 - 2x$$
$$x + 2x = 195 - 2x + 2x$$
$$3x = 195$$
$$\frac{3x}{3} = \frac{195}{3}$$
$$x = 65$$

5. *Check:* Measure $\angle A = 65°$, and measure of complement of $\angle A = (90 - 65)° = 25°$.

Is 65 15 more than $2 \cdot 25$?
$$65 \overset{?}{=} 15 + 50$$
$$65 = 65 \checkmark$$

$\therefore \angle A$ measures $65°$.   Answer.

Oral Exercises

State the degree measure of the angle that is the complement of the angle with the given measure.

EXAMPLE $46°$

SOLUTION $90 - 46 = 44;$ $44°$.

1. $76°$	**2.** $9°$	**3.** $25°$	**4.** $60°$
5. $x°$	**6.** $(2x)°$	**7.** $(y + 1)°$	**8.** $(n - w)°$

State the degree measure of the angle that is the supplement of the angle with the given measure.

9. $105°$	**10.** $70°$	**11.** $8°$	**12.** $\frac{1}{2}°$
13. $y°$	**14.** $(2y)°$	**15.** $(y + 1)°$	**16.** $(2y + 1)°$

Written Exercises

In Exercises 1–12 the measures of two angles of a triangle are given. Find the measure of the third angle.

EXAMPLE   47°, 115°

SOLUTION   Let x = the measure of the third angle. Then:

$$x + 47 + 115 = 180$$
$$x = 180 - 162 = 18$$

∴ the third angle measures 18°.

A   **1.** 30°, 45°   **2.** 25°, 70°   **3.** 30°, 90°

4. 12°, 78°   **5.** 115°, 35°   **6.** 140°, 20°

7. 13°, 29°   **8.** $x°$, $4x°$ **9.** $\dfrac{n°}{2}$, $n°$

10. $z°$, 80°   **11.** $k°$, $(2k - 10)°$ **12.** $3w°$, $(w + 20)°$

Exercises 13–20 refer to the Law of Reflection: $i = r$. That is, when a light ray strikes a reflecting surface, the measure, i, of the angle of incidence is equal to the measure, r, of the angle of reflection.

B   **13.**

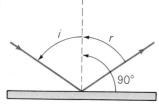

14.

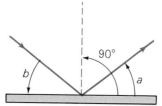

$i = (2x - 10)°$
$r = (50 - x)°$
Find x.

$a = (3y)°$
$b = (7y - 48)°$
Find y.

15. $i = (3x - 5)°$
$r = (65 + x)°$
Find x.

16. $i = (4y)°$
$r = (5y - 53)°$
Find y.

17. $i = (39 - 2n)°$
$r = (3 + 4n)°$
Find n.

18. $i = (50 - 3w)°$
$r = (2w - 25)°$
Find w.

19. $i = (7k - 19)°$
$r = (3k + 37)°$
Find k.

20. $i = (17 - 4g)°$
$r = (49 + 2g)°$
Find g.

Problems

A

1. An angle is 28° greater in measure than its supplement. What is the degree measure of its supplement?

2. Find the measure of an angle that has a measure 19° less than that of its complement.

3. Find the measure of two complementary angles if one measures 12° more than twice the other.

4. Find the measure of two supplementary angles if one measures 18° more than 3 times the other.

5. Find the measure of an angle whose supplement measures 8° more than twice its complement.

6. Find the measure of an angle for which the sum of the measures of its complement and supplement is 194°.

7. One angle of a triangle measures 3 times a second angle, and the third angle measures 4° less than the sum of the measures of the other two. Find the measure of each angle.

8. The supplement of $\angle x$ measures 25° more than 4 times its complement. What is the measure of $\angle x$?

9. An equiangular triangle is one in which the three angles have the same measure. What is the measure of each angle?

10. One angle of a triangle measures 20° more than the second angle. The third angle measures 30° less than twice the sum of the measures of the first two angles. Find the measure of each angle in the triangle.

11. In any isosceles triangle, two angles have equal measures (the base angles). If the third angle measures 12° more than the sum of the measures of the base angles, find the measure of each angle in the triangle.

vertex angle

base angles

12. The degree measures of two angles of a triangle are consecutive even integers. The third angle measures 10° more than the smallest angle of the triangle. What are the degree measures of the angles?

B

13. Explain how you can show that the sum of the degree measures of the angles of a parallelogram is 360°.

14. The degree measure of $\angle w$ is twice the degree measure of $\angle x$ and one less than the degree measure of $\angle y$. If the sum of the degree measures is at least 61°, what is the least possible degree measure of $\angle w$?

15. The degree measure of $\angle r$ is 5 times the degree measure of $\angle s$. If $\angle r$ measures at least 240° more than $\angle s$, what is the least possible degree measure of $\angle s$?

16. In $\triangle ABC$ (read "triangle A, B, C"), $\angle A$ is a right angle and the measure of $\angle B$ is at least 4 times the measure of $\angle C$.

   a. What is the least possible degree measure of $\angle B$?
   b. Is there a greatest possible degree measure for $\angle B$?

17. The degree measure of $\angle A$ is 20° less than the degree measure of $\angle B$ and one half the degree measure of $\angle C$. If the sum of the degree measures of the three angles is at most 90°, what is the greatest possible degree measure of $\angle A$?

18. The second angle of a triangle is $\frac{2}{3}$ of the first angle, and the third angle is 30° less than $\frac{5}{6}$ of the first angle. Find the three angles.

C   19. An angle of a triangle measures at least $\frac{2}{3}$ of its supplement and at most 5 times its complement. Find the possible measures of the angle.

5-8 Problems of Uniform Motion

An object that moves without changing speed, or **rate,** is in **uniform motion.** The formula of uniform motion is:

$$\text{rate} \cdot \text{time} = \text{distance}$$

r	t	$r \cdot t = d$
3 m/s	7 s	$3 \cdot 7$, or 21, m
2 km/min	8 min	$2 \cdot 8$, or 16, km
1050 km/h	3 h	$1050 \cdot 3$, or 3150, km

When motion is not actually uniform, this formula may be applied by considering *average* rate.

EXAMPLE 1   (Motion in the same direction) Two cars heading for Juneau, Alaska, along the same route left a gas station at the same time. Their average speeds differed by 12 km/h. The faster car reached Juneau after 6 h of travel time. The slower car took an hour longer. Find the average speed of each car.

SOLUTION   1. The problem asks for the average speed of each car.

   2.   Let x = the rate of the slow car in km/h.
          Then $x + 12$ = the rate of the fast car in km/h.

Make a sketch illustrating the given facts and arrange the facts in a chart:

The fast car's rate is 12 km/h greater than the slow car's.

The fast car travels for 6 h. The slow car travels for 7 h.

The cars travel the same distance.

Fast car
Slow car

$6(x + 12)$
$7x$

	r (km/h)	t (h)	$r \cdot t = d$ (km)
Fast car	$x + 12$	6	$6(x + 12)$
Slow car	x	7	$7x$

3. Distance of fast car = Distance of slow car

4.
$$
\begin{aligned}
6(x + 12) &= 7x \\
6x + 72 &= 7x \\
6x + 72 - 6x &= 7x - 6x \\
72 &= x
\end{aligned}
$$

Then:
$$x + 12 = 72 + 12 = 84$$

5. To check the speeds of 72 km/h and 84 km/h, you must answer the question: Did the cars travel the same distance?

Fast car traveled $6(84) = 504$ km
Slow car traveled $7(72) = 504$ km
$504 = 504$ ✓

∴ the average speed of the fast car was 84 km/h; the average speed of the slow car was 72 km/h. **Answer.**

EXAMPLE 2 (Motion in opposite directions) A fishing trawler radioed the Coast Guard for a helicopter to pick up an injured crewman. At the time the helicopter left the Coast Guard station, the trawler was 660 km away and heading directly toward the station.

If the average speed of the trawler was 30 km/h and that of the helicopter was 300 km/h, how long did it take the helicopter to reach the trawler?

(*Solution on next page.*)

SOLUTION

1. The problem asks for the number of hours the helicopter and the trawler traveled to reach one another.

2. Let t = the number of hours each traveled. Make a sketch showing the given facts.

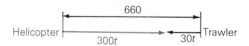

	r (km/h)	t (h)	$r \cdot t = d$ (km)
Helicopter	300	t	$300t$
Trawler	30	t	$30t$

3. Helicopter's distance + Trawler's distance = Original distance between them

$$300t + 30t = 660$$

Solving the equation and checking the results are left to you.

EXAMPLE 3 (Round trip) Using the butterfly stroke to swim the length of the pool, Eva averaged 1.5 m/s. Swimming back, she used the breast stroke and averaged 1.2 m/s. The entire swim took 1 min and 15 s. How long was the pool?

SOLUTION

1. The problem asks for the length of the pool.

2. The given facts are these: (a) The total time was 1 min 15 s, or 75 s. (b) The butterfly-stroke rate was 1.5 m/s. (c) The breast-stroke rate was 1.2 m/s. (d) The distances swum in each direction were the same.

3. Let n = the number of seconds of butterfly-stroke swimming. Then $75 - n$ = the number of seconds of breast-stroke swimming.

	r (m/s)	t (s)	$r \cdot t = d$ (m)
Butterfly stroke	1.5	n	$1.5n$
Breast stroke	1.2	$75 - n$	$1.2(75 - n)$

Distance by butterfly stroke = Distance by breast stroke

$$1.5n = 1.2(75 - n)$$

Completing the solution is left to you. Notice that when you have found the proper value of n, you must use it to find the length of the pool.

Oral Exercises

June and Carol started out from the same location at the same time. June drove due south at an average speed of 75 km/h, and Carol drove due north at an average speed of 85 km/h. They drove for t hours. In Exercises 1–4 represent the required number in terms of t.

1. The number of kilometers that June drove.

2. The number of kilometers that Carol drove.

3. The number of kilometers between June and Carol after the trip.

4. If they had driven in the same direction, how far apart would they be after t hours?

A freight train, averaging 56 km/h, takes x hours to go from Pedro Passe to Warners Fall. On the same run, a passenger train averages 80 km/h and makes the trip 2 hours faster. In Exercises 5–8 represent the required number in terms of x.

5. The number of kilometers traveled by the freight train between Pedro Passe and Warners Fall.

6. The number of hours traveled by the passenger train.

7. The number of kilometers traveled by the passenger train.

8. An open sentence based on the fact that the freight and passenger trains travel the same distance between Pedro Passe and Warners Fall is "$56x = \underline{\ ?\ }$."

9. Two people start out from the same location at the same time. If one travels due west at an average rate of 72 km/h, and the other travels due east at an average rate of 80 km/h, how far apart will they be after y hours?

Problems

A 1. Two cars leave Coon Rapids at the same time, one traveling north at an average of 70 km/h and the other south at 85 km/h. After how many hours will the cars be at least 542.5 km apart?

2. A jet and a light plane leave the same airport at the same time and travel in opposite directions at 960 km/h and 560 km/h, respectively. In how many hours will they be 4560 km apart?

3. The Espinola family drove their car for 2 hours at 80 km/h to get to O'Hare Airport. Then they took an airplane to Dallas, flying for 2 h. If the total distance traveled was 1460 km, what was the average speed of the airplane?

4. Two airplanes start toward each other at the same time from airports located 1792 km apart. One plane flies at the average rate of 400 km/h. At what average rate must the second plane fly if they pass each other no more than 2 h later?

5. Lionel left Reedsville at 9:00 A.M. one day and drove to Fitzwilliam at an average rate of 80 km/h. At 10:00 A.M. the same day, State Trooper Collins left Reedsville, following the same route. If both men arrived in Fitzwilliam at 3:00 P.M., at what average rate had State Trooper Collins traveled? How far is Reedsville from Fitzwilliam?

6. Nancy Evans flew from Stoddard to Brookline and back. She maintained an average rate of 700 km/h to Brookline and of 800 km/h back to Stoddard. If the actual flying time for the round trip was $7\frac{1}{2}$ h, how long did the flight to Brookline take? How far is Stoddard from Brookline?

7. A commuter train traveling due north at 80 km/h passes a freight train traveling due south at 70 km/h. How long after they pass each other will the trains be 25 km apart?

8. Carla rode her bicycle from her home to the bicycle shop and then walked back home. She averaged 10 km/h riding and 5 km/h walking. How far is it from her home to the bicycle shop if her traveling time totaled 75 min?

9. Running at the average rate of 400 m/min, an athlete dashed to the end of the course and then walked back to her starting position at an average rate of 80 m/min. How long was the course if it took her 6 min to make the dash and walk back?

10. A ski lift carries a skier up a slope at the rate of 40 m/min. The skier returns from the top to the bottom on a path parallel to the lift at an average rate of 880 m/min. How long is the lift if the round-trip traveling time is 23 min?

11. The average rate of a passenger train is 32 km/h faster than that of a freight train. In 6 h the passenger train travels the same distance as the freight train travels in 9 h. Find the rate of the passenger train.

12. Irene made a certain trip by moped in 2 h. It took her 2.5 h to make the return trip, since her average speed returning was 12 km/h less than her speed going. Find the two speeds at which she traveled.

13. A helicopter traveling at an average speed of 225 km/h left San Jose 1 h after a train which had departed at 7:00 A.M. If the helicopter overtook the train in $\frac{4}{5}$ h, find the average rate of the train.

14. A westbound bus left Sanders City at 8:00 A.M. An eastbound bus traveling at the same average rate left the same terminal at 10:00 A.M. At 3:00 P.M. the two buses were at least 612 km apart. What is the least possible rate of the westbound bus?

B 15. At 9:00 A.M. Mark and Tom get on their motorbikes to meet each other from towns located 45 km apart. Mark travels 5 km/h faster than Tom. At what rate must each travel if they are to meet at 10:12 A.M.?

16. A car traveled 462 km in 7 h. For the last 3 h of the trip its average rate was 20 km/h less than twice its average rate for the first 4 h. Find the two rates at which the car traveled.

17. Calvin drove from Greenfield to Munsonville at an average speed of 64 km/h. By traveling at an average rate of 78 km/h, he could have arrived 10 min earlier. How far is it from Greenfield to Munsonville?

18. In still water Meg can row 6 km/h. However, on the Branch River it took her 4 h to row upstream to her friend's cabin and only 2 h to return. Find the rate at which the river was flowing.

19. An airplane had been flying for 4 h when a change in wind decreased the plane's ground speed by 75 km/h. If the entire trip of 1536 km took 6 h, how far did the plane travel during the first 4 h?

20. A cargo ship must average 45 km/h to make its 14-hour run on schedule. During the first 4 h, bad weather forced the captain to reduce speed to 30 km/h. What should the speed of the ship be for the rest of the trip to keep on schedule?

21. The Donovan family hiked to an overnight campsite at the rate of 5 km/h. The following morning they returned on horseback over the same route at 16 km/h. The total time spent going and returning was 8.4 h. Find the distance to the campsite.

22. Rita started out on a bicycle ride. She traveled at 12 km/h until she had a flat tire. She left her bike and returned home, walking at 5 km/h. She arrived home 1.5 h after she started out. How far had she traveled on her bike?

23. A private airplane had been flying for 1 h when a change of wind direction doubled the rate of speed. If the entire trip of 384 km took 2.5 h, how far did the plane travel in the first hour?

24. Terry averages 12 km/h on a bicycle trip. One hour after she leaves, her mother sets out to catch her in the car. What is the least speed the mother can average if she is to catch Terry in no more than 20 min?

C 25. Two honeybees leave two locations 180 m apart and fly, without stopping, continually back and forth between these two places at the rates of 4 m/s and 5 m/s, respectively. When do they meet for the first time? the second time?

26. A turbojet train traveling 120 km/h took 10 s to pass a boy riding a bicycle in the same direction at 6 km/h. How long was the train?

27. Barbara and Thelma jog around an oval track that has a circumference of 190 m. Barbara's rate is 207 m/min while Thelma's rate is only 180 m/min. In how many minutes will Barbara be one full lap ahead?

28. An eastbound airplane departs from an airport at the same time as a westbound plane. If the eastbound plane flies at least 450 km/h and the westbound plane at least 550 km/h, what is the latest possible time they can be 2750 km apart?

29. Ralph and Stefan start from the same point and walk in opposite directions for 4 h. They are then 36 km apart. If, instead, they had walked in the same direction for 6 h they would be only 5 km apart. Find their rates of walking.

30. A train running at the rate of 96 km/h overtakes another train, 144 m long, running 72 km/h in the same direction on a parallel track, and passes it in 30 s. In what time would the first train pass another of its own length going in the opposite direction at 48 km/h?

31. A passenger airplane landing at Crystal City between the hours of 6:00 P.M. and 8:00 P.M. must expect to go into a "holding pattern" (circle airport to await a landing clearance) for an average time of 20 min. Northern Airlines Flight 364 from Wedgewood arrives at Crystal City airport 1 h late due to head winds which reduced its ground speed from 800 to 640 km/h. If the flight left Wedgewood on schedule at 1:10 P.M. Crystal City time, when can it expect to land at Crystal City?

32. If Mary Lou were able to increase her cycling speed by 8 km/h, she would be able to cover in 2 h a distance at least as great as that which now takes her 3 h. What is the best speed she achieves at present?

5-9 Mixture Problems; Problems without Solutions

The following example illustrates the kind of problem in which two or more ingredients of different values are combined to form a mixture with a given value. To solve such a problem use this rule:

The sum of the values of the original ingredients must equal the value of the mixture.

EXAMPLE 1 How much gold selling at $8 per gram on the world market must be mixed with silver at $0.50 per gram to obtain 15 g of an alloy worth $4 per gram?

SOLUTION 1. The problem asks for the number of grams of gold to use in 15 grams of the alloy.
2. Let n = the number of grams of gold.
   Then $15 - n$ = the number of grams of silver.

	Number of grams	Dollars per gram	Value
Gold	n	8	$8n$
Silver	$15 - n$	0.5	$0.5(15 - n)$
Alloy	15	4	$4(15)$

3.

The value of the gold	added to	the value of the silver	must equal	the value of the alloy
$8n$	$+$	$0.5(15 - n)$	$=$	$4(15)$

Steps 4 and 5 are left to you. You should find that the amount of gold is 7 g.

Does every problem have a solution? Consider this example.

EXAMPLE 2 Find three consecutive integers whose sum is 85.

SOLUTION
1. The problem asks for three integers that are consecutive and have 85 as their sum.

2. Let $x =$ the first integer.
Then $x + 1 =$ the second integer;
$x + 2 =$ the third integer.

3. The sum of the integers is 85.

$$x + (x + 1) + (x + 2) = 85$$

4.
$$3x + 3 = 85$$
$$3x = 82$$
$$x = 27\tfrac{1}{3}$$

Then: $x + 1 = 28\tfrac{1}{3}$
$x + 2 = 29\tfrac{1}{3}$

5. The numbers obtained, $27\tfrac{1}{3}$, $28\tfrac{1}{3}$, and $29\tfrac{1}{3}$ are *not integers;* therefore, they do *not* satisfy the problem's requirements.

What is the trouble? The trouble is that the conditions of the problem do not fit with one another; they are **inconsistent.** If x is an integer, then

$$x + (x + 1) + (x + 2) = 3x + 3 = 3(x + 1),$$

and $3(x + 1)$ is a multiple of 3. But 85 is not a multiple of 3. Therefore, the sum of three consecutive integers cannot be 85.

In reading problems, you should be on the lookout for inconsistent conditions. You should also be able to recognize problems in which not enough facts are given for a solution. Most of the problems in the following set can be solved. But some of them fail to have solutions either because their conditions are inconsistent or because they give too few facts.

Oral Exercises

State a variable expression for the value in cents of the given item.

EXAMPLE $(3k + 1)$ quarters

SOLUTION $25(3k + 1)$ cents, or $75k + 25$ cents

1. x dimes
2. p nickels
3. $(z + 2)$ fifteen-cent stamps
4. $(3m + 1)$ five-dollar tickets
5. k nickels and $(10 + k)$ dimes
6. $(t - 6)$ ten-dollar bills

Problems

A
1. The registration fee at a convention of the National Council of Teachers of Mathematics was $10 for members and $23 for non-members. If receipts at the convention were $13,505 and 1040 persons attended, how many members were registered?

2. Pamela has an equal number of dimes and quarters. If she has $3.85 in all, how many coins of each kind does she have?

3. Airline fares from Red Cactus to Snow City are $98 for first class and $75 for tourist class. If a flight carried 63 passengers for a total fare of $5139, how many first-class passengers were on the flight?

4. Patrick bought some 12¢ and 15¢ stamps. He bought 35 stamps in all and paid $4.77 for them. How many stamps of each kind did he buy?

5. Seth has 10 more quarters than nickels. How many of each kind of coin has he, if the total is $5.80?

6. Tickets to a school play cost $3 in advance and $4 at the door. In all, 684 tickets were sold for a total of $2457. How many tickets were sold at the door?

7. In a certain triangle, the measure of the first angle is twice that of the second angle, and the second and third angles are complementary. Find the measures of the angles.

8. A small-appliance dealer sold 32 tape recorders. Some of the tape recorders were reel-to-reel models, each selling at $275. The others were tape-cassette recorders at $215 each. The sale brought in $7720. How many tape-cassette recorders were sold?

9. On a certain test each question in Part A was worth 4 points and each question in Part B was worth 6 points. George answered 15 questions correctly and had a point total of 76. How many questions did he answer correctly in each part of the test?

10. How many kilograms of Brand A tea worth $7.50/kg must be added to 10 kg of Brand B tea worth $5.50/kg to form a mixture worth $6.50/kg?

11. Janice Costa made 225 kg of concrete by mixing cement and water with sand and crushed rock. The mass of the sand and crushed rock in the mixture was 10 times that of the water. The mass of the cement exceeded twice the mass of the water by 4 kg. How much cement did Janice use?

12. Chico took $35 in bills to the bank to get change. He asked for twice as many dimes as nickels and 3 times as many quarters as nickels. How many quarters did he get?

B   13. Sue Wood saved all the nickels, dimes, and quarters from her paper route for one week. She found that she had twice as many quarters as dimes, half as many nickels as dimes, and $3.75 in all. How many coins of each kind did she save?

14. Find two integers such that their sum is 18 and 3 times the lesser differs from 4 times the greater by −37.

15. Beth earned $8 more than Edie arranging flowers. After Beth was given $1 and Edie spent $3, together they had $12. How much money did Edie have in the beginning?

16. The Cutrona twins, Cathy and Carrie, together have 19 pails of blackberries. If Cathy sells 3 pails of berries, and Carrie doubles her supply, the twins together will have 40 pails of berries. How many pails of berries does each have now?

17. During the year Len's monthly salary was raised from $950 to $1130, and he received a bonus of $1500. If his total earnings for the year were $14,340, how many months did he work at each salary that year?

18. A piggy bank has 3 fewer quarters than dimes, and 2 more than 3 times as many nickels as dimes. If the bank contains $12.35, how many coins of each kind does it contain?

19. Floyd bought 3 times as many 12¢ stamps as he did 10¢ stamps and 6 more 10¢ stamps than he did 15¢ stamps. How many of each kind of stamp did he buy with $4.26?

20. Find two consecutive odd integers whose sum is 157.

C   21. A projectile is fired at a target and the sound of its impact is heard less than 7 s later. If sound travels at the average rate of 330 m/s, at most how far away is the target?

22. Helen Todd has $10,000 invested in stocks and bonds. From the stocks, she receives $2\frac{1}{2}\%$ interest annually and from the bonds $4\frac{1}{2}\%$ interest annually. Her annual income from these stocks and bonds is $300. How much has she invested in stocks?

23. Juan flies a light plane at an average rate of 160 km/h. Due to a severe storm, he returns on a course 128 km longer than the course he took on the previous trip. His rate is 128 km/h, and the trip takes 3 h longer. How far does he fly on the return trip?

24. Find the least three consecutive integers such that the sum of the first and the last equals twice the second.

25. Sally worked at the Lovely Lamp Land packing light bulbs. She received 5¢ for each bulb she packed successfully and was fined 12¢ for each bulb she broke. If she handled 187 bulbs and was paid $8.16, how many bulbs did she break?

26. A recipe for 100 kg of plate glass calls for 8 times as much silica as lime. Also, the mass of silica required is 12 kg more than 4 times the mass of soda. How many kg of silica, lime, and soda are used in the recipe, if other ingredients have a mass of 6 kg?

27. After washing a rectangular wall hanging whose length was 4 times its width, Freda found it to be unchanged in width, but to be 2 cm shorter than it was before. If the wall hanging decreased 24 cm² in area, find its original dimensions.

Self-Test 2

VOCABULARY   consecutive multiples (p. 160)   supplementary angles (p. 164)
                 complementary angles (p. 164)

1. Find three consecutive odd integers whose sum is 16 more than   *Obj. 1, p. 159*
   the greatest of the three integers.

2. In a certain triangle, the degree measure of the first angle is 12 less than twice that of the second angle. If the third angle measures 42°, what are the degree measures of the first and the second angle?

3. A freight train left Bonds Corner at 6:00 A.M. traveling 72 km/h. At 8:00 A.M. an express train traveling 96 km/h left the same station. At what time did the express overtake the freight?

4. A store sold 45 notebooks on the first day of school, some at 35¢ each and the rest at 50¢ each. The total receipts for notebooks were $20.10. How many of each kind were sold that day?

DIVERSION

Find the least two positive numbers whose sum is an even integer and whose difference is an odd integer.

Symbolic Logic: Boolean Algebra

Most people think of algebra as the study of operations with numbers and variables. Another kind of algebra that is used in the design of electronic digital computers involves operations with logical statements. This "algebra of logic" is called *Boolean algebra* in honor of its origina-tor, George Boole (British, 1815–1864).

In Boolean algebra you use letters such as p, q, r, s, and so on, to stand for statements. For example, you might let p represent the statement "3 is an odd integer," and q, the statement "5 is less than 3." In this case, the statement p has the truth value T (the statement is true), whereas q has the truth value F (the statement is false).

The table below shows the operations that are used in Boolean algebra to produce compound statements from any given statements p and q.

Operation	How Read	Symbols
Conjunction	p and q	$p \wedge q$
Disjunction	p or q	$p \vee q$
Conditional	If p, then q	$p \rightarrow q$
Equivalence	p if and only if q	$p \leftrightarrow q$
Negation	not p	p'

The rules for assigning truth values to compound statements are shown in the five *truth tables* appearing below and on page 180.

Conjunction

p	q	$p \wedge q$
T	T	T
T	F	F
F	T	F
F	F	F

Disjunction

p	q	$p \vee q$
T	T	T
T	F	T
F	T	T
F	F	F

Notice that the conjunction $p \wedge q$ is true when *both* p and q are true; otherwise, $p \wedge q$ is false. On the other hand, the disjunction $p \vee q$ has the value T provided *at least one* of the statements p, q is true.

Conditional				Equivalence			
p	q	$p \rightarrow q$		p	q	$p \leftrightarrow q$
T	T	T		T	T	T
T	F	F		T	F	F
F	T	T		F	T	F
F	F	T		F	F	T

When is the conditional $p \rightarrow q$ false? Only when p is true and q is false.

The equivalence $p \leftrightarrow q$ is a brief way of stating the conjunction

$$(p \rightarrow q) \wedge (q \rightarrow p).$$

The equivalence is true when p and q are both true or both false.

Negation

p	p'
T	F
F	T

The negation p' is the denial of p. Therefore, it is reasonable to agree that p' is false when p is true, and true when p is false.

EXAMPLE 1 Let r stand for "$1 > 2$," and s, for "$2 < 4$." Read each of the following statements. Then by referring to the truth tables above and on page 179, give the truth value of the statement and a reason for your answer.

    a. $r \wedge s$ **b.** $r \vee s$ **c.** $r \rightarrow s$ **d.** $r \leftrightarrow s$ **e.** r'

SOLUTION **a.** $r \wedge s$: $1 > 2$ and $2 < 4$. F; the truth value of r is F.

    b. $r \vee s$: $1 > 2$ or $2 < 4$. T; the truth value of s is T.

    c. $r \rightarrow s$: If $1 > 2$, then $2 < 4$. T; the truth value of r is F, and that of s is T.

    d. $r \leftrightarrow s$: $1 > 2$ if and only if $2 < 4$. F; r and s have different truth values.

    e. r': not $(1 > 2)$, that is, $1 \leq 2$. T; the truth value of r is F.

EXAMPLE 2 Show that for all truth values of r, p, and q, the truth value of

$$r \vee (p \wedge q) \leftrightarrow (r \vee p) \wedge (r \vee q)$$

is T.

SOLUTION Plan: Construct a truth table containing the *eight* combinations of truth values for r, p, and q.

Can you tell how the table on page 181 was constructed? Entries in column 4 were obtained by "*and*-ing" the entries in columns 2 and 3. The entries in column 5 result from "*or*-ing" the entries in columns 1 and 4. The rest of the table is obtained similarly.

r	p	q	$p \wedge q$	$r \vee (p \wedge q)$	$r \vee p$	$r \vee q$	$(r \vee p) \wedge (r \vee q)$
T	T	T	T	T	T	T	T
T	T	F	F	T	T	T	T
T	F	T	F	T	T	T	T
T	F	F	F	T	T	T	T
F	T	T	T	T	T	T	T
F	T	F	F	F	T	F	F
F	F	T	F	F	F	T	F
F	F	F	F	F	F	F	F

When you compare the fifth and last columns, you can see that $r \vee (p \wedge q)$ and $(r \vee p) \wedge (r \vee q)$ have the same array of truth values. Therefore,

$$r \vee (p \wedge q) \leftrightarrow (r \vee p) \wedge (r \vee q)$$

is a true statement no matter what truth values are assigned to $r, p,$ and q.

A compound statement that is true for all truth values of its component statements is called a **tautology**.

Exercises

In Exercises 1–9 assume that r and p are true statements and that q is a false statement. Determine the truth value of each statement.

1. $q \rightarrow r$
2. $r' \wedge p$
3. $p \vee q'$
4. $r \wedge (p \vee q)$
5. $p \vee (q \wedge r)$
6. $(p \vee r) \rightarrow q$
7. $p' \rightarrow (q \vee r')$
8. $r \rightarrow (q \rightarrow r)$
9. $(p \rightarrow q) \leftrightarrow (p' \vee q)$

Given that $p \rightarrow q$ is false, show that each statement in Exercises 10–12 is true.

10. $q \rightarrow p$
11. $p \wedge q'$
12. $(p \vee q) \wedge p$

Construct a truth table for each of the following statements. Is the statement a tautology?

13. $(p \vee q) \rightarrow p$
14. $(q \wedge q') \rightarrow p$
15. $(p \vee q) \leftrightarrow (q \vee p)$
16. $(p \rightarrow q) \leftrightarrow (p' \vee q)$
17. $r \wedge (p \vee q) \leftrightarrow (r \wedge p) \vee (r \wedge q)$
18. $[(p \rightarrow q) \wedge (q \rightarrow r)] \rightarrow (p \rightarrow r)$

Chapter Summary

Statements 1–4 are true for all real values of each variable unless noted otherwise.

1. **Axiom of comparison:** One and only one of the following statements is true:

$$a < b, \ a = b, \ b < a.$$

2. **Transitive axiom of inequality:**

   a. If $a < b$ and $b < c$, then $a < c$.

   b. If $a > b$ and $b > c$, then $a > c$.

Axioms 3 and 4 are used to produce equivalent inequalities.

3. **Additive axiom of inequality:**

   a. If $a < b$, then $a + c < b + c$ and $a - c < b - c$.

   b. If $a > b$, then $a + c > b + c$ and $a - c > b - c$.

4. **Multiplicative axiom of inequality:**

   a. If $a < b$ and $c > 0$, then $ac < bc$ and $\dfrac{a}{c} < \dfrac{b}{c}$.

   b. If $a > b$ and $c > 0$, then $ac > bc$ and $\dfrac{a}{c} > \dfrac{b}{c}$.

   c. If $a < b$ and $c < 0$, then $ac > bc$ and $\dfrac{a}{c} > \dfrac{b}{c}$.

   d. If $a > b$ and $c < 0$, then $ac < bc$ and $\dfrac{a}{c} < \dfrac{b}{c}$.

5. If S and T are any sets, then the set whose members are the elements belonging to both S and T is called the *intersection* of S and T and is denoted by $S \cap T$.

6. Sets having no members in common are *disjoint sets*.

7. If S and T are any sets, then the set whose members are the elements belonging to S or T, or to both S and T, is called the *union* of S and T and is denoted by $S \cup T$.

8. A sentence which is formed by joining two sentences by the word *and* is called a *conjunction* of sentences. For a conjunction to be true, both of the joined sentences must be true.

9. A sentence formed by joining two sentences by the word *or* is called a *disjunction* of sentences. For a disjunction to be true, *at least one* of the joined sentences must be true.

10. The guidelines suggested earlier for solving word problems can be applied to situations involving inequalities.

Chapter Review

5-1
1. If $a < -2$, then which statement is *not* true?

   a. $a + 5 < 3$ **b.** $-3a < 6$ **c.** $a^2 > 4$ **d.** $5a < -10$

2. If $a < b$ and $b < -4$, then which statement is *not* true?

   a. $-4 < a$ **b.** $a - 3 < b - 3$ **c.** $a < -4$ **d.** $ab > b^2$

5-2
3. Solve: $5 - 3x > 14$

   a. $x > -3$ **b.** $x < 3$ **c.** $x > 3$ **d.** $x < -3$

4. Solve: $3x + 6 \leq -9$

   a. $x \leq -1$ **b.** $x \leq -5$ **c.** $x \leq 5$ **d.** $x \leq 9$

5-3
5. If $M \cap N = \emptyset$, which statement must be true?

   a. $N = \emptyset$ **b.** $M \cup N = M$

   c. M and N are disjoint sets   **d.** $M \subset N$

5-4
6. Solve: $x + 4 > 7$ or $-2x > 4$

   a. $x < -2$ or $x > 3$ **b.** $-2 < x < 3$ **c.** $x > -2$ **d.** $\emptyset$

5-5
7. Graph the solution set of $|3m - 9| \leq 6$

   a.

   b.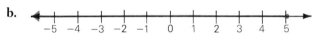

   c.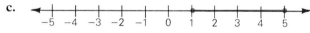

   d.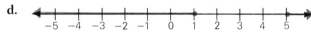

5-6
8. Translate into an inequality. The sum of three consecutive even numbers is at least 228.

   a. $3n \geq 228$ **b.** $3n + 3 \geq 228$

   c. $n + 6 \geq 228$ **d.** $3n + 6 \geq 228$

9. Five times the smallest of three consecutive integers is at least 18 more than the sum of the integers. What are the least possible values of the integers?

   a. 9, 10, 11   **b.** 14, 15, 16   **c.** 11, 12, 13   **d.** 10, 11, 12

5-7
10. In an isosceles triangle, each base angle measures $10°$ less than twice the vertex angle. Find the measure of a base angle.

    a. $40°$ **b.** $70°$ **c.** $47.5°$ **d.** $54°$

5-8
11. A tugboat travels at 24 km/h in still water. If it takes 5 h to make a trip downstream, and 7 h to return, how fast is the river flowing?

    a. 4 km/h   **b.** 5 km/h   **c.** 8 km/h   **d.** 2 km/h

12. Ace Appliances sells black-and-white TV sets for $140 and color sets for $350. If they sell 8 more black-and-white sets than color sets, and n is the number of black-and-white sets, which expression is the value in dollars of the color sets sold?

 a. $350(8 - n)$ b. $350(n + 8)$ c. $350n$ d. $350(n - 8)$

5-9

Chapter Test

1. If $x < -2$, which of the following statements are true?
 $3x < -6$, $x - 5 < -7$, $-2x < 4$, or $x + 2 > 0$

5-1

2. Solve, and graph the solution set: $5 - 2x \geq 17 + x$

5-2

3. If $A = \{$the integers between -5 and $0\}$ and
 $B = \{$the even numbers between -5 and $3\}$,
 then $A \cup B = \underline{\ ?\ }$ and $A \cap B = \underline{\ ?\ }$.

5-3

4. Solve, and graph the solution set: $5 - 3a < 11$.

5-4

5. Solve, and graph the solution set: $|5 - 2k| > 1$.

5-5

6. Find four consecutive even numbers such that twice the largest is the same as the sum of the other three numbers.

5-6

7. In $\triangle ABC$, $\angle A$ and $\angle B$ are complementary, and the measure of $\angle B$ is twice the measure of $\angle A$. Find the measures of all the angles in the triangle.

5-7

8. A runner leaves first base, running at 7.5 m/s; the ball is thrown from first base 1 s later at a speed of 30 m/s. How long does it take for the ball to catch up with the runner?

5-8

9. An elevator has a safety rating of 1500 kg. At most, how many men and women can be carried safely on the elevator if there are three more women than men, and if women average 55 kg and men average 70 kg?

5-9

DIVERSION

The Queen in a game of chess may move in any direction: up, down, right, left, or diagonally. She may move any distance as long as she is not blocked. Show how the Queen in the diagram at the right can pass through the 9 shaded squares in just 4 moves.

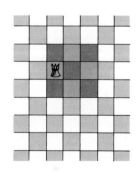

Amelia Earhart
1898–1937

Amelia Earhart's career in aviation ended in early July, 1937. In June of that year, accompanied by Lieutenant Commander Fred Noonan, she took off from Miami, Florida, in an attempt to make the first round-the-world flight near the equator. After negotiating two thirds of the distance safely, their plane, a twin-engined Lockheed, vanished near Howland Island in the Pacific Ocean. A great naval search was conducted but no trace of the plane was ever found.

Amelia Earhart had not always been a flyer. After a tour of duty as a military nurse in Canada during World War I, she did social work in Boston. Her decision to fly was made over strong protests by her family. In 1932 she became the first woman to fly solo across the Atlantic, flying from Harbour Grace, Newfoundland, to Ireland. In 1935 she made a solo flight from Hawaii to the mainland. In the 1930's she took a great interest in the development of commercial aviation.

Careers

in Architecture

An architect, working together with draftsmen and engineers, designs and supervises the construction of a wide variety of projects, ranging from individual houses to entire towns. In designing a structure, the architect must ensure that it is safe, useful, attractive, and compatible with other buildings in the area.

All phases of the project are the architect's responsibility, from the original idea for a design to the final construction. A bachelor's or master's degree in architecture plus experience in an architect's office are prerequisites to becoming a licensed architect.

Voyager I took this photo of Io, one of Jupiter's moons, at a distance of 490,000 km. A volcanic explosion, spewing material to an altitude of 160 km, can be seen silhouetted against dark space.

Functions, Relations, and Graphs

Ordered Pairs and Their Graphs

OBJECTIVES *for Sections 6-1 through 6-3:*
1. *Graph ordered pairs of numbers, and state the coordinates of points, in a coordinate plane.*
2. *Identify the domain and range of a relation specified by a set of ordered pairs and draw a mapping diagram for the relation.*
3. *Find the set of ordered pairs corresponding to a mapping diagram for a relation.*
4. *Determine whether or not a given relation with a finite domain is a function.*
5. *Graph functions and relations with finite domains in a coordinate plane.*

6-1 Coordinates in a Plane

In Section 1-4 you learned that a number line assigns a real number, or *coordinate*, to every point on the line and contains the *graph* of every real number.

$$\begin{array}{ccccccccccccc} & -6 & -5 & -4 & -3 & -2 & -1 & 0 & 1 & 2 & 3 & 4 & 5 & 6 \end{array}$$

Figure 1

In this section you will learn how to set up a *number plane,* or **coordinate plane.** Here are the steps to take:

1. Draw two perpendicular number lines that intersect at the origin of each. This intersection point is called the **origin** in the coordinate plane. As in Figure 2, one line is usually horizontal (the *x*-**axis**) and the other is vertical (the *y*-**axis**). A single arrowhead on each of these two **coordinate axes** shows the positive direction on the axis. The axes are usually directed as shown.

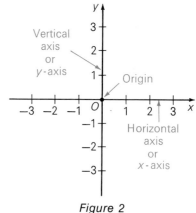

Figure 2

2. Each point *P* in the plane is assigned a unique *pair of coordinates* as follows:

   a. Draw a vertical line from *P* to the *x*-axis. The coordinate of the intersection point on the *x*-axis is the **first coordinate,** or *x*-**coordinate,** or **abscissa** (ab-*siss*-a) of *P*. In Figure 3, the abscissa of *P* is 2.

   b. Draw a horizontal line from *P* to the *y*-axis. The coordinate of the intersection point on the *y*-axis is the **second coordinate,** or *y*-**coordinate,** or **ordinate** (*ord*-n-it) of *P*. In Figure 3, the ordinate of *P* is 3.

   c. The numbers in the **ordered pair** (2, 3) or, in general,

   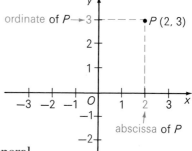

   (*abscissa, ordinate*) or (*x-coordinate, y-coordinate*)

   are the **coordinates** of *P*.

Figure 3

3. Conversely, each ordered pair of numbers, for example, (−2, 4), is assigned a unique point in the plane as follows (see Figure 4):

   a. Draw a *vertical line* through the graph of −2 on the *x*-axis.

   b. Draw a *horizontal line* through the graph of 4 on the *y*-axis.

   c. The *point T* of intersection of these lines is the **graph** of (−2, 4). When you locate *T* in a coordinate plane, you say that you have "plotted, or graphed, the point (−2, 4)."

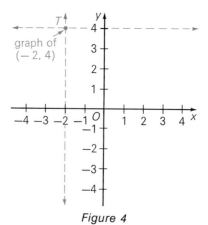

Figure 4

The two axes of a coordinate plane separate the rest of the plane into four regions called **quadrants** which are numbered as shown in Figure 5. Notice the range of values of the coordinates of points in each quadrant.

Graph paper with a printed grid is helpful in finding the coordinates of a point or in graphing an ordered pair of numbers.

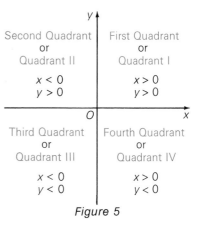

Figure 5

EXAMPLE 1   Graph $(1, 3)$, $(-1, 3)$, $(-1, -3)$, and $(1, -3)$.

SOLUTION

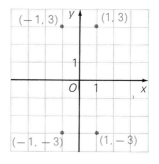

EXAMPLE 2   State the coordinates of each lettered point in the figure below.

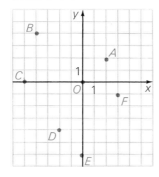

SOLUTION $A(2, 2)$
$B(-4, 4)$
$C(-5, 0)$
$D(-2, -4)$
$E(0, -6)$
$F(3, -1)$
$O(0, 0)$

In working with a coordinate plane, you take the following facts for granted:

1. There is exactly one point in the coordinate plane paired with each ordered pair of real numbers.
2. There is exactly one ordered pair of real numbers paired with each point in the coordinate plane.

Oral Exercises

Name the coordinates of each of the following points.

1. A 2. B 3. C 4. D

5. H 6. P 7. M 8. origin

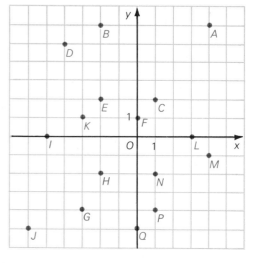

Name by letter the point having the given
coordinates.

9. $(0, 1)$ 10. $(-3, 1)$
11. $(-5, 0)$ 12. $(3, 0)$
13. $(0, -5)$ 14. $(-3, -4)$
15. $(1, -2)$ 16. $(-6, -5)$

17. Name the ordinate of every point on the
x-axis.

18. Name the abscissa of every point on the
y-axis.

In Exercises 19–26 name the quadrant(s) or axis in which the point
(x, y) would lie under the given conditions.

19. $x > 0, y > 0$ 20. $x < 0, y < 0$ 21. $x < 0, y > 0$ 22. $x > 0, y < 0$
23. $x = 0$ 24. $y = 0$ 25. $xy > 0$ 26. $xy < 0$

Written Exercises

Plot each ordered pair of numbers.

A 1. $(2, -5)$ 2. $(-2, -3)$ 3. $(0, 5)$ 4. $(2, 4)$
   5. $(0, -1)$ 6. $(-4, 0)$ 7. $(-6, 3)$ 8. $(-2, 2)$
   9. $(-1, -1)$ 10. $(5, -5)$ 11. $(3, -6)$ 12. $(5, 0)$

Plot three points in at least two different quadrants whose coordinates
are integers satisfying the given requirement.

B 13. The ordinate equals the abscissa.

14. The ordinate is the additive inverse of the abscissa.

15. $y = x + 3$ 16. $y = \frac{1}{4}x$ 17. $y = \frac{1}{2}x + 1$ 18. $y = 3x - 2$
19. $y = 2 - x$ 20. $y = |x|$ 21. $y < x$ 22. $y > x$

190 | Chapter 6

Each of Exercises 23–28 lists three vertices of a rectangle. Graph these, sketch the rectangle, and determine the coordinates of the fourth vertex.

23. $(2, 3), (2, -3), (-2, -3)$ **24.** $(4, 3), (0, 0), (4, 0)$

25. $(-3, -4), (4, 2), (-3, 2)$ **26.** $(-1, 7), (-4, 7), (-4, -4)$

27. $(5, -\frac{3}{2}), (-1, -6), (5, -6)$ **28.** $(-1, -1), (-1, -5), (-\frac{1}{2}, -1)$

For any three given points not all on one line, there are three possible ways of choosing a fourth point so that the four points are vertices of a parallelogram. In Exercises 29 and 30 what are the coordinates of the three possible points?

C **29.** $(4, 2), (2, -1), (7, 1)$ **30.** $(-1, 0), (1, -4), (0, 5)$

31. Three vertices of an isosceles trapezoid are the points with coordinates $(0, 0)$, $(-5, 0)$, and $(-6, -2)$. Find the coordinates of two possible points for the fourth vertex.

32. The base of an isosceles triangle has endpoints $(-6, 0)$ and $(0, -4)$. Using only integers as coordinates, name the coordinates of six points that can serve as the third vertex.

6-2 Functions and Relations

The table in Figure 6 lists the lengths of three lobsters of different ages. The contents of the table can also be presented in a *mapping diagram* (Figure 7) or as a *set of ordered pairs* of numbers (Figure 8).

Lobster Lengths

Age in Months	Length in mm
0.5	14
14	65
300	495

Figure 6

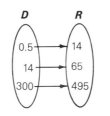

Figure 7

Figure 8

The table, the mapping diagram, and the set of ordered pairs all assign to each member of the set

$$D = \{0.5, 14, 300\}$$

exactly one member of the set

$$R = \{14, 65, 495\}.$$

An assignment such as this is called a *function*.

A **function** consists of two sets, D and R, together with a rule that assigns to each element of D *exactly* one element of R. Each element of R is assigned to *at least* one element of D. The set D is called the **domain** of the function, and R is called the **range** of the function. In the function described above, the domain is {0.5, 14, 300} and the range is {14, 65, 495}.

Two ordered pairs of numbers are **equal** when their first coordinates are equal *and* their second coordinates are equal.

$$\overset{\text{equal}}{(12, 0)} = (4 \cdot 3, \underset{\text{equal}}{1} - 1) \qquad \overset{\text{unequal}}{(8, 7)} \neq (5, 7)$$

Thus, in terms of ordered pairs, you can describe a function whose domain and range are sets of numbers in this way:

> A *function* is a set of ordered pairs of numbers in which no two different ordered pairs have the same first coordinate.

EXAMPLE State whether or not the given set of ordered pairs is a function, and draw a mapping diagram.

  a. {(0, 1), (0, −3), (1, 5), (2, 8)}
  b. {(−2, −1), (1, 1), (5, 1)}

SOLUTION **a.** This is *not* a function because the different ordered pairs (0, 1) and (0, −3) have the same first coordinate.

 b. This is a function.

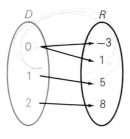

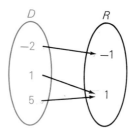

Each of the sets of ordered pairs in the preceding example is a *relation*. A **relation** is any set of ordered pairs of elements. The set of first coordinates of the ordered pairs is the **domain** of the relation; the set of second coordinates is the **range**. For the example above:

Domain	Range
a. $D = \{0, 1, 2\}$	$R = \{-3, 1, 5, 8\}$
b. $D = \{-2, 1, 5\}$	$R = \{-1, 1\}$

Note that a function is a special kind of relation. In fact:

> A *function* is a relation in which each member of the domain is paired with *exactly one* element of the range.

Oral Exercises

a. State the domain and range of each relation.
b. Is the relation a function?

1. $\{(1, 0), (2, 1), (3, 2)\}$ **2.** $\{(4, 2), (4, -2), (0, 0)\}$

3. $\{(0, 1), (5, 1), (-5, -1)\}$ **4.** $\{(9, -3), (1, -1), (1, 1), (9, 3)\}$

Tell whether or not each mapping diagram specifies a function.

5.   **6.**   **7.**

8. 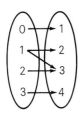   **9.**   **10.**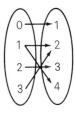

State the value of each variable for which the given equation is a true statement.

11. $(3, y) = (3, 6)$ **12.** $(x, -5) = (-2, -5)$ **13.** $(x, -4) = (8, y)$

Written Exercises

Find the set of ordered pairs for each mapping diagram. In each case tell whether or not the relation is a function.

A **1.**   **2.**   **3.**   **4.**

5.   **6.**   **7.**   **8.**

Functions, Relations, and Graphs | **193**

Find all real values of m and n for which the given equation is true.

EXAMPLE $(m + 4, 3n + 6) = (2m, n - 6)$

SOLUTION $(m + 4, 3n + 6) = (2m, n - 6)$ if and only if

$$m + 4 = 2m \quad \text{and} \quad 3n + 6 = n - 6$$
$$4 = m \qquad\qquad 2n = -12$$
$$m = 4 \qquad\qquad n = -6$$

Check: $[4 + 4, 3(-6) + 6] \overset{?}{=} (2 \cdot 4, -6 - 6)$
$(8, -12) = (8, -12)$ ✓

∴ the values of m and n are 4 and -6. Answer.

9. $(3m - 8, n) = (4, -8)$

10. $(6m, n - 2) = (-36, -2)$

11. $(4m + 9, 4n) = (3m + 13, 2n - 6)$

12. $(3m - 7, 5 - 2n) = (m + 7, 2n - 1)$

B 13. $(1 - m, 2) = (2 + m, |n|)$

14. $(|m| + 2, n) = (3, -6n)$

15. $(1, |n| - 6) = (3 - |m|, -6)$

16. $(2m, 3n - 4) = (5 - 2m, n - 6)$

State the domain and range of each relation and draw a mapping diagram. Is the relation a function?

17. $\{(1, 2), (0, 1), (1, 3), (2, 1)\}$

18. $\{(6, 4), (12, 6), (9, 3), (3, 2)\}$

19. $\{(2, 1), (1, 0), (2, 0), (1, 2)\}$

20. $\{(1, 4), (3, 8), (0, 3), (2, 5)\}$

21. $\{(3, -6), (2, -3), (1, -1), (4, -10)\}$

22. $\{(0, 0), (-1, -1), (2, 2), (3, 3)\}$

23. $\{(6, 9), (12, 18), (3, 4\frac{1}{2}), (9, 13\frac{1}{2})\}$

24. $\{(3, 2.1), (6, 4.2), (9, 6.3), (12, 8.4)\}$

In each of Exercises 25 and 26, there is a mapping diagram. Find the set of ordered pairs, and state whether it is a function. Give a rule for calculating the number that is associated with each given ordered pair. (*Hint:* The first coordinate of an ordered pair may itself be an ordered pair.)

C 25.

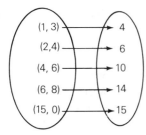

26.

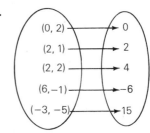

6-3 Graphing Relations and Functions

Figure 9 shows the graphs in a coordinate plane of the ordered pairs in the relation given by the table below.

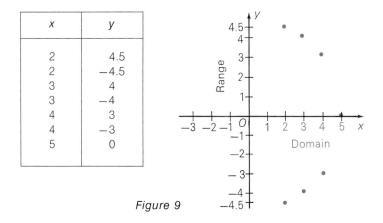

x	y
2	4.5
2	−4.5
3	4
3	−4
4	3
4	−3
5	0

Figure 9

We call this set of points the **graph** of the relation. Notice that domain elements are named on the horizontal axis and range elements on the vertical axis.

Some functions can be computed by evaluating variable expressions. For example, consider the variable expression $3x - 2$ and the domain for x, $\{0, 1, 2\}$. To each value x of the domain the function assigns a single value of the variable expression $3x - 2$ as shown below in the table and the set of ordered pairs formed.

D x	R $3x - 2$
0	−2
1	1
2	4

$\{(0, -2), (1, 1), (2, 4)\}$

Figure 10   Figure 11

Letters such as f, h, G, S, and so on, are often used to name functions. The arrow notation

$$f: x \rightarrow 3x - 2$$

is read "the function f that assigns $3x - 2$ to x." For the domain $\{0, 1, 2\}$, as shown above, the range of this function is $\{-2, 1, 4\}$. Members of the range are called **values** of the function. The values of the function f, above, are −2, 1, and 4.

To state that

$$f: x \rightarrow 3x - 2$$

assigns to the number 0 the number -2, you use the equation

$$f(0) = -2.$$

This may be read: "f at zero equals -2," or "f of zero equals -2," or "the value of f at zero equals -2." Notice that $f(0)$ does not name the product of f and 0. It names the number that f assigns to 0.

EXAMPLE Draw the graph of $f: x \rightarrow 3x + 1$
if $x \in \{-2, -1, 0, 1, 2\}$.

SOLUTION 1. Construct a table giving the value of f for each value of x.

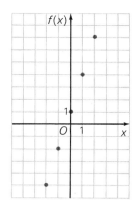

x	$f(x) = 3x + 1$
-2	$3(-2) + 1 = -5$
-1	$3(-1) + 1 = -2$
0	$3 \cdot 0 + 1 = 1$
1	$3 \cdot 1 + 1 = 4$
2	$3 \cdot 2 + 1 = 7$

2. Plot the ordered pairs obtained in the table, as shown above.

Notice that the vertical axis has been labeled $f(x)$ as a reminder that for a given first coordinate x, the second coordinate of a point of the graph is the value of f at x.

Another way to specify the function in the preceding example is to write

$$f = \{(x, f(x)): x \in \{-2, -1, 0, 1, 2\} \text{ and } f(x) = 3x + 1\}$$

and say "f equals the set of ordered pairs x, $f(x)$ such that $x \in \{-2, -1, 0, 1, 2\}$ and $f(x) = 3x + 1$."
You can also specify f like this:

$$f = \{(x, 3x + 1): x \in \{-2, -1, 0, 1, 2\}\}$$

The relations graphed in this section are finite sets. Later you will learn how to graph some relations that are infinite sets.

Oral Exercises

Each of Exercises 1–9 pictures the graph of a relation. State whether the relation is a function.

1.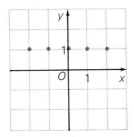

2.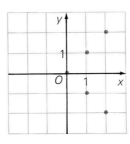

3.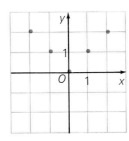

4.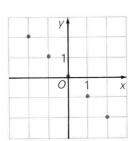

5.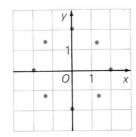

6.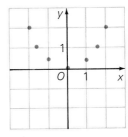

7.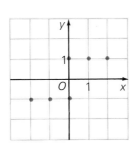

8.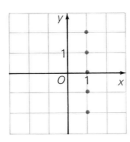

9.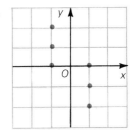

10. How can a straight line be used as a test to determine whether the graph of a relation is a function?

◻IVERSION

Copy the nine dots in the figure at the right. Then draw four straight lines that pass through all nine dots without taking your pencil off the paper.

Written Exercises

In Exercises 1–9 specify each function in two other ways. Assume the domain is $D = \{-2, -1, 0, 1, 2\}$. Then compute the table of ordered pairs of the function.

EXAMPLE $f: x \to 3x - 2$

SOLUTION $f = \{(x, f(x)): x \in D \text{ and } f(x) = 3x - 2\}$
               $f = \{(x, 3x - 2): x \in D\}$

x	-2	-1	0	1	2
$3x - 2$	$3(-2) - 2$	$3(-1) - 2$	$3(0) - 2$	$3(1) - 2$	$3(2) - 2$
$f(x)$	-8	-5	-2	1	4

A   **1.** $f: x \to 2x$ **2.** $g: x \to \dfrac{x}{4}$ **3.** $h: x \to 2x + 1$

4. $m: x \to 1 - x$ **5.** $s: x \to 2 - x^2$ **6.** $r: x \to x^2 + 1$

7. $w: x \to (x + 1)^2$ **8.** $t: x \to x^2 - 2x + 1$ **9.** $u: x \to x^3$

10–18. Graph the functions specified in Written Exercises 1–9 above.

Graph each of the following relations. Is the relation a function?

19. $\{(2, 1), (0, 0), (3, 2), (2, -1), (3, -2)\}$ **20.** $\{(1, 0), (-1, 2), (0, 1), (3, 2), (2, 1)\}$

21. $\{(4, 10), (3, 5), (1, 1), (0, 2), (-1, 5), (-2, 10), (2, 2)\}$

22. $\{(0, 2), (5, 3), (-3, -1), (0, -2), (-4, 0), (-3, 1), (5, -3)\}$

B   **23.** $\{(x, x^2): x \in \{-4, -2, 0, 2, 4\}\}$ **24.** $\{(x^2, x): x \in \{-4, -2, 0, 2, 4\}\}$

25. $\{(x^4, x): x \in \{-2, -1, 0, 1, 2\}\}$ **26.** $\{(x, x^4): x \in \{-2, -1, 0, 1, 2\}\}$

27. $\{(x, f(x)): x \in \{0, 1, 2\} \text{ and } f(x) = x^2 - x - 6\}$

28. $\{(x, g(x)): x \in \{-2, -1, 0\} \text{ and } g(x) = x^2 - 2x - 4\}$

29. $\{(x, h(x)): x \in \{-1, 0, 1\} \text{ and } h(x) = x^2 - 1\}$

30. $\{(x, r(x)): x \in \{-2, -1, 1, 2\} \text{ and } r(x) = x^3\}$

31. $\{(x, s(x)): x \in \{5, 6, 7, 8\} \text{ and } s(x) = \frac{1}{2}x\}$

32. $\{(x, t(x)): x \in \{-3, -1, 0, 1\} \text{ and } t(x) = x^2 - 2x - 2\}$

In Exercises 33–40 let the function described have domain $D = \{\text{the integers between, but not including, } -4 \text{ and } 4\}$. In each case graph the function.

C   **33.** g assigns 1 to each nonnegative member of D and -1 to each negative member of D.

34. f assigns x^2 to each member x of D.

35. l assigns to each member of D its absolute value.

36. p assigns to each member of D the negative of its absolute value.

37. $\{(x, r(x)): r(x) = |2x|\}$

38. $\{(x, s(x)): s(x) = -|2x|\}$

39. $\{(x, t(x)): t(x) = 2|x - 1|\}$

40. $\{(x, w(x)): w(x) = |2x - 1|\}$

Self-Test 1

VOCABULARY   abscissa (p. 188)   ordinate (p. 188)   quadrant (p. 189)

1. Graph $(1, -2)$, $(3, 4)$, $(-2, 3)$, $(-1, -3)$ and $(-5, 0)$ in a coordinate plane.

2. State the coordinates of points A, B, C, and D.

3. State the domain and range of

$$\{(0, -1), (1, 0), (-1, -2), (-2, -3)\}$$

and draw a mapping diagram.

Obj. 1, p. 187

Obj. 2, p. 187

4. Find the set of ordered pairs for the mapping diagram.

5. Which of the following specifies a function?
  a. $\{(1, 0), (-2, 1), (0, 3), (2, 1)\}$
  b. $\{(5, 2), (6, 3), (5, 3)\}$

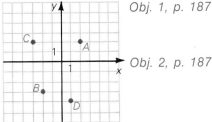

Obj. 3, p. 187

Obj. 4, p. 187

Graph each of the following.

6. $f: x \rightarrow 2x - 1$ with $x \in \{-2, -1, 0, 1, 2\}$

7. $\{(1, -2), (1, 3), (2, 0), (3, -2)\}$

Check your answers with those at the back of the book.

Obj. 5, p. 187

programming in BASIC

You can use the computer to print out ordered pairs. Suppose that you want to find the values of

$$3x + 1$$

if $x \in \{-2, -1, 0, 1, 2\}$. You can start with

$$x = -2$$

and add 1 to the value of x each time through in order to find the values of $3x + 1$ for $x = -1$, $x = 0$, and so on. You use a LET statement to give x its initial value.

You can also print the values in columns, with the elements of each ordered pair side by side on the same line, by using a comma. Try the following program on your computer.

10	PRINT "X", "3X+1"	The computer copies all the
20	LET X=−2	letters, numerals, and spaces
30	PRINT X, 3*X+1	between quotation marks. The
40	IF X=2 THEN 70	comma makes the second col-
50	LET X=X+1	umn begin printing in the six-
60	GOTO 30	teenth space from the left mar-
70	END	gin on the same line.

Lines 30 to 60 form a loop. Line 40 gives the test that will end the loop. Notice that line 50 tells the computer to take the value labeled X, add 1 to it, and store the changed value under the label X.

You can use FOR and NEXT statements to give successive values to a variable. To run through a succession of integral values of X, you use the form shown in this program:

```
10   PRINT "X", "3X+1"
20   FOR X=−2 TO 2
30   PRINT X, 3*X+1
40   NEXT X
50   END
```

When you RUN this program, you will get the same results as you did with the preceding longer program.

Exercises

Print out the ordered pairs of each function using the domain

$$\{-5, -4, -3, -2, -1, 0, 1, 2, 3, 4, 5\}.$$

(*Hint:* Use FOR $X = -5$ TO 5)

1. $x \rightarrow 2x$

2. $x \rightarrow 2x + 1$

3. $x \rightarrow 2x - 1$

4. $x \rightarrow 2(x + 1)$

5. $x \rightarrow 0.5x^2$

6. $x \rightarrow 0.5x^2 + 1$

7. $x \rightarrow 0.5x^2 - 1$

8. $x \rightarrow 0.5(x + 1)^2$

9. $x \rightarrow 2 \times 3.14159x$

10. $x \rightarrow 3.14159x^2$

11. $x \rightarrow 4(3.14159x^2)$

12. $x \rightarrow 4(3.14159/3)x^3$

13. $x \rightarrow |x|$

14. $x \rightarrow |x| + 1$

15. $x \rightarrow 2x|x| + 1$

16. $x \rightarrow 2x|x + 1|$

17. $x \rightarrow |x| + |-x|$

18. $x \rightarrow |x| - |-x|$

19. $x \rightarrow |x^2 - 2|$

20. $x \rightarrow |x^3 - 3|$

21. $x \rightarrow |x^4 - 4|$

Open Sentences in Two Variables

OBJECTIVES for Sections 6-4 through 6-6:
1. *Solve and graph open sentences in two variables over specified replacement sets.*
2. *Graph linear equations in two variables in a coordinate plane.*
3. *Use the concepts of proportion and direct variation to solve problems.*

6-4 Solving Open Sentences in Two Variables

A solution of an open sentence in *one* variable, such as

$$5x - 3 = 7,$$

is a value of the variable for which the sentence is a true statement. The sentence above, of course, has just one solution, namely, 2.

A **solution of an open sentence in two variables**, such as

$$2x + 3y = 11,$$

is an *ordered pair* (x, y) of values of the variables for which the sentence is a true statement. You can check that $(4, 1)$ is a solution of the given open sentence:

$$2x + 3y = 11$$
$$2 \cdot 4 + 3 \cdot 1 \stackrel{?}{=} 11$$
$$8 + 3 \quad = 11 \checkmark$$

You can also verify that $(1, 3)$ and $(-2, 5)$ are two other solutions.

To **solve an open sentence in two variables,** you find all its solutions. To do this, you look for all ordered pairs of numbers chosen from the replacement sets of the variables that **satisfy** the sentence, that is, make it a true statement. The set of all solutions of the open sentence is called the **solution set** over the given replacement sets of the variables.

EXAMPLE 1 Given that the set of whole numbers is the replacement set of each variable, find the solution set of

$$3x + 5y = 30.$$

Then graph the solution set in a coordinate plane.

SOLUTION 1. Transform the sentence into an equivalent one having y as its left member.

$$3x + 5y = 30$$
$$5y = 30 - 3x$$
$$y = \tfrac{1}{5}(30 - 3x)$$
$$y = 6 - \tfrac{3}{5}x$$

(Solution continued on next page.)

2. For x substitute the name of each member of its replacement set in turn and find the corresponding value of y. Since y must be a whole number, $\frac{3}{5}x$ must also be a whole number. Therefore, x must be a multiple of 5.

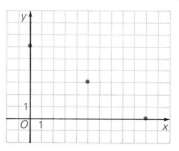

x	$y = 6 - \frac{3}{5}x$	y
0	$6 - \frac{3}{5} \cdot 0$	6
5	$6 - \frac{3}{5} \cdot 5$	3
10	$6 - \frac{3}{5} \cdot 10$	0
15	$6 - \frac{3}{5} \cdot 15$	-3

3. If the value of y found in Step 2 belongs to the replacement set of y, then the ordered pair (x, y) satisfies the sentence. As you see in the table, values of x greater than 10 yield negative numbers for y; therefore, x cannot exceed 10.

$\therefore$ the solution set is

$\{(0, 6), (5, 3), (10, 0)\}$. **Answer.**

Notice that the solution set of the equation in Example 1 is a function with domain $\{0, 5, 10\}$ and range $\{0, 3, 6\}$. You can specify that function in this way:

$$f: x \rightarrow y = 6 - \tfrac{3}{5}x \quad \text{with } x \in \{0, 5, 10\}.$$

Another way to specify f is the following:

$$f = \{(x, y): x \in \{0, 5, 10\} \text{ and } y = 6 - \tfrac{3}{5}x\}$$

In general, the solution set of any open sentence in two variables is a relation whose domain is the set of first coordinates and whose range is the set of second coordinates of the ordered pairs satisfying the sentence.

In the next example, the solution set is a relation, but *not* a function.

EXAMPLE 2   Find the solution set of $y + 2x \leq 1$ if $x \in \{-1, 0, 1\}$ and $y \in \{$the integers greater than $-2\}$.

SOLUTION $y + 2x \leq 1$

$y \leq 1 - 2x$

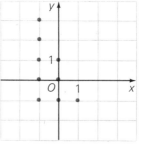

x	$1 - 2x$	$y \leq 1 - 2x$	y
-1	$1 - 2(-1)$	$y \leq 3$	$-1, 0, 1, 2, 3$
0	$1 - 2 \cdot 0$	$y \leq 1$	$-1, 0, 1$
1	$1 - 2 \cdot 1$	$y \leq -1$	-1

Figure 12

$\therefore$ the solution set is

$\{(-1, -1), (-1, 0), (-1, 1), (-1, 2), (-1, 3),$
$(0, -1), (0, 0), (0, 1), (1, -1)\}$. **Answer.**

The graph of the solution set of an open sentence in two variables is usually called the **graph of the open sentence.** Figure 12 shows the graph of the inequality in Example 2.

Oral Exercises

Is the given ordered pair of numbers a solution of the open sentence? Why? Assume that $\mathcal{R}$ is the replacement set of each variable.

EXAMPLE 1 $2x - 5y = -1;$ $(2, -1)$

SOLUTION   Not a solution, because $2 \cdot 2 - 5(-1) \neq -1$.

1. $2x + 4y = 10;$ $(1, 2)$
2. $y + 3x \geq 2;$ $(-2, -3)$
3. $y > 5x - 3;$ $(-5, -27)$
4. $x \leq 4y - 5;$ $(10, 20)$
5. $y + x^2 = 27;$ $(-5, 2)$
6. $xy - y = 4;$ $(3, -2)$

Transform each open sentence into an equivalent one having y as one member.

EXAMPLE 2 $3x - 4y = 7$ **SOLUTION** $y = \frac{3}{4}x - \frac{7}{4}$

7. $4x - y = 3$
8. $x - y = -3$
9. $3x + 2y = 6$
10. $6x - 3y = 1$
11. $x + y \geq 2$
12. $2x + 3y < 7$
13. $4x - y \leq 6$
14. $-x + y \geq 0$

Name three ordered pairs belonging to the given set if $x, y \in \{$the integers$\}$.

15. $\{(x, y): y = -2x\}$
16. $\{(x, y): y = -x + 4\}$
17. $\{(x, y): y > -x + 4\}$

Written Exercises

Find the solution set of each equation given that $\{-2, -1, 0, 1, 2\}$ is the replacement set of x, and $J = \{$the integers$\}$ is the replacement set of y. Graph the solution set.

A
1. $y = 3x$
2. $y = -3x$
3. $y = 2x + 3$
4. $y = x - 2$
5. $y = 4 - 2x$
6. $y = -4x + 2$
7. $y = \frac{1}{2}x$
8. $y = -\frac{1}{2}x$
9. $2x + 3y = 2$
10. $-3x + y = -4$
11. $3x + 2y = -6$
12. $4x + 5y = 0$
13. $y = x^2 - 1$
14. $y = 1 - x^2$
15. $y = x^3 - 2$
16. $y = 2 - x^3$

Graph each function. Assume that the domain is $\{-3, 0, 3\}$.

17. $\{(x, y): y = -2x + 1\}$
18. $\{(x, y): y = 3 - 2x\}$

Functions, Relations, and Graphs | **203**

Find the solution set of each inequality given that $\{-1, 0, 1\}$ is the replacement set of x and $\{-2, -1, 0, 1, 2\}$ is the replacement set of y. Graph the solution set.

B **19.** $y < x$ **20.** $y > x$ **21.** $y \le x$ **22.** $x + y > 0$

23. $x + y < 0$ **24.** $x - y \ge 0$ **25.** $y + 1 \le 2x$ **26.** $y + 1 \ge 2x$

27. $1 - y < -2x$ **28.** $3x - y < 3$ **29.** $2y - 4 \ge 4x$ **30.** $4x + 4 \le 2y$

31. $x + y < x - y$ **32.** $3 - y > 2x$ **33.** $2x + y \ge 2y$ **34.** $3x < 2x - 3y$

Graph the solution set of each open sentence for the given replacement set.

C **35.** $|x| + 2y = y + 4;\quad x \in \{-3, -2, -1, 0, 1, 2, 3\},\ y \in \{\text{the positive integers}\}$

36. $|3x| > 3x + y;\quad x \in \{-3, -2, -1, 0, 1, 2, 3\},\ y \in \{\text{the nonnegative integers}\}$

37. $|x| + |y| = 4;\quad x, y \in \{\text{the whole numbers less than } 7\}$

38. $|x| \le |y|;\quad x, y \in \{-3, -2, -1, 0, 1, 2, 3\}$

39. $y < x^2 + 1;\quad x \in \{-2, -1, 0, 1, 2\},\ y \in \{1, 2, 3, 4\}$

40. $x^2 + y^2 \le 10;\quad x, y \in \{\text{the positive integers}\}$

41. $|x| + x \ge y;\quad x \in \{-3, -2, -1, 0, 1, 2, 3\},\ y \in \{0, 1, 2, 3, 4, 5, 6\}$

42. $y = x^2 + 6x;\quad x \in \mathcal{R},\ y \in \{-9, 0, 7\}$

programming in BASIC

You can use a computer to find out whether or not a given ordered pair is a solution of an open sentence in two variables. Try this program for several values of X and Y:

```
10   INPUT X
20   INPUT Y
30   IF 2*X+3*Y=11 THEN 60
40   PRINT "NO"
50   GOTO 70
60   PRINT "YES"
70   END
```

You can change line 30 to test values in other sentences. For example:

```
30   IF Y<5*X+2 THEN 60
```

The line

```
50   GOTO 70
```

can be replaced with the simpler statement:

```
50   STOP
```

You can use the computer to test the values of (X, Y) for all the ordered pairs of numbers in given replacement sets. Suppose that the replacement set of both X and Y is

$$(-3, -2, -1, 0, 1, 2, 3).$$

The computer should test the ordered pairs:

$$(-3, -3), (-3, -2), (-3, -1),$$
$$(-3, 0), (-3, 1), (-3, 2), (-3, 3),$$
$$(-2, -3), (-2, -2), (-2, -1),$$
$$(-2, 0), (-2, 1), (-2, 2), (-2, 3),$$

and so on.

You can do this by using two loops, one *inside* the other. Such loops are called *nested loops*.

Let us find solutions of $3 * X - Y > 0$. We shall want to print the ordered pairs that are solutions, but skip to the NEXT Y when the ordered pair tested is *not* a solution. Thus, to find solutions of

$$3*X - Y > 0$$

we transfer to NEXT Y when

$$3*X - Y <= 0$$

is true.

```
10   FOR X=-3 TO 3
20   FOR Y=-3 TO 3
30   IF 3*X-Y<=0 THEN 50
40   PRINT "(";X;",";Y;")",
50   NEXT Y
60   NEXT X
70   END
```

(Outside loop, Inside loop)

The comma at the *end* of line 40 will make the results print in 5 columns.

Exercises

1–6. Using $\{-1, 0, 1\}$ as the replacement set of x and $\{-2, -1, 0, 1, 2\}$ as the replacement set of y, check your solutions of Exercises 19–34 on page 204.

Taking the set of integers from -10 to 10 as the replacement set for x and y, find the solution set of:

7. $x^2 + y^2 = 100$

8. $x^2 + 4y^2 = 100$

9. $4x^2 + y^2 = 100$

10. $|x| + |y| = 10$

6-5 The Graph of a Linear Equation in Two Variables

Which of the following figures shows the graph of the equation

$$x + y = 2?$$

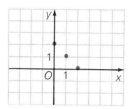

Figure 13

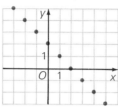

Figure 14

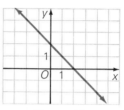

Figure 15

The answer depends on the replacement sets of x and y.

Replacement Set of x and y	Graph of $x + y = 2$
$W = \{$the whole numbers$\}$	Figure 13 The graph is a set of 3 points.
$J = \{$the integers$\}$	Figure 14 (a partial graph) The graph is an infinite set of isolated points.
$\mathcal{R} = \{$the real numbers$\}$	Figure 15 The graph is a straight line.

We call the line shown in color in Figure 15 "the graph of the equation $x + y = 2$," or "the line whose equation is $x + y = 2$," or simply "the line $x + y = 2$."

Any equation which can be written equivalently in the form

$$ax + by = c,$$

where a, b, and c are real numbers with a, the **coefficient of x**, and b, the **coefficient of y**, *not both* 0, is called a **linear equation in two variables**, x and y. Thus $5x - 2y = 8$ and $3y = 4$ are linear, but $x^2 + y = 3$, $xy = 7$, and $\frac{1}{x} - y = 1$ are not.

Figure 15 illustrates the following fact that we take for granted.

> If $\mathcal{R}$ is the replacement set of each variable in a linear equation in two variables, then the graph of the equation in a coordinate plane is a straight line.

Hereafter, unless otherwise stated, you should assume that the replacement set of each variable in a linear equation is $\mathcal{R}$.

Only two points are needed to determine a line. Therefore, to graph a linear equation in two variables, you need find only two of its solutions.

EXAMPLE 1 Graph $3x - 2y = 6$ in a coordinate plane.

SOLUTION The two solutions that are easiest to obtain are those of the form $(0, y)$ and $(x, 0)$.

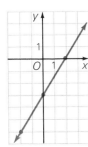

1. Substitute 0 for x and solve for y. Then plot the point $(0, -3)$.

$$3x - 2y = 6$$
$$3(0) - 2y = 6$$
$$y = -3$$

2. Substitute 0 for y and solve for x. Then plot the point $(2, 0)$.

$$3x - 2(0) = 6$$
$$x = 2$$

3. As a check, determine a third solution; for example, substitute -2 for x and solve for y. Then plot the point $(-2, -6)$.

$$3(-2) - 2y = 6$$
$$-6 - 2y = 6$$
$$-2y = 12$$
$$y = -6$$

4. Draw the line containing the three points plotted.

EXAMPLE 2 In a coordinate plane, graph:

a. $x = -3$

b. $4y = 8$

SOLUTION

a. Every point whose abscissa is -3 satisfies $x = -3$. Therefore, the graph is the vertical line 3 units to the left of the y-axis.

b. $4y = 8$ is equivalent to $y = 2$. Therefore every point whose ordinate is 2 satisfies $4y = 8$. Hence, the graph is the horizontal line 2 units above the x-axis.

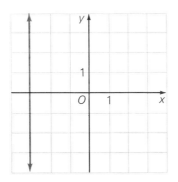

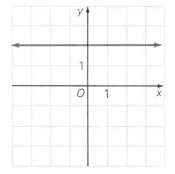

A function whose ordered pairs satisfy a linear equation is called a **linear function.** Thus,

$$f = \{(x, y): x \in \mathcal{R} \text{ and } y = 2 - x\}$$

is a linear function. Its graph is the straight line shown in color in Figure 15 (page 206).

Oral Exercises

State whether the equation is linear.

1. $3x + 6y = 4$ **2.** $y = x^2 + 2$ **3.** $\dfrac{y}{2} = 3x - 4$ **4.** $5x - y = 0$ **5.** $xy = 3$

State whether or not the graph of the given relation is a line. Assume x, $y \in \mathcal{R}$.

6. $\{(x, y): x - 3 = 5\}$ **7.** $\{(x, y): y + 1 = 2\}$

8. $\{(x, y): y = x^2 - x + 2\}$ **9.** $\{(x, y): 2y - x = y + 2x\}$

10. $\{(x, y): y = \dfrac{1}{x}\}$ **11.** $\{(x, y): y = |x|\}$

In Exercises 12–15 state in words the relationship between the ordinate and abscissa of points on the graph of the given equation.

EXAMPLE $y = 2x + 3$

SOLUTION The ordinate is 3 more than two times the abscissa.

12. $y = -5x$ **13.** $y = \tfrac{1}{2}x - 1$ **14.** $y = x^2 + 6$ **15.** $y = -3x + \tfrac{1}{2}$

Since no two different ordered pairs in a function have the same x-coordinate, no vertical line intersects the graph of a function in more than one point.

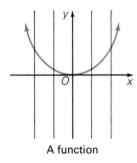

A function

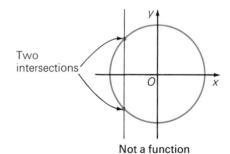

Two intersections

Not a function

Use this "vertical-line test" to determine whether the relations graphed in Exercises 16–21 are functions.

16.

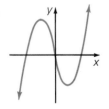

17.

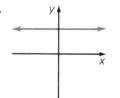

18.

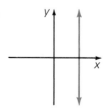

19.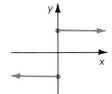

20.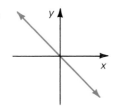

21.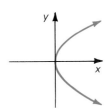

Written Exercises

Graph each equation in a coordinate plane.

A
1. $x = 3$
2. $y = 2$
3. $x = -1$
4. $y = -2$

5. $y = x + 2$
6. $y = x - 2$
7. $2y - x = 0$
8. $x + 2y = 0$

9. $4x - 2y = 6$
10. $-x + 2y = 10$
11. $-4x + 2y = 8$
12. $6x - 3y = -12$

Find the coordinates of the point where the graph of each equation crosses (a) the x-axis and (b) the y-axis.

13. $3x - 5y = 15$
14. $-3x + 2y = 24$
15. $5x = 12 - y$
16. $6x = -36 + 3y$

17. $3x = 4y$
18. $5y = -7x$
19. $-y = 3x + 2$
20. $y = \frac{2}{3}x + 6$

In Exercises 21–26 use a formula to specify, and then graph, the function defined by each rule. In each case, let the domain be $D = \{\text{nonnegative real numbers}\}$.

EXAMPLE The force of gravity on Earth on a given mass is 2.6 times the force of gravity on Mars on the same mass.

SOLUTION Let m = the force of gravity on Mars
e = the force of gravity on Earth
$\{(m, e): e = 2.6m \text{ and } m \in D\}$

m	e
0	0
1	2.6
2	5.2

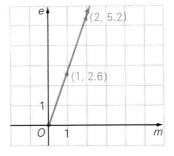

B 21. Cost for telephone service in a certain city is $22 for installation and $4.50 a month for service.

22. In water there is one oxygen atom for every two hydrogen atoms.

23. A certain recipe calls for four times as many grams of stone-ground wheat flour as enriched white flour.

Use a formula to specify, and then graph, the function defined by each rule.

24. When enlarging a certain photograph, its width is to be two fifths of its length.

25. Currency in West German marks can be exchanged for 2.3 times as many French francs.

26. A certain alloy of nickel and silver contains three fourths as much silver as nickel.

Graph the function with the indicated domain.

27. $f: x \to \dfrac{x}{2} - 4, x \in \mathcal{R}$

28. $h: x \to \dfrac{x - 2}{2}, x \in \mathcal{R}$

29. $i: x \to \frac{1}{2}x - 2, x \in \{\text{the even whole numbers}\}$

30. $j: x \to -2(1 - x), x \in \{\text{the nonnegative integers}\}$

31. $\{(x, y): y = -3(x + 2)\}, x \in \mathcal{R}$

32. $\{(x, y): y = 2(x - 2)\}, x \in J$

33. $C: r \to 2\pi r, 0 \le r \le 6$ (Use $\frac{22}{7}$ for π.)

34. $C: d \to \pi d, 0 \le d \le 6$ (Use $\frac{22}{7}$ for π.)

Graph each function. Is it a linear function?

C   35. $\{(x, y): y = |x| + 1, x \in \mathcal{R}\}$

36. $\{(x, y): y = 1 - |x|, x \in \mathcal{R}\}$

37. $\{(x, y): y = |x - 2|, x \in \mathcal{R}\}$

38. $\{(x, y): y = -|x - 2|, x \in \mathcal{R}\}$

39. $\{(x, y): y = |x + 2|, x \in \mathcal{R}\}$

40. $\{(x, y): y = \left|\dfrac{x}{3}\right|, x \in \mathcal{R}\}$

41. The function which assigns the number 1 to each positive number, -1 to each negative number, and 0 to 0.

42. The function which assigns the number 0 to each negative number x, and the number x itself to each nonnegative number x.

6-6 Direct Variation and Proportion

The table below shows the distance traveled in kilometers for a given length of time at a constant rate of speed, 85 km/h.

t (h)	d (km)	$\dfrac{d}{t}$
1	85	$\frac{85}{1} = 85$
2	170	$\frac{170}{2} = 85$
3	255	$\frac{255}{3} = 85$
4	340	$\frac{340}{4} = 85$

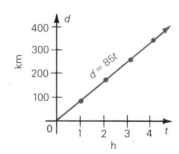

Figure 16

Note that the constant rate of speed may be shown by the formula

$$\frac{d}{t} = 85, \text{ or } d = 85t.$$

Such a formula describes a function called a *linear direct variation*.
A **linear direct variation** (or simply, a **direct variation**) is a function f such that

$$f: x \to kx \quad \text{where } k \text{ is a nonzero constant.}$$

Other ways to specify f are

$$f: f(x) = kx, \quad f: x \to y = kx, \quad \text{or} \quad f = \{(x, y): y = kx\}.$$

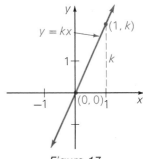

Figure 17

You say, "*y, or* $f(x)$*, varies directly as x,*" or "*y is directly proportional to x;*" k is called the **constant of variation** or the **constant of proportionality**.
The graph of a linear direct variation with $\Re$ as domain and range is a straight line (Figure 17), two points of which are $(0, 0)$ and $(1, k)$. (See Exercise 24, page 214.)

EXAMPLE 1   Given that u varies directly as v, and that the value of u is -12 when the value of v is 4, find

   a. the constant of variation.

   b. the value of u if 7 is the value of v.

SOLUTION   Let $u = kv$.
   a. Substituting -12 for u and 4 for v, you find:

$$-12 = k(4); \quad \therefore k = \frac{-12}{4} = -3. \quad \textbf{Answer.}$$

   b. $u = -3v$
      If $v = 7$, then
         $u = -3(7) = -21.$
      $\therefore$ if 7 is the value of v, then -21 is the value of u.   **Answer.**

Suppose that one ordered pair of a direct variation specified by

$$\{(x, y): y = kx\}$$

is (x_1, y_1) [read "x sub one, y sub one"] and another is (x_2, y_2), where both (x_1, y_1) and $(x_2, y_2) \neq (0, 0)$. Then

$$y_1 = kx_1 \quad \text{and} \quad y_2 = kx_2, \quad \text{so that} \quad \frac{y_1}{x_1} = k \quad \text{and} \quad \frac{y_2}{x_2} = k.$$

Therefore,

$$\frac{y_1}{x_1} = \frac{y_2}{x_2}.$$

This last equation, $\dfrac{y_1}{x_1} = \dfrac{y_2}{x_2}$, is called a **proportion** and may be read:

means

y_1 is to x_1 as y_2 is to x_2.

extremes

Note the names **means** given to x_1 and y_2 and **extremes** to y_1 and x_2. Example 2 on page 215 proves the following theorem.

Theorem.   If $\dfrac{y_1}{x_1} = \dfrac{y_2}{x_2}$, then $x_1y_2 = y_1x_2$. That is:

In a proportion, the product of the means equals the product of the extremes.

EXAMPLE 2   A mass of 12 g stretches a spring 3 cm. If the distance D a spring is stretched is directly proportional to the mass m:

a. How far will an 8-g mass stretch the spring?

b. What mass will stretch the spring 8 cm?

SOLUTION $\dfrac{D_1}{m_1} = \dfrac{D_2}{m_2}$,   or $D_1m_2 = D_2m_1$.

a. $D_1 = 3$, $m_1 = 12$, $m_2 = 8$. Find D_2.

$$\frac{3}{12} = \frac{D_2}{8}, \quad \text{or} \quad 3(8) = D_2(12)$$

$$12D_2 = 24$$
$$D_2 = 2$$
$$\therefore \ 2 \text{ cm.} \quad \text{Answer.}$$

b. $D_1 = 3$, $m_1 = 12$, $D_2 = 8$. Find m_2.

$$\frac{3}{12} = \frac{8}{m_2}, \quad \text{or} \quad 3m_2 = 12(8)$$

$$3m_2 = 96$$
$$m_2 = 32$$
$$\therefore \ 32 \text{ g.} \quad \text{Answer.}$$

Notice that in solving Example 2 it was not necessary to find a constant of proportionality because a proportion was used.

Oral Exercises

State whether the set of ordered pairs of real numbers (x, y) satisfying the given equation, or specified by the given table, is a direct variation. For each direct variation, state the constant of proportionality.

1. $y = -2x$ **2.** $y = \frac{1}{3}x$ **3.** $xy = 1$ **4.** $\dfrac{y}{x} = 4$ **5.** $3x = y$ **6.** $y = \dfrac{1}{2x}$

7.

x	1	2	3	4
y	2	4	6	8

8.

x	1	2	4	8
y	4	2	1	$\frac{1}{2}$

9.

x	0	1	-1	2
y	0	-3	3	-6

a. Read each proportion in two ways.
b. In each case, state an equation obtained by applying the theorem on page 212.

EXAMPLE $\dfrac{n}{4} = \dfrac{13}{7}$

SOLUTION   **a.** n divided by 4 equals $\frac{13}{7}$; n is to 4 as 13 is to 7.
b. $7n = 52$.

10. $\dfrac{x}{5} = \dfrac{4}{7}$ **11.** $\dfrac{36}{3n} = \dfrac{2}{3}$ **12.** $\dfrac{x_1}{y_1} = \dfrac{y_1}{y_2}$ **13.** $\dfrac{x_1}{x_2} = \dfrac{y_1}{y_2}$ **14.** $\dfrac{k+1}{4} = \dfrac{k-1}{5}$

Written Exercises

A   **1.** y varies directly as x. y is 24 when x is 6. Find y when:
   a. x is 1   **b.** x is -2 **c.** x is 25
   2. x is directly proportional to t. x is 8 when t is 16. Find x when:
   a. t is 20   **b.** t is -8 **c.** t is -3

In these direct variations, find the value of the indicated variable.

3. $x_1 = 8, y_1 = 18, x_2 = -4, y_2 = \underline{\ ?\ }$ **4.** $x_2 = -3, y_1 = -4, y_2 = 6, x_1 = \underline{\ ?\ }$
5. $y_1 = 2, x_1 = 3, y_2 = -6, x_2 = \underline{\ ?\ }$ **6.** $r_1 = 25, s_1 = 6, r_2 = -25, s_2 = \underline{\ ?\ }$
7. $s_2 = -1, r_2 = -4, r_1 = 9, s_1 = \underline{\ ?\ }$ **8.** $r_2 = -8, s_1 = 5, s_2 = -10, r_1 = \underline{\ ?\ }$

Find the value of the variable for which each proportion is true.

9. $\dfrac{4}{7} = \dfrac{x-3}{x}$ **10.** $\dfrac{y-4}{y} = \dfrac{5}{9}$ **11.** $\dfrac{z}{z-16} = \dfrac{5}{3}$ **12.** $\dfrac{w}{w-3} = \dfrac{7}{4}$

13. $\dfrac{3p+2}{12} = \dfrac{5}{9}$ **14.** $\dfrac{3k-2}{27} = \dfrac{11}{6}$ **15.** $\dfrac{7}{10r+9} = \dfrac{2}{3r+2}$ **16.** $\dfrac{c-1}{c+3} = \dfrac{2}{3}$

B 17. If x varies directly as $y - 2$, and $x = 6$ when $y = -1$, find x when $y = 20$.

18. If x varies directly as $y + 4$, and $x = -2$ when $y = -5$, find y when $x = -16$.

19. If y is directly proportional to $2z + 3$, and $y = -14$ when $z = 2$, find y when $z = -3$.

20. If $s + 2$ varies directly as $3r - 1$, and $s = 4$ when $r = 1$, find s when $r = -2$.

21. If $2k - 1$ is directly proportional to $2t + 1$, and $k = 5$ when $t = -2$, find k when $t = -\frac{1}{2}$.

22. If v varies directly as r^3, and $v = \frac{9}{2}\pi$ when $r = \frac{3}{2}$, write the formula for v in terms of r.

23. If g varies directly as $\dfrac{h^2}{2}$, and $g = 1$ when $h = -2$, write the formula for g in terms of h.

24. Show that $(0, 0)$ and $(1, k)$ are ordered pairs of the function $f: x \rightarrow kx$, $k \neq 0$.

25. Show that $(0, 0)$, $(1, 2c)$, and $(-1, -2c)$ are ordered pairs of the function $g: x \rightarrow 2cx$, $c \neq 0$.

Find (a) a formula and (b) a proportion for the direct variation described.

EXAMPLE 1   The circumference of a circle varies directly as its radius. A circle whose radius is 1 cm has a circumference of 2π cm.

SOLUTION   Let C = circumference in centimeters and r = radius in centimeters. Then:

a. $C = kr$, $2\pi = k(1)$, $k = 2\pi$; $C = 2\pi r$ b. $\dfrac{C_1}{r_1} = \dfrac{C_2}{r_2}$

26. The total cost of a given number of kilograms of apples varies directly as the number of kilograms. One kilogram of apples costs $2.25.

27. The distance traveled at a constant speed of 90 km/h varies directly as the time traveled.

28. The yearly interest on a loan varies directly as the amount of the loan. A loan of 1 dollar calls for a yearly interest of 12 cents or 0.12 dollar.

29. Distance on a map varies directly with actual distance. On a certain map 0.5 cm represents 1 km.

30. Under constant pressure, the volume in cubic centimeters of a dry gas is directly proportional to its Kelvin temperature. The constant of proportionality for a certain gas is 1.5.

31. The resistance in wire is directly proportional to its length. For a certain type of wire, the constant of proportionality is 0.005.

In Exercises 32–35 prove that each equation is a true statement.

EXAMPLE 2   If $\dfrac{y_1}{x_1} = \dfrac{y_2}{x_2}$, and if $x_1,\ x_2 \neq 0$, then $y_1 x_2 = x_1 y_2$.

PROOF

1. $\dfrac{y_1}{x_1} = \dfrac{y_2}{x_2}$	1. Given
2. $(x_1 x_2)\left(\dfrac{y_1}{x_1}\right) = (x_1 x_2)\left(\dfrac{y_2}{x_2}\right)$	2. Multiplicative property of equality
3. $y_1 x_2 \left(\dfrac{x_1}{x_1}\right) = x_1 y_2 \left(\dfrac{x_2}{x_2}\right)$	3. Commutative and associative axioms of multiplication
4. $\therefore\ y_1 x_2 = x_1 y_2$	4. Identity axiom of multiplication

C   **32.** If $\dfrac{y_1}{x_1} = \dfrac{y_2}{x_2}$, and $x_1,\ x_2,\ y_2 \neq 0$, then $\dfrac{y_1}{y_2} = \dfrac{x_1}{x_2}$.

33. If $\dfrac{y_1}{x_1} = \dfrac{y_1}{x_2}$, and if $x_1,\ x_2,\ y_1 \neq 0$, then $x_1 = x_2$.

34. If $\dfrac{y_1}{x_1} = \dfrac{y_2}{x_2}$, and if $x_1,\ x_2 \neq 0$, then $\dfrac{y_1 + x_1}{x_1} = \dfrac{y_2 + x_2}{x_2}$.

35. If $\dfrac{y_1}{x_1} = \dfrac{y_2}{x_2}$, and if $x_1,\ x_2 \neq 0$, then $\dfrac{y_1 - x_1}{x_1} = \dfrac{y_2 - x_2}{x_2}$.

36. If $\dfrac{x + 2}{y - 2} = \dfrac{7}{4},\ \dfrac{x - 2}{y + 2} = \dfrac{3}{8}$, and $y \neq 2,\ y \neq -2,\ y \neq 0$, find the value of $\dfrac{x}{y}$.

Problems

Solve each problem. Assume that direct variation is involved in each case.

A   **1.** In a certain class election Rhonda received 6 out of every 8 votes cast. How many votes did she get if 320 students voted?

2. Twenty-seven grams of hydrochloric acid neutralize 30 g of lye. How many grams of lye are neutralized by 1620 g of hydrochloric acid?

3. To make ammonia, a chemist combines 28 parts by mass of nitrogen with 6 parts by mass of hydrogen. How much hydrogen must be combined with 224 g of nitrogen to produce ammonia?

4. When an electric current is 5 amperes (A), the electromotive force is 25 volts (V). Find the force when the current is 95 A.

5. How long is a 25-kg roll of wire which is 0.9 kg/m?

6. A mass of 24 kg causes a beam to bend 4 cm. If the amount of bending varies directly as the mass, what mass will cause the beam to bend 7 cm?

7. A mass of 15 g stretches a spring 6 cm. If the distance a spring is stretched is directly proportional to the mass, what mass will stretch the spring 10 cm?

8. If a credit union paid a total dividend of $28 on 200 shares of its stock, how much of a dividend did it pay on 350 shares?

9. If 0.75 m² of a fabric has a mass of 840 g, what is the mass of 1.5 m² of the same fabric?

10. To prepare a nutrient broth for bacteria, a laboratory technician adds to distilled water 15 g of peptone for every 9 g of beef extract. If she used 90 g of beef extract in making several batches of the broth, how much peptone did she use?

B 11. The distance from an observer to a lightning bolt is directly proportional to the time elapsed from the instant the observer sees the bolt to the instant she hears the thunder. If an observer hears the thunder 7 s after she sees the bolt, then the bolt is approximately 2.4 km away. If an observer hears thunder 9 s after seeing the bolt, approximately how far away is the bolt?

12. The mass of a uniform iron rod varies directly as its length. If a rod 60 cm long has a mass of approximately 675 g, find the approximate mass of a rod whose length is 234 cm.

13. A certain precious jewel's price varies as the square of its mass. If a jewel has a mass of 1.75 g and is worth $825, find the cost of a similar jewel with a mass of 3.25 g.

14. On a map, 1 cm represents an actual distance of 80 m. Find the actual area of a parcel of land which is represented on the map by a rectangle measuring 13.75 cm by 17 cm.

15. The surface area of a human body is directly proportional (approximately) to the square of the person's height. How many times as much skin does a giant 3 m tall have compared to a person 2 m tall?

16. The distance it takes to stop after you have applied the brakes of a car is directly proportional to the square of the speed of the car. A car going 64 km/h stops in 27 m. If you are going 90 km/h, how far will the car travel after you have applied the brakes?

17. An architect is designing a house which will be 20 m long and 16 m wide. A scale drawing of the floor plan is 30 cm × 24 cm. In the drawing, find the dimensions representing a 4.2 m × 5.5 m dining room.

C 18. The average number of red cells in 1 mm³ of blood is 4,500,000. In order to conduct a test on a sample of blood, a laboratory technician dilutes 1 part of blood with 299 parts of a salt solution. How many red blood cells would the technician expect to find in 0.75 mm³ of the diluted solution?

19. The heat loss due to radiation from a hot metal is directly proportional to the fourth power of the temperature. If the temperature is tripled, how many times as much is the resulting radiation?

20. A certain ocean liner displaces approximately 45,000 t and is almost 300 m long. For a 45-kg model of the ocean liner what should be the scale and the length, if the mass varies directly as the cube of the length?

21. The surface area of a sphere varies directly as the square of its radius. If the surface is 64π m² when the radius is 4 m, what is the surface area when the radius is 9 m?

22. If x varies directly as y, and y varies directly as z, prove that x varies directly as z.

Self-Test 2

VOCABULARY   solution of an open sentence in two variables (p. 201)
linear equation in two variables (p. 206)
direct variation (p. 211)   constant of variation (p. 211)

1. Find the graph and the solution set of   *Obj. 1, p. 201*
$$2x - y = -1$$
if $x \in \{-1, 0, \frac{1}{4}, \frac{1}{2}\}$ and $y \in$ {the positive integers}.

2. Graph $-2x + y = -1$ in a coordinate plane.   *Obj. 2, p. 201*

3. If w is directly proportional to p, and $w = -3$ when $p = 6$, what   *Obj. 3, p. 201*
is the value of w when $p = -8$?

4. A photograph measuring 4 cm high and 3 cm wide is to be enlarged. The height of the enlargement is to be 6 cm. How wide should it be to keep the same proportions as the original?

Check your answers with those at the back of the book.

Lines in a Coordinate Plane

***OBJECTIVES** for Sections 6-7 through 6-10:*

1. *Find the slope and y-intercept of a nonvertical line given its equation.*
2. *Graph, and find an equation of a nonvertical line given its slope and y-intercept.*
3. *Find an equation of a line given the slope of the line and the coordinates of a point on it.*
4. *Find the slope and an equation of a line given the coordinates of two points on the line.*
5. *Graph linear inequalities in two variables.*

6-7 Slope and y-Intercept of a Line

Like many other everyday terms, the familiar word "slope," as in "ski slope," has a special mathematical meaning. To discover that meaning, let us look at some examples.

The table below shows the coordinates of a few of the points (Figure 18) on the line

$$y = 2x + 3.$$

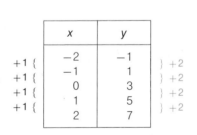

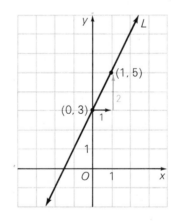

Figure 18

Notice that when the x-coordinates of two points on the line differ by 1, their y-coordinates differ by 2. We call the number 2 the *slope* of the line.

If you study the graph of the equation

$$y = 4x + 3$$

(Figure 19), you see that when the x-coordinates of two points on that line differ by 1, their y-coordinates differ by 4. The number 4 is the slope of the line.

x	y	
−2	−5	} +4
−1	−1	} +4
0	3	} +4
1	7	} +4
2	11	

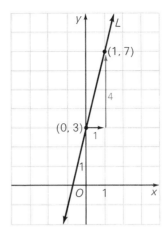

Figure 19

Notice that the numbers 2 and 4, the slopes of the lines in Figures 18 and 19, indicate the "steepness" of the graphs.

In general, if L is the graph of

$$y = mx + b,$$

then we call m the **slope** of the line. If you move from one point on L to another whose x-coordinate is 1 more (Figure 20), then the y-coordinate changes by m.

Notice that the coordinates of the point where L crosses the y-axis are $(0, b)$. The y-coordinate of this point, b, is called the **y-intercept** of the line.

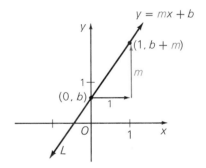

Figure 20

EXAMPLE 1   Find the slope and y-intercept of the graph of $3x + 4y = 8$.

SOLUTION   Transform the equation into an equivalent one in slope-intercept form, $y = mx + b$, and then read the values of the slope m and the y-intercept b.

$$3x + 4y = 8$$
$$3x + 4y - 3x = 8 - 3x$$
$$4y = 8 - 3x$$
$$\frac{4y}{4} = \frac{8 - 3x}{4}$$
$$y = 2 - \tfrac{3}{4}x$$
$$y = -\tfrac{3}{4}x + 2$$

∴ the slope is $-\tfrac{3}{4}$ and the y-intercept is 2.   **Answer.**

EXAMPLE 2   Draw the line with slope -3 and y-intercept -2; then find an equation for the line.

SOLUTION   1. The y-intercept is -2. Therefore, you plot the point $(0, -2)$.

2. Since the slope is -3, you move from $(0, -2)$ 1 unit to the right and 3 units down to locate a second point on the line.

3. Draw the line containing the two points.

4. $y = mx + b$
$\quad\quad\;\downarrow\quad\quad\;\downarrow$
$y = -3x + (-2)$, or $3x + y = -2$.

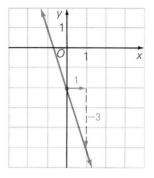

Figures 18 and 19 and the diagram accompanying Example 2 illustrate this fact: Lines that rise from left to right have positive slope, and lines that fall from left to right have negative slope.

The equation of any horizontal line (Figure 21) is equivalent to one of the form

$$y = b, \quad \text{or} \quad y = 0 \cdot x + b.$$

Thus, *the slope of every horizontal line is* 0.

Figure 21

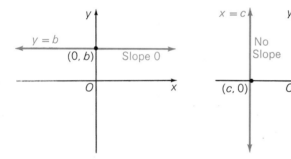

Figure 22

Any vertical line (Figure 22) has an equation of the form

$$x = c, \quad \text{or} \quad 0 \cdot y = x - c.$$

You cannot transform such an equation into slope-intercept form, because you cannot divide by 0. Thus, *vertical lines have no slope.*

Oral Exercises

State the slope and y-intercept (if any) of the line whose equation is given.

1. $y = 3x - 2$ **2.** $y = x + 3$ **3.** $y = -4x - 1$ **4.** $y = -\frac{2}{3}x$

5. $y + 2 = 0$ **6.** $x - 3 = 0$ **7.** $2y = 8x - 4$ **8.** $7 + 5x = y$

Written Exercises

In Exercises 1–6, P and Q are points on a line with slope m. Find the y-coordinate of Q.

EXAMPLE $P(4, 6)$, $Q(6, \underline{?})$; $m = -4$

SOLUTION   Note that: x-coordinate of $Q = x$-coordinate of $P + 2$

since $6 = 4 + 2$.

$\therefore y$-coordinate of $Q = y$-coordinate of $P + 2m$.

$6 + 2(-4) = -2$.   Answer.

A

1. $P(0, 0)$, $Q(2, \underline{?})$; $m = 4$ **2.** $P(0, -4)$, $Q(1, \underline{?})$; $m = -6$

3. $P(3, 6)$, $Q(2, \underline{?})$; $m = 9$ **4.** $P(7, -4)$, $Q(-5, \underline{?})$; $m = -\frac{1}{2}$

5. $P(-2, -1)$, $Q(2, \underline{?})$; $m = 0$ **6.** $P(1, -3)$, $Q(7, \underline{?})$; $m = \frac{5}{6}$

a. Graph the line whose slope and y-intercept are given.
b. Write an equation of the line. Give your answer in the form $Ax + By = C$ where A, B, and C are integers.

7. $m = 3$, $b = 2$ **8.** $m = -4$, $b = -1$ **9.** $m = -2$, $b = -3$ **10.** $m = 6$, $b = -2$

11. $m = \frac{1}{4}$, $b = 1$ **12.** $m = -\frac{2}{3}$, $b = 0$ **13.** $m = 0$, $b = -3$ **14.** $m = -\frac{7}{4}$, $b = -\frac{1}{2}$

6-8 Determining an Equation of a Line: Slope and One Point

Given the slope of a line and the coordinates of any point through which it passes, you can use the slope-intercept form of a linear equation to find an equation of the line.

EXAMPLE 1   Find an equation of the line through the point $(3, 1)$ with slope $-\frac{7}{4}$.

SOLUTION   1. The slope-intercept form of the equation of the line is

$$y = -\tfrac{7}{4}x + b.$$

2. Since the point $(3, 1)$ lies on the line, its coordinates must satisfy this equation:

$$1 = -\tfrac{7}{4}(3) + b$$
$$1 = -\tfrac{21}{4} + b$$
$$\tfrac{25}{4} = b$$

3. $\therefore$ an equation of the line is

$$y = -\tfrac{7}{4}x + \tfrac{25}{4}. \quad \text{Answer.}$$

An equivalent form of this equation is $7x + 4y = 25$.

EXAMPLE 2 Find an equation of the vertical line through the point $(2, -3)$.

SOLUTION The x-coordinate of every point on this vertical line must be the same as the x-coordinate of $(2, -3)$. Thus, an equation of the line is

$$x = 2. \quad \text{Answer.}$$

Written Exercises

Find an equation of the line having the given slope (if any) and passing through the given point.

A
1. $1; (2, 4)$
2. $3; (4, 2)$
3. $-2; (-3, 0)$
4. $-1; (2, -1)$

5. $\frac{3}{4}; (-5, 0)$
6. $-\frac{1}{2}; (-1, -3)$
7. $0; (-\frac{1}{2}, 2)$
8. Vertical line (no slope); $(-5, -\frac{1}{2})$

9. $-\frac{2}{3}; (0, \frac{1}{2})$
10. $\frac{5}{9}; (0, 0)$

Determine the value of k so that the graph of the given equation contains the given point.

B
11. $2x + ky = 4; (3, 1)$
12. $kx + 3y = 5; (1, 4)$

13. $4x - ky = 2; (-3, -2)$
14. $5x - 2y + k = 0; (-1, 3)$

15. $kx - 5 = -2y; (-1, 2)$
16. $3y - 1 = -kx; (4, -1)$

17. $7x + ky = -3; (-3, -9)$
18. $4x - ky + 6 = 0; (-1, 1)$

19. $5x - 6y = k; (4, 3)$
20. $x + k = 4y; (-5, -1)$

Determine an equation of the line satisfying the given conditions.

21. With y-intercept 2 and having the same slope as the graph of $y = 2x + 4$.

22. With y-intercept -1 and having the same slope as the graph of $6y - 3x = 12$.

23. Through $(6, -2)$ and having the same slope as the graph of $4y - 3x = -8$.

24. Through $(-2, -3)$ and having the same slope as the graph of $2x - 3y = -3$.

25. Through $(-1, -2)$ and vertical.

26. Through $(-4, 6)$ and having the same slope as the x-axis.

Find an equation of the line that contains the given point and is parallel to the given line. (Two lines in a plane are parallel if they do not intersect. Parallel lines have the same slope.)

27. $(2, 4); 2x + y = 6$
28. $(0, -2); x - y = -4$

29. $(2, -3); x + 2y = 6$
30. $(-6, -2); x - 3y = 6$

Find the value of r and the value of s for which the graphs of the given equations are the same line.

C 31. $6x - 2y = -r$; $sx - y = -1$ 32. $2s - rx = y$; $y = 3x - 10$

33. Show that the y-intercept of the line with slope m through a given point (x_1, y_1) is $y_1 - mx_1$.

34. Use the result of Exercise 33 to show that an equation of the line with slope m through the point (x_1, y_1) is $y - y_1 = m(x - x_1)$.

35. If (x, y), (x_1, y_1), and (x_2, y_2) are coordinates of three points on a line, show that:

$$\frac{y_1 - y_2}{x_1 - x_2} = \frac{y - y_1}{x - x_1}.$$

6-9 Determining an Equation of a Line: Two Points

The points $P(-2, -7)$ and $T(4, 5)$ lie on the line

$$y = 2x - 3,$$

whose slope is 2 (Figure 23).

    What are the differences between the ordinates (y-coordinates) of P and T and the abscissas (x-coordinates) of P and T? What is the quotient of these differences?

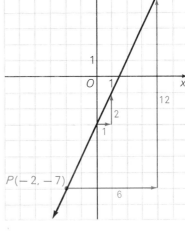

1. ordinate of T − ordinate of P
$$= 5 - (-7) = 12$$

2. abscissa of T − abscissa of P
$$= 4 - (-2) = 6$$

3. $\dfrac{\text{ordinate of } T - \text{ordinate of } P}{\text{abscissa of } T - \text{abscissa of } P}$

$$= \tfrac{12}{6} = 2 \qquad\qquad \textit{Figure 23}$$

Notice that the quotient computed in Step 3 is the *slope* of the line. This example illustrates the following fact.

> If P and T are any two different points of a nonvertical line, then a formula for the slope m of the line is:
>
> $$m = \frac{\text{ordinate of } T - \text{ordinate of } P}{\text{abscissa of } T - \text{abscissa of } P}.$$

You can use this to find an equation of a line through two given points.

EXAMPLE Find an equation of the line through the points (4, 1) and (1, −2).

SOLUTION Let P be the point (4, 1) and T be (1, −2).

1. Find the slope: $m = \dfrac{-2-1}{1-4} = \dfrac{-3}{-3} = 1$

2. The slope-intercept form of the equation is:
$$y = mx + b$$
$$y = 1 \cdot x + b, \quad \text{or} \quad y = x + b$$

3. Choose one of the points, say $P(4, 1)$. Since it lies on the line:
$$1 = 4 + b$$
$$-3 = b$$
Thus $y = x + (-3)$ is an equation of the line.

4. To check, show that the coordinates of the other point, $T(1, -2)$, satisfy the equation:
$$y = x - 3$$
$$-2 \overset{?}{=} 1 - 3$$
$$-2 = -2 \ \checkmark$$
$\therefore$ an equation of the line is $y = x - 3$. **Answer.**

In solving the preceding example, you could have given P the coordinates (1, −2) and T the coordinates (4, 1). You should check that the slope still turns out to be 1.

Oral Exercises

Name the slope m (if any) of each line pictured.

EXAMPLE

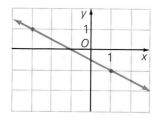

SOLUTION

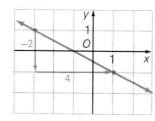

Counting squares as shown, you find:
$$m = \frac{-2}{4} = -\frac{1}{2}.$$

1.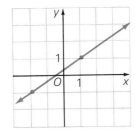

2.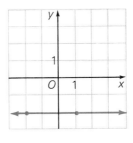

3.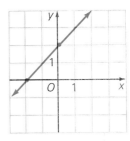

4.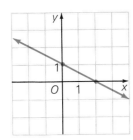

5.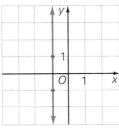

6.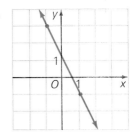

7. Describe the graph of a line having slope $-\frac{2}{3}$.

8. Describe the graph of a line having slope 0 and y-intercept -1.

Written Exercises

Find the slope of the line joining the given points.

A 1. $(1, 4)$, $(4, 7)$ 2. $(2, 3)$, $(2, 1)$ 3. $(5, -2)$, $(-3, 2)$

4. $(-4, -2)$, $(0, 0)$ 5. $(5, 6)$, $(3, 2)$ 6. $(-1, 3)$, $(2, 3)$

7. $(6, -5)$, $(1, 10)$ 8. $(-5, -3)$, $(0, 0)$ 9. $(-3, 4)$, $(3, -4)$

10–18. Find an equation of the line through the points given in Written Exercises 1–9.

Find the linear function f with domain $\mathfrak{R}$ satisfying the given conditions.

EXAMPLE $f(1) = -3$, $f(-2) = 5$

SOLUTION The graph of f is the line through $(1, -3)$ and $(-2, 5)$. Thus, using the method of the example on page 224, you can show that the equation of the graph is

$$y = -\tfrac{8}{3}x - \tfrac{1}{3}.$$

$\therefore f = \{(x, y): y = -\tfrac{8}{3}x - \tfrac{1}{3}\}.$ **Answer.**

B 19. $f(0) = -2$, $f(2) = 12$ 20. $f(0) = -1$, $f(-3) = 6$

21. $f(-2) = -1$, $f(1) = 2$ 22. $f(-3) = 5$, $f(2) = -6$

23. $f(-1) = -1$, $f(3) = -3$ 24. $f(0) = -\tfrac{1}{2}$, $f(\tfrac{1}{4}) = -2$

25. $f(-3) = 5$ and the slope of the graph of f is $\frac{1}{2}$.

26. $f(-2) = -3$ and the slope of the graph of f is $-\frac{2}{3}$.

27. $f(-4) = 0$ and the slope of the graph of f is 0.

The vertices of a right triangle are given. Find the slopes of the three sides.

28. $(-1, 0), (-4, 1), (1, 6)$ **29.** $(5, 0), (-5, -2), (1, -6)$

30. $(1, 4), (3, 0), (-5, 1)$ **31.** $(0, 4), (-2, -2), (-5, -1)$

C   **32.** Show that if (x_1, y_1) and (x_2, y_2) are different points on the graph of $y = 2x + 1$, then

$$\frac{y_2 - y_1}{x_2 - x_1} = 2.$$

33. Show that if (x_1, y_1) and (x_2, y_2) are different points on the graph of $y = mx + b$, then

$$m = \frac{y_2 - y_1}{x_2 - x_1}.$$

34. Use the result of Exercise 33, above, and Exercise 34, page 223, to show that an equation of the line through two different points (x_1, y_1) and (x_2, y_2) [not on the same vertical line] is

$$y - y_1 = \frac{y_2 - y_1}{x_2 - x_1}(x - x_1).$$

35. Use the result of Exercise 33, above, to show that if f is any linear function and x_1 and x_2 are two different real numbers, then

$$\frac{f(x_2) - f(x_1)}{x_2 - x_1}$$

is the slope of the graph of f.

36. If a function f is a direct variation, prove that for any real numbers x and y, $f(x + y) = f(x) + f(y)$.

programming in BASIC

In BASIC you may also use variables such as X1, X2, where a single letter is followed by a single digit.

You may also INPUT several values on one line. Since the computer will print only one question mark, it is useful to precede the INPUT statement with a PRINT statement reminding you of the values you are to INPUT.

The following program for finding the slope of a line through two points (X1, Y1) and (X2, Y2) uses these ideas:

```
10   PRINT "INPUT X1, Y1"
20   INPUT X1, Y1
30   PRINT "INPUT X2, Y2"
40   INPUT X2, Y2
50   IF X1=X2 THEN 80
60   PRINT "SLOPE ="; (Y2−Y1)/(X2−X1)
70   STOP
80   PRINT "NO SLOPE"
90   END
```

Notice that instead of using 70 GOTO 90 we have used the simpler statement 70 STOP.

The computer will print the value of the slope as a decimal, correct to 6 digits.

Exercise

1. Write a program that will accept the coordinates of four points (X1, Y1), (X2, Y2), (X3, Y3), and (X4, Y4) and tell you whether or not the line through (X1, Y1) and (X2, Y2) is parallel to the line through (X3, Y3) and (X4, Y4). What exceptional cases do you have to watch out for in your program?

6-10 The Graph of a Linear Inequality in Two Variables

A line like the graph of

$$y = -\tfrac{1}{2}x + 1$$

separates the rest of the coordinate plane into two regions called half-planes, one above the line (colored shading in Figure 24) and the other below it (gray shading). The line itself is called the boundary (or bounding line) of each half-plane.

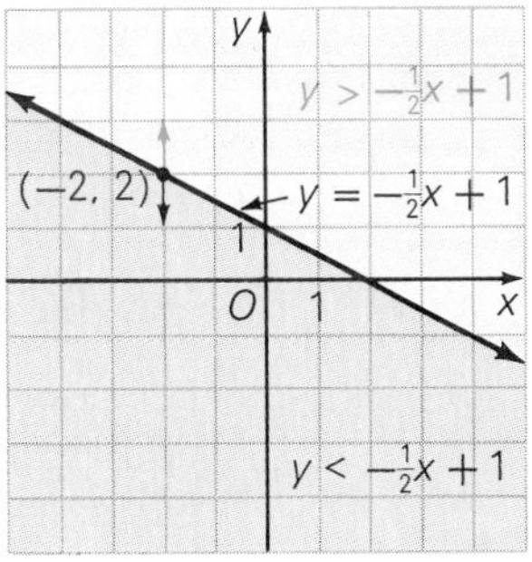

If you start at any point of the line, say $(-2, 2)$, and move vertically above it, the y-coordinates of points increase. Thus, the half-plane above the line is the graph of

$$y > -\tfrac{1}{2}x + 1.$$

Figure 24

Because it does *not* include its boundary, this half-plane is called an open half-plane.

If you move vertically downward from any point of the line, the y-coordinates of points decrease. Thus, the open half-plane below the line is the graph of

$$y < -\tfrac{1}{2}x + 1.$$

Colored shading in Figure 25 shows the graphs of the four inequalities:

a. $x < 1$ b. $x \leq 1$ c. $x > 1$ d. $x \geq 1$

Notice that a dashed line is used for the boundary in parts (a) and (c) of the figure to show that the graphs are open half-planes. In parts (b) and (d), however, the graphs include the boundary line and are called **closed half-planes.** Drawing the boundary as a solid line shows this.

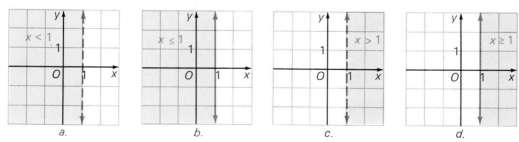

Figure 25

Given any linear equation in two variables

$$ax + by = c,$$

there are four **linear inequalities in two variables** associated with it. The following chart describes their graphs.

Open Sentence	Graph
$ax + by = c$	A straight line
$ax + by > c$	An open half-plane
$ax + by \geq c$	A closed half-plane
$ax + by < c$	An open half-plane
$ax + by \leq c$	A closed half-plane

EXAMPLE   Graph the inequality $3x - y > 0$.

SOLUTION   1. Transform the given inequality into an equivalent one (page 148) having y as a left member:

$$3x - y > 0$$
$$3x > y$$
$$y < 3x$$

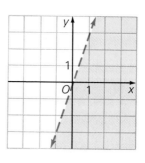

2. Graph $y = 3x$ and show it as a *dashed* line.

3. Shade the half-plane *below* the line.

228 | *Chapter 6*

Oral Exercises

Transform each open sentence into an equivalent one having y as one member.

1. $4x + y < -4$
2. $-2x + y \geq 1$
3. $y + x > 0$
4. $3y < x$
5. $3x + 5y \leq 0$
6. $2x - y < -2$
7. $0 < y - \frac{x}{3}$
8. $4x < 5 - 2y$
9. $8x + 4y < -12$

Name which, if any, of the given points belong to the graph of the given inequality.

10. $y - x \leq 0$; $(-3, 2)$, $(1, -2)$
11. $2y + x \geq 4$; $(-1, -3)$, $(-2, 0)$
12. $x - y > 0$; $(6, 7)$, $(-3, -7)$
13. $y - 3x > 5$; $(-2, 0)$, $(5, -10)$

Written Exercises

Graph each inequality.

A
1. $y < 4$
2. $y > -2$
3. $x > -3$
4. $x < 0$
5. $y \leq x$
6. $y \geq 2x$
7. $x + y > 0$
8. $y - x < 0$
9. $y \geq x - 1$
10. $y \leq 2 - x$
11. $3y + x > 6$
12. $y + 2x \leq 3$
13. $4x - 2y \leq 10$
14. $7x - 5y > -15$
15. $8y + 3x \leq 16$
16. $2x - 3y \leq 9$
17. $3y \geq 2x - 9$
18. $3y - 2x < -9$

Self-Test 3

VOCABULARY   slope-intercept form of a linear equation (p. 219)
open and closed half-planes (pp. 227–228)
boundary of a half-plane (p. 227)

1. Find the slope and y-intercept of the graph of $3x + 2y = -4$.   *Obj. 1, p. 218*
2. Graph, and find an equation of, the line having slope -3 and y-intercept 2. Give your answer in the form "$Ax + By = C$."   *Obj. 2, p. 218*
3. Find an equation of the line which   *Obj. 3, p. 218*
   a. has slope -1 and passes through the point $(2, -4)$.
   b. is vertical and contains the point $(-2, 2)$.
4. Find the slope and an equation of the line passing through the points $(-4, -2)$ and $(3, 5)$.   *Obj. 4, p. 218*
5. Graph the inequality $2y + x \leq 4$.   *Obj. 5, p. 218*

Check your answers with those at the back of the book.

Transformations of the Plane: Translations

The figure at the left below shows coordinate axes in the plane, together
with a square with vertices (1, 1), (2, 1), (2, 2), and (1, 2).

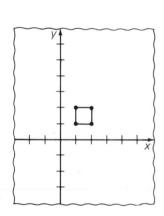

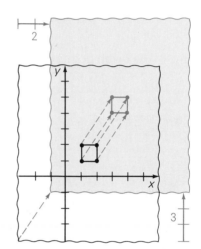

Now, imagine that the axes remain fixed while the plane, like a sheet
of glass, slides rigidly into a new position, such as the one shown at the
right above. Do you see that the new coordinates of the vertices of the
square after this "sliding" are (3, 4), (4, 4), (4, 5), and (3, 5)?

In fact, every point in the plane now has a new pair of coordinates,
(x', y'), where x' is 2 greater than the original x-coordinate and y' is 3
greater than the original y-coordinate. Thus for every point (x', y'), you
have

$$x' = x + 2 \quad \text{and} \quad y' = y + 3.$$

Such a "sliding" of the plane is an example of a **transformation,** or
mapping, of the plane called a **translation.** In a translation of the plane,
if the point originally coinciding with the origin slides to a new location
with coordinates (h, k), then every point in the plane has new coordinates (x', y') which are related to the original coordinates (x, y) by the
equations of translation

$$x' = x + h \quad \text{and} \quad y' = y + k.$$

EXAMPLE A translation maps the point with coordinates (5, 4) into the point with
coordinates (8, −2).

    a. Find the equations of translation.

    b. Find the new coordinates of the point with old coordinates (−4, 0).

SOLUTION   **a.** Let $(x, y) = (5, 4)$ and $(x', y') = (8, -2)$, and substitute in the general translation equations

$$x' = x + h \qquad \text{and} \qquad y' = y + k.$$
$$8 = 5 + h \qquad \qquad -2 = 4 + k$$
$$h = 3 \qquad \qquad k = -6$$

∴ the equations of translation are $x' = x + 3$ and $y' = y - 6$.

Answer.

b. Use the equations $x' = x + 3$ and $y' = y - 6$, and substitute -4 for x and 0 for y.

$$x' = -4 + 3 \qquad \qquad y' = 0 - 6$$
$$x' = -1 \qquad \qquad y' = -6$$

∴ the new coordinates of $(-4, 0)$ are $(-1, -6)$.   Answer.

The following result is proved in more advanced courses.

Under a translation:

1. Every line is mapped (translated) into a line parallel to the original line.

2. Every line segment is mapped (translated) into a line segment of equal length.

3. Every angle is mapped (translated) into an angle of equal measure.

Exercises

In Exercises 1–8 (a) find equations of translation that map (translate) the first point into the second, and (b) find the new coordinates under this translation of the point whose original coordinates are $(-7, 3)$.

1. $(5, 5), (6, 2)$ 2. $(3, 8), (5, 1)$

3. $(-7, 2), (6, -1)$ 4. $(3, -2), (8, 1)$

5. $(-7, 0), (0, 2)$ 6. $(0, -3), (5, 0)$

7. $(-5, -2), (4, -7)$ 8. $(-6, -3), (-7, 1)$

9. Under what kind of translation of the plane will (a) a vertical line be mapped into itself? (b) a horizontal line be mapped into itself?

10. Use the slope formula (page 223) to show that the slope of the line segment with endpoints (a, b) and (c, d), with $c \neq a$, is equal to the slope of the line segment with endpoints $(a + h, b + k)$ and $(c + h, d + k)$, and hence prove that under a translation a nonvertical line segment is mapped into a parallel line segment.

Chapter Summary

1. A *coordinate plane* can be set up using two perpendicular number lines whose point of intersection is the origin of each. One line, called the *x*-axis, is usually horizontal and the other, the *y*-axis, is usually vertical.

2. Each point in the plane is assigned a unique pair of coordinates. The first coordinate is called the *x-coordinate*, or *abscissa*, and the second coordinate is called the *y-coordinate*, or *ordinate*. Conversely, each pair of coordinates is assigned a unique point in the coordinate plane.

3. A *relation* is any set of ordered pairs of elements. The set of first coordinates of the ordered pairs is the *domain* of the relation; the set of second coordinates is the *range*.

4. A *function* is a relation in which each element of the domain is paired with exactly one element of the range.

5. **a.** Arrow notation such as $f: x \to 5 - 3x$ is used for functions to show that the function f assigns $5 - 3x$ to each value of x in the domain of f.
   b. If $f: x \to 5 - 3x$, the notations "$f(0) = 5$" and "$f(3) = -4$" are used to show that the value of f at 0 is 5 and the value of f at 3 is -4.
   c. If $f(a) = b$, then (a, b) is one of the number pairs of f.

6. A *solution* of an open sentence in two variables is an ordered pair of values of the variables for which the sentence is a true statement. The set of all such ordered pairs is called the *solution set* over the given replacement sets of the variables.

7. For all real numbers a, b, and c, with a and b not both zero, any equation which can be written equivalently in the form

$$ax + by = c$$

is called a *linear equation in two variables*. If the replacement set of each variable is $\Re$, the graph of such an equation in a coordinate plane is a straight line.

8. A *linear direct variation* is a function f such that

$$f: x \to kx,$$

where k is a nonzero constant; k is called the *constant of variation*, or the *constant of proportionality*. In a *proportion*, the product of the *means* equals the product of the *extremes*.

9. The *slope* of a line is the change in the *y*-coordinate as you move from one point on the line to another whose *x*-coordinate is one more than the *x*-coordinate of the first point. If L is the graph of "$y = mx + b$," then m is the slope of the line L.

The ratio

$$\frac{\text{ordinate of } T - \text{ordinate of } P}{\text{abscissa of } T - \text{abscissa of } P},$$

where T and P are points on a line, also yields the slope m of the line.

10. A line with slope m and y-intercept b is the graph of the equation $y = mx + b$. This *slope-intercept form* of a linear equation is very useful in finding an equation of a line from information given about the line, and vice versa.

11. Any linear equation in two variables of the form $ax + by = c$ has four *linear inequalities in two variables* associated with it. The graph of each of these inequalities is a half-plane.

Chapter Review

1. For the point $P(-2, 5)$, -2 is called the: 6-1

   a. slope   b. y-intercept   c. abscissa   d. ordinate

2. The point $P(5, -3)$ is in Quadrant _?_.

   a. I   b. II   c. III   d. IV

3. For all points on the y-axis, the x-coordinate is _?_.

   a. positive   b. 0   c. negative   d. undefined

4. Which of the following is *not* a function?   6-2

   a. $\{(-3, 2), (-1, 1), (1, 1), (3, 2)\}$
   b. $\{(2, 2), (3, 3), (4, 4)\}$
   c.   d.

5. The domain of the relation $\{(-3, 9), (0, 0), (2, 4)\}$ is _?_.

   a. $\{-3, 0, 2\}$ b. $\{9, 0, 4\}$ c. $\{-3, 0, 2, 4, 9\}$ d. $\mathcal{R}$

6. Which function is graphed below?   6-3

   a. $f: x \rightarrow 2x - 3, x \in \mathcal{R}$

   b. $f: x \rightarrow \dfrac{x + 3}{2}, x \in \{-3, -1, 1, 3\}$

   c. $f: x \rightarrow -3 + 2x, x \in \{\text{the integers}\}$
   d. $f: x \rightarrow 2x - 3, x \in \{0, 1, 2, 3\}$

   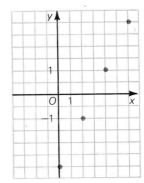

7. For the function $f\colon x \to 4 - x^2$, $x \in \{-1, 0, 1, 2\}$, which statement is *not* true?

   a. $f(1) = f(-1)$ **b.** $f(2) < f(1)$

   c. If $f(a + 3) = 4$, then $a = -3$.   **d.** $f(0) = f(2)$

8. Which ordered pair is *not* a member of the solution set of the open sentence $2y - 3x \geq 9$?   *6-4*

   a. $(-1, 3)$ **b.** $(1, 6)$ **c.** $(2, 7)$ **d.** $(-2, 2)$

9. The equation $3x + 2y = 6$ is equivalent to:

   a. $y = -3x + 6$ **b.** $y = -\frac{3}{2}x + 3$

   c. $y = -\frac{3}{2}x + 6$ **d.** $y = \frac{3}{2}x + 3$

10. If the graphs of $y = 2x - 5$ and $-3x + y = 6$ cross the x-axis at points A and B respectively, then: *6-5*

   a. A is to the right of B.   **b.** A is to the left of B.

   c. A is above B.   **d.** A is below B.

11. w varies directly as h, and w is -2 when h is 6. Then when h is -9, w is ___?___.   *6-6*

   a. $\frac{4}{3}$ **b.** 3   **c.** 27   **d.** -3

12. The value of x for which the proportion $\dfrac{x + 8}{x} = \dfrac{5}{9}$ is true is ___?___.

   a. 10   **b.** -2 **c.** $\frac{36}{7}$ **d.** -18

13. A line with slope $-1\frac{1}{2}$ which passes through the point $(-1, 4)$ also passes through the point ___?___.   *6-7*

   a. $(3, -2)$ **b.** $(2, 2)$ **c.** $(1, 7)$ **d.** $(-3, 1)$

14, 15. Write the letter of the equation which has been graphed in each diagram below.

   a. $y = -2$ **b.** $2y - x = 2$

   c. $2x - y = 1$ **d.** $x = -2$

14. 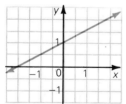   **15.**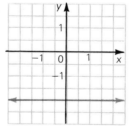

16. A line through $(3, 5)$ has slope $\frac{1}{2}$. An equation of this line is ___?___.   *6-8*

   a. $y = \frac{1}{2}x + 5$ **b.** $y = \frac{1}{2}x + 6\frac{1}{2}$

   c. $y = \frac{1}{2}x - 1$ **d.** $2y = x + 7$

17. An equation for the horizontal line through $(-2, -4)$ is ___?___.

   a. $y = -4$ **b.** $y = x - 4$ **c.** $x = -2$ **d.** $y = 2x$

18. A line passes through $(-2, 2)$ and $(10, -1)$. The y-intercept of this line is ___?___.

6-9

   a. 2   **b.** $\frac{5}{2}$ **c.** $\frac{3}{2}$ **d.** 6

19. Which open sentence is graphed here?

6-10

   a. $x - 2y > 4$
   b. $x - 2y \leq 4$
   c. $2x - y \geq 2$
   d. $x - 2y \geq 4$

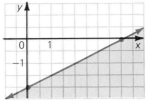

Chapter Test

1. Graph the following ordered pairs of numbers in a coordinate plane.

6-1

   a. $(-2, -3)$ **b.** $(5, 0)$ **c.** $(-4, 2)$ **d.** $(4, -2)$

2. State the domain and range of each relation. Is the relation a function?

6-2

   a. $\{(-4, 2), (-2, 1), (2, -1), (4, -2)\}$ **b.**

3. For the function $f: x \rightarrow 3x - 5$, $x \in \{-2, -1, 0, 1, 2\}$:

6-3

   a. Complete the table.
   b. Graph the function.

x	-2	-1	0	1	2
$f(x)$					

4. Name four ordered pairs belonging to the set $\{(x, y): y < x + 1\}$ if $x, y \in \{0, 1, 2\}$.

6-4

5. Is the equation $xy + x = 3$ linear?

6-5

6. Graph the following equations in a coordinate plane.
   a. $2x - y = 5$ **b.** $y + 4 = 1$

7. m varies directly as n, and m is 10 when n is -5.

6-6

   a. Find m when n is 3.   **b.** Find n when m is 11.

8. The slope of the line with equation $2y = -4x + 3$ is ___?___. The y-intercept of this line is ___?___.

6-7

9. A line with slope 3 contains the point $(-2, -4)$. Find the coordinates of three more points on this line.

10. Find an equation of the line passing through the point $(-3, 4)$ and having slope $\frac{2}{3}$.

6-8

11. Find an equation of the line through $(2, -1)$ which is parallel to the line $2x + y = 5$.

12. Find an equation of the line through the points $(-2, 4)$ and $(-6, 6)$.

6-9

13. Graph $3x - 4y > 12$.

6-10

Functions, Relations, and Graphs | **235**

At the top left of the organ's keyboard is the control panel of a microprocessor register control system, which allows up to 40 register combinations to be stored. Stops for whole recitals can be stored, leaving the organist free to concentrate on the keys and pedals.

7

Systems of Open Sentences

Solving Systems of Linear Equations in Two Variables

OBJECTIVES *for Sections 7-1 through 7-4:*
1. *Determine whether a system of two linear equations in two variables is consistent by comparing the graphs of the equations and the slope-intercept forms of the equations.*
2. *Determine the solution set of a system of two linear equations in two variables by using graphs, linear combinations, and substitution.*

7-1 Using Graphs

The three parts of Figure 1 show how the graphs of two linear equations in two variables can be related to each other. The possibilities are:

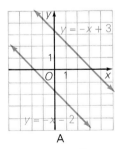

A

$y = -x + 3$
$y = -x - 2$

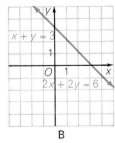

B

$x + y = 3$
$2x + 2y = 6$

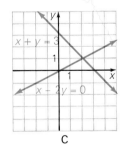

C

$x + y = 3$
$x - 2y = 0$

Figure 1

A. The graphs are *parallel* lines. (**Parallel lines** are lines that lie in the same plane, but have *no* point in common.)

B. The graphs are *coincident* lines. They have *all* their points in common.

C. The graphs are *intersecting* lines. (In this book we shall consider intersecting lines to be *different*, that is, *noncoincident*, lines that intersect. Thus, intersecting lines have *exactly one* point in common, called their **point of intersection.**)

Together, two or more equations such as

$$x + y = 3 \quad \text{and} \quad x - 2y = 0$$

form a **system of linear equations,** or a **set of simultaneous linear equations, in two variables.** To **solve** such a system, you find the ordered pairs of numbers that satisfy *all* the equations in the system. Each such ordered pair is called a **solution of the system;** the set of all solutions is called the **solution set of the system.**

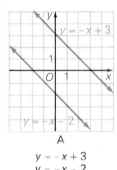 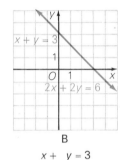 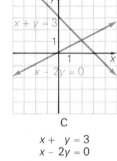

Figure 1

A	B	C
$y = -x + 3$	$x + y = 3$	$x + y = 3$
$y = -x - 2$	$2x + 2y = 6$	$x - 2y = 0$

You can use the graphs in Figure 1 to solve the three given systems:

A	**B**	**C**
$y = -x + 3$	$x + y = 3$	$x + y = 3$
$y = -x - 2$	$2x + 2y = 6$	$x - 2y = 0$

The graphs have no common point; therefore, the system has no solution, and the solution set is ∅.

The graphs coincide; therefore, there are an unlimited number of solutions, and the solution set is $\{(x, y): x + y = 3\}$.

The graphs intersect at (2, 1) but nowhere else; therefore, (2, 1) is the one solution of the system, and the solution set is $\{(2, 1)\}$.

You should check that (2, 1) satisfies both equations of System C:

$$x + y = 3 \qquad 2 + 1 \overset{?}{=} 3 \qquad 3 = 3 \checkmark$$
$$x - 2y = 0 \qquad 2 - 2 \cdot 1 \overset{?}{=} 0 \qquad 0 = 0 \checkmark$$

> To solve a system of linear equations in two variables graphically:
>
> **1.** Graph the equations in the same coordinate plane.
>
> **2.** Determine the coordinates of all points common to the graphs.

A system of equations having no solution, such as System A above, is called an **inconsistent** system. Because Systems B and C do have solutions, they are called **consistent** systems.

You can tell whether or not a system of two linear equations in two variables representing nonvertical lines is consistent by comparing the slope-intercept forms of the equations of the system. For example, consider the slope-intercept forms of the equations in Systems A, B, and C.

A	**B**	**C**
$y = -1x + 3$	$y = -1x + 3$	$y = -1x + 3$
$y = -1x + (-2)$	$y = -1x + 3$	$y = \frac{1}{2}x + 0$

A	**B**	**C**
1. Same slope, but different y-intercepts.	1. Same slope and same y-intercept.	1. Different slopes.
2. The graphs are parallel.	2. The graphs coincide.	2. The graphs intersect in a single point.
3. The system is inconsistent.	3. The system is consistent.	3. The system is consistent.
4. The solution set is $\varnothing$.	4. The solution set is $\{(x, y): y = -x + 3\}$.	4. The solution set is $\{(2, 1)\}$.

Oral Exercises

Determine the coordinates of the point of intersection.

1.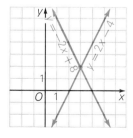

2.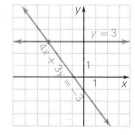

3.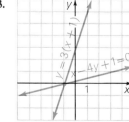

Determine the coordinates of the point of intersection.

4. 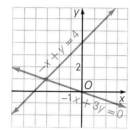 **5.** 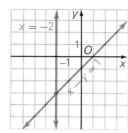 **6.**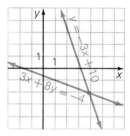

Tell whether $(-2, -7)$ is a solution of the given system.

7. $y + 2x = 3$
  $x - y = -9$

8. $2x - y = 3$
  $x + y = -9$

9. $y - 2x = -3$
  $x = y + 9$

The sentences in Exercises 10–13 refer to lines in the same plane. In each exercise tell which one of the given replacements for _?_ makes the given sentence a true statement.

EXAMPLE Lines having the same slope are either coincident or _?_.
  a. intersecting **b.** vertical **c.** parallel **d.** none of these

SOLUTION **c.** parallel

10. Nonvertical lines that intersect in exactly one point have _?_.
  a. the same slope **b.** no slope **c.** different slopes **d.** none of these

11. Parallel lines either have the same slope or are _?_.
  a. coincident lines **b.** vertical lines **c.** horizontal lines **d.** none of these

12. If a vertical line intersects another line in exactly one point, the second line cannot be _?_.
  a. vertical **b.** horizontal **c.** of negative slope **d.** of positive slope

13. If a system of two linear equations in x and y has at least two solutions, the graphs of the equations are _?_.
  a. parallel lines **b.** coincident lines **c.** intersecting lines. **d.** none of these

Written Exercises

Graph the given system, and from your graph determine the solution set of the system.

A

1. $y = x$
  $y + x = 4$

2. $y = 3x$
  $x + y - 8 = 0$

3. $x + y = 7$
  $x - y = 3$

4. $y - x = -2$
  $x = 6 - y$

5. $3x + 6y = 0$
  $y = 9 - 5x$

6. $3x - 2y = 6$
  $4y = 6x - 12$

7. $x + 2y = 6$
  $x = 2 - 2y$

8. $3x + 4y = 12$
  $2y = -3x$

9. $y = 4 - x$
    $x = 4 + y$

10. $2x + 3y = 2$
     $x + y = 0$

11. $2x - 3y = -3$
     $2x + 3y = 15$

12. $2y - 3x = -6$
     $x = 4$

In Exercises 13–18 determine whether the graphs of the equations of the given system are parallel lines, the same line, or intersecting lines. Then state whether the solution set of the system is the empty set, an infinite set, or a set with one member.

EXAMPLE $3y + 2x = 12$
           $6x + 9y = -9$

SOLUTION   Transform the equations into slope-intercept form:

$$3y + 2x = 12 \qquad y = -\tfrac{2}{3}x + 4$$
$$6x + 9y = -9 \qquad y = -\tfrac{2}{3}x - 1$$

The graphs have the same slope, $-\tfrac{2}{3}$, but different y-intercepts, 4 and -1. $\therefore$ the graphs are parallel lines, and the solution set of the system is the empty set. Answer.

13. $x - 2y = -7$
     $x - 5 = 2y$

14. $3x + 3y = -9$
        $-4x - 4y = 12$

15. $y - 5x = -5$
     $10x - 2y = -8$

16. $2x + 3y = 12$
     $3x - 2y = 5$

17. $3x = 4y - 8$
     $6x - 8y + 16 = 0$

18. $2x + 8 = y$
     $2x + 3y = -2$

B **19.** Find the point on the graph of $2x - 3y = 2$ where the ordinate is equal to the abscissa.

20. Find the point on the graph of $2x + 5y = 6$ where the ordinate is twice the abscissa.

21. Find the point on the graph of $7x + 2y = 10$ where the abscissa is the opposite of the ordinate.

22. Find the point on the graph of $5x + 6y = 9$ where the ordinate is two thirds of the abscissa.

Find the area of the triangle whose vertices are the points of intersection of the graphs of the given equations.

23. $y - x = 3$
     $y = 11 - x$
     $y = 2$

24. $2y - x = 5$
     $y = -2x$
     $x = 3$

25. $y = x + 4$
     $x + y = 10$
     $y = 2$

Find the area of the parallelogram whose vertices are the points of intersection of the graphs of the given equations.

C **26.** $y = 2x + 2$
       $y = 2$
       $y = 2x - 6$
       $y = 0$

27. $y = x + 4$
     $y = -x + 6$
     $y = x$
     $y = -x + 2$

28. $y - x = -2$
     $y + x = 2$
     $y - x = -6$
     $y + x = -4$

7-2 Using Addition or Subtraction

You do not need the graphs in Figure 2 to see that $\{(-3, 2)\}$ is the solution set of the system

$$x = -3$$
$$y = 2. \quad (1)$$

However, the graphs in Figure 3 do help you to see that $\{(-3, 2)\}$ is also the solution set of the system

$$-x + y = 5$$
$$4x + y = -10. \quad (2)$$

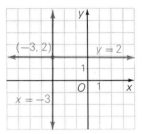

Figure 2

Systems of equations in x and y (or any other two variables) that have the same solution set are called **equivalent systems**. Thus, Systems (1) and (2) above are equivalent.

To solve a system such as System (2) algebraically, you use number properties to find an equivalent system that can be solved by inspection. The next two examples show how you can use addition or subtraction to solve a system of equations in x and y in which the coefficients of x (or y) have the same absolute value.

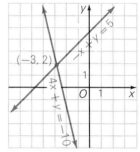

Figure 3

EXAMPLE 1 Solve: $x + 2y = 14$
$$x - 3y = -11$$

SOLUTION Use the additive property of equality to obtain the equivalent system made up of the equations of the horizontal and vertical lines through the point of intersection. The drawings picture the steps below.

1. To obtain an equation in which the coefficient of x is zero, subtract each member of the second given equation from the corresponding member of the first equation.

$$
\begin{array}{r}
x + 2y = 14 \\
x - 3y = -11 \\
\hline
5y = 25 \\
y = 5
\end{array}
$$

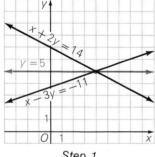

Step 1

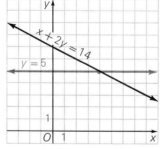

Step 2

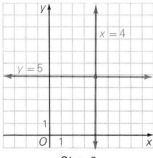

Step 3

2. Replace one of the given equations (say the second) by $y = 5$ to obtain the equivalent system:

$$x + 2y = 14$$
$$y = 5$$

3. Substitute 5 for y in the first equation of Step 2 and solve for x.

$$x + 2y = 14$$
$$x + 2(5) = 14$$
$$x + 10 = 14$$
$$x = 4$$

Check:
$$x + 2y = 14 \qquad\qquad x - 3y = -11$$
$$4 + 2(5) \overset{?}{=} 14 \qquad\qquad 4 - 3(5) \overset{?}{=} -11$$
$$14 = 14 \ \checkmark \qquad\qquad -11 = -11 \ \checkmark$$

$\therefore$ the solution set is $\{(4, 5)\}$. **Answer.**

When you replace one equation of a system of equations by another equation in which the coefficient of one of the variables is 0 (Steps 1 and 2 in Example 1), you often say that you have **eliminated** that variable.

EXAMPLE 2   Solve: $6r - 5s = -8$
$\qquad\qquad\qquad\quad 2r + 5s = -16$

SOLUTION   1. Add the corresponding members of the given equations.

$$6r - 5s = -8$$
$$\underline{2r + 5s = -16}$$
$$8r \qquad = -24$$
$$r \qquad = -3$$

Check:

$$6r - 5s = -8$$
$$6(-3) - 5(-2) \overset{?}{=} -8$$
$$-18 + 10 \overset{?}{=} -8$$
$$-8 = -8 \ \checkmark$$

2. Substitute -3 for r in one of the given equations.

$$6r - 5s = -8$$
$$6(-3) - 5s = -8$$
$$-18 - 5s = -8$$
$$-5s = 10$$
$$s = -2$$

$$2r + 5s = -16$$
$$2(-3) + 5(-2) \overset{?}{=} -16$$
$$-6 - 10 \overset{?}{=} -16$$
$$-16 = -16 \ \checkmark$$

$\therefore$ the solution set is $\{(-3, -2)\}$. **Answer.**

In solving problems involving systems of linear equations in two variables, it is customary to give the coordinates of the solution(s) of a system in alphabetical order of the variables involved unless the conditions of the problem indicate otherwise. Thus in Example 2 the solution set was given as $\{(-3, -2)\}$, not $\{(-2, -3)\}$.

Written Exercises

Use addition or subtraction to solve each system.

A   1. $x + y = 7$
         $x - y = 3$

2. $x - y = -9$
   $x + y = 17$

3. $x - y = 4$
   $4x - y = -2$

4. $w - 2z = -20$
   $2z + w = 48$

5. $3h - i = 18$
   $3h + i = 60$

6. $3k - l = 10$
   $2k - l = 7$

7. $s + 2t = 5$
   $3s + 2t = 17$

8. $2m + 5n = -10$
   $4m - 5n = 10$

9. $3y - 5x = -19$
   $2x + 3y = -5$

10. $2w - 5z = 12$
    $2w - 3z = 10$

11. $3x - 2y = -1$
    $3x - 4y = 9$

12. $2m - 3n = -4$
    $4m - 3n = -4$

13. $2 - i = -3j$
    $12 = i - j$

14. $6 = x - 3y$
    $28 = x - y$

15. $3p - 5q = -36$
    $3p + 5q = 12$

B   16. $2y = 21 - 5x$
      $7x = 2y + 39$

17. $(x + y)7 = 21$
    $x - 7y = -6$

18. $5w = 4z$
    $3w - 4z = 8$

19. $y - x = -3$
    $6y + x = 4$

20. $-8 = 2y - 3x$
    $16 = 6y - 3x$

21. $12x - 2 = 7y$
    $12x = 7y$

C   22. If $a \neq 0$ and $b \neq 0$, under what conditions will the system

$$ax + by = c$$
$$bx + ay = c$$

have no solution?

23. Prove: If $a \neq 0$ and $b \neq 0$, then the system of equations

$$ax + by = c$$
$$ax - by = d$$

has exactly one ordered pair in its solution set.

7-3 Using Linear Combinations

Sometimes you have to use the multiplicative property of equality to transform one or both of the given equations in a system of two linear equations. You do this to obtain an equivalent system in which a coefficient in one equation has the same absolute value as the corresponding coefficient in the other.

EXAMPLE 1   Solve: $3x + y = 7$
                    $2x - 3y = 12$

SOLUTION

1. Multiply both members of the first equation by 3.

$$9x + 3y = 21$$
$$2x - 3y = 12$$

2. Add the second equation to the "new" first equation and solve for x.

$$11x = 33$$
$$x = 3$$

3. Substitute 3 for x in one of the given equations, for example, the first, and solve for y.

$$3x + y = 7$$
$$3(3) + y = 7$$
$$9 + y = 7$$
$$y = -2$$

4. Checking $(3, -2)$ in the two original equations is left to you.

∴ the solution set is $\{(3, -2)\}$.   **Answer.**

EXAMPLE 2   Solve: $7a + 6b = -44$
$3a - 14b = 64$

SOLUTION

1. Multiply the first equation by 3 and the second by -7.

$$3(7a + 6b) = 3(-44) \rightarrow \quad 21a + 18b = -132$$
$$-7(3a - 14b) = -7(64) \rightarrow \quad -21a + 98b = -448$$

2. Add the resulting equations.

$$116b = -580$$
$$b = -5$$

3. Substitute -5 for b in one of the given equations and solve for a.

$$7a + 6b = -44$$
$$7a + 6(-5) = -44$$
$$7a - 30 = -44$$
$$7a = -14$$
$$a = -2$$

4. Checking $(-2, -5)$ in the two given equations is left to you.

∴ the solution set is $\{(-2, -5)\}$.   **Answer.**

When you multiply one equation of a system by a nonzero constant and another equation of the system by another nonzero constant, and add the two resulting equations, the equation that you obtain is called a linear combination of the given equations. Thus, the first equation found in Step 2 of Examples 1 and 2 is a linear combination of the equations given in each example.

Oral Exercises

In Exercises 1–6 you can eliminate one variable in the system by addition if you first multiply both members of one of the equations by some number. Which equation would you choose, and by what number would you multiply both members? Which variable would be eliminated?

EXAMPLE $2x - 5y = 21$
$4x + 3y = 6$

SOLUTION   Multiply both members of the first equation by -2, and then add the resulting equations to eliminate x.

1. $x + y = 7$
$2x + 3y = -3$

2. $w - z = 4$
$5w - 3z = 7$

3. $3m + 2n = 10$
$2m + n = 5$

4. $r - 3s = 6$
$3r + 2s = 0$

5. $5i + 12j = 24$
$9i + 4j = 8$

6. $9x + 5y = 13$
$3x + 4y = 5$

To solve each of the following systems of equations, you would first choose a variable to eliminate. Then you would multiply both members of each equation by some number.
a. Which variable would you eliminate?
b. By what number would you multiply both members of the first equation?
c. By what number would you multiply both members of the second equation?

7. $19x - 4y = 1$
$4x - 3y = 20$

8. $8c + 3d = 21$
$13c - 8d = -2$

9. $5w - 13z = 19$
$3w + 11z = 7$

10. $5p - 7q = -2$
$6p - 5q = 0$

11. $7a + 2b = 4$
$5a + 7b = -1$

12. $8x + 5y = -12$
$5x + 8y = 27$

Written Exercises

Solve each system.

A
1. $x + 2y = 1$
$3x + y = 8$

2. $w + 11z = -6$
$2w + z = 9$

3. $8c - 3d = -32$
$d - c = -1$

4. $3x + 5y = 9$
$9x + 2y = -12$

5. $8x + 9y = -45$
$x + 6y = 9$

6. $12x - 10y = 30$
$2x + 5y = 15$

7. $5p - 4q = -4$
$4p - 7q = 12$

8. $4m + 20n = 8$
$5m + 8n = 27$

9. $4h + 3k = 14$
$9h - 2k = 14$

10. $20x - 9z = 6$
$12x - 7z = 2$

11. $7y + 10z = 17$
$8y + 15z = 23$

12. $5m + 6n = 16$
$6m - 5n = 7$

13. $21r - 5s = 80$
$8r + 3s = 55$

14. $-5a + 3c = 25$
$4a + 2c = 2$

15. $2e + 3f = 0$
$5e - 2f = -19$

B **16.** $\dfrac{2}{x} + \dfrac{3}{y} = 20$ (*Hint:* Rewrite $\dfrac{2}{x} + \dfrac{3}{y} = 20$ as $2\left(\dfrac{1}{x}\right) + 3\left(\dfrac{1}{y}\right) = 20$. Let

$\dfrac{3}{x} + \dfrac{2}{y} = 15$ $a = \dfrac{1}{x}$ and $b = \dfrac{1}{y}$. Solve for a and b. Then $x = \dfrac{1}{a}$ and $y = \dfrac{1}{b}$.)

17. $\dfrac{9}{x} + \dfrac{1}{y} = 36$ **18.** $\dfrac{2}{x} + \dfrac{3}{y} = 5$ **19.** $\dfrac{1}{x} - \dfrac{2}{y} = -1$

$\dfrac{2}{x} - \dfrac{3}{y} = -21$ $\dfrac{1}{x} + \dfrac{1}{y} = 2$ $\dfrac{2}{y} + \dfrac{1}{x} = 11$

Note: The equations in Exercises 16–19 are not linear in the original
variables.

Solve for x and y in terms of the other variable(s).

20. $x - 5k = -2y$ **21.** $4cx - 5dy = 24cd$ **22.** $7ay - 12x = 39a$
   $k = x - 2y$ $3cx + 3cd = 2dy$ $3ay - 8x = -9a$

C **23.** Prove: If $a \neq 0$, $b \neq 0$, and $c \neq 0$, then the points at which the graphs
of

$$ax + by = c \quad \text{and} \quad by - ax = c$$

cross the axes are the vertices of an isosceles triangle.

24. Prove: If a, b, c, and d are real numbers such that $a \neq 0$, $b \neq 0$, and
$c \neq d$, then the system of equations

$$ax + by = c$$
$$ax + by = d$$

has no solution.

25. a. Solve the system: $9x - 8y = 1$
                    $6x + 12y = 5.$

b. Show that for every real number e and every real number f, the
solution of the system in part **a** also satisfies the equation

$$e(9x - 8y - 1) + f(6x + 12y - 5) = 0,$$

thus proving that the solution of the system in part **a** is also a
solution of every linear combination of the equations in that
system.

7-4 Using Substitution

The following method is sometimes easier to use than the linear-combination method. Given two equations in x and y, you can transform one
of them to express x (or y) in terms of the other variable. You can then
use the substitution principle to replace *one* of the given equations by a
third equation involving only one variable.

EXAMPLE   Solve: $2x + 4y = -4$
$x - 6y = 30$

SOLUTION   1. Solve for x in the second equation.
$$x - 6y = 30$$
$$x = 30 + 6y$$

2. Substitute this expression for x in the other equation.
$$2x + 4y = -4$$
$$2(30 + 6y) + 4y = -4$$

3. Solve for y.
$$60 + 12y + 4y = -4$$
$$16y = -64$$
$$y = -4$$

4. Solving for x and checking are left to you. You should find that the solution set is $\{(6, -4)\}$.

The transformations used in solving systems of linear equations are summarized below.

Transformations That Produce an Equivalent System of Linear Equations

1. Replacing any equation of the system with an equivalent equation in the same variables.

2. Replacing any equation with a linear combination of itself and another equation of the system.

3. In any equation substituting for one variable (a) its value, if known, or (b) an equivalent expression for that variable obtained from another equation of the system.

The graph of an equation formed from a given pair of equations by using one or more of these transformations is a line through the point of intersection of the graphs of the given pair if these lines intersect.

Oral Exercises

To use the substitution method, you first select one equation. Using that equation, you find an expression for one variable in terms of the other. In Exercises 1–6:
a. Which equation would you select?
b. For which variable would you find an expression?

1. $x + y = 12$
$3x - y = 4$

2. $4y + 3x = 12$
$y - 4x = 3$

3. $x + 4y = 10$
$3x + 5y = -6$

4. $4x - y = -15$
$7x + 3y = 7$

5. $x - 2y = 5$
$5x - 2y = 16$

6. $5x + 3y = 45$
$x - 6y = 40$

Written Exercises

Use substitution to solve each system of equations.

A 1. $y = 2x$
   $3x + y = 5$

2. $3x + 2y = 5$
   $x + y = 2$

3. $x + 5y = 2$
   $x = -3y$

4. $5r - 3s = -1$
   $r + s = 3$

5. $h - k = 1$
   $k + h = -5$

6. $p + 3q = 2$
   $2p + 3q = 7$

7. $x - 3y = -4$
   $2x + 6y = 5$

8. $x + 2y = 15$
   $3x - 2y = 5$

9. $2x - y = 2$
   $3x - 2y = 3$

10. $3m - 5n = 8$
    $m + 2n = -1$

11. $3i - 4j = 5$
    $i + 7j = 10$

12. $4w - 3z = 15$
    $w - 2z = 0$

Solve each system.

B 13. $2x - 3y = 3$
   $x - \dfrac{y}{2} = 2$

14. $6x + 44y = -60$
    $15x = 12 - 2y$

15. $\dfrac{m}{4} - n = -6$
    $4m + 7n = -4$

16. $\dfrac{c}{2} + \dfrac{d}{2} = 37$
    $c - d = 16$

17. $x - \dfrac{y}{2} = 4$
    $3x - 2y = -15$

18. $\dfrac{r}{5} - \dfrac{s}{5} = -2$
    $\dfrac{r}{3} + \dfrac{s}{3} = -4$

In Exercises 19 and 20 let m_1, m_2, b_1, and b_2 denote real numbers. Prove the given theorems.

C 19. If $m_1 \neq m_2$, then the one and only ordered pair (x, y) that satisfies the system

$$y = m_1x + b_1$$
$$y = m_2x + b_2$$

is $\left(\dfrac{b_2 - b_1}{m_1 - m_2}, \dfrac{m_1b_2 - m_2b_1}{m_1 - m_2} \right)$.

20. If $m_1 = m_2$, but $b_1 \neq b_2$, then the system

$$y = m_1x + b_1$$
$$y = m_2x + b_2$$

has no solution.

21. Prove that any solution of the system $ax + by + c = 0$
$$y = mx + k$$
is also a solution of the equation $ax + b(mx + k) + c = 0$.

22. Use the results of Exercises 19 and 20 to prove that two lines are parallel if and only if the lines have no slope and different x-intercepts or both lines have the same slope and different y-intercepts.

Self-Test 1

VOCABULARY   simultaneous linear equations (p. 238)
linear combination of equations (p. 245)

1. Tell whether the given system is consistent or inconsistent: *Obj. 1, p. 237*

$$x - 3y = 7$$
$$-5x + 15y = -5$$

2. Solve the given system graphically: *Obj. 2, p. 237*

$$3x - y = 2$$
$$4x - 2y = -6$$

3. Use linear combinations to solve the system:

$$2x + 3y = 6$$
$$3x + 2y = 5$$

4. Use substitution to solve the system:

$$3x + y = 0$$
$$4x - 3y = -26$$

Check your answers with those at the back of the book.

programming in BASIC

In order to solve systems of linear equations on a computer, you must arrange the solution in steps that a computer can handle. One way is to write the solution in formula form.

Exercises

1. Using either the linear combination or the substitution method, verify that the solution of

$$ax + by = c$$
$$dx + ey = f$$

   is $\left(\dfrac{ce - bf}{ae - bd}, \ \dfrac{af - cd}{ae - bd} \right)$ provided $ae - bd \neq 0$.

2. Write a program in which you INPUT values of A, B, C, D, E, F, and then find the solution set. Be sure to include a special print-out in case $A * E - B * D$ is 0.

Use your program to solve exercises in the text.

Solving Problems

OBJECTIVES for Sections 7-5 through 7-7:
1. *Use two variables to solve certain word problems, such as problems involving prices and ages.*
2. *Use two variables to solve motion problems and digit problems.*

7-5 Using Two Variables to Solve Problems

In solving word problems, it is often helpful to use two equations in two variables. Here are two examples.

EXAMPLE 1   If you rent a standard-model sedan for a week from Thrifty Rent-A-Car, the charge is $79 plus 7¢ per kilometer. Airways Rental System charges $59 plus 9¢ per kilometer for a one-week rental of a similar car.

    a. For what distance will the charges of the two companies be equal?

    b. If cost is the deciding factor and you know the number of kilometers to be driven, tell how to choose the company to patronize.

SOLUTION   (Part a)

1. The problem asks for the number of kilometers for which Thrifty and Airways will have equal charges.

2. Let k = the number of kilometers for which the charges will be equal,

    C = the charge in dollars for driving either car k km.

3. Thrifty's charge is $79 (7900 cents) plus 7 cents a kilometer.

$$C \;=\; 7900 \;+\; 7k$$

    Airways' charge is $59 (5900 cents) plus 9 cents a kilometer.

$$C \;=\; 5900 \;+\; 9k$$

4. Use substitution to solve the system of equations in Step 3:

$$C = 7900 + 7k$$
$$C = 5900 + 9k \;\rightarrow\; 7900 + 7k = 5900 + 9k$$
$$2000 = 2k$$
$$1000 = k$$

5. *Check:* If the number of kilometers is 1000, will

$$\text{Thrifty's charge} = \text{Airways' charge?}$$
$$7900 + 7(1000) \overset{?}{=} 5900 + 9(1000)$$
$$7900 + 7000 \overset{?}{=} 5900 + 9000$$
$$14900 = 14900 \;\checkmark$$

∴ the charges are equal when the number of kilometers is 1000.

SOLUTION   (Part b)
Graph the equations in Step 3, part (a). The graphs show that:

(1) if the planned trip is less than 1000 km, then Airways is the more economical choice.

(2) if the planned trip is greater than 1000 km, then choosing Thrifty is more economical.

Answer.

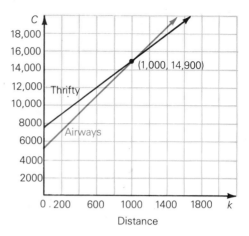

EXAMPLE 2   An antique dealer has two valuable chairs. Ten years ago the older chair was twice as old as the other one was. Ninety years from now, the older one will be only 1.5 times as old as the other will be. How old are the chairs?

SOLUTION   1. The problem asks for the age of each chair now.

2. Let x = the age in years of the older chair, and
   y = the age in years of the other chair.

3.

Time	Older Chair	Other Chair	
10 years ago	$x - 10$	$y - 10$	$x - 10 = 2(y - 10)$
Now	x	y	
90 years hence	$x + 90$	$y + 90$	$x + 90 = 1.5(y + 90)$

4–5. Solving the system of equations and checking are left to you. It turns out that the chairs are 210 and 110 years old, respectively.

Oral Exercises

Translate into a system of two equations in two variables.

1. The sum of two numbers is 25. One number is 4 times the other.

2. One number is 7 less than another. Their sum is 35,

3. The sum of two numbers is 50. Their difference is 26.

4. A 30-m cable is cut into two pieces. One piece is twice the length of the other.

5. Four pears and 2 apples cost $1.96. Three apples and 3 pears cost $2.10.

6. The perimeter of a rectangle is 100 cm. The rectangle is twice as long as it is wide.

7. Two angles are complementary (page 164). The degree measure of one is 22 more than the degree measure of the other.

Tell what expressions for ages belong in the blank spaces.

8.

Time	Cal	Len
Now	x	y
4 years ago	?	?
7 years hence	?	?

9.

Time	Marilyn	Sandra
In 1960	k	$2h - 5$
In 1948	?	?
In 1985	?	?

Problems

Solve each problem.

A

1. Bill's house lot has 3 more trees than Fran's. The two lots together have 27 trees. How many trees does each lot have?

2. In a game, Pearl scored 2 points more than twice the number of points Lydell scored. If a total of 26 points were scored, how many points did each score?

3. The perimeter of a rectangular lot which is three times as long as it is wide is 184 m. What are the dimensions of the lot?

4. The difference between three times one number and a lesser one is 23. The sum of the lesser and twice the greater is 27. Find the numbers.

5. Dolly cashed a check for $265 and asked the teller to give her only 5- and 10-dollar bills. She received 36 bills in all. How many of each denomination did the teller give her?

6. The charge for admission to the aquarium is $2.75 for adults and $1.50 for children. On a day when 589 people paid to visit the aquarium, the receipts were $1151. Find the number of adults and the number of children who paid for their tickets that day.

7. Carlos invested $10,000, part at 5% and the rest at 6%. His total annual income from these investments was $575. How much did he invest at each rate?

8. Two angles are complementary. If the degree measure of one of them is 28 greater than the degree measure of the other, what is the degree measure of each angle?

9. If Kurt were 10 years older, he would be twice as old as his sister, and their combined ages would be 36. What are the ages of Kurt and his sister?

10. The base of an isosceles triangle is one half as long as each of the congruent sides. The perimeter is 55 cm. How long is the base?

11. A large box of raisins sells for 82¢, and a medium box for 56¢. Reggie buys several boxes for a total of $3.58. If he spent $1.34 more for the large boxes than for the medium boxes, how many boxes of each size did he buy?

12. In 5 years a girl will be two thirds the age of her aunt. Three years ago she was half as old as the aunt is now. What is the age of the girl?

13. The degree measure of one of two supplementary angles (page 164) is 6 more than one half that of the other. What are the degree measures of the angles?

B 14. On a grand jury there are 4 fewer men than twice the number of women. If there were 2 more women on the jury and 2 fewer men, the jury would be equally divided between the sexes. How many women are on the jury?

15. A shipment of 19 cars, some 1300 kg apiece and the others 2200 kg each, has a total mass of 31,000 kg. Find the number of each kind of car in the shipment.

16. One amount at 4% and another at 6% yield $57. If the investments were interchanged, their income would increase by $6. What are the amounts?

17. Each side of a square frame is 30 cm long. The frame encloses two matching rectangular prints. The length of each print is 4 cm more than twice the width. The perimeter of the frame is 16 cm less than the combined perimeters of the prints. Find the dimensions of each print.

18. The ages of Janet's mother and father total 68 years. If her mother's age were doubled, the difference between their ages would be 40 years. How old are Janet's mother and father?

19. Mike and Carrie live 18 blocks apart in opposite directions from Alumni Field. If Mike lives 3 more blocks than one fourth as far from the field as Carrie does, how far does Mike live from the field?

20. A rental company has a fixed charge for the first three days and an additional charge for each day thereafter. Rosa paid $2.70 for an article she kept 7 days. Sal paid $2.10 for one he kept 5 days. Find the fixed charge and the charge for each extra day.

21. The average of two numbers is $\frac{7}{36}$. One third of their difference is $\frac{2}{9}$. Find the numbers.

C 22. Leanne is twice as old as her brother will be when Leanne is 8 times as old as her brother is now. What is the relationship between Leanne's present age and her brother's?

7-6 Motion Problems

To solve motion problems about airplanes flying with or against the wind, you have to know what these words mean:

tail wind: a wind blowing in the same direction as the one in which the airplane is heading.

head wind: a wind blowing in the direction opposite to the one in which the airplane is heading.

wind speed: the speed of the wind.

air speed: the speed of the airplane in still air.

ground speed: the speed of the airplane relative to the ground.

You also have to know these facts:

1. With a *tail wind:* ground speed = air speed + wind speed

2. With a *head wind:* ground speed = air speed − wind speed

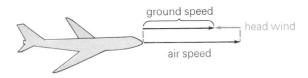

EXAMPLE   With a tail wind, a light plane can fly 960 km in 3 h. Going against the wind, the plane can fly the same distance in 4 h. What are the wind speed and the air speed of the plane?

SOLUTION   1. The problem asks for the wind speed and the air speed of the plane.

      2. Let x = the air speed of the plane in kilometers per hour, and y = the wind speed in kilometers per hour.

      3.

With a	ground speed r (km/h)	time t (h)	distance $d = rt$ (km)
tail wind	$x + y$	3	$3(x + y)$
head wind	$x - y$	4	$4(x - y)$

$$\begin{array}{l} 3(x + y) = 960 \\ 4(x - y) = 960 \end{array} \quad \text{or} \quad \begin{array}{l} x + y = 320 \\ x - y = 240 \end{array}$$

(Solution continued on next page.)

4–5. Solving the system of equations and checking are left to you. You should find that:

the air speed of the plane is 280 km/h;
the wind speed is 40 km/h.   Answer.

The method of solution used in this example can also be used to solve problems involving boats moving in a current.

Oral Exercises

In Exercises 1–4 Sharon rows at the rate 4 km/h in still water, and the rate of the current is 2 km/h.

1. How fast does Sharon move rowing upstream?
2. How fast does Sharon move rowing downstream?
3. How long would it take her to row 6 km upstream and 6 km back?
4. What would happen if she tried to row upstream in a current flowing at 4.5 km/h?

In Exercises 5–8, the plane's air speed is 400 km/h.

5. What is the plane's ground speed on a windless day?
6. What is its ground speed if it has a 40 km/h tail wind?
7. What is its ground speed when it encounters head winds of 24 km/h?
8. How long will the plane take to fly 1980 km
   a. with a 40 km/h tail wind?
   b. with a 40 km/h head wind?

Problems

Solve each problem.

A   1. With a tail wind, a plane flew 280 km in half an hour. With no change in the wind, the return trip took 40 min. Find the speed of the wind and the plane's rate in still air.

2. A boat travels 6 km in 45 min. The return trip takes 1.5 h. Find the boat's speed in still water.

3. The air speed of a light plane was 250 km/h. If the plane traveled the same distance in 3 h with the wind as it flew in 4 h against the wind, what was the wind speed?

4. Flying against a head wind, a plane could fly 2100 km in 6 h, but it would require only 5 h for the return trip with the wind. Find the wind speed and the air speed of the plane.

5. A salmon swims 24 km from its birthplace to the ocean in 3 h. The return trip upstream to spawn takes 4 h. How fast does the salmon swim in still water?

6. Charlie flew 720 km against the wind in 90 min. The return trip took 1 h 15 min, with no wind change. What was the wind speed?

7. Carol took 36 min to row 5 km. When she returned, she took 2 h. Find the speed of the current.

8. A cyclist rode 8 km in $\frac{1}{4}$ h with the wind, and returned in $\frac{1}{3}$ h against the same wind. Find the speed without a wind of the cyclist.

B 9. A swimmer took 75 min to swim 1.6 km. On the return trip against the current she swam 0.2 km in the time that she swam 0.8 km on the trip downstream. Find her average rate. What was the speed of the current?

10. A crew rowed a boat 10 km with the tide in 0.5 h. Later, when the tide was flowing at half its previous rate, the same crew rowed 10 km against the tide in 1 h. Find the original rate of the tide.

11. A river steamer sails a distance up a river in the time it sails twice that distance downstream. If the speed of the steamer is s and that of the current is c, find the relationship between s and c.

12. Larry can row a certain distance with the current in 3.5 h. To cover the same distance against the current, he takes 4.5 h. How many times faster is his rate of rowing in still water than the rate of the current?

13. A plane can fly a certain distance in 4 h with an 80 km/h head wind. Flying in the opposite direction with the same wind, it can fly that distance in 3 h. What is the plane's air speed?

C 14. A round-trip flight of 2508 km took 7.5 h. There was a head wind at first and then a tail wind. The part of the flight with the wind took an hour and a half less than the other part of the trip. Flying with the wind gave the plane a 25% increase in ground speed over its ground speed flying against the wind. Find the speed of the plane in still air and the wind speed.

7-7 Digit Problems

You can write a two-digit decimal number like 83 in expanded form:

$$\text{tens digit} \quad \text{units digit}$$
$$\downarrow \qquad \downarrow$$
$$83 = 10 \cdot 8 + 3$$

All two-digit decimal numbers have this kind of expanded form:

$$\text{tens digit} \quad \text{units digit}$$
$$\downarrow \quad \downarrow$$
$$10t + u,$$

where $t \in \{1, 2, 3, 4, 5, 6, 7, 8, 9\}$ and $u \in \{0, 1, 2, 3, 4, 5, 6, 7, 8, 9\}$.

To represent a number with the same digits in reverse order, you write

$$10u + t.$$

In each case, the sum of the values of the digits is represented by $t + u$.

EXAMPLE A clerk mistakenly reversed the two digits in the price of a marking pen and overcharged the customer 27¢. If the sum of the digits was 15, what was the correct price of the pen?

SOLUTION 1. The problem asks for the correct price of the pen.

2. Let $t = $ the tens digit of the correct price, and
   $u = $ the units digit of the correct price. Then
$$10t + u = \text{the correct price, and}$$
$$10u + t = \text{the price mistakenly charged.}$$

3. The price charged was 27 cents more than the correct price:
$$10u + t = 10t + u + 27$$
$$10u + t - (10t + u) = 27$$
$$9u - 9t = 27$$
$$u - t = 3$$

The sum of the digits was 15:
$$t + u = 15, \quad \text{or} \quad u + t = 15$$

4. $u - t = 3$
   $\underline{u + t = 15}$
   $\quad 2u = 18 \qquad \longrightarrow 9 - t = 3$
   $\quad\; u = 9 \qquad\qquad\quad t = 6$

5. *Check:* When you reverse the digits of 69, is the resulting number 27 more? $96 = 69 + 27$ ✓
   Is the sum of the digits 15? $6 + 9 = 15$ ✓

$\therefore$ the correct price was 69 cents.   **Answer.**

Problems

Solve each problem.

A   1. The sum of the digits of a two-digit number is 11. The value of the number is 13 times the units digit. Find the number.

2. The sum of the digits of a two-digit number is 9. The number with the digits reversed is 9 times the original tens digit. Find the original number.

3. The sum of the digits of a two-digit number is 9. If the order of the digits is reversed, the result is a number exceeding the original number by 9. Find the original number.

4. The units digit of a two-digit number is 3 times the tens digit. The sum of the digits is 12. Find the number.

5. The sum of the digits of a two-digit number is 14. If 18 is subtracted from the number, the result is the number with its digits reversed. Find the number.

6. The units digit of a two-digit number exceeds 2 times the tens digit by 2. The sum of the digits is 11. Find the number.

B 7. Find a three-digit number whose units digit is 3 times its hundreds digit and 2 times its ten digit, and the sum of whose digits is 11.

8. A three-digit number is 198 more than itself reversed. The sum of the digits is 19. The hundreds digit is 3 times the tens digit. Find the original number.

9. Show that the difference between a two-digit number and the number with the order of the digits reversed is always divisible by 9.

10. Show that the difference between a three-digit number and the number with the order of the digits reversed is always divisible by 99.

C 11. If a two-digit number is divided by its tens digit, the quotient is 12 and the remainder is 1. If the number with its digits interchanged is divided by its original units digit, the quotient is 10 and the remainder is 4. Find the original number.

12. The sum of the digits of a two-place decimal number is 13. When its digits are reversed, the new number exceeds the original by 0.45. Find the original number.

13. The sum of the digits of a two-place decimal number is 9. The number with its digits reversed is 0.09 more than 4 times the original number. Find the original number.

14. Show that the difference between a two-digit number and the number obtained when the tens digit is decreased by 1 and the units digit is increased by 1 is always 9.

15. Find all three-digit numbers, if there are any, that satisfy all of the requirements **a** through **c**.
    a. The units digit is one half the tens digit.
    b. The hundreds digit is 2 less than 4 times the units digit.
    c. The difference between the number and the number reversed is 396.

Self-Test 2

VOCABULARY   air speed (p. 255)   ground speed (p. 255)

1. Five years ago, Jake was 7 times as old as Jenny was at that time.   *Obj. 1, p. 251*
   Ten years from now, Jake will be twice as old as Jenny will be.
   Find the age of Jenny now.

2. Ruth needed 1 h to row 2 km upstream, but only 10 min to row   *Obj. 2, p. 251*
   back downstream. Find her rowing rate in still water and the
   rate of the current.

3. The sum of the digits of a two-digit number is 9. When the digits
   are interchanged, the new number exceeds the original number
   by 27. Find the original number.

Check your answers with those at the back of the book.

Additional Topics (Optional)

***OBJECTIVES** for Sections 7-8 through 7-10:*
1. *Graph the solution set of a system of two or three linear inequalities in two variables.*
2. *Find the maximum (or minimum) value, if it exists, of a linear expression over a convex polygonal region in a coordinate plane.*
3. *Use the mastery of Objective 2 to solve linear-programming problems that involve two variables.*
4. *Solve a system of three linear equations in three variables.*

7-8 Graphs of Systems of Linear Inequalities

You can use graphs to determine the solution set of a system of linear inequalities in two variables.

EXAMPLE   Draw the graph of the solution set of this system of inequalities:

$$2x + y \leq 1$$
$$x - 2y < 2$$

SOLUTION   1. Transform each inequality into an equivalent one with y as one member.

$$2x + y \leq 1 \quad \rightarrow \quad y \leq 1 - 2x$$
$$x - 2y < 2 \quad \rightarrow \quad y > \tfrac{1}{2}x - 1$$

2. Draw the graph of $y = 1 - 2x$. The graph of $y \leq 1 - 2x$ is the closed half-plane (Section 6–10) on and below the line $y = 1 - 2x$ (colored shading).

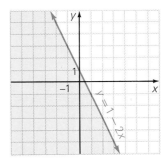

3. Draw the graph of $y = \frac{1}{2}x - 1$. The graph of $y > \frac{1}{2}x - 1$ is the open half-plane above the line $y = \frac{1}{2}x - 1$ (gray shading).

4. The intersection of the half-planes found in Steps 2 and 3 (double shading) is the graph of the given system:

$$2x + y \leq 1$$
$$x - 2y < 2$$

(Of course, the figure shows only a partial graph.)

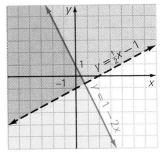

Written Exercises

Graph each system of inequalities. Use shading to show the solution set of the system.

A

1. $y \geq 0$
   $x \leq 0$

2. $y \geq -1$
   $x \leq 1$

3. $y < 1$
   $x > 2$

4. $y > x$
   $x < 3$

5. $y < \frac{1}{2}x$
   $x > -1$

6. $y \geq x$
   $y \geq x + 2$

7. $y \leq x - 1$
   $y \leq 1 - x$

8. $y > 2x + 3$
   $y > -2x + 1$

9. $3x + y \leq 0$
   $2x - 3y > 6$

10. $x + 3y \geq -1$
    $x - 2y \leq 2$

11. $2x + 4y \leq 4$
    $2x + 4y \geq -4$

12. $3y \geq x - 27$
    $x - 2y \geq -2$

EXAMPLE $y \leq x$
$y \geq -1 - x$
$x \leq 4$

SOLUTION   1. The graph of $y \leq x$ consists of the points on and below the line $y = x$ (red).

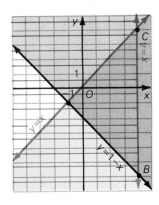

2. The graph of $y \geq -1 - x$ consists of the points on and above the line $y = -1 - x$ (gray).

3. The graph of $x \leq 4$ consists of the points on and to the left of the vertical line $x = 4$ (horizontal shading).

4. The graph of the solution set of the given system is the intersection of the shaded regions, that is, triangle ABC, including the points on its sides and in its interior.

Graph each system of inequalities.

B **13.** $x + y \geq 2$
         $x \leq 0$
         $y \geq 0$

14. $y \geq 3 - x$
     $y \leq x + 3$
     $y \leq -1$

15. $y \geq -x$
     $y \leq x$
     $x \geq -2$

16. $2y \geq x + 2$
     $2y + x \leq 6$
     $y > 0$

17. $2y - x \leq 0$
     $2x - y \geq 0$
     $x + y > 0$

18. $x \leq y + 1$
     $2x \geq -2y + 2$
     $x \geq 3 - y$

19. $x + y > 1$
     $x > -1$
     $x < 2$

20. $3x - y \leq 6$
     $x \geq 0$
     $y > -2$

C **21.** $y \geq 3x + 1$
         $y \geq -x - 1$
         $y \leq 2 - 2x$

22. $2y \leq x + 4$
     $y \geq x + 2$
     $y \geq -x + 1$

23. $x - 2y \leq 2$
     $y < -\frac{1}{2}x + 3$
     $x \leq 0$

24. $2x + 3y + 9 \geq 1$
     $x + y < 0$
     $y \geq 0$

Show the solution set of the system in a four-way shaded region.

25. $x \leq 0$
        $y \geq 0$
    $x + y > -1$
    $x + y < 1$

26. $2x + y < 3$
     $2y + x > -3$
     $x - y \leq 0$
     $x + y \geq 0$

7-9 Linear Programming

Many practical problems in business, science, and industry can be solved using the techniques of *linear programming*. This branch of mathematics is concerned with problems in which a quantity represented by a linear equation—often profit or cost—is to be maximized or minimized subject to conditions expressed by a system of linear inequalities. The following example illustrates the method.

During an illness Mr. Gates supplements his daily diet with vitamin pills. Each day he needs at least 4 mg of thiamine, 6 mg of riboflavin, and 80 mg of niacin. To meet these needs, he can buy either Brand X pills at 3¢ apiece or Brand Y pills at 4¢ apiece. These brands contain the following amounts of the vitamins:

	Brand X	Brand Y
Thiamine	1 mg	4 mg
Riboflavin	3 mg	5 mg
Niacin	30 mg	40 mg

What combination of pills will provide his minimum daily needs for the three vitamins at the lowest cost?

Let $x =$ the number of Brand X pills used daily,
    $y =$ the number of Brand Y pills used daily, and
    $C =$ the daily cost in cents for the pills.

Then:

$$C = 3x + 4y.$$

Mr. Gates wants to minimize C subject to the following inequalities (*constraints*):

$$\left.\begin{array}{l} x + 4y \geq 4 \\ 3x + 5y \geq 6 \\ 30x + 40y \geq 80 \end{array}\right\}$$ The total daily amount of each vitamin must equal at least the daily need.

$$\left.\begin{array}{l} x \geq 0 \\ y \geq 0 \end{array}\right\}$$ He cannot use a negative amount of either pill.

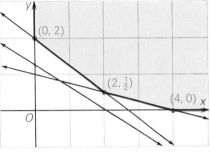

Figure 4

The graph of the solution set of this system of inequalities is indicated by colored shading in Figure 4. (Notice that $3x + 5y \geq 6$ has no effect on the result.) This shaded region (called the *feasible region*) has the following characteristics:

1. Where it is bounded, the boundary is determined by straight lines. The points of the region where such boundary lines intersect are called **corner points.** In this case the corner points are $(0, 2)$, $(2, \frac{1}{2})$, and $(4, 0)$.
2. Every point on the boundary is a part of the region.
3. The region is *convex*.

(A region is said to be **convex** if whenever you choose any two points in the region and draw the segment joining them, the segment is contained in the region. For example, the shaded region in part (*a*) of Figure 5 is convex, but the shaded region in part (*b*) is not.)

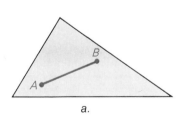

a.

Figure 5

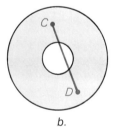

b.

Any plane region which is the intersection of a finite number of closed half-planes has the three characteristics listed above and is called a **convex polygonal region.**

A remarkable result, not proved here, is that:

> Over a convex polygonal region, any maximum or minimum values of a *linear expression ax + by*, where a and b are real numbers, occur for the coordinates of a corner point of the region.

It is possible that maximum or minimum values may also occur at other points besides the corner points. What is important is that by testing values at the corner points, you can, in a finite number of steps, find the greatest and least values (if they exist) of any linear expression over any closed polygonal region.

Thus the minimum value of C can be found by evaluating $3x + 4y$ at each corner point:

$$(0, 2): \quad 3(0) + 4(2) = 8$$
$$(2, \tfrac{1}{2}): \quad 3(2) + 4(\tfrac{1}{2}) = 8$$
$$(4, 0): \quad 3(4) + 4(0) = 12$$

In this case both $(0, 2)$ and $(2, \tfrac{1}{2})$ yield the minimum value, 8; however it is inconvenient to take half a pill.

Therefore the most economical and practical choice for Mr. Gates is 2 Brand Y pills and no Brand X pills.

Written Exercises

Which of the following shaded regions are convex?

A   1.   2.   3.   4.

5. Evaluate the linear expression $3x + y$ at each corner point of the convex polygonal region shown at the right.

6. What is the minimum value of $3x + y$ over the region shown?

7. Does $3x + y$ have a maximum value over the region shown? If so, what is the value? If not, explain why.

8. Explain why $2x + y$ does not have a maximum value over the region shown.

9. Explain why $x - y$ does not have a minimum value over the region shown.

10. Shade the region that represents the solution set of the system:
$$x \geq 1,\ y \geq -x + 6,\ y \geq -\tfrac{1}{2}x + 4,\ y \geq -2x + 8,\ y \geq 0.$$

11. Evaluate the expression $3x + 4y$ for the values of x and y given in each of the following ordered pairs:

    a. $(1, 6)$ b. $(2, 4)$ c. $(4, 2)$ d. $(8, 0)$

    Notice that the graphs of these four ordered pairs are the corner points of the region you shaded in Exercise 10.

In Exercises 12–15:

a. Graph the solution set of the system of inequalities and label the corner points with their coordinates.

b. Find the minimum value of the expression $2x + 3y$ over the region graphed, by evaluating it at each of the corner points.

B

12. $x \geq 0$
$y \geq -x + 4$
$y \geq -\frac{1}{2}x + 3$
$y \geq 0$

13. $x \geq 0$
$y \geq -2x + 8$
$y \geq -x + 3$
$y \geq 1$

14. $x \geq 0$
$y \geq -3x + 10$
$y \geq -2x + 9$
$y \geq -\frac{1}{2}x + 6$
$y \geq 1$

15. $x \geq 1$
$y \geq -3x + 9$
$y \geq -x + 5$
$y \geq -\frac{1}{4}x + 2$
$y \geq 0$

In Exercises 16–19, use the following information: A zoo is given a large sum of money provided that (1) the zoo spend at least $200 for rabbits and (or) monkeys, (2) the zoo buy at least 14 rabbits and (or) monkeys. Each rabbit costs $4 and each monkey costs $40.

16. Introduce variables for the number of rabbits and the number of monkeys to be purchased. Write four inequalities that express the conditions of the problem. (Remember that the number of each kind of animal must be at least zero.)

17. Graph the solution set of the system of inequalities. State the coordinates of the corner points.

18. Suppose that it costs 10 cents per day to feed each rabbit and 50 cents per day to feed each monkey. Daily feeding costs are to be held to a minimum. How many rabbits and monkeys should be bought?

19. Suppose that it costs 8 cents per day to feed each rabbit and 80 cents per day to feed each monkey. Find five ways the zoo can meet the conditions while holding feeding costs to a minimum.

C

20. A machine shop needs to manufacture at least 1920 bolts and 1700 screws. A small machine can make 80 bolts and 75 screws per hour. A larger machine can make 120 bolts and 100 screws per hour. The cost of running the small machine is $2 per hour, and of running the large machine is $3.10 per hour. How long should each machine be used for the most economical production?

21. Each day the Yellow Sands Post Office handles at least 3120 first-class letters and 1890 second-class advertising flyers. The cancelling machine now in use can handle 390 letters and 315 flyers per hour at an hourly cost of $2.10. The manufacturer of the machine is about to release an improved model which can do 520 letters and 210 flyers each hour at a cost of $2.25 per hour. Neither machine can be operated for more than 10 hours due to internal heating.

a. How would the expense of handling the mail be affected if the post office could use both machines?

b. Would the use of both machines be practical if the post office reduced the cost of operating the old machine by 30¢ an hour?

programming in BASIC

A computer can assist you in solving a problem in linear programming.

Here is an elementary program, using the example in Section 7-9.

First, sketch the graph of the solution set of the inequalities. (Recall that $3x + 5y \geq 6$ has no effect on the result.)

Then, after studying the graph, write the equations of the lines in the order in which you will want to find their intersections. Rewrite each in the form $ax + by = c$:

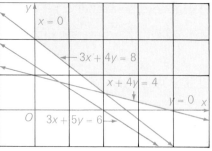

$$\begin{cases} x = 0 \longrightarrow 1x + 0y = 0 \\ 3x + 4y = 8 \longrightarrow 3x + 4y = 8 \\ x + 4y = 4 \longrightarrow 1x + 4y = 4 \\ y = 0 \longrightarrow 0x + 1y = 0 \end{cases}$$

The following program finds the corner points and evaluates the expression $3x + 4y$ at each. Notice that the (" ",) in lines 110 and 130 moves the print-out over 15 spaces.

```
10   PRINT "HOW MANY EQUATIONS";
20   INPUT N
30   FOR I=1 TO N−1
40   PRINT "INPUT A,B,C"
50   INPUT A,B,C
60   PRINT "INPUT D,E,F"
70   INPUT D,E,F
80   LET Q=A*E−B*D
90   LET X=(C*E−B*F)/Q
100  LET Y=(A*F−C*D)/Q
110  PRINT "  ","CORNER PT.",
120  PRINT "VALUE OF 3X + 4Y"
130  PRINT "  ","(";X;",";Y;")",
140  PRINT 3*X+4*Y
150  NEXT I
160  END

RUN

HOW MANY EQUATIONS? 4
INPUT A,B,C
?1,0,0
INPUT D,E,F
?3,4,8
```

	CORNER PT.	VALUE OF 3X + 4Y
	(0, 2)	8

```
INPUT A,B,C
?3,4,8
INPUT D,E,F
?1,4,4
```

	CORNER PT.		VALUE OF 3X + 4Y
	(2, .5)		8

```
INPUT A,B,C
?1,4,4
INPUT D,E,F
?0,1,0
```

	CORNER PT.		VALUE OF 3X + 4Y
	(4, 0)		12

```
END
```

This verifies the result in the text that (0, 2) and (2, 0.5) yield the minimum value, 8.

You can use this program for other problems by changing lines 120 and 140.

Exercises

1. Write a program that will find coordinates of points on $x + 4y = 4$ when x has the values 2, 2.5. 3, 3.5, 4, and evaluate $3x + 4y$ at each of these points. (Use LET $X = X + .5$ and a GOTO statement. Use a test to end the loop after $X = 4$.)

2. Use the computer to check your answers to Exercises 5–6 on page 264.

3. Use the computer to check your answers to Exercises 10–11 on page 264.

7-10 Systems of Linear Equations in Three Variables

You have seen that a solution of an equation in *one* variable, such as

$$2x + 1 = 0,$$

is a *single* real number, and that a solution of an equation in *two* variables, such as

$$3x - 5y = 3,$$

is an *ordered pair* of real numbers.

A **solution** of an equation in *three* variables, such as

$$x + 2y - 4z = 9,$$

is an **ordered triple** (x, y, z) of real-number values of the variables for which the equation is a true statement; the set of all such ordered triples is the **solution set** of the equation.

EXAMPLE 1   Which of the following ordered triples are solutions of the equation

$$3x + y + 2z = 13?$$

a. $(2, 1, 3)$ **b.** $(1, 3, 2)$ **c.** $(1, 2, 4)$

SOLUTION   **a.** $3(2) + 1 + 2(3) = 13$
b. $3(1) + 3 + 2(2) = 10 \neq 13$
c. $3(1) + 2 + 2(4) = 13$

$\therefore$ $(2, 1, 3)$ and $(1, 2, 4)$ are solutions.   **Answer.**

Any equation of the form

$$ax + by + cz = d, \quad x, y, \text{ and } z \in \mathcal{R},$$

where a, b, and c are constants (not all 0), is called a **first-degree,** or **linear, equation in three variables.** (In later courses you will learn that the graph of such an equation is a plane in space.) You can find as many solutions for such equations as you wish by choosing values for two of the variables and determining the corresponding value for the third variable.

EXAMPLE 2   Find three solutions of $x - 2y + z = 6$.

SOLUTION   1. Transform the given sentence into an equivalent sentence having z as one member.

$$x - 2y + z = 6$$
$$z = 6 - x + 2y$$

2. Replace x and y with convenient numbers and determine the corresponding values of z.

x	y	$6 - x + 2y$	z
0	0	$6 - 0 + 2(0)$	6
1	0	$6 - 1 + 2(0)$	5
0	1	$6 - 0 + 2(1)$	8

$\therefore$ three solutions are $(0, 0, 6)$, $(1, 0, 5)$, and $(0, 1, 8)$.   **Answer.**

In Sections 7-2 through 7-4 you learned to solve systems of two linear equations in two variables. Similar methods can be used to solve systems of three linear equations in three variables since the transformations listed on page 248 are also valid for these systems. A **solution** of a system of three equations in three variables is an ordered triple satisfying *all* the equations in the system; the **solution set** of the system is the set of all such triples.

EXAMPLE 3 Solve: $\begin{aligned} x + y + 2z &= -1 \quad (1) \\ x - 2y + z &= 0 \quad (2) \\ 3x + y - 2z &= 5 \quad (3) \end{aligned}$

SOLUTION

1. Subtract (2) from (1), that is, subtract the members of Equation (2) from the corresponding members of Equation (1), to eliminate x.

$$\begin{aligned} x + y + 2z &= -1 \\ x - 2y + z &= 0 \\ \hline 3y + z &= -1 \end{aligned}$$

2. Multiply (2) by 3 and subtract (3) from this equation to eliminate x.

$$\begin{aligned} x - 2y + z &= 0 \\ 3x - 6y + 3z &= 0 \\ 3x + y - 2z &= 5 \\ \hline -7y + 5z &= -5 \end{aligned}$$

3. Next, eliminate z between the equations obtained in Steps 1 and 2. For example, multiply the equation in Step 1 by -5 and add this equation to the equation in Step 2. Then solve for y.

$$\begin{aligned} 3y + z &= -1 \\ -15y - 5z &= 5 \\ -7y + 5z &= -5 \\ \hline -22y &= 0 \\ y &= 0 \end{aligned}$$

4. Substitute 0 for y in the equation obtained in Step 1 (or the equation obtained in Step 2) and solve for z.

$$\begin{aligned} 3(0) + z &= -1 \\ z &= -1 \end{aligned}$$

5. Substitute 0 for y and -1 for z in one of the original equations, say (1), and solve for x.

$$\begin{aligned} x + 0 + 2(-1) &= -1 \\ x &= 1 \end{aligned}$$

6. Checking that $(1, 0, -1)$ satisfies the three original equations is left to you.

$\therefore$ the solution set is $\{(1, 0, -1)\}$. **Answer.**

Written Exercises

A
1. Which of the following are solutions of $2x - y - 3z = 0$?
   a. $(2, -\frac{1}{3}, 3)$ b. $(-1, -5, 1)$ c. $(2, 3, \frac{1}{3})$
2. Which of the following are solutions of $2y - 3x - 4z = 6$?
   a. $(-5, 0, -4)$ b. $(0, -5, -4)$ c. $(-5, -4, 0)$

Find three solutions of each of the following equations.

3. $x - y + z = 1$ 4. $y - 2x - z = 1$ 5. $2x - 3y - z = -12$

Find the solution set of each system.

6. $\begin{aligned} -2x + y - z &= -4 \\ 3x + y + 4z &= 1 \\ x + 2y + z &= 5 \end{aligned}$
7. $\begin{aligned} x - 2y + 3z &= 2 \\ x - y - z &= 0 \\ 2x - 3y - 2z &= -2 \end{aligned}$
8. $\begin{aligned} 2x + 5y + z &= 5 \\ x + y + z &= 1 \\ x + 2y - z &= -3 \end{aligned}$

Solve each problem using three variables.

EXAMPLE The sum of the length, width, and height of a rectangular box is 17 cm. The width is half the height. Twice the length exceeds the sum of the width and height by 7 cm. Find the length, width, and height of the box.

SOLUTION 1. The problem asks for the length, width, and height of the box.

2. Let x = the length in cm,
y = the width in cm,
z = the height in cm.

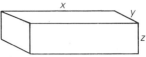

3. Then: $x + y + z = 17$ or $x + y + z = 17$
$y = \frac{1}{2}z$ $y - \frac{1}{2}z = 0$
$2x = y + z + 7$ $2x - y - z = 7$

4–5. Solving this system and checking are left to you. You should find that the length is 8 cm, the width is 3 cm, and the height is 6 cm.

9. The sum of three numbers is 7. One of the numbers is 1 more than the sum of the other two numbers. It is also 4 times the difference between the other two numbers. What are the three numbers?

10. Rachel has $1.30 in change in her purse, consisting entirely of quarters, dimes, and nickels. She has the same number of nickels as quarters, and one more dime than nickels. How many of each kind of coin does she have?

11. A certain triangle has a perimeter of 44 mm. The length of the longest side is 4 mm less than the sum of the lengths of the other sides. Twice the length of the shortest is 9 mm more than the difference between the lengths of the other sides. What is the length of each side of the triangle?

12. Carlos has one-dollar, five-dollar, and ten-dollar bills, totaling $171. He has the same number of five-dollar bills as one-dollar and ten-dollar bills put together. If he has 30 bills in all, how many bills of each kind does he have?

Solve each system.

B 13. $3(x + y + z) = 2$
$y = 1 - 3x - 6z$
$27x + 9y + 3z = -8$

14. $x + 2y - z = 0$
$x - y + \frac{2}{3}z = \frac{1}{2}$
$2x + y + 4z = -6$

15. $2y + z = 11 - x$
$3z = 19 - x - 3y$
$2x + y + z = 13$

16. $z - y - x = -2$
$y = 4x + z$
$y + z = 4 - 6x$

17. $-x - y + z = 0$
$y = -2x - 7$
$x + 3y + z = 2$

18. $3x = 4 - 5z$
$3y = -1 - 4z$
$2z = 2y - 5x$

Find a, b, and c so that the following ordered triples will be solutions of the given equation.

C **19.** $ax + by + cz = 7$; $(0, -1, 3)$, $(2, -1, 0)$, $(1, 0, 2)$
  20. $ax - 2y + bz = c$; $(2, 0, -2)$, $(-2, -7, 0)$, $(0, 1, 2)$
  21. $ax + by - z = c$; $(4, 0, 9)$, $(1, -3, 0)$, $(0, 2, 9)$

Find a, b, and c so that the following ordered pairs will be solutions of the given equation.

  22. $y = ax^2 + bx + c$; $(0, 3)$, $(1, 3)$, $(-1, 5)$
  23. $y = ax^2 + bx + c$; $(0, -1)$, $(1, 2)$, $(2, 9)$
  24. $y = ax^2 + bx + c$; $(-1, -1)$, $(0, -2)$, $(-2, 0)$

Self-Test 3

VOCABULARY   linear programming (p. 262)   solution of an equation in
                 constraints (p. 263)   three variables (p. 268)

1. Graph the solution set of the system: $x + y \le 1$ *Obj. 1, p. 260*
                                                $2x - y \le 4$

2. Find the minimum value of the linear expression $60x + 75y$ over   *Obj. 2, p. 260*
the convex polygonal region shown below.

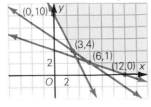

3. Cheshire Transportation needs at least 650 barrels of diesel fuel,   *Obj. 3, p. 260*
324 barrels of gasoline, and 48 barrels of oil to keep its trucks in
operation. Coos Oil can deliver 130 barrels of diesel fuel, 36
barrels of gasoline, and 4 barrels of oil for a bulk wholesale rate
of $3060. A similar plan, costing $2620, is available from Sullivan
Oil for 65 barrels of diesel fuel, 54 barrels of gasoline, and 12
barrels of oil. How many standing orders should Cheshire
Transportation place with each firm in order to meet its petro-
leum needs at the least cost?

4. Solve: $y = 2x + 3z + 3$ *Obj. 4, p. 260*
              $x = 1 - y - 2z$
              $z = 2 + x - 2y$

Check your answers with those at the back of the book.

Matrices

You can find the solution of a system such as
$$x + 4y = 9$$
$$2x + y = 4$$

by working only with the coefficients of x and y and the constant terms. To do this, you represent the coefficients and constants by means of an ordered array of numbers called a **matrix** (plural: **matrices**). Parentheses or brackets are used to group the **elements** of a matrix. Thus

$$\begin{pmatrix} 1 & 4 \\ 2 & 1 \end{pmatrix}, \quad \text{or} \quad \begin{bmatrix} 1 & 4 \\ 2 & 1 \end{bmatrix},$$

is the **coefficient matrix** of the given system, and

$$\begin{pmatrix} 1 & 4 & 9 \\ 2 & 1 & 4 \end{pmatrix}$$

is the **augmented matrix** of the system.

Now, compare the sequence of steps used to solve the given system by linear combinations, as shown at the left, with the corresponding sequence of matrices shown at the right.

$$\begin{array}{ll} x + 4y = 9 & \begin{pmatrix} 1 & 4 & 9 \\ 2 & 1 & 4 \end{pmatrix} \end{array} \quad \begin{array}{l} \text{Row 1} \\ \text{Row 2} \end{array}$$

Multiplying each member of the first equation by -2 and adding the result to the second equation produces the equivalent system:

$$\begin{array}{ll} x + 4y = 9 & \begin{pmatrix} 1 & 4 & 9 \\ 0 & -7 & -14 \end{pmatrix} \end{array} \quad \begin{array}{l} (-2 \times \text{Row 1}) + \text{Row 2} \\ \text{in preceding matrix} \end{array}$$

Dividing the second equation by -7, or multiplying by $-\frac{1}{7}$, produces the equivalent system:

$$\begin{array}{ll} x + 4y = 9 & \begin{pmatrix} 1 & 4 & 9 \\ 0 & 1 & 2 \end{pmatrix} \end{array} \quad \begin{array}{l} -\frac{1}{7} \times \text{Row 2 in} \\ \text{preceding matrix} \end{array}$$

Adding -4 times the second equation to the first equation yields the equivalent system:

$$\begin{array}{ll} x + 0y = 1 & \begin{pmatrix} 1 & 0 & 1 \\ 0 & 1 & 2 \end{pmatrix} \end{array} \quad \begin{array}{l} \text{Row 1} + [(-4) \times \text{Row 2}] \\ \text{in preceding matrix} \end{array}$$

The only solution of the system, $(1, 2)$, is evident by inspection from this last set of equations, and from the right-hand column of the corresponding matrix. Do you see that, at each step, the matrix shown is obtained from the preceding matrix in the sequence in exactly the same way as the corresponding system of equations is obtained from its preceding system?

Each of the matrices in this example is said to be **row-equivalent** to each of the other matrices in the example. In general:

> Two matrices are **row-equivalent** if one can be obtained from the other by means of one or more of the following **row transformations:**
>
> 1. Interchanging two rows.
> 2. Multiplying each entry in a row by the same nonzero real number.
> 3. Multiplying each entry in a row by a nonzero real number and adding the resulting product to the corresponding entry in another row.

As illustrated by the example, you can use the concept of equivalent matrices to solve a system of linear equations in two variables.

> ## Steps in Solving Systems by Matrices
>
> 1. Represent the system by its augmented matrix.
> 2. Use row transformations to obtain an equivalent matrix of the form
> $$\begin{pmatrix} 1 & 0 & p \\ 0 & 1 & q \end{pmatrix}.$$
> 3. Read the solution, (p, q), from this matrix.

EXAMPLE   Solve: $3x - 4y = 5$
$\qquad\qquad\qquad\quad 2x + 5y = -12$

SOLUTION   1. $\begin{pmatrix} 3 & -4 & 5 \\ 2 & 5 & -12 \end{pmatrix}$

2. $\begin{pmatrix} 1 & -\frac{4}{3} & \frac{5}{3} \\ 2 & 5 & -12 \end{pmatrix}$ $\frac{1}{3} \times$ Row 1

3. $\begin{pmatrix} 1 & -\frac{4}{3} & \frac{5}{3} \\ 0 & \frac{23}{3} & -\frac{46}{3} \end{pmatrix}$ Row 2 + $[(-2) \times$ Row 1$]$

4. $\begin{pmatrix} 1 & -\frac{4}{3} & \frac{5}{3} \\ 0 & \frac{1}{3} & -\frac{2}{3} \end{pmatrix}$ $\frac{1}{23} \times$ Row 2

(Solution continued on next page.)

5. $\begin{pmatrix} 1 & 0 & -1 \\ 0 & \frac{1}{3} & -\frac{2}{3} \end{pmatrix}$ Row 1 + (4 × Row 2)

6. $\begin{pmatrix} 1 & 0 & -1 \\ 0 & 1 & -2 \end{pmatrix}$ 3 × Row 2

The solution set is $\{(-1, -2)\}$.   Answer.

If, at any step in the solution process, you obtain a matrix of the form

$\begin{pmatrix} a & b & c \\ 0 & 0 & 0 \end{pmatrix}$ or $\begin{pmatrix} 0 & 0 & 0 \\ a & b & c \end{pmatrix}$, the two equations in the system are equivalent;

$\begin{pmatrix} a & b & c \\ 0 & 0 & d \end{pmatrix}$ or $\begin{pmatrix} 0 & 0 & d \\ a & b & c \end{pmatrix}$, $d \neq 0$, the two equations in the system are inconsistent.

Exercises

Use matrices to solve each system.

1. $2x - y = 0$
   $x + 3y = 7$

2. $2x - 3y = -5$
   $5x - 8y = -13$

3. $-2x + y = -6$
   $x - 3y = -2$

4. $4x - 3y = -15$
   $x + 2y = -1$

5. $3x - 5y = 17$
   $-5x + y = -21$

6. $x - 3y = 5$
   $2x - 6y = 7$

Chapter Summary

1. The graphs of two linear equations in two variables are either *parallel lines, coincident lines,* or *intersecting lines,* that is, *noncoincident* lines that intersect.

2. The *solution set* of a *system of linear equations in two variables* is the set of all ordered pairs of numbers that satisfy *all* the equations in the system.

3. To solve a system of linear equations in two variables graphically, graph the equations in the same coordinate plane and determine the coordinates of all points common to the graphs.

4. A system of equations having no solution is called an *inconsistent system.* A system having one or more solutions is called a *consistent system.*

5. Either the *linear-combination method* or the *substitution method* may be used to solve a system of linear equations in two variables.

6. A *system of linear inequalities* may be solved graphically by finding the intersection of the half-planes which are the graphs of the inequalities of the system.

7. Systems of three or more equations in *three or more variables* can be solved using the methods developed for systems of two equations in two variables.

Chapter Review

1. If the system $2x - 4y = 5$ is consistent, then: 7-1
$nx - y = 3$

  a. $n = 2$ **b.** $n = \frac{1}{2}$ **c.** $n < 0$ **d.** $n \neq \frac{1}{2}$

2. In which quadrant do the graphs of the equations in the   7-2
system $3x - y = -15$ intersect?
$y = x + 1$

  a. I   **b.** II   **c.** III   **d.** IV

3. In the system $3x - 5y = 8$ the first equation   7-3
$3y + 9 = -2x$
is multiplied by a and the second equation by b. Which choice of values for a and b will eliminate one of the variables when the resulting equations are added?

  a. $a = 1, b = -1$ **b.** $a = 2, b = 3$
  c. $a = 3, b = 5$ **d.** $a = 3, b = -5$

4. The system $3x + y = 3$ is *not* equivalent to   7-4
$2y - 3x = -12$

  a. $3x + y = 3$ **b.** $6x + 2y = 6$
     $10y - 5x = -50$ $2y - 3x = -12$
  c. $3x + y = 3$ **d.** $y = -3x + 3$
     $y = \frac{3}{2}x - 6$ $2y - 3x = -12$

5. Two angles are supplementary. The degree measure of one angle is   7-5
30 more than twice the degree measure of the other. Which system correctly expresses this information?

  a. $x + y = 180$ **b.** $x + y = 90$
     $x + 30 = 2y$ $x - 30 = 2y$
  c. $y = 180 - x$ **d.** $x + y = 180$
     $2x = y + 30$ $x - 30 = 2y$

6. If a plane's air speed is s km/h and there is a 50 km/h head wind,   7-6
how many hours will it take to fly 650 km?

  a. $\dfrac{650}{s - 50}$ **b.** 13   **c.** $\dfrac{700}{s}$ **d.** $\dfrac{650}{s + 50}$

Systems of Open Sentences | **275**

7. The tens digit of a two-digit number is twice the units digit. If the digits of the number are reversed, the original number is 12 less than twice the result. What is the number?

7-7

    a. 96   **b.** 84   **c.** 63   **d.** 42

8. Which system has the solution set shaded in the graph shown?

7-8

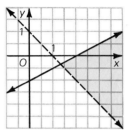

    a. $\quad x \le 2y + 2$
           $x + y > 1$

    b. $\quad y < 1 - x$
           $2 + 2y \le x$

    c. $x - 2y \ge 2$
           $x + y > 1$

    d. $x - 2y > 2$
           $x > -y + 1$

9. At which point of the convex polygonal region shown does the minimum value of $5x + 3y$ occur?

7-9

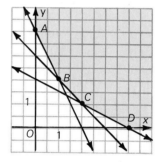

    a. A **b.** B **c.** C **d.** D

10. Tell which of the following is a solution of the system

7-10

$$2x - 2y + z = -4$$
$$x + y - 3z = -9$$
$$2y - x - z = 1$$

    a. $(3, 0, 2)$ **b.** $(3, 1, -8)$
    c. $(-8, -7, -2)$ **d.** $(-3, 0, 2)$

Chapter Test

1. Solve the system "$x + 2y = 11$ and $2x - 3y = -6$" graphically.

7-1

2. Is the system $\quad 2x - y = 5$ consistent?
                $2y - 1 = 4x + 1$

In Exercises 3–6, solve each system using the method of your choice.

7-2, 7-3, 7-4

    3. $\quad 2x - 3y = 5$
          $-5x + 3y = 1$

    4. $\quad x + 4y = 8$
          $2x + 3y = 1$

    5. $3a - 4b = -1$
          $a = 2b - 3$

    6. $2m - 5n = 21$
         $7m - 4n = 6$

7. Julio bought some apples and some oranges. He spent $4.77 and bought 26 pieces of fruit. If apples cost 23¢ each and oranges cost 12¢ each, how many of each kind of fruit did he buy? 7-5

8. Flying with the wind, a plane takes 6 h for a trip. Returning against the wind, the plane takes 7 h. If the air speed of the plane is 480 km/h more than the wind speed, find the air speed of the plane. 7-6

9. The tens digit of a two-digit number is one more than the units digit. If the digits are reversed, the original number exceeds the resulting number by one more than the tens digit of the original number. Find the original number. 7-7

10. Graph the solution set of the system: 7-8

$$x \geq 0$$
$$y \geq \tfrac{1}{3}x$$
$$x + 2y \leq 5$$

11. Find the maximum value of the expression $5x + 8y$ over the region graphed in Exercise 10. 7-9

12. Solve the system: 7-10

$$x + y + z = 3$$
$$x + 2y - z = -1$$
$$-2x - y + 3z = 11$$

================================ R. A. Fisher
1890–1962

After graduation from Cambridge University in England, Fisher was offered the job of sorting and assessing sixty-six years' data on agricultural trials for £200, and stayed on for fourteen years! During this period he became recognized as the foremost statistician of his time, and a brilliant geneticist as well. A hallmark of his work throughout his career was his blending of theory and experiment. In order to aid in understanding whether experimental results supported a theory or not, he formulated a complete theory of hypothesis testing. His work comprised a virtual revolution in theoretical statistics. Fisher passed many important ideas to others, for them to develop and take the credit, so that it's difficult to be aware of his full contribution to knowledge.

Besides statistics, Fisher made major advances in the emerging science of genetics. His research showed that the recently discovered work of Mendel was not incompatible with the theory of natural selection of Charles Darwin. He was to reach his conclusions by using new mathematical approaches, and through his characteristic combining of theory and painstaking experimentation.

Located in the Pyrenees mountains of France, this ten-story parabolic mirror is part of the world's largest solar furnace, capable of generating temperatures up to 2000°C.

8 Polynomials and Their Factors

Products

OBJECTIVES *for Sections 8-1 through 8-4:*
1. *Add or subtract polynomials.*
2. *Simplify expressions for products of powers.*
3. *Simplify expressions for products of polynomials.*

8-1 Polynomials

Expressions such as -2, $5x^2$, and $9y$ are called **monomials.** You say that:

$$-2 \quad \text{is a constant monomial}$$
$$5x^2 \quad \text{is a monomial in } x$$
$$9y \quad \text{is a monomial in } y$$

In general, a monomial in x is an expression of the form

where a denotes a nonzero real number and n denotes a positive integer.

As suggested on page 279, you call the number denoted by a the **numerical coefficient** (or simply, the coefficient) of the monomial and the number denoted by n the **degree** of the monomial. Thus:

in $5x^2$, the coefficient is 5 and the degree is 2.

in $9y$, the coefficient is 9 and the degree is 1.

A monomial like $1x^5$, in which the coefficient is 1, is usually written x^5.

Sometimes an expression like $\dfrac{2}{x}$ is called a monomial in x. However, this book will use the words "monomial in x" with the meaning given above.

A nonzero constant monomial, such as -2, is said to have **degree zero**. The zero monomial, 0, has **no degree**.

More than one variable may appear in a monomial. For instance, $12x^2y^3$ is a monomial in x and y. Its coefficient is 12 and its *degree* is $2 + 3$, the sum of the exponents of x and y. The degree of $6xy^3$ is $1 + 3$, or 4. In general, a monomial in x and y is an expression of the form

$$\underset{\text{coefficient}}{a}\overset{\text{degree} = n + m}{x^n y^m},$$

where a is a nonzero constant and n and m denote positive integers. Its coefficient is the number denoted by a, and its degree is the sum of the integers denoted by n and m. Monomials in three or more variables can be defined similarly.

Two monomials are said to be **similar**, or **like**, if they are exactly the same (except for the order of their factors) or if they differ only in their coefficients. For example, the expressions

$$9xy^4, \quad (-13)xy^4, \quad 9y^4x, \quad 178xy^4$$

are *similar*, or *like*, monomials, but

$$9xy^4, \quad 9x^4y, \quad 9y^5, \quad 9xw^4$$

are *unlike monomials*.

To describe the expression

$$3y^4x + 2y^2x^2 + y - 7$$

you can think of it as

$$3y^4x + 2y^2x^2 + y + (-7).$$

Thus you can say that it is a string of monomials connected by plus ($+$) signs. Such an expression is called a **polynomial**. The monomials in the expression are called the **terms** of the polynomial, and the coefficients of the terms are called the **coefficients** of the polynomial. The coefficient of a nonzero constant monomial is defined to be the constant itself.

Thus, the terms of

$$3y^4x + 2y^2x^2 + y - 7$$

are $3y^4x$, $2y^2x^2$, y, and -7; its coefficients are 3, 2, 1, and -7.

It is sometimes convenient to say that a monomial is a polynomial of *one* term. A polynomial of two terms is called a **binomial**; a polynomial of three terms is a **trinomial**. Thus, $x^2 + y^2$ is a binomial, and $x^2 + 2xy + y^2$ is a trinomial.

A polynomial is in **simple form** when no two of its terms are similar. For example

$$3y^4x + 2y^2x^2 + 3y - 7$$

is in simple form, but

$$7x^3 + 5x^2 + 4x^2 - x + 3$$

is *not* in simple form. As you learned in Chapter 2, you may simplify it by using the distributive axiom to add like terms:

$$7x^3 + (5 + 4)x^2 - x + 3 = 7x^3 + 9x^2 - x + 3$$

To add polynomials, you simply add the like terms of the polynomials. Thus,

$$\begin{aligned} (4x^3 + 5x - 3) + (2x^2 - 3x + 4) &= 4x^3 + 5x - 3 + 2x^2 - 3x + 4 \\ &= 4x^3 + 2x^2 + 2x + 1 \end{aligned}$$

Similarly, *to subtract one polynomial from another,* you subtract the similar terms:

$$\begin{aligned} (4x^3 + 5x - 3) - (2x^2 - 3x + 4) &= 4x^3 + 5x - 3 - 2x^2 + 3x - 4 \\ &= 4x^3 - 2x^2 + 8x - 7 \end{aligned}$$

For a polynomial *in simple form* we define the **degree** of the polynomial to be the greatest of the degrees of its terms. The degrees of the terms of

$$3y^4x + 2y^2x^2 + y - 7$$

are, in order, 5, 4, 1, and 0. The degree of this polynomial is, therefore, 5.

Oral Exercises

State whether the given expression is a monomial. If it is a monomial, give its coefficient and degree.

1. -5 **2.** r^3 **3.** $2x^2y$ **4.** $\dfrac{4}{w}$ **5.** $t^2 + 1$ **6.** $\frac{2}{3}m + n$

In Exercises 7–12 identify each polynomial as a monomial, binomial, trinomial, or as none of these. Give the coefficients and, if the polynomial is in simple form, its degree.

7. $3x^2 - 5x + 2$ **8.** $k^3 - 2k^2$ **9.** $2p^2 + 3q^2 + p^2$

10. $-7y^4$ **11.** $3xyz - 2x^2y^2z^2$ **12.** $-3t + 6t^2 - 2t^3 + 5t^4$

13. The lengths of the sides of a certain triangle are $3c^2 - 2cd + d^2$, $c^2 - d^2$, and $3d^2 + 2cd - 2c^2$. What is its perimeter?

14. If two polynomials of degree 3 are added, must their sum be a polynomial of degree 3? Explain your answer.

Written Exercises

Copy the polynomial and underline similar terms (monomials) in the same way. Simplify the polynomial.

EXAMPLE $6r^2s - 4rs^2 + 2r^2s - rs^2 + 5$

SOLUTION $\underline{6r^2s} - \underline{\underline{4rs^2}} + \underline{2r^2s} - \underline{\underline{rs^2}} + 5 = 8r^2s - 5rs^2 + 5$

A   **1.** $xy - 7 + 4xy$ **2.** $-2u + 5w - 6u$

3. $-4a^2 + 5a - 4a + 2a^2$ **4.** $-5r^2s - rs^2 + 3r^2s - 4rs$

5. $7k^4 - 2k^2 + 3k^4 - 5k$ **6.** $7abc^2 - 3a^2bc - 2abc^2 + 3ab^2c$

7. $6 + 4i^2j^2 - 3ij - 8$ **8.** $-x^3 - x^4 - 3x^4 + 2x^3 + x^4$

9. $3ef^2 - 5ef + 7e^2f - 4e^2f + ef$ **10.** $2s^2t^2 - 3s^3t + 2s^2t^2 - 4s^3t - 3st^3$

11. $9x^2yz - 2xy^2z + 3xyz - 7x^2yz + xy^2z - 1$

12. $2ab - 5abc - 3ac + 6ab + abc - ac$

13. $-3uv^2w + uvw^2 + 2u^2vw - 4uvw^2 + 3u^2vw + 9uv^2w$

14. $11j^2k^2t^2 - 21j^2t^2 + 13k^2t^2 - 4j^2k^2t^2 + 12k^2t^2 - 5j^2t^2$

15. $-3^2x^3y + 2^4x^3y - 2^2x^3y + 2^3x^3y$

16. $2^5k + 3^5p - 2^5kp + 5^2k - 5^3p + 5^2kp$

In Exercises 17–24 supply the missing exponent(s) or variable(s) to make a pair of similar terms.

17. $2a^2b^3$; $5a^?b^?$ **18.** $6rx^4$; $2\,\underline{?}\,x^4$ **19.** $-6s^2t$; $4s^?\,\underline{?}$ **20.** k^2m^3n; $5\,\underline{?}\,{}^2m^3\,\underline{?}$

21. efz^4; $\underline{?}\,fz^?$ **22.** a^2bw^4; $\underline{?}\,a^2w^?$ **23.** $15h^5t$; $\underline{?}\,h^5$ **24.** $3m^6$; $-2\,\underline{?}\,{}^?$

In Exercises 25–36 add or subtract the polynomials as indicated.

25. $(5p^2 - p - 9) + (p^3 + 2p^2 - 5)$ **26.** $(\frac{1}{2}x - \frac{3}{2}) + (\frac{3}{2}x + \frac{1}{2})$

27. $(3xy + 4x^2y) + (3xy^2 - 6xy)$ **28.** $(r^2 + 2rm + m^2) + (r^2 - m^2)$

29. $(k^2 - g^2) - (k^2 + 2g^2)$ **30.** $(m^2n + 3mn) - (7mn - mn^2)$

B **31.** $(5c^3 - 2c^2 + 3c + 2) + (c^3 + 3c^2 - 7c - 2)$

32. $(k^2 + 5k - 2k^3 - 3) + (2 - k^2 + 3k^3 - 5k)$

33. $(2g - 4 - 3g^2 + g^4) + (g^2 - g + 2g^4 - g^3 - 1)$

34. $(2m^4 + 3m^3 - m^2 + 2m - 3) - (2m^5 + m^4 - 2m + 3)$

35. $(4z^6 - 2z^4 + 3z^3 - 2z^2 - 1) - (3z^6 - 2 + z - 2z^2 + 3z^3 + z^5 - 2z^4)$

36. $(2t^6 - 5t^2 - 7 + 2t^4 - 3t^5) - (t^6 - 3t^3 + 2t^5 + 5)$

C **37.** What polynomial must be added to $2ab - 5b^2 - a^2$ to give $2a^2 - 3ab + 2b^2$?

38. What polynomial must be subtracted from $k^2 - k - 1$ to give $k - k^2$?

39. Subtract the sum of $3x^2 - 2xy + 5y^2$ and $-4x^2 + 5xy - 2y^2$ from $2x^2 + 7xy - 3y^2$.

40. From the sum of $5s - 2r$ and $3r - 5s$ subtract the sum of $7s - 2r$ and $4r + 3s$.

DIVERSION

If you were offered the following choices of a salary, which would you take?
A. $4000 at the end of 6 months and a semi-annual raise of $400.
B. $8000 at the end of the year and an annual raise of $1600.

8-2 Multiplication and Powers

Compare the two equations below.

$$\overbrace{3^2 \cdot 3^3}^{2 + 3 = 5 \text{ factors}} = \underbrace{(3 \cdot 3)}_{2 \text{ factors}} \cdot \underbrace{(3 \cdot 3 \cdot 3)}_{3 \text{ factors}} = 3^5 \qquad \leftarrow \textit{Special case}$$

$$\overbrace{a^m \cdot a^n}^{m + n \text{ factors}} = \underbrace{(a \cdot a \ldots a)}_{m \text{ factors}} \underbrace{(a \cdot a \ldots a)}_{n \text{ factors}} = a^{m+n} \qquad \leftarrow \textit{General case}$$

Do you see that, in general, you have the following *law of exponents* for products of powers?

For every real number a, and all positive integers m and n:

$$a^m \cdot a^n = a^{m+n}$$

Similarly, by comparing the special case and general case in each of the next two pairs of equations, you can discover two more laws of exponents.

$$\overbrace{}^{\text{3 factors}} \qquad \overbrace{}^{\text{3 terms}} \quad \overbrace{}^{\text{6 factors}}$$

1. $(3^2)^3 = \overbrace{(3^2)(3^2)(3^2)}^{} = \overbrace{3^{2+2+2}}^{} = \overbrace{3^{2\cdot3}}^{}$ ← *Special case*

$$\overbrace{}^{n \text{ factors}} \qquad\qquad \overbrace{}^{n \text{ terms}} \quad \overbrace{}^{mn \text{ factors}}$$

2. $(a^m)^n = \overbrace{(a^m)(a^m)\ldots(a^m)}^{} = \overbrace{a^{m+m+\cdots+m}}^{} = \overbrace{a^{mn}}^{}$ ← *General case*

> For every real number a and all positive integers m and n:
>
> $$(a^m)^n = a^{mn}$$

$$\overbrace{}^{\text{3 factors}} \qquad\qquad \overbrace{}^{\text{3 factors}} \quad \overbrace{}^{\text{3 factors}}$$

1. $(2 \cdot 5)^3 = \overbrace{(2 \cdot 5)(2 \cdot 5)(2 \cdot 5)}^{} = \overbrace{(2 \cdot 2 \cdot 2)(5 \cdot 5 \cdot 5)}^{} = 2^3 \cdot 5^3$ ← *Special case*

$$\overbrace{}^{m \text{ factors}} \qquad\qquad \overbrace{}^{m \text{ factors}} \quad \overbrace{}^{m \text{ factors}}$$

2. $(ab)^m = \overbrace{(ab)(ab)\ldots(ab)}^{} = \overbrace{(a \cdot a \ldots a)(b \cdot b \ldots b)}^{} = a^m b^m$ ← *General case*

> For all real numbers a and b and each positive integer m:
>
> $$(ab)^m = a^m b^m$$

The three preceding laws are called the laws of exponents for multiplication.

By using these laws of exponents, along with the associative and commutative properties of multiplication, you can **multiply** monomials; that is, **simplify expressions for products** of monomials.

EXAMPLE   Simplify: $(-3a^3b^2)(5ab^4)^2$

SOLUTION $(-3a^3b^2)(5ab^4)^2 = (-3a^3b^2)[5^2a^2(b^4)^2]$
$= (-3a^3b^2)(25a^2b^8)$
$= (-3 \cdot 25)(a^3 \cdot a^2)(b^2 \cdot b^8)$
$= -75a^5b^{10}$

Oral Exercises

Simplify each expression.

1. $e^2 \cdot e^3$ 2. $f^5 \cdot f^3$ 3. $c \cdot c^3$ 4. $(xy)^2$

5. $(k^3)^2$ **6.** $(2p^2)^2$ **7.** $n \cdot n^2 \cdot n^4$ **8.** $rm^2 \cdot rm \cdot r^3m$

9. $(-2x^2)^3$ **10.** $-(2x^2)^3$ **11.** $(-x^2)(x)(x^2)^4$ **12.** $p^c \cdot p^d$

Give the square of each expression.

13. $-3q^2$ **14.** $2z^3$ **15.** $3k^2j$ **16.** $-5x^4y^3$

Give the cube of each expression.

17. x^2 **18.** $2m$ **19.** $3a^3$ **20.** $-2mn^2$

Written Exercises

Simplify.

A

1. $(2x^2)(-3x)$ **2.** $(4y^3)(3y^4)$ **3.** $(5c)(-3c^2)$

4. $(7kl^2)(-l^2)$ **5.** $(x^2y)(xy^2)$ **6.** $(-3ab^2)(a^2b^3)$

7. $(2p)(-p^4)(7p^3)$ **8.** $(5m^2)(-3m^3)(m^2)$ **9.** $(0.2n^2)(4n^2)(5n)$

10. $(6e^2)(0.5e)(10e^3)$ **11.** $(\frac{1}{4}s^2)(3s)(\frac{4}{3}s^3)$ **12.** $(\frac{1}{6}a^3)(12a)(-2a^3)$

13. $(2k)^2(-2k^2)$ **14.** $(4n^2)(-4n)^2$ **15.** $(-r)(rs^2)^2(r^2s)^3$

16. $(3ef)^2(2e^3f)^3(-2e)$ **17.** $(-3m^2n)^2(5mn^2)(-2mn)$ **18.** $(4p^2q)^2(3pq^3)(p^3q^2)$

Find a polynomial equivalent to the given expression.

B

19. $5x^3 - x(-3x^2)$ **20.** $(y^2)(2y) + (y)(y^2)$ **21.** $(3x^2y)(2x^2y^3) + (-xy)^4$

22. $(2z^3)(2z)^4 - (4z)^2(2z^5)$ **23.** $a^5(a^2) + 3a(a^2) - a^2(a)$

24. $3ab(ab) + 8a(ab^2) - 2(ab)^2$ **25.** $(-3x)(2x^2y) + (5x)[xy(-3y)]$

26. $(\frac{1}{2}x)^2(24x^3) + (-2x)^3(\frac{1}{4}x^2)$ **27.** $(-4s^2)(3rt^2)(-r^2) + (-2rst)(4r^2st)$

28. $(6x^2yz^3)(-3xy^2z) + (-2z^2)(-9x^3y^2)(yz^2)$

29. $-(-5c)(2^2a^5b^3)(5ab^7) + (3^3b^6a^3c)(-a^2b^2)(ab^2)$

30. $(5^2x^3y^2z)(-3x^4y^4z^5) - (-z^3x^2y^3)(7^2x^5y^4z^2)$

C

31. $(2xy)(3x)^3(-y) + (2y)(3xy)(2x)^3 - (-2xy)(y)(xy)^2$

32. $(7y^2)(2x^2) - (-4y)(-3y)(-5x^2) + (5y^2)(-2x)(3x)$

33. $(4xyz)(-3x^2yz) - (2x)^2(3xy^4) - (5y^2)(2xy^2)(-x^2)$

34. $[(2xy^2)(-3x^2y) + (2y^3)(5x^3)] - (-2y^3)(-3x)(5x^2y)$

In Exercises 35–40 assume each expression in an exponent denotes a positive integer, and simplify.

35. $k^n \cdot k^{2n}$ **36.** $c^{3n} \cdot c^{n-1}$ **37.** $x^{n-2} \cdot x^{2n+2}$

38. $a^{2n} \cdot a^{2-2n}$ **39.** $10^{n-1} \cdot 10^{5-n}$ **40.** $y^n \cdot y^{2n-1} \cdot y$

8-3 Multiplying Polynomials

Can you express the product

$$3a(2a + 5)$$

as a polynomial in simple form? You can by using the distributive axiom, together with the commutative and associative axioms for multiplication. Thus, for all real numbers a:

$$3a(2a + 5) = 3a(2a) + 3a(5) = 6a^2 + 15a$$

You call $6a^2 + 15a$ the **product** of the monomial $3a$ and the polynomial $2a + 5$.

EXAMPLE 1 Express $-2t^3(3t^2 - 2t + 5)$ as a polynomial in simple form.

SOLUTION $-2t^3(3t^2 - 2t + 5) = -2t^3(3t^2) + (-2t^3)(-2t) + (-2t^3)(5)$
$$= -6t^5 + 4t^4 - 10t^3.$$

To simplify an expression such as

$$(a - 2)(2a^2 + 3a - 1),$$

you apply the distributive axiom successively, treating $2a^2 + 3a - 1$ as a number to be multiplied by $a - 2$. Thus:

$$(a - 2)(2a^2 + 3a - 1) = a(2a^2 + 3a - 1) - 2(2a^2 + 3a - 1)$$
$$= a(2a^2) + a(3a) + a(-1)$$
$$\quad -2(2a^2) - 2(3a) - 2(-1)$$
$$= 2a^3 + 3a^2 - a - 4a^2 - 6a + 2$$
$$\therefore (a - 2)(2a^2 + 3a - 1) = 2a^3 - a^2 - 7a + 2$$

When you write $(a - 2)(2a^2 + 3a - 1)$ as $2a^3 - a^2 - 7a + 2$, you say you have **simplified the expression** $(a - 2)(2a^2 + 3a - 1)$, or that you have **multiplied the polynomials** $a - 2$ and $2a^2 + 3a - 1$ to obtain the product $2a^3 - a^2 - 7a + 2$. You can also use a vertical arrangement to obtain products of polynomials, as illustrated in Example 2.

EXAMPLE 2 Simplify: $(y + 3)(y^3 - 2y^2 + 3)$

SOLUTION

$$
\begin{array}{llll}
& y^3 - 2y^2 & & + 3 \\
& \underline{y + 3} & & \\
y(y^3 - 2y^2 + 3) \rightarrow & y^4 - 2y^3 & & + 3y \\
3(y^3 - 2y^2 + 3) \rightarrow & \underline{\quad\quad 3y^3 - 6y^2} & & + 9 \\
(y + 3)(y^3 - 2y^2 + 3) \rightarrow & y^4 + \quad y^3 - 6y^2 & + 3y & + 9
\end{array}
$$

Notice that in the first line of the vertical arrangement a space is left for the "missing term" that you could show by writing $0y$ between $-2y^2$ and $+3$. The expressions at the left of the arrows above are usually omitted.
The preceding examples suggest the following rule.

Rule for Multiplying Polynomials

To multiply two polynomials, multiply each term of one of the polynomials by each term of the other, and add the resulting products.

In solving a word problem, you sometimes have to multiply polynomials.

EXAMPLE 3   One rectangle is twice as long as it is wide. A second rectangle is 3 cm longer and 1 cm narrower than the first rectangle, but has the same area. Find the dimensions of the first rectangle.

SOLUTION   1. The problem asks for the dimensions of the first rectangle.

2.   Let x = the width of the first rectangle, so that
               $2x$ = the length of the first rectangle.
         Then $x - 1$ = the width of the second rectangle, and
               $2x + 3$ = the length of the second rectangle.

3. Measured in square centimeters:

<u>Area of first rectangle</u>   is equal to   <u>area of second rectangle</u>
         $2x(x)$ $=$ $(2x + 3)(x - 1)$

4. $2x^2 = 2x^2 + x - 3$
         $2x^2 - 2x^2 = 2x^2 + x - 3 - 2x^2$
               $0 = x - 3$
         $\therefore x = 3$ and $2x = 6$

5. The check is left to you.
   The dimensions of the first rectangle are 3 cm and 6 cm. **Answer.**

Oral Exercises

Simplify the given expression.

1. $4(x - 3)$ **2.** $-3(2y + 5)$ **3.** $-1(-5f + 8)$

4. $k(k^2 - 2k + 3)$ **5.** $-a^2(3a^2 - 2a^4)$ **6.** $-5t^3(4 - 3t + 7t^2)$

7. Use the distributive axiom to show that $(2x - 1)(3x - 2) = (1 - 2x)(2 - 3x)$.

Written Exercises

Simplify.

A
1. $k^2(k^3 + 2k - 1)$
2. $x^4(x^3 - 2x^2 + 3)$
3. $-y^3(y^2 - 4y + 2)$
4. $n^2(n^3 - 6n - 4)$
5. $rs(r^2 + rs - s^2)$
6. $4x^2y(3x^2y^2 - 4xy + 3)$
7. $5p^2(-2p^3 + 4p^2 + 2p - 1)$
8. $3r^2s(2rs^2 - 5r^2s^2 + 4r^2s)$
9. $-x^2y^3(-5xy^2 - 2x^2y + xy^3)$
10. $9xk^3(6x^3k - 4x^2k^2 + xk^3)$
11. $(y + 2)(y + 1)$
12. $(x + 3)(x + 2)$
13. $(z - 3)(z + 2)$
14. $(w + 6)(w - 3)$
15. $(s - 4)(s - 5)$
16. $(r - 9)(r - 1)$

B
17. $(3m + 1)(2m - 3)$
18. $(5n + 2)(2n - 1)$
19. $(2h + 3)^2$
20. $(6i - 2)^2$
21. $(5r - 3)(3r - 5)$
22. $(5x - 2)(5x + 2)$
23. $(x - 3)(x^2 + 3x + 9)$
24. $(z + 5)(2z^2 - 5z + 3)$
25. $(x - y)(x^2 + xy + y^2)$

Find the solution set of the given open sentence over $\mathcal{R}$.

EXAMPLE $(x - 3)2x + x(9 - 2x) \leq -27$

SOLUTION $2x^2 - 6x + 9x - 2x^2 \leq -27$
$$3x \leq -27$$
$$x \leq -9$$
$\therefore$ the solution set is $\{x: x \leq -9\}$.

26. $x(x + 2) - x(x - 3) = -7$
27. $y(3 - 2y) + 2y(y - 2) = 3$
28. $(m + 1)^2 = m^2 + 5$
29. $p^2 - (p + 1)^2 = -8$
30. $(k - 1)^2 - k^2 \geq -3$
31. $7 > 4r^2 - (1 - 2r)^2$
32. $(x + 1)(x + 2) - (x + 1)^2 < -9$
33. $(s + 3)^2 - (s + 5)(s - 3) \leq -4$

Simplify. Assume that all variable exponents denote positive integers.

C
34. $(2g + 5)(g^2 - g - 1) + (g^2 + 7g + 6)(g - 3)$
35. $(j - 3)^2 - (j^2 + j - 3)(j^2 + 5)$
36. $(2k + 1)(k^2 + k + 1) - (3k - 2)(2k^2 - k - 3)$
37. $(a + b)(a^2 - ab + b^2) - (a - b)(a^2 + ab + b^2)$
38. $c^n(3c^n - 2c + 5)$
39. $e^{n+1}(e^{2n} + f)$
40. $(3m^n + 1)(m^n - 3)$
41. $(r^{2n} - 6)(r^{2n} + 6)$

42. Explain why the degree of the product of two nonzero polynomials is the sum of the degrees of the polynomials.

43. Give an example to show that the degree of the sum of two polynomials may be less than the degree of either of the polynomials.

Problems

Solve each problem.

A

1. A side of one square is 2 cm longer than a side of a second square, and the area of the second square is 36 cm² less than that of the first square. Find the length of a side of each square.

$(s + 2)^2$

$s + 2$

s^2

s

2. One rectangle is 3 cm longer and 4 cm wider than a second rectangle, and their areas differ by 56 cm². The smaller rectangle is 4 cm longer than it is wide. Find the dimensions of the rectangle with the smaller area.

3. The difference of the squares of two consecutive positive integers is 55. Find the integers.

4. The difference of the squares of two consecutive odd integers is 72. Find the integers.

B

5. The length of a rectangular box is 6 cm greater than the width. The height of the box is 15 cm less than the width. The sum of the areas of the top and bottom is 1092 cm² greater than the sum of the areas of the ends. Find the dimensions of the box.

6. A rectangular box is 4 cm longer and 3 cm narrower than a certain cube. The rectangular box and the cube have equal heights and equal surface areas. Find the length of an edge of the cube.

7. The product of two consecutive even positive integers is 16 greater than the square of the lesser of the integers. Find the integers.

8. The product of three consecutive integers is 15 less than the cube of the middle integer. Find the integers.

C

9. A rectangular picture is 3 cm longer than it is wide. The picture is surrounded by a frame 2 cm wide. The area of the picture and frame is 100 cm² greater than that of the picture. Find the width of the picture.

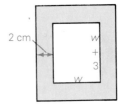

10. A guidebook says that a certain castle, twice as long as it is wide, is surrounded by a moat 6 m wide. The area of the castle floor (the area of the rectangle inside the moat) is 864 m² less than the area of the region bounded by the outer edge of the moat. Find the dimensions of the castle.

11. A concrete walk around a circular fishpond is 1 m wide. At $2/m², the total cost of the walk came to $81.64. Find the area of the surface of the pond. (Use $\pi \approx 3.14$.)

12. Georgia wants to build three boxes, each of them a cube. The length of an edge of one box is to be the average of the lengths of the edges of the other two boxes, and the volume of that box is to be the average of the volumes of the other two. Write an equation that describes this situation. Can you discover a solution?

8-4 Finding Products at Sight

You can ordinarily multiply two binomials such as $2x + 1$ and $3x - 2$ mentally. The procedure for obtaining each term of the product is outlined below.

$$
\begin{array}{r}
2x + 1 \\
3x - 2 \\
\hline
6x^2 + 3x \\
- 4x - 2 \\
\hline
6x^2 - x - 2
\end{array}
$$

1. Multiply the first terms of the binomials.
2. Multiply the first term of each binomial by the second term of the other; add the products.
3. Multiply the second terms of the binomials.

The following display shows the process horizontally.

$$
(2x + 1)(3x - 2) = \overset{1}{6x^2} + \overset{2}{3x} - \overset{3}{4x} - 2 = 6x^2 - x - 2
$$

A special case of this procedure arises when you *square a binomial*:

$$(a + b)^2 = a^2 + 2ab + b^2$$

1. Square the first term of the binomial.
2. Double the product of the two terms.
3. Square the second term of the binomial.

Similarly: $(a - b)^2 = a^2 - 2ab + b^2$

Other examples of squaring a binomial are as follows:

$$(a + 2)^2 = a^2 + 4a + 4$$
$$(3b - 5)^2 = 9b^2 - 30b + 25$$
$$(2x - 3y)^2 = 4x^2 - 12xy + 9y^2$$

When you know how to square a binomial at sight, you can use your knowledge to find the cube (the third power) of the binomial, and even higher powers of the binomial:

$(a + b)^3 = (a + b)(a + b)^2$
$ = (a + b)(a^2 + 2ab + b^2)$

To complete the multiplication, you can use the vertical arrangement shown at the right.

$$
\begin{array}{r}
a^2 + 2ab + b^2 \\
a + b \\
\hline
a^3 + 2a^2b + ab^2 \\
a^2b + 2ab^2 + b^3 \\
\hline
a^3 + 3a^2b + 3ab^2 + b^3
\end{array}
$$

Thus you can write $(a + b)^3 = a^3 + 3a^2b + 3ab^2 + b^3$.

Another special case of the procedure for multiplying binomials occurs when *the first terms of the binomials are the same, but the second terms are opposites of each other*:

$$(a + b)(a - b) = a^2 + (ab - ab) - b^2$$
$$= a^2 + 0 - b^2$$
$$= a^2 - b^2$$

Thus, $\qquad (a + b)(a - b) = a^2 - b^2$.

This special case is used in the following examples:

$$(x + 2)(x - 2) = x^2 - 4$$
$$(3t - 1)(3t + 1) = 9t^2 - 1$$
$$(2y^2 + 6)(2y^2 - 6) = 4y^4 - 36$$

Oral Exercises

Simplify.

1. $(a + 3)(a + 2)$
2. $(b - 3)(b + 2)$
3. $(c - 2)(c - 3)$
4. $(d - 4)(d - 6)$
5. $(e - 5)(e + 5)$
6. $(7 - f)(7 + f)$
7. $(2g - 1)(4g - 2)$
8. $(3h - 4)(2h + 5)$
9. $(2i + 1)^2$
10. $(3j - 2)^2$
11. $(4k^2 - 3)(k^2 + 5)$
12. $(7l + 4)(4l - 7)$
13. Show that $(2x - 3)^2 \neq 4x^2 + 9$.
14. Show that $(4x - 3)(2x - 5) = -(4x - 3)(5 - 2x)$.

Written Exercises

In Exercises 1–10 simplify the expression by using the fact that one factor is the sum and the other factor is the difference of two particular real numbers.

EXAMPLE 48(52) SOLUTION $48(52) = (50 - 2)(50 + 2) = 50^2 - 2^2$
$$= 2500 - 4 = 2496$$

A
1. 18(22)
2. 17(23)
3. 28(32)
4. 39(41)
5. 58(62)
6. 45(55)
7. 47(53)
8. 66(74)
9. 87(93)
10. 196(204)

Simplify each expression.

11. $(2a - 1)(5a + 2)$
12. $(3b - 5)(2b + 3)$
13. $(5c + 1)(7c + 8)$
14. $(4d - 5)(3d - 7)$
15. $(7 + 9e)(5 - 2e)$
16. $(3 + 2f)(2 - 3f)$
17. $(4g + 9h)(3g - 5h)$
18. $(4i - 3j)(2i + 7j)$
19. $(tx + c)^2$
20. $(x - 3y)^2$
21. $(13k - 8)(13k + 8)$
22. $(11l - 12)(11l + 12)$

Simplify.

23. $(8 + 3m^2)(3m^2 - 8)$ **24.** $(6n^2 - 7)(7 + 6n^2)$ **25.** $(-4p^2 + 3)(4p^2 + 3)$

B **26.** $r(r + 4)(2r - 3)$ **27.** $2s(4s - 3)(s - 4)$ **28.** $(t + 8)(t - 4)^2$

29. $(u - 3)(3u - 2)^2$ **30.** $(v - 2)(3v)(2v - 5)^2$ **31.** $(w + 3)(4w - 1)^2(-2w)$

32. $(4x - 3)^3$ **33.** $(5y + 2)^3$ **34.** $(1 - x)^3$

Solve and check the given open sentence.

35. $(2x + 5)(x - 3) = (x + 7)(2x + 3)$ **36.** $(2y - 1)(3y + 2) = (6y + 5)(y - 3)$

37. $(3t + 3)(4t - 3) = (3 - 4t)(-3t - 2)$ **38.** $(2w + 4)(4w + 5) = (4w - 1)(2w + 6)$

39. $(2x + 5)(6x - 1) = (4x + 3)(3x - 2)$ **40.** $(8y + 3)(y - 2) = (4y - 1)(2y + 5)$

In Exercises 41–50 assume that all variables in exponents denote positive integers, and simplify the expression.

C **41.** $(2x^2 + x - 2)(x^2 + 3x + 5)$ **42.** $(c + 2d + 3e)(2c - d + 2e)$

43. $(p + 1)(p^4 - p^3 + p^2 - p + 1)$ **44.** $(q - 1)(q^4 + q^3 + q^2 + q + 1)$

45. $(x^n - 3)(2x^n + 1)$ **46.** $(y^{2n} - 4)(y^{2n} + 4)$ **47.** $(r^{2n} - 2s^n)(r^{2n} + s^n)$

48. $(5w^n - z^n)(w^n + 5z^n)$ **49.** $(e^n + e^m)(e^n - e^{2m})$ **50.** $(f^{2n} - f^n)^3$

Self-Test 1

VOCABULARY monomial (p. 279) polynomial (p. 280)
numerical coefficient (p. 280) degree of a polynomial (p. 281)

Add or subtract the polynomials.

1. $(6a^2 + 4a - 5) + (a^2 - 7a + 5)$ *Obj. 1, p. 279*

2. $(5x^3 + 9x^2 + 9) + (6x^2 - 5x + 3)$

3. $(4x^3 - x) - (3x^3 + 6x^2 + x)$

4. $(7x^3 - x^2 + 9) - (8x^2 - 10)$

Simplify.

5. $3x^2 \cdot x^3$ **6.** $(2x)^2 x^5$ **7.** $(r^2s)^3 \cdot rs^2$ *Obj. 2, p. 279*

Simplify.

8. $3k(k^2 - 5)$ **9.** $(m + 2)(m - 5)$ *Obj. 3, p. 279*

10. $(4n - 3)(3n + 4)$ **11.** $(2p + 3)^2$

12. $(6r + 5)(6r - 5)$ **13.** $(3s^2 - 1)(1 + 3s^2)$

Check your answers with those at the back of the book.

Factors

OBJECTIVES *for Sections 8-5 through 8-9:*
1. *Factor monomials and find the greatest common factor of two or more given monomials.*
2. *Completely factor quadratic polynomials.*

8-5 Factoring Monomials

To **factor a number over a given set of numbers** means to express the number as a product of two or more members of the given set (called the **factor set**). For example, over the set J of integers, 96 can be factored as

$$96 = 48(2), \quad 96 = 16(6), \quad 96 = 2^5(3),$$

or in a number of other ways. However, if you write $96 = \frac{1}{2}(192)$, you have not factored 96 over J because $\frac{1}{2}$ is not an integer.

An important subset of J, which is often used as the factor set, is the set of *prime numbers*. A **prime number** is an integer *greater than 1* that has no positive integral factors other than itself and 1. The first ten primes are 2, 3, 5, 7, 11, 13, 17, 19, 23, 29.

To express a positive integer as a product of primes, you can sometimes proceed in more than one way. For instance:

$$
\begin{aligned}
96 &= 12 \cdot 8 \\
&= 4 \cdot 3 \cdot 8 \\
&= 2 \cdot 2 \cdot 3 \cdot 8 \\
&= 2 \cdot 2 \cdot 3 \cdot 4 \cdot 2 \\
&= 2 \cdot 2 \cdot 3 \cdot 2 \cdot 2 \cdot 2
\end{aligned}
\qquad
\begin{aligned}
96 &= 32 \cdot 3 \\
&= 16 \cdot 2 \cdot 3 \\
&= 8 \cdot 2 \cdot 2 \cdot 3 \\
&= 4 \cdot 2 \cdot 2 \cdot 2 \cdot 3 \\
&= 2 \cdot 2 \cdot 2 \cdot 2 \cdot 2 \cdot 3
\end{aligned}
$$

Either way, you obtain $96 = 2^5(3)$, which is the one and only factorization of 96 over the set of prime numbers. Other prime factorizations merely vary the order in which the prime factors 2 and 3 appear.

Once you have the prime factorization of a positive integer, you can easily list all its positive integral factors. Let A be the set of positive integral factors of $96 = 2^5(3)$. Then:

$$
\begin{aligned}
A &= \{1, 2, 2^2, 2^3, 2^4, 2^5, 3, 2(3), 2^2(3), 2^3(3), 2^4(3), 2^5(3)\} \\
&= \{1, 2, 4, 8, 16, 32, 3, 6, 12, 24, 48, 96\}.
\end{aligned}
$$

Now, let B be the set of positive integral factors of 60. Thus:

$$B = \{1, 2, 3, 4, 5, 6, 10, 12, 15, 20, 30, 60\}.$$

Then the common positive integral factors of 96 and 60 are the members of the intersection of A and B.

$$A \cap B = \{1, 2, 3, 4, 6, 12\}$$

The greatest member of $A \cap B$, 12, is called the *greatest common factor* of 96 and 60. In general, the **greatest common factor** (G.C.F.) of two or more integers is the greatest integer that is a factor of each of the given integers.

It is not necessary to search the roster of $A \cap B$ to find the G.C.F. of 96 and 60. You can proceed in the following way:

1. Factor each number over the set of primes.

$$96 = 2^5 \cdot 3; \ 60 = 2^2 \cdot 3 \cdot 5$$

2. Compare the powers of each prime which is a factor of *both* of the numbers, and for each such prime choose the lesser power.

$2^2 < 2^5$; choose 2^2
$3 = 3$; choose 3

3. The G.C.F. is the product of the powers chosen in Step 2.

$\therefore$ the G.C.F. of 96 and 60 is
$$2^2 \cdot 3 = 12.$$

To **factor a monomial** whose numerical coefficient is an integer, you seek other monomials with integral coefficients whose product is the given monomial. Thus, over the set of monomials with integral coefficients, the factors of $-2c^2$ are

$$1, -1, 2, -2, c, -c, 2c, -2c, c^2, -c^2, 2c^2, \text{ and } -2c^2.$$

If you write a similar list of factors of $6cd$, you will find that the common factors of $-2c^2$ and $6cd$ are

$$1, -1, 2, -2, c, -c, 2c, \text{ and } -2c.$$

We call $2c$ the greatest common factor of $-2c^2$ and $6cd$. The **greatest common factor** of two or more monomials whose numerical coefficients are integers is the monomial with the greatest numerical coefficient and the greatest degree that is a factor of each of the given monomials.

The following example shows an efficient way to determine the G.C.F. of monomials.

EXAMPLE Find the G.C.F. of $-102x^2y^4$ and $42x^7y^3$.

SOLUTION

1. Find the G.C.F. of the numerical coefficients of the monomials.

$-102 = (-1) \cdot 17 \cdot 3 \cdot 2$;
$42 = 7 \cdot 3 \cdot 2$

$\therefore$ G.C.F. of the coefficients is $3 \cdot 2$, or 6.

2. Compare the powers of each variable which is a factor of *both* monomials, and choose the power in which the exponent is *least*.

Compare x^2 and x^7; choose x^2.
Compare y^4 and y^3; choose y^3.

3. The G.C.F. is the product of the constant determined in Step 1 and the powers chosen in Step 2.

$\therefore$ the G.C.F. of $-102x^2y^4$ and $42x^7y^3$ is $6x^2y^3$. **Answer.**

Oral Exercises

In Exercises 1–8 tell why the given statement is true or why it is false.

EXAMPLE   **a.** 5 is a factor of 0 over J.

             b. 6 is a factor of 14 over J.

SOLUTION   **a.** True, because $0 \in J$ and $5 \cdot 0 = 0$.

             b. False, because there is no integer n such that $6n = 14$.

1. 5 is a factor of 15 over J.

2. 1 is a factor of every integer.

3. 4 is a factor of 10 over J.

4. 0 is a factor of 7 over J.

5. 17 has no prime factors.

6. 2 is not a factor of 8 over J.

7. 0 is a factor of every integer.

8. Every integer has a prime factor.

In Exercises 9–16 name the greatest common factor of the given monomials.

9. 8, 18

10. -16, 32

11. $6x$, $15x^2$

12. $15ab^2$, $-10a^2b$

13. $30c^3d$, $35c^2d^2$

14. $6e^3f$, $11g^2$

15. $20s^3t^2$, $25st^3$

16. $36x^4y^3$, $54x^2y^5$

Written Exercises

Factor each integer over the set of primes.

A   **1.** 36

2. 75

3. 29

4. 200

5. 480

6. 500

7. 3640

8. 1800

Specify the set of positive integral factors of the given number by roster.

9. 24

10. 32

11. 58

12. 70

13. 90

14. 100

15. 121

16. 169

Find the greatest common factor of the given monomials.

17. 16, 42

18. 24, 108

19. 325, 350

20. 693, 882

21. $80x^2y^3$, $52x^3y$

22. $35a^3n$, $105a^2n^3$

23. $-38r^3s^2$, $114r^5s^7$

24. $-108w^2t^5$, $114w^3t^7$

25. $176d^3e^3f^2$, $208d^2e^3f^4$

Name the monomial factor, if there is any, by which the first monomial must be multiplied to obtain the second monomial.

B   **26.** $6x$, $6x^2y$

27. $12xy$, $24x^2y$

28. $-7a^2b$, $-56a^4b^3$

29. $5a^2b^3$, $125a^8b^9$

30. $2r^2s^2$, $8r^6s^6$

31. 0, $32efg^5$

32. $-12ef^2g$, $-48e^3f^3g$

33. $7m^2np$, $-28m^5n^2p^3$

34. $19m^2n^3$, 1

Polynomials and Their Factors | **295**

Name the greatest common factor of the terms in the polynomial.

35. $12x^3 + 72x^2 - 120x$

36. $-18y^6 + 12y^5 - 6y^4$

37. $27e^2f^3 - 36e^3f^4 + 9e^4f^5$

38. $48e^4f^2 - 24e^2f^3 - 72e^5f^2$

39. $9x^{2n} + 12x^{3n} + 15x^{4n}$
    (*Note: n* is a positive integer.)

40. $a^2b^3c^2 - ab^2c + a^3b^2c^3$

programming in BASIC

In this section you were shown a method for listing all the positive integral factors of a positive integer. This method used the prime factorization of the integer. Another method is to divide the given integer systematically by 2, 3, 4, . . . to see whether or not the division is exact. That is a good job for a computer.

Try this program:

```
10  LET N=84
20  FOR F=2 TO N/2
30  LET Q=N/F
40  PRINT F; Q,
50  NEXT F
60  END
```

When you RUN this program, you will be able to pick, out of all the quotients listed, those that are integers. These indicate the factors 2, 3, 4, 6, 7, 12, 14, 21, 28, 42.

But we would like a program that will print out *only* the values of F for which Q is an integer. The computer has a special function that will help here:

INT(X) will find the greatest integer less than or equal to X.

In the program above, if Q = INT(Q) then Q is an integer and F is a factor of N.

Try this revised program for N=84:

```
10   PRINT "N=";
20   INPUT N
30   PRINT "THE FACTORS OF";N; "ARE: 1";
40   FOR F=2 TO N/2
50   LET Q=N/F
60   IF Q<>INT(Q) THEN 80
70   PRINT F;
80   NEXT F
90   PRINT N
100  END
```

Exercises

1. Try the preceding program for 85, 86, 87, 88, 89.
2. What does it mean when the computer prints out only 1 and the value of N?
3. Use the computer to list the factors of 108 and 60. Check off the common factors and find the GCF.
4. Write a program that will print out the prime numbers less than 500.

8-6 Factoring Monomials from Polynomials

You **factor a polynomial** by expressing it as a product of polynomials belonging to a given set. In this book, unless otherwise stated, *the factor set for a polynomial whose terms have integral coefficients will be the set of all polynomials whose terms have integral coefficients.*

To factor a common monomial factor from the terms of a polynomial, you simply use the distributive axiom. For example, to factor

$$48x^3 - 42x^2,$$

note that $6x^2$ is the greatest common factor of the terms. In fact:

$$48x^3 - 42x^2 = 6x^2(8x) - 6x^2(7)$$

Therefore, by the distributive axiom:

$$48x^3 - 42x^2 = 6x^2(8x - 7).$$

We call $6x^2$ the **greatest monomial factor** of $48x^3 - 42x^2$ because it is the greatest common factor of the terms of the polynomial.

Notice that the degree of $48x^3 - 42x^2$ is 3, which is greater than 2 (the degree of $6x^2$) and also greater than 1 (the degree of $8x - 7$). (In fact, $3 = 2 + 1$.)

A polynomial, like $48x^3 - 42x^2$, that can be expressed as a product of two or more nonconstant polynomials of lower degree is said to be **reducible**. Is $6x^2$ reducible? Yes, it is, because $6x^2 = 6x \cdot x$. But $8x - 7$ is *not* reducible, and so is called an **irreducible** polynomial. Notice that 1 is the greatest monomial factor of $8x - 7$.

When you write

$$48x^3 - 42x^2 = 6x^2(8x - 7)$$

you have *factored* the given polynomial *completely*. The **factorization** of a polynomial is **complete** when the polynomial is expressed as the product of a monomial and one or more irreducible polynomials, each of which has 1 as its greatest monomial factor. If the polynomial itself is irreducible and has 1 as its greatest monomial factor, then it is already factored completely. Thus $2x - 6$ is *irreducible,* but to *factor* it *completely* you write $2x - 6 = 2(x - 3)$. On the other hand, $8x - 7$ is irreducible and is itself factored completely.

Is there more than one complete factorization of $48x^3 - 42x^2$? Except for unimportant changes in the factors, such as changing their order as in $6x(8x - 7)x$ or introducing -1 as a factor as in $-6x^2(-8x + 7)$, the complete factorization is unique.

Oral Exercises

State the greatest monomial factor of the polynomial. If the polynomial is irreducible, so state.

1. $2a - 8$
2. $kb^2 + |k$
3. $2c^3 - 2c^2 + 5$
4. $px - py + qz$
5. $x^2 - x + xy$
6. $d + de + d^2$
7. $15r^2 - 18r + 45$
8. $2\pi^2 + 2\pi rh$

Written Exercises

A 1–8. In Oral Exercises 1–8 factor the polynomial completely.

In Exercises 9–32, factor completely.

9. $12x^2y + 16xy^2 - 4xy$
10. $5x^2y - 10xy^2 + 35xy$
11. $15a^2b^3 - 18a^5b^2 + 24ab^4$
12. $30a^2b^3 + 48a^2b^6 - 36a^2b^2$
13. $3p^4 - 12p^3 + 6p^2 - 21p$
14. $14p^4 + 28p^2 - 35p - 7p^5$
15. $30m^2n - 24mn^2 + 36m^3n$
16. $75m^5n^2 - 225m^2n^5 + 375m^3n^3$

EXAMPLE $p(p - 2) - 7(p - 2)$

SOLUTION   Consider $(p - 2)$ to be a single number.
$$p(p - 2) - 7(p - 2) = (p - 2)(p - 7)$$

17. $a(a - 3) + 5(a - 3)$
18. $b(b - 5) - 2(b - 5)$
19. $2c(3c + 1) - (3c + 1)$
20. $4d(2d - 3) - 3(2d - 3)$
21. $4x(3x - 7) + 5(3x - 7)$
22. $y(y^2 - 1) - (y^2 - 1)$

B 23. $(x + 3)^2 + 2(x + 3)$
24. $(y - 1)^2 - 3(y - 1)$
25. $4(z + 5)^2 - 3(z + 5)$
26. $7(z - 3)^2 + 5(z - 3)$
27. $5(2w + 3)^2 - 4(2w + 3)$
28. $6(w + 4)^2 - (w + 4)$
29. $e^2(e - f) + f(e - f)$
30. $3e(e^2 + 1) + 4(e^2 + 1)$
31. $5x(y^3 + z) - (y^3 + z)$
32. $x(x + 2y) - y(x + 2y)$

Supply the missing binomial to make the sentence a true statement. Assume that variable exponents represent positive integers.

C 33. $a^{2n} - 2a^n = a^n(\underline{\ ?\ })$
34. $b^{3n} + b^n = b^n(\underline{\ ?\ })$

35. $4c^{n+1} - 2c^n = 2c^n(\underline{\ ?\ })$

36. $d^{n+1} - 9d^n = d^n(\underline{\ ?\ })$

37. $e^{n-1} + 4e^n = e^{n-1}(\underline{\ ?\ })$

38. $f^{n-1} - 5f^n = f^{n-1}(\underline{\ ?\ })$

If $n \in J$, find the value of n that satisfies the equation.

39. $2^{2n+1} = 2^n(4)$

40. $\left(\frac{1}{2}\right)^{2n-3} = \frac{1}{32}$

41. $(3^{2n-1})(3^{n+2}) = 9^2$

42. $4(2^{2n-1}) = 8 \cdot 2^4$

8-7 Factoring Special Polynomials

Your experience with products of binomials (Section 8-4) will help you to factor many polynomials at sight. Below are three *factor patterns* that occur frequently.

1. $a^2 - b^2 = (a + b)(a - b)$

In this pattern, $a^2 - b^2$ is called a *difference of squares*.

2. $a^2 + 2ab + b^2 = (a + b)^2$

3. $a^2 - 2ab + b^2 = (a - b)^2$

In patterns 2 and 3, the left member is called a *trinomial square* because it has *three* terms and is the square of a binomial.

The three factor patterns above are used in the following examples:

$$y^2 - 9 = (y + 3)(y - 3)$$
$$16t^2 - 1 = (4t + 1)(4t - 1)$$
$$m^2 + 6m + 9 = m^2 + 2 \cdot 3m + 3^2 = (m + 3)^2$$
$$9k^2 - 12k + 4 = (3k)^2 - 2 \cdot 2 \cdot 3k + 2^2 = (3k - 2)^2$$

Sometimes a common factor of the terms of a polynomial conceals a familiar factor pattern. Therefore, the first step in factoring a polynomial is to express it as a product of its greatest monomial factor (or the negative of the greatest monomial factor) and a polynomial whose greatest monomial factor is 1.

EXAMPLE 1   Factor $15r^4 - 60r^2$ completely.

SOLUTION
$$15r^4 - 60r^2 = 15r^2(r^2 - 4)$$
$$= 15r^2(r + 2)(r - 2)$$

EXAMPLE 2   Factor $4a - 4 - a^2$ completely.

SOLUTION
$$4a - 4 - a^2 = -a^2 + 4a - 4$$
$$= -1(a^2 - 4a + 4)$$
$$= -(a - 2)^2$$

By rearranging and grouping the terms of a polynomial, you can sometimes recognize factors. In the following example, the red parentheses indicate a useful grouping of the terms of the polynomial.

EXAMPLE 3   Factor $r^3 + 2r^2 - 9r - 18$ completely.

SOLUTION $r^3 + 2r^2 - 9r - 18 = (r^3 + 2r^2) + (-9r - 18)$
$$= r^2(r + 2) - 9(r + 2)$$
$$= (r^2 - 9)(r + 2)$$
$$= (r + 3)(r - 3)(r + 2)$$

The process used in Example 3 is called **factoring by grouping**. In the next example, the first three terms of the polynomial form a trinomial square. By grouping these terms, you can see that the polynomial itself can be expressed as a difference of squares.

EXAMPLE 4   Factor $a^2 - 4a + 4 - b^2$ completely.

SOLUTION $a^2 - 4a + 4 - b^2 = (a^2 - 4a + 4) - b^2$
$$= (a - 2)^2 - b^2$$
$$= [(a - 2) + b][(a - 2) - b]$$
$$= (a + b - 2)(a - b - 2).$$

Oral Exercises

Factor the given polynomial completely. If the polynomial is not a difference of squares or a trinomial square, so state.

1. $w^2 - t^2$ 2. $x^2 - 2y^2$ 3. $j^2 + k^2$ 4. $p^2 - 12p + 36$
5. $x^2y^2 - m^2$ 6. $4m^2 - n^2$ 7. $9a^2 + 12a + 4$ 8. $k^2 + 16k - 64$

Written Exercises

Factor completely.

A   1. $16a^2 - 1$ 2. $9 - b^2$ 3. $25 - 4c^2$

4. $49 - 36d^2$ 5. $75e^2 - 147f^2$ 6. $32g^2 - 18h^2$

7. $i^2 - 18i + 81$ 8. $j^2 - 10j + 25$ 9. $98k^2 - 28k + 2$

10. $36m^2 + 12m + 1$ 11. $4n^2 + 12n + 9$ 12. $18p^3 - 60p^2 + 50p$

13. $6x + xy + 6y + y^2$ 14. $4 + 4z + z + z^2$ 15. $w^2 - w + 3wz - 3z$

16. $3s + r - 6s^2 - 2rs$ 17. $6t^3 - 3t - 4t^2 + 2$ 18. $3a^2 - 3ab + 2a - 2b$

B   19. $x^2y^3 + 2x^2 - y^3 - 2$ 20. $8 - 2f^2 - 4e^3 + e^3f^2$ 21. $25a^2 + 10a + 1 - b^2$

22. $k^2 - 12k + 36 - h^2$ 23. $c^2 + 4cd + 4d^2 - 25$ 24. $(2s - 1)^2 - (s + 2)^2$

25. $(r - 1)^2 - 8(r - 1) + 16$ 26. $2(x - 1) - 4 - (x - 1)^2$

27. $a^2 - (b + c)^2$ 28. $c^2(c - 1) - 4d^2(c - 1)$

29. $d(r + s) - e(r + s) - f(r + s)$ 30. $(a - b)^3 + 2(a - b)^2 + (a - b)$

C 31. $16 - k^2 + 14kh - 49h^2$ 32. $25 - 4m^2 - 20nm - 25n^2$

33. $a^2 - b^2 - c^2 + 2bc$ 34. $4b - 1 - 4b^2 + 4a^2$

Assume that all variable exponents represent positive integers.

35. $y^{2n} - 1$ 36. $x^{2n} - 4y^{2n}$ 37. $x^{2n} - 8x^n + 16$

38. $x^{4n} - 4x^{2n} + 4$ 39. $x^{2n}(y + m) - w^{2n}(y + m)$ 40. $x^{8n} - y^{2n}$

8-8 Factoring Quadratic Trinomials (1)

A polynomial in simple form that has degree two and that contains a single variable is a quadratic polynomial. For example,

$$x^2 - 5x + 4, \quad 3z^2 + 9z, \quad \text{and} \quad -5y^2$$

are quadratic polynomials.

The term of degree two in a polynomial in simple form, such as x^2 in $x^2 - 5x + 4$, is called the quadratic term. The term of first degree, $-5x$, is called the linear term, and 4, the numerical term or term of zero degree, is the constant term. In the polynomial $3z^2 + 9z$, the "missing" constant term is 0, which has no degree.

It is easy to see that $3z^2 + 9z$ is reducible over the set of polynomials with integral coefficients because its terms have the common factor $3z$. Thus:

$$3z^2 + 9z = 3z(z + 3).$$

Recognizing that $x^2 - 5x + 4$ is also reducible (and is the product of two binomials each of degree 1) is harder. However, your experience in multiplying binomials at sight will help you. Recall that for binomial products in which the coefficients of the linear terms are both 1, you have:

$$(x + a)(x + b) = x^2 + ax + bx + ab = x^2 + (a + b)x + ab.$$

This process leads to the following clues to the factorization of $x^2 - 5x + 4$:

Clue 1. The product (1) of the linear terms in the factors must be x^2. Thus, you have: $(x \quad)(x \quad)$

Clue 2. The product (3) of the constant terms in the factors must be 4. Thus, the only possibilities are those shown at the right.
(Note that the constant terms must both be positive or both be negative because their product, 4, is positive.)

$(x - 4)(x - 1)$
$(x + 4)(x + 1)$
$(x - 2)(x - 2)$
$(x + 2)(x + 2)$

Clue 3. The coefficient of the linear term in $x^2 - 5x + 4$ is the sum of the two constant terms in the factors. Since $-4 + (-1) = -5$, the only possibility for factors is: $(x - 4)(x - 1)$

A check by multiplication verifies that: $x^2 - 5x + 4 = (x - 4)(x - 1)$.

If the quadratic term in a trinomial has a negative coefficient, it is usually helpful first to factor -1 from each term in the trinomial.

EXAMPLE   Factor $-z^2 + 3z + 10$ completely.

SOLUTION   First, factor -1 from each term:

$$-z^2 + 3z + 10 = -1(z^2 - 3z - 10) = -(z^2 - 3z - 10)$$

Now, use the three clues to factor $z^2 - 3z - 10$.

Clue 1. $(z \quad)(z \quad)$

Clue 2.
(Note that one constant term must be positive and one negative because their product, -10, is negative.)

$(z - 10)(z + 1)$
$(z + 10)(z - 1)$
$(z + 2)(z - 5)$
$(z - 2)(z + 5)$

Clue 3. $2 + (-5) = -3$ $(z + 2)(z - 5)$

Check: $(z + 2)(z - 5) = z^2 - 3z - 10$.

Hence $z^2 - 3z - 10 = (z + 2)(z - 5)$, and

$$-z^2 + 3z + 10 = -(z + 2)(z - 5). \quad \text{Answer.}$$

Oral Exercises

When looking for binomial factors of the given polynomial, which of the following patterns would you try?

a. $(x + \quad)(x + \quad)$ b. $(x - \quad)(x - \quad)$ c. $(x + \quad)(x - \quad)$

EXAMPLE 1 $x^2 - 7x - 8$

SOLUTION   Pattern c

EXAMPLE 2 $x^2 - 14x + 49$

SOLUTION   Pattern b

1. $x^2 + 5x + 6$
2. $x^2 - 9x + 8$
3. $x^2 + 10x + 24$
4. $x^2 - 2x - 8$
5. $x^2 + 2x - 35$
6. $x^2 + 3x - 4$
7. $x^2 - 10x + 21$
8. $x^2 - 11x + 18$
9. $x^2 - 4x - 21$

Written Exercises

A   **1–9.** Factor each trinomial in Oral Exercises 1–9.

Factor each trinomial completely. If the trinomial is irreducible, so state.

EXAMPLE 1 $-21k + 3k^2 - 24$

SOLUTION
$$-21k + 3k^2 - 24 = 3k^2 - 21k - 24$$
$$= 3(k^2 - 7k - 8)$$
$$= 3(k - 8)(k + 1)$$

10. $6a^2 - 24a - 72$
11. $6 - b - b^2$
12. $-3c^2 - 15c + 18$
13. $2d^3 + 16d^2 - 40d$
14. $-e^4 + 21e^2 - 4e^3$
15. $f^4 + 30f^2 + 11f^3$
16. $2k^3 - 14k^2 - 20k$
17. $h^3 - 16h - 6h^2$
18. $3x^8 + 108 - 36x^4$

EXAMPLE 2 $9x^3 - 18x^2y + 9xy^2$

SOLUTION
$$9x^3 - 18x^2y + 9xy^2 = 9x(x^2 - 2xy + y^2)$$
$$= 9x(x - y)(x - y)$$

B   19. $a^2 + ab - 6b^2$
20. $c^2 + 4cd - 5d^2$
21. $e^2 + 15ef + 5f^2$
22. $k^2 - kg - 3g^2$
23. $px^2 - 4p^2x - 45p^3$
24. $7xy - y^2 + 18x^2$
25. $ry^2 + 14r^2y + 45r^3$
26. $8s^2 - 9st + t^2$
27. $18xy^2 - 36x^2y + 18x^3$
28. $x^2 - 6y^2 - xy$
29. $cy^3 + 24c^3 + 10c^2y$
30. $35x^2 + 2xy - y^2$

C   31. $a^2b^2 + ab - 2$
32. $c^2d^2 - cd - 2$
33. $4x^2y^2 + 20xy + 24$
34. $6r^2s^2 - 6rs - 36$
35. $-16e^2f - 10e^2fg - e^2fg^2$
36. $-27kz - 42k - 3kz^2$

8-9 Factoring Quadratic Trinomials (2)

The clues you learned in Section 8-8 for factoring quadratic trinomials whose quadratic terms have a coefficient of 1 or -1 can also be used to factor quadratic trinomials whose quadratic terms have coefficients other than 1 or -1. To see how this is done, study the following example.

Polynomials and Their Factors | **303**

EXAMPLE   Factor $4x^2 + 23x - 6$ completely.

SOLUTION   *Clue 1.* The product of the linear terms in the fac- $(2x\ \ )(2x\ \ )$
tors must be $4x^2$. Thus, the possibilities are: $(4x\ \ )(x\ \ )$

Clue 2. The product of the constant terms in the $(2x - 6)(2x + 1)$
factors must be -6. This leads to the possi- $(2x - 1)(2x + 6)$
bilities shown at the right. $(2x - 3)(2x + 2)$
(Note that one of the constant terms must be $(2x - 2)(2x + 3)$
positive and one negative because their $(4x - 6)(x + 1)$
product, -6, is negative.) $(4x + 6)(x - 1)$
     $(4x - 1)(x + 6)$
     $(4x + 1)(x - 6)$
     $(4x - 3)(x + 2)$
     $(4x + 3)(x - 2)$
     $(4x - 2)(x + 3)$
     $(4x + 2)(x - 3)$

Clue 3. The linear term of the polynomial, $23x$, must
be the sum of the products of the first term of
each binomial factor with the last term of the
other. Therefore the only possibility is: $(4x - 1)(x + 6)$

Check: $(4x - 1)(x + 6) = 4x^2 + 23x - 6$

$$\therefore 4x^2 + 23x - 6 = (4x - 1)(x + 6). \quad \text{Answer.}$$

Of course, with practice, you will be able to check the possibilities
arising from Clue 2 mentally and will not have to write out all the
combinations.

Oral Exercises

In Exercises 1–12, when looking for binomial factors of the given
polynomial, name the expressions you would try as linear terms and
constant terms in the binomials.

EXAMPLE $6x^2 - 23x + 20$ **SOLUTION** $2x, 3x;\ 6x, x.$ $4, 5;\ 2, 10;\ 20, 1$

1. $3a^2 + 4a + 1$ 2. $2b^2 + 3b + 1$ 3. $2c^2 - 3c + 1$

4. $2d^2 + 5d + 3$ 5. $4e^2 + 3e - 1$ 6. $6f^2 + 5f + 1$

7. $3k^2 - 2k - 5$ 8. $4s^2 + 25s - 21$ 9. $8r^2 + 3r - 5$

10. $12x^2 - 7x - 12$ 11. $16y^2 - 50y + 25$ 12. $18t^2 - 9t - 14$

Written Exercises

A   **1–12.** Factor each trinomial in Oral Exercises 1–12 completely.

Factor each trinomial completely. If the trinomial is irreducible, so state.

13. $16x^2 - 16x - 5$ 14. $6y - 4 + 7y^2$ 15. $16z^2 - 11z - 5$

16. $9w^2 + 9w - 4$ 17. $4u^2 + 16u + 15$ 18. $-8v^2 + 6v + 2$

B 19. $15a^2 + 16ab + 4b^2$ 20. $9c^2 - 3cd - 2d^2$ 21. $6e^2 - 3ef - 9f^2$

22. $12g^2 + 42gh + 18h^2$ 23. $4r^3s - 6rs^3 - 10r^2s^2$ 24. $5m^2n + m^3 + 4mn^2$

25. $9u^2 + 6u - 27u^3$ 26. $-4v^3 + 10v^2 + 6v$ 27. $15l^2 - 18l^3 + 18l$

C 28. $9n^4 + 8n^2 - 1$ 29. $8m^4 + 14m^2 - 4$ 30. $27cd^2 - 12d^3 + 27c^2d$

31. $24ey^2 + 7ey - 6e$ 32. $ay^4 - 8ay^2 + 15a$ 33. $6cnx^2 - 5cnx - 6cn$

34. $y^4 + y^2 + 1$ (*Hint:* $y^4 + y^2 + 1 = y^4 + 2y^2 + 1 - y^2$)

35. $x^4 - 7x^2 + 9$ (*Hint:* $x^4 - 7x^2 + 9 = x^4 - \underline{\ ?\ } + 9 - \underline{\ ?\ }$)

36. $k^4 - 7k^2 + 1$ 37. $c^4 + c^2d^2 + d^4$

38. Find integral values of a, b, and c such that $ax + b$ is a factor of both $2x^2 - 5x + c$ and $4x^2 + 4x + 1$.

39. Show that there is an infinite set of integral values of c for which the trinomial $x^2 + x + c$ can be factored over the set of polynomials with integral coefficients.

Self-Test 2

VOCABULARY   factor over a set (p. 293)   irreducible polynomial (p. 297)

                   prime number (p. 293)   quadratic polynomial (p. 301)

                   greatest common factor

                      (p. 294)

1. Factor 224 over the set of prime numbers.   *Obj. 1, p. 293*

2. Find the greatest common factor of $24a^2b^3$ and $36a^3b^2$.

Factor completely.

3. $6x - 3x^2$ 4. $4y^2 - 16$ *Obj. 2, p. 293*

5. $y^2 - 10y + 25$ 6. $7w - 5 + 6w^2$

7. $5v^2 - 80$ 8. $m^2 - 2m - 63$

9. $9p^2 + 52p - 12$ 10. $-12n^2 + 29n - 15$

Check your answers with those at the back of the book.

Applications of Factoring

OBJECTIVES for Sections 8-10 and 8-11:
1. *Solve quadratic equations by factoring.*
2. *Use quadratic equations to solve word problems.*

8-10 Using Factoring to Solve Equations

The multiplicative property of 0 guarantees that the following theorem is true.

Theorem.   For all real numbers a and b,

if $a = 0$ or $b = 0$, then $ab = 0$.

Compare the preceding theorem with the following one, whose proof is outlined in Exercises 38–41 on page 310.

Theorem.   For all real numbers a and b,

if $ab = 0$, then $a = 0$ or $b = 0$.

Do you see that the hypothesis, $a = 0$ or $b = 0$, of the first theorem is the conclusion of the second theorem? Moreover, the hypothesis of the second theorem, $ab = 0$, is the conclusion of the first. We say that each of these theorems is the *converse* of the other. Two statements are called **converses** when the hypothesis of each statement is the conclusion of the other statement.

Alternative ways of expressing a statement and its converse are shown in the following display.

Statement	**Converse**
If $a = 0$ or $b = 0$, then $ab = 0$.	If $ab = 0$, then $a = 0$ or $b = 0$.
Or: $ab = 0$ if $a = 0$ or $b = 0$.	*Or:* $ab = 0$ only if $a = 0$ or $b = 0$.

Thus, a statement combining the first theorem and its converse is:

Zero-Product Property of Real Numbers

For all real numbers a and b, $ab = 0$ if and only if $a = 0$ or $b = 0$.

Notice that a theorem using the words "if and only if" involves a *conjunction* of two statements that are converses of each other. For the theorem to be true, *both* statements must be true.

EXAMPLE 1   Reword the following statement as a conjunction of statements, and tell whether the given statement is true:

"For each real number r: $r > 0$ if and only if $r^2 > 0$."

SOLUTION   For each real number r:

if $r > 0$, then $r^2 > 0$, *and* if $r^2 > 0$, then $r > 0$.   **Answer.**

The statement "If $r > 0$, then $r^2 > 0$" is a true statement for *every* real number r, but the converse, "If $r^2 > 0$, then $r > 0$," is false for some real values of r. For example, when r is replaced by -2, the converse asserts "If $(-2)^2 > 0$ (that is, if $4 > 0$), then $-2 > 0$," which is false.

Therefore the given statement is false.   **Answer.**

The zero-product property of real numbers is often stated as follows:

A product of real numbers is zero if and only if at least one of the factors is zero.

You can use this property to solve an equation having one member 0 and the other member a product of binomials of degree 1.

EXAMPLE 2   Solve $(2x + 3)(x - 4) = 0$.

SOLUTION   The zero-product property implies that this equation is equivalent to the *disjunction*

$$2x + 3 = 0 \quad \text{or} \quad x - 4 = 0$$
$$2x = -3$$
$$x = -\tfrac{3}{2} \quad \text{or} \quad x = 4$$

$Check:$ $[2(-\tfrac{3}{2}) + 3](-\tfrac{3}{2} - 4) \overset{?}{=} 0$ $[2(4) + 3](4 - 4) \overset{?}{=} 0$

$0 \cdot (-\tfrac{11}{2}) \overset{?}{=} 0$ $11 \cdot 0 \overset{?}{=} 0$

$0 = 0 \ \checkmark$ $0 = 0 \ \checkmark$

$\therefore$ the solution set is $\{-\tfrac{3}{2}\} \cup \{4\} = \{-\tfrac{3}{2}, 4\}$.   **Answer.**

You can also use the zero-product property to solve an equation that is equivalent to one in which one member is 0 and the other is a polynomial that can be expressed as a product of factors with linear terms.

EXAMPLE 3 Solve $z^3 - 3z^2 = 10z$.

SOLUTION

1. Transform the given equation into an equivalent one with 0 as the right member.

$$z^3 - 3z^2 = 10z$$
$$z^3 - 3z^2 - 10z = 0$$

2. Factor the left member completely.

$$z(z^2 - 3z - 10) = 0$$
$$z(z - 5)(z + 2) = 0$$

3. Solve the equivalent disjunction.

$$z = 0 \text{ or } z - 5 = 0 \text{ or } z + 2 = 0$$
$$z = 0 \text{ or } \qquad z = 5 \text{ or } z = -2$$

4. The check is left for you.

The solution set is $\{0, 5, -2\}$.   Answer.

Oral Exercises

State a disjunction equivalent to each of the following open sentences.

1. $x(x - 2) = 0$

2. $3y(y + 4) = 0$

3. $(x + 3)(x - 4) = 0$

4. $(z - 5)(z - 5) = 0$

5. $(w - 2)(3w - 4) = 0$

6. $r(r - 6)(r + 6) = 0$

State the converse of each of the following statements. Tell whether (a) the statement or (b) its converse is true for all real values of the variables. When both statement and converse are true, restate them in combined form using the words "if and only if."

7. If $a + c = b + c$, then $a = b$.

8. If $a = b$, then $ac = bc$.

9. If $x = 3$ and $y = 4$, then $x + y = 7$.

10. If $x^2 = 2x$, then $x^2 - 2x = 0$.

11. If $rs = 1$, then $r = 1$ and $s = 1$.

12. If $a < b$, then $a - c < b - c$.

13. If $-a > 0$, then $a < 0$.

14. If $a > b$, then $ac > bc$.

Reword each of the following theorems in a form using a conjunction.

15. For all real numbers a and b, $a > b$ if and only if $a - b > 0$.

16. For all real numbers a and b, $ab > 0$ if and only if either $a > 0$ and $b > 0$ or $a < 0$ and $b < 0$.

17. For all positive real numbers a and b, $a < b$ if and only if $\frac{1}{a} > \frac{1}{b}$.

18. For all positive real numbers c and d and all real numbers a and b, $\frac{a}{c} < \frac{b}{d}$ if and only if $ad < bc$.

Written Exercises

Solve each equation.

A
1. $(r + 6)(r - 3) = 0$
2. $(s - 5)(s - 2) = 0$
3. $(t + 8)(t + 11) = 0$
4. $(a - 3)(a - 10) = 0$
5. $2b(b - 2)(b + 2) = 0$
6. $(3d - 4)(2d + 1) = 0$
7. $(2e - 3)(e + 6) = 0$
8. $x^2 + 11x + 28 = 0$
9. $x^2 - 12x + 36 = 0$
10. $6y^2 - 12y = 0$
11. $4z^2 - 16z = 0$
12. $f^2 - 25 = 0$
13. $9k^2 - 49 = 0$
14. $2x^2 + 3 = -5x$
15. $4y^2 - 3y - 1 = 0$
16. $23m - 6 = -4m^2$
17. $4p^3 - 10p^2 = 6p$
18. $0 = 9t^2 + 6t - 27t^3$

EXAMPLE 1 $(2x - 3)(x - 2) = 21$

SOLUTION

1. Simplify the left member.
2. Rewrite with the right member zero.
3. Factor the left member.
4. Write as a disjunction.
5. The check is left for you.

$\therefore$ the solution set is $\{-\frac{3}{2}, 5\}$. **Answer.**

$$(2x - 3)(x - 2) = 21$$
$$2x^2 - 7x + 6 = 21$$
$$2x^2 - 7x - 15 = 0$$
$$(2x + 3)(x - 5) = 0$$
$$2x + 3 = 0 \quad \text{or} \quad x - 5 = 0$$
$$x = -\tfrac{3}{2} \text{ or } x = 5$$

B
19. $(a + 2)(a - 3) = 6$
20. $(3x + 1)(x + 5) = -11$
21. $(c + 1)(c + 8) = -12$
22. $(d + 3)(d + 1) = 8$
23. $(3e + 1)^2 - 25 = 0$
24. $(f - 4)(f + 3) = -10$
25. $(2x - 1)^2 = 16$
26. $-12 = (y - 5)(y + 2)$
27. $z^2 + 2z(z - 6) = -5(z - 2)$
28. $(k + 2)^2 = k(2 + 3k)$
29. $4(w + 5) = -2w(5 + w)$
30. $(p + 3)^2 - (p + 3) - 12 = 0$
31. $(q - 1)^2 - (q - 1) - 6 = 0$
32. $6(m + 2)^2 - 5(m + 2) - 6 = 0$

In Exercises 33–37, using *x* as the variable, write an equation which has the given solution set and which has 0 as one member and a quadratic polynomial as the other member.

EXAMPLE 2 $\{3, -8\}$

SOLUTION The equation must be equivalent to $\quad x = 3 \quad$ or $\quad x = -8$;
that is $\quad x - 3 = 0 \quad$ or $\quad x + 8 = 0.$

Hence the equation is equivalent to $(x - 3)(x + 8) = 0.$

$\therefore x^2 + 5x - 24 = 0.$ **Answer.**

33. $\{-4, 2\}$
34. $\{0, -5\}$
35. $\{-2\}$
36. $\{6, -6\}$
37. $\{-9, -4\}$

Polynomials and Their Factors | 309

Supply the missing reason in each step of the following proof of the second theorem stated on page 306.

Case 1. $a = 0$. In this case, the conclusion is true whether or not $b = 0$. Therefore, the theorem holds.

Case 2. $a \neq 0$. In this case, the axiom of multiplicative inverses ensures that $\dfrac{1}{a}$ is a real number. Therefore, you may reason as follows:

C **38.** $\qquad ab = 0 \qquad$ ___?___

39. $\quad \therefore \dfrac{1}{a}(ab) = \dfrac{1}{a} \cdot 0 \qquad$ ___?___

40. But $\dfrac{1}{a}(ab) = b \qquad$ ___?___

41. Also $\dfrac{1}{a} \cdot 0 = 0 \qquad$ ___?___

$\qquad \therefore b = 0 \qquad$ Substitution principle

Therefore, given that $ab = 0$, it follows that $a = 0$ or $b = 0$.

42. Prove: For all real numbers a and b, $ab > 0$ if and only if either $a > 0$ and $b > 0$ or $a < 0$ and $b < 0$.

8-11 Using Factoring to Solve Problems

When you write an open sentence expressing the relationships in a given problem, you may obtain an equation having more than one root. Each root must be tested to see whether it gives a meaningful answer to the problem. The solution set of the equation contains only *possible solutions* of the problem. *Actual solutions* are found by checking each possible solution with the words of the problem.

EXAMPLE 1   The square of a positive integer is 79 less than twice the square of the next consecutive integer. Find the integers.

SOLUTION   1. The problem asks for two consecutive positive integers.

2. $\qquad$ Let $x = $ the first integer.

$\qquad$ Then $x + 1 = $ the next consecutive integer.

3. $2(x + 1)^2 - x^2 = 79$

4. $2(x^2 + 2x + 1) - x^2 = 79$
$\qquad 2x^2 + 4x + 2 - x^2 = 79$
$\qquad\qquad x^2 + 4x - 77 = 0$
$\qquad\qquad (x + 11)(x - 7) = 0$

$\qquad x + 11 = 0 \qquad$ or $\quad x - 7 = 0$
$\qquad\qquad x = -11 \quad$ or $\qquad\quad x = 7$

5. Since the problem asks for consecutive *positive* integers, you reject -11. Check that 7 and 8 meet the conditions of the problem.

$$2(8)^2 - 7^2 \overset{?}{=} 79$$
$$128 - 49 \overset{?}{=} 79$$
$$79 = 79 \ \checkmark$$

The two consecutive positive integers are 7 and 8. **Answer.**

EXAMPLE 2 The Sun City swimming pool measures 25 m by 20 m. Plans are made to pave an area of 250 m² around the pool. The paved area is to be of uniform width. How wide should it be?

SOLUTION 1. The problem asks for the width of the paved area.

2. Let x = width of paved area in meters. Then
$20 + 2x$ = total width in meters, and the area to be paved is
$$2x(20 + 2x) + 2x(25).$$

3. $2x(20 + 2x) + 2x(25) = 250$

4. $40x + 4x^2 + 50x = 250$
$4x^2 + 90x - 250 = 0$
$(4x - 10)(x + 25) = 0$

$4x - 10 = 0$ or $x + 25 = 0$
$4x = 10$ $x = -25$
$x = 2.5$ Since the width must be positive,
                      -25 is rejected.

Check: $2(2.5)[20 + 2(2.5)] + 2(2.5)(25) \overset{?}{=} 250$
$125 + 125 \overset{?}{=} 250$
$250 = 250 \ \checkmark$

The width of the paved area is 2.5 m. **Answer.**

Problems

A   1. Find two positive consecutive integers whose product is 42.

2. The square of a positive integer exceeds 3 times the integer by 54. Find the integer.

3. Find two negative consecutive integers such that the sum of their squares is 61.

4. Find two integers which differ by 5, while their squares differ by 55.

5. The dimensions of a rectangular flower garden in a park are 12 m × 8 m. Surrounding the garden is a paved walk of uniform width. The combined area of the garden and the walk is 192 m². Find the width of the walk.

6. The length of a rectangle is 3 m less than twice its width. If the area is 77 m², find the length of the rectangle.

7. The sum of two numbers is 17 and their product is 42. Find the numbers.

8. The sum of two integers is 4. The sum of the squares of the integers is 16 greater than the product of the integers. Find the integers.

9. The height of a photograph is 40 cm greater than the width, and the area of the photograph is 896 cm². Can the photograph be hung in a 35 cm × 25 cm wall space? How wide is it?

10. A rectangle is 2 m longer than it is wide. The area of the rectangle is 99 m². Find its length and width.

B 11. The perimeter of a rectangle is 60 cm and the area is 161 cm². Find the dimensions of the rectangle.

12. The area of an asphalt walk around a rectangular fishpond is equal to the area of the fishpond. The walk is of uniform width. If the fishpond measures 6 m × 9 m, find the width of the walk.

13. The sum S of the first n consecutive natural numbers is given by the formula $S = \frac{1}{2}n(n + 1)$. How many such natural numbers must be added to give a sum of 276?

14. Let h be a function whose domain is the set of positive numbers. Given that $h: x \rightarrow 6x^2 + x - 5$, find the value of x such that $h(x) = 0$.

15. What is the error in the following argument? The equation

$x^2 + 4x + 3 = 1$ is equivalent to $(x + 1)(x + 3) = 1 \cdot 1$.

Hence, the given equation is equivalent to:

$$x + 1 = 1 \quad \text{or} \quad x + 3 = 1$$
$$x = 0 \qquad\qquad x = -2$$

∴ the solution set is $\{0, -2\}$.

16. Consider a set of k points no three of which lie on a line. The number n of segments that can be drawn connecting all possible pairs of these points is given by the formula

$$n = \frac{k(k - 1)}{2}, \text{ where } k \text{ is a positive integer.}$$

How many points are there in the particular set for which the total number of segments joining pairs of points is 28?

17. The length of one leg of a right triangle is 4 cm less than twice the length of the other leg. The area of the triangle is 24 cm². Find the length of each leg.

18. The lengths of the sides of two cubes differ by 2 cm and the volumes differ by 152 cm³. Find the length of a side of the smaller cube.

19. Jessica made a rectangular pen for her dog using a side of the garage for one side and 10 m of fencing for the other three sides. If the area enclosed by the fence was 12 m², find the dimensions of the pen.

20. A side of a house is in the shape of a triangle on top of a rectangle. The rectangle is 4 times as long as it is high, and the altitude of the triangular part is 1 m more than the height of the rectangle. The total area of the side is 60 m². Find the height of the house.

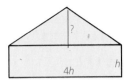

21. Jim is k years old. His brother Joe is k^2 years old. In 8 years, Joe will be twice as old as Jim is then. How old is Joe now?

22. Jack has a garden 10 m × 17 m. He makes part of it into a badminton court by subtracting equal amounts from its length and width. The area of the garden that is left is 92 m². By how much does he reduce each dimension of his garden?

23. An open rectangular box with length 4 times its width is made from a rectangular piece of metal by cutting out a 2 cm square from each corner and turning up the sides. If the volume of the box is 128 cm³, find the length of a side of the original piece of metal.

C 24. Show that the sum of the squares of any two consecutive integers is 1 more than a multiple of 4. (*Hint:* Let x and $x + 1$ denote the integers. Also use the fact that one of every two consecutive integers is even.)

25. Show that the square of an odd integer is 1 more than a multiple of 8. (*Hint:* If x denotes an integer, then $2x + 1$ denotes an odd integer. Square $2x + 1$ and use the fact that one of every two consecutive integers is even.)

Self-Test 3

VOCABULARY   converse (p. 306)
zero-product property of real numbers (p. 307)

Solve.

1. $a(2a - 5) = 0$ 2. $b^2 = -4b$ *Obj. 1, p. 306*

3. $c^2 - 3c = 28$ 4. $12d^2 = 35d - 18$

5. Find two consecutive positive integers such that the sum of their squares is 113.   *Obj. 2, p. 306*

Check your answers with those at the back of the book.

Transformations of the Plane: Reflections

On pages 230–231, we discussed translations. Another kind of transformation (mapping) of the plane is a *reflection*. In a reflection, each point in the plane is transformed into its mirror image across an axis of reflection. For example, if the axis of reflection is the *x*-axis (diagram at the left below), then the square with vertices (1, 1), (2, 1), (2, 2), and (1, 2) is mapped into the square with vertices (1, −1), (2, −1), (2, −2), and (1, −2).

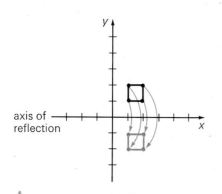
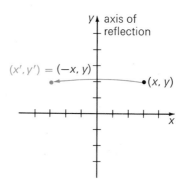

A reflection about the *x*-axis maps each point (x, y) in the plane into a new point (x', y') whose coordinates are related to those of the point (x, y) by the reflection equations

$$x' = x \quad \text{and} \quad y' = -y.$$

Similarly, as the figure at the right above suggests, the equations for a reflection in the *y*-axis are

$$x' = -x \quad \text{and} \quad y' = y.$$

EXAMPLE Sketch the line segment with endpoints $(-1, 3)$ and $(2, 1)$, and then sketch the reflections of this segment in the *x*-axis and in the *y*-axis, respectively.

SOLUTION

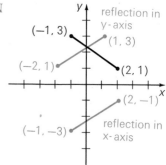

The following result is proved in more advanced courses.

Under a reflection:

1. Every line in the plane is mapped (reflected) into a line in the plane.

2. Every line segment is mapped (reflected) into a line segment of equal length.

3. Every angle is mapped (reflected) into an angle of equal measure.

Translations and reflections of the plane are called *rigid transformations* because they preserve the size and shape of geometric figures in the plane.

Another rigid transformation of the plane which you will study in later courses is a *rotation* of the plane. The diagram at the right shows the effect of a 45° rotation of the plane on the square with vertices (1, 1), (2, 1), (2, 2), and (1, 2).

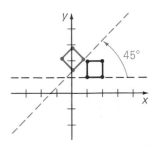

Exercises

In Exercises 1–8 the coordinates of the endpoints of a line segment are given. In each case (a) sketch the segment and label its endpoints with their coordinates and (b) in the same coordinate plane, sketch the reflections of the segment in the *x*-axis and in the *y*-axis, respectively. Label the endpoints of each reflected segment with their coordinates.

1. $(3, 1)$, $(5, 0)$
2. $(0, 2)$, $(3, 1)$
3. $(-2, 3)$, $(1, 4)$
4. $(4, -2)$, $(1, 1)$
5. $(-1, -1)$, $(1, 1)$
6. $(-1, 1)$, $(1, -1)$
7. $(1, -4)$, $(3, 2)$
8. $(-3, 2)$, $(-1, -2)$

9. What must be true of a line (other than an axis) if it is transformed into itself by a reflection in the *y*-axis? the *x*-axis?

10. What must be true of a line (other than an axis) if it is transformed into a parallel line by a reflection in the *y*-axis? the *x*-axis?

Chapter Summary

1. Expressions such as $2x^3$ and $-5a^3b^2$ are called *monomials*. Two or more monomials connected by plus (+) signs form a *polynomial*.

2. For all real numbers a and b, and all positive integers m and n:
   $$a^m \cdot a^n = a^{m+n} \qquad (a^m)^n = a^{mn} \qquad (ab)^m = a^m b^m$$

3. Polynomials may be added, subtracted, and multiplied by using the laws of exponents and the commutative, associative, and distributive axioms.

4. To factor a number over a given set of numbers means to express the number as the product of numbers from the given set.

5. To factor a polynomial with integral coefficients means to express it as the product of polynomials with integral coefficients.

6. A polynomial is *factored completely* when none of its factors can be factored further.

7. Three frequently occurring factor patterns are:
$$a^2 - b^2 = (a + b)(a - b) \qquad a^2 + 2ab + b^2 = (a + b)^2$$
$$a^2 - 2ab + b^2 = (a - b)^2$$

8. The use of *factoring* to solve equations is based on the zero-product property of real numbers: For all real numbers a and b, $ab = 0$ if and only if $a = 0$ or $b = 0$.

Chapter Review

1. The degree of the polynomial $2a^2b^3 - 3a^3b + a^2b^2$ is: 8-1

   a. 2   **b.** 3   **c.** 4   **d.** 5

2. $(3a^3 - 4a^2b + 7b^2) + (2a^3 + 8ab^2 - 2b^2) = \underline{\ ?\ }$.

   a. $5a^3 + 4a^2b^2 + 5b^2$ **b.** $5a^3 - 4a^2b + 8ab^2 + 5b^2$

   c. $5a^6 + 4a^3b^3 + 5b^4$ **d.** $5a^3 - 4a^2b + 8ab^2 + 9b^2$

3. $2c(2c^3)^2 = \underline{\ ?\ }$.   8-2

   a. $4c^7$ **b.** $8c^6$ **c.** $8c^7$ **d.** $4c^6$

4. $3a(4a + 5) = \underline{\ ?\ }$.   8-3

   a. $12a^2 + 15a$ **b.** $12a^2 + 5$

   c. $27a^2$ **d.** $60a^2$

5. $(2t + 3)(t - 4) = \underline{\ ?\ }$.

   a. $2t^2 - 11t - 12$ **b.** $2t^2 - 5t - 12$

   c. $2t^2 - 8t - 1$ **d.** $2t^2 + 11t - 12$

6. $(2m - 3n)^2 = \underline{\ ?\ }$.   8-4

   a. $4m^2 + 9n^2$ **b.** $4m^2 - 9n^2$

   c. $4m^2 + 12m + 9n^2$ **d.** $4m^2 - 12mn + 9n^2$

7. $(2r - 7)(2r + 7) = \underline{\ ?\ }$.

   a. $4r^2 - 49$ **b.** $4r^2 - 28r - 49$

   c. $4r^2 + 28r - 49$ **d.** $4r^2 - 28r + 49$

8. Which of the following is *not* a prime number? 8-5
   a. 59   b. 181   c. 87   d. 47

9. The greatest common factor of $24x^4y^2$ and $-60x^3y^2$ is _?_.
   a. $6x^3y^2$ b. $12xy$ c. $12x^3y^2$ d. $120x^4y^2$

In Exercises 10–15, factor completely.

10. $18x - 12y + 6 =$ _?_. 8-6
    a. $3(6x - 4y + 2)$ b. $6(3x - 2y + 1)$
    c. $6(3x - 2y)$ d. The polynomial is already factored completely.

11. $18a^3b^2 - 27a^2b^3 =$ _?_.
    a. $9ab(2a^2b - 3ab^2)$ b. $9(2a^3b^2 - 3a^2b^3)$
    c. $3a^2b^2(6a - 9b)$ d. $9a^2b^2(2a - 3b)$

12. $4m^2 + 24m + 36 =$ _?_. 8-7
    a. $4(m + 3)^2$ b. $(2m + 6)^2$
    c. $4(m + 3)(m - 3)$ d. $4m(m + 15)$

13. $x^2 - 10x - 24 =$ _?_. 8-8
    a. $(x - 6)(x - 4)$ b. $(x - 12)(x + 2)$
    c. $(x - 8)(x + 3)$ d. $(x + 12)(x - 2)$

14. $2t^2 - 22t + 36 =$ _?_.
    a. $2(t - 3)(t - 6)$ b. $(2t - 18)(t - 2)$
    c. $2(t - 9)(t - 2)$ d. $(t - 9)(2t - 4)$

15. $2x^2 - 5x - 12 =$ _?_. 8-9
    a. $(2x + 3)(x - 4)$ b. $2(x - 1)(x + 6)$
    c. $(2x - 3)(x + 4)$ d. The polynomial is already factored completely.

16. If $rs = 0$, then: 8-10
    a. $r = 0$ b. $s = 0$
    c. $r = 0$ and $s = 0$ d. $r = 0$ or $s = 0$

17. The solution set of $x^3 = 4x$ is:
    a. $\{2\}$ b. $\{2, -2\}$ c. $\emptyset$ d. $\{-2, 0, 2\}$

18. The solution set of $2x^2 - 24x + 72 = 0$ is:
    a. $\{-6\}$ b. $\{6\}$ c. $\{0, 6\}$ d. $\{4, 9\}$

19. Three consecutive multiples of 3 are such that the product of the first two is 93 more than the third one. Which equation expresses this relationship?

8-11

   a. $x(x + 3) + 93 = (x + 6)$ **b.** $x(3x) = 6x + 93$

   c. $x(x + 3) = (x + 6) + 93$ **d.** $x(x + 3) = 93(x + 6)$

Chapter Test

1. State the degree of the polynomial $-3p^5s - 5p^3s^3 + 7p^2s^3$.

8-1

Simplify.

2. $(-3a^3 + 2a^2b + 4b^2) - (2a^3 - 3ab^2 - 5b^2)$

3. $3y(-2x^2)^3 + 2y(3x^3)^2$ 4. $(2c^2d)^3(3cd)^2$

8-2

5. $2r(r^2 + 3r - 5)$ 6. $(x - 3)(x + 4)$

8-3

7. $(m - 9)(m + 9)$ 8. $(2a - 5)^2$ 9. $(3w + 5)(2w - 3)$

8-4

10. Find the greatest common factor of $30a^2bc^4$ and $-45abc^2$.

8-5

Factor completely.

11. $75v^2z^4 + 30v^2z^2 - 45vz^3$.

8-6

12. $9n^2 - 4$ 13. $8x^2 - 24x + 18$

8-7

14. $c^2 - 6c + 8$ 15. $m^3 - 7m^2 - 18m$

8-8

16. $6x^2 + x - 40$

8-9

17. Solve the equation $h^2 - 15 = 2h$.

8-10

18. The length of a room is 5 m less than twice its width. If 52 m² of wall-to-wall carpeting is needed to cover the floor, find the dimensions of the room.

8-11

ON THE CALCULATOR

An exclamation point is used in the following way to show certain multiplications:

$$2! = 2 \times 1, \quad 3! = 3 \times 2 \times 1, \quad 4! = 4 \times 3 \times 2 \times 1, \text{ and so on.}$$

The symbol 2! is read "factorial 2."

Using a calculator, find the value of $(n - 1)! + 1$ when n is an integer and $2 \leq n < 12$. In each case, determine whether or not n is a factor of $(n - 1)! + 1$. Make a table like the one at the right. What do you notice about n when it is a factor of $(n - 1)! + 1$?

n	$(n - 1)! + 1$	Is n a factor?
2	2	Yes
3	3	Yes
4	7	No
. . .	. . .	. . .

Cumulative Review—Chapters 5–8

1. Which inequality is equivalent to $5 - 3m > -1$?

   a. $m > 2$ **b.** $m < -2$ **c.** $m < 2$ **d.** $m > -2$

2. If $R = \{1, 2, 3, 4, 5\}$ and $S = \{1, 3, 5\}$, which statement is *not* true?

   a. $R \cup S = R$ **b.** $S \subset R$ **c.** $R \cap S = S$ **d.** $S \cap R = R$

3. If A is the solution set of $x \leq 4$ and B is the solution set of $x \geq 1$, then which set is graphed here?

   a. A **b.** B **c.** $A \cup B$ **d.** $A \cap B$

4. Which of these is the graph of $x - 2 \geq 1$ or $2x + 3 \geq 1$?

   a. ![graph] $-2\ -1\ \ 0\ \ 1\ \ 2\ \ 3\ \ 4$

   b. ![graph] $-2\ -1\ \ 0\ \ 1\ \ 2\ \ 3\ \ 4$

   c. ![graph] $-2\ -1\ \ 0\ \ 1\ \ 2\ \ 3\ \ 4$

   d. ![graph] $-2\ -1\ \ 0\ \ 1\ \ 2\ \ 3\ \ 4$

5. The sum of three consecutive integers is at least 97 more than the smallest of the integers. Which of these inequalities does *not* correctly express this relationship?

   a. $x + (x + 1) + (x + 2) \leq x + 97$ **b.** $3x + 3 \geq x + 97$

   c. $x + (x + 1) + (x + 2) \geq x + 97$ **d.** $[x + (x + 1) + (x + 2)] - x \geq 97$

6. In which quadrant is the point $(-3.8, -1.7)$?

   a. I **b.** II **c.** III **d.** IV

7. Which set is *not* a function?

   a. $\{(-2, 3), (-1, 1), (0, 0), (1, 1), (2, 3)\}$ **b.** $\{(2, 5), (0, 7), (2, 8), (4, 9)\}$

   c. $\{(-2, -4), (0, -4), (2, -4)\}$ **d.** $\{(2, 2), (3, 3)\}$

8. For the function $f: x \rightarrow 2 + (x - 1)^2, x \in \mathcal{R}$, which statement is *not* true?

   a. $f(3) = 6$ **b.** $f(-2) = f(4)$

   c. $f(-1) > f(1)$ **d.** $f(0) = 0$

9. Which equation is graphed?

   a. $3x + 2y = 6$ **b.** $3x - 2y = 6$

   c. $2x + 3y = 6$ **d.** $-3x + 2y = -6$

10. If a line contains $(-4, 2)$ and is parallel to $4x - 2y = 7$, it must also contain:

    a. $(0, 0)$ **b.** $(-2, 6)$ **c.** $(-2, 3)$ **d.** $(-3, 6)$

11. Which inequality is graphed here?

    a. $x + 3y \leq -3$ **b.** $x + 3y \geq -3$

    c. $x + 3y < -3$ **d.** $x + 3y > -3$

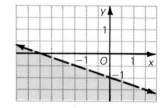

12. The graphs of the system $3x - 2y = 5$ are:
$$7 + 6x = 4y$$

    a. parallel lines   **b.** the same line   **c.** intersecting lines

13. Which system can be solved by the addition method?

    a. $3x - 5y = 8$ **b.** $2x + 7y = 4$ **c.** $x - 3y = 1$ **d.** $2x + 7y = 8$
        $2x + 3y = 1$ $-5x + 6y = 2$ $2x + y = 3$ $3x - 7y = 2$

14. To solve the system $5x - 6y = 8$ by the linear combination method,
$$3x + 4y = 9$$
find numbers to multiply the first and second equations by, respectively, to eliminate y.

    a. $3, -5$ **b.** $6, 4$ **c.** $2, 3$ **d.** $9, 8$

15. Solve the system $x - 2y = -8$
$$3x + 4y = 6$$

    a. $\{(2, 5)\}$ **b.** $\{(-2, 3)\}$ **c.** $\{(0, 4)\}$ **d.** $\{(-2, 5)\}$

16. The vertex angle of an isosceles triangle is $20°$ more than each base angle. Which system correctly represents this triangle?

    a. $x = y + 20$ **b.** $x = y + 20$ **c.** $x - y = 20$ **d.** $y = x - 20$
        $x + 2y = 180$ $2x + y = 180$ $x + y = 180$ $2x + y = 180$

17. Flying with the wind, a plane travels 3000 km in 4 h. Flying against the wind, it takes 5 h to cover the same distance. Find the wind speed.

    a. 50 km/h   **b.** 150 km/h   **c.** 100 km/h   **d.** 75 km/h

18. Simplify: $(-3a^2bc)^2(abc^2)^3$

    a. $9a^2b^5c^7$ **b.** $81a^7b^8c^8$ **c.** $9a^7b^5c^8$ **d.** $9a^{10}b^8c^8$

19. Simplify: $(2m - 3)^2$

    a. $4m^2 + 9$ **b.** $4m^2 + 12m + 9$ **c.** $4m^2 - 12m + 9$ **d.** $2m^2 - 9$

20. Factor completely: $12m^3n^3 - 8m^2n + 4mn$

    a. $4mn(3m^2n^2 - 2m + 1)$ **b.** $4mn(3m^2n^2 - 8m)$

    c. $4m(3m^2n^3 - 2mn + n)$ **d.** $(3mn + 4)(4m^2n^2 + mn)$

21. Factor completely: $4c^2 + 12c + 9$

   a. $(2c - 3)^2$ **b.** $(2c + 3)^2$

   c. $(2c + 3)(2c - 3)$ **d.** $(4c + 9)(c + 1)$

22. Factor completely: $6n^2 + 7n - 10$

   a. $(3n - 2)(2n + 5)$ **b.** $(6n + 5)(n - 2)$

   c. $(3n + 10)(2n - 1)$ **d.** $(6n - 5)(n + 2)$

23. Solve: $x^2 + 4x = 12$

   a. $\{-4, 3\}$ **b.** $\{-6, 2\}$ **c.** $\{12, 8\}$ **d.** $\{6, -2\}$

Wernher von Braun
1912–1977

Born in Wirsitz, Germany, von Braun was the person chiefly responsible for both the V-2 weapon used by Germany in World War II, and the rocket that sent men to the moon. His father was a secretary of agriculture in the Weimar Republic and his mother was an amateur astronomer. Her interest in astronomy no doubt stimulated the young Wernher's curiosity about space.

At the end of World War II, von Braun and some of his colleagues came to the United States to begin work on a succession of missiles that led to the development of the Jupiter-C, the first rocket to carry a U.S. satellite, Explorer 1, into orbit. That was in 1958, and was the result of a crash program directed by von Braun in response to the launching by the U.S.S.R. in 1957 of Sputnik 1, the first artificial satellite to be placed into Earth orbit.

Subsequently, von Braun was engaged in developing the Saturn series of rockets, the rockets which dispatched the Apollo spacecraft on their historic trips to the moon.

Anchored several miles off the coast, weather station Alpha records wind velocity, temperature, and movement of ocean currents.

Polynomials and Rational Expressions

Division of Polynomials

OBJECTIVES *for Sections 9-1 through 9-3:*
1. *Use the laws of exponents for division to simplify quotients of two monomials and quotients of other polynomials divided by monomials.*
2. *Use the algorithm for division to simplify quotients of polynomials.*

9-1 Laws of Exponents for Division

Numerical computations often reveal properties of numbers. Consider the following example.

EXAMPLE 1   Show that $\dfrac{9 \cdot 8}{3 \cdot 2} = \dfrac{9}{3} \cdot \dfrac{8}{2}$.

SOLUTION $\dfrac{9 \cdot 8}{3 \cdot 2} = \dfrac{72}{6} = 12$

$\dfrac{9}{3} \cdot \dfrac{8}{2} = 3 \cdot 4 = 12$

$\therefore \dfrac{9 \cdot 8}{3 \cdot 2} = \dfrac{9}{3} \cdot \dfrac{8}{2}$

Example 1 suggests the following theorem:

Basic Property of Quotients

For all real numbers r and s, and all nonzero real numbers t and u,

$$\frac{rs}{tu} = \frac{r}{t} \cdot \frac{s}{u}.$$

PROOF

1. $\dfrac{r}{t} \cdot \dfrac{s}{u} = \left(r \cdot \dfrac{1}{t}\right) \cdot \left(s \cdot \dfrac{1}{u}\right)$
   Relationship between multiplication and division

2. $\phantom{\dfrac{r}{t} \cdot \dfrac{s}{u}} = (rs) \cdot \left(\dfrac{1}{t} \cdot \dfrac{1}{u}\right)$
   Commutative and associative axioms of multiplication

3. $\phantom{\dfrac{r}{t} \cdot \dfrac{s}{u}} = (rs) \cdot \left(\dfrac{1}{tu}\right)$
   Property of the reciprocal of a product

4. $\phantom{\dfrac{r}{t} \cdot \dfrac{s}{u}} = \dfrac{rs}{tu}$
   Relationship between multiplication and division

5. $\dfrac{r}{t} \cdot \dfrac{s}{u} = \dfrac{rs}{tu}$
   Transitive property of equality

You can obtain the three following useful consequences of the basic property of quotients by (1) replacing t with 1, (2) replacing r with 1, and (3) replacing r with 1 and s with 1. This latter is simply the property of the reciprocal of a product given on page 86.

---

1. For all real numbers r and s and all nonzero numbers u,

$$\frac{rs}{u} = r \cdot \frac{s}{u}.$$

2. For all real numbers s and all nonzero real numbers t and u,

$$\frac{s}{tu} = \frac{1}{t} \cdot \frac{s}{u}.$$

3. For all nonzero real numbers t and u,

$$\frac{1}{tu} = \frac{1}{t} \cdot \frac{1}{u}.$$

Using these properties you can write the following examples:

$$\frac{-8 \cdot 34}{17} = -8 \cdot \frac{34}{17} = -8 \cdot 2 = -16$$

$$\frac{32}{8 \cdot (-7)} = \frac{32}{8} \cdot \frac{1}{(-7)} = 4 \cdot \frac{1}{(-7)} = -\frac{4}{7}$$

$$\frac{1}{7} \cdot \frac{1}{11} = \frac{1}{7 \cdot 11} = \frac{1}{77}$$

The examples below show how you can simplify quotients of powers by using the laws of exponents for multiplication along with the basic property of quotients and the three results mentioned above. Note that in each example, a denotes any *nonzero* real number. Also, each example uses the fact that the quotient of any power of a nonzero number and the power itself is 1. For instance,

$$\frac{a^2}{a^2} = 1, \quad \text{since} \quad \frac{a^2}{a^2} = a^2 \cdot \frac{1}{a^2} = 1.$$

Thus,

$$\frac{a^7}{a^4} = \frac{a^3 \cdot a^4}{a^4} = a^3 \cdot \frac{a^4}{a^4} = a^3 \cdot 1 = a^3$$

$$\frac{a^2}{a^6} = \frac{a^2}{a^4 \cdot a^2} = \frac{1}{a^4} \cdot \frac{a^2}{a^2} = \frac{1}{a^4} \cdot 1 = \frac{1}{a^4}$$

The preceding examples suggest a pattern for simplifying any quotient of the form $\frac{a^m}{a^n}$, where $a \neq 0$ and m and n are positive integers.

$\frac{a^m}{a^n}$ $(m = n)$	$\frac{a^m}{a^n}$ $(m > n)$	$\frac{a^m}{a^n}$ $(m < n)$
$\frac{a^m}{a^n} = \frac{a^n}{a^n}$	$\frac{a^m}{a^n} = \frac{a^{m-n} \cdot a^n}{a^n}$	$\frac{a^m}{a^n} = \frac{a^m}{a^{n-m} \cdot a^m}$
$= a^n \cdot \frac{1}{a^n}$	$= a^{m-n} \cdot \frac{a^n}{a^n}$	$= \frac{1}{a^{n-m}} \cdot \frac{a^m}{a^m}$
$= 1$	$= a^{m-n} \cdot 1$	$= \frac{1}{a^{n-m}} \cdot 1$
	$= a^{m-n}$	$= \frac{1}{a^{n-m}}$

The following theorem summarizes the results in the preceding table.

Laws of Exponents for Division

For all positive integers m and n and every nonzero real number a:

1. If $m = n$, then $\frac{a^m}{a^n} = 1$.   **2.** If $m > n$, then $\frac{a^m}{a^n} = a^{m-n}$.

3. If $m < n$, then $\frac{a^m}{a^n} = \frac{1}{a^{n-m}}$.

EXAMPLE 2 Use the basic property of quotients and the laws of exponents for division to show that for all real numbers x and y such that $x \neq 0$ and $y \neq 0$,

$$\frac{18x^6y^4}{3x^3y} = 6x^3y^3.$$

SOLUTION $\dfrac{18x^6y^4}{3x^3y} = \dfrac{18}{3} \cdot \dfrac{x^6}{x^3} \cdot \dfrac{y^4}{y} = 6x^{6-3}y^{4-1} = 6x^3y^3$

The result of Example 2 means that for any *nonzero* values of x and y, the expression $\dfrac{18x^6y^4}{3x^3y}$ has the same value as the expression $6x^3y^3$. We say that these expressions are *equivalent, subject to the restrictions $x \neq 0$ and $y \neq 0$*. When we replace $\dfrac{18x^6y^4}{3x^3y}$ by $6x^3y^3$, we say that we have *divided* $18x^6y^4$ by $3x^3y$, and we call $6x^3y^3$ the *quotient* of the monomials. We also say that we have *simplified* $\dfrac{18x^6y^4}{3x^3y}$.

EXAMPLE 3 Find the quotient when $10a^3b^4$ is divided by $5a^4b^3$, and use the result to state a fact about real numbers.

SOLUTION $\dfrac{10a^3b^4}{5a^4b^3} = \dfrac{10}{5} \cdot \dfrac{a^3}{a^4} \cdot \dfrac{b^4}{b^3} = 2 \cdot \dfrac{1}{a^{4-3}} \cdot b^{4-3} = 2 \cdot \dfrac{1}{a} \cdot b = \dfrac{2b}{a}.$

For all nonzero real numbers a and b,

$$\frac{10a^3b^4}{5a^4b^3} = \frac{2b}{a}.$$

Oral Exercises

In Exercises 1–4 use the basic property of quotients to state a numeral to replace the question mark in the sentence so that the resulting statement is true.

1. $\dfrac{24}{32} = \dfrac{3}{8} \cdot \dfrac{?}{4}$

2. $\dfrac{36}{28} = \dfrac{4}{7} \cdot \dfrac{9}{?}$

3. $\dfrac{1}{-21} = \dfrac{1}{?} \cdot \dfrac{1}{3}$

4. $\dfrac{-6}{64} = \dfrac{?}{32} \cdot \dfrac{2}{2}$

5. Complete the following statement of the basic property of quotients: For all real numbers a and b, and all nonzero real numbers c and d, $\dfrac{ab}{cd} = \underline{\ ?\ }$. (There is more than one way to complete the statement correctly.)

6. Why is it necessary to specify that $t \neq 0$ and $u \neq 0$ in the basic property of quotients stated on page 324?

Simplify each expression. Assume that variable expressions appearing as exponents denote positive integers and that no denominator has zero as a value.

7. $\dfrac{a^3}{a}$ 8. $\dfrac{b^5}{b^2}$ 9. $\dfrac{c^6}{c^9}$ 10. $\dfrac{d^7}{d^{11}}$ 11. $\dfrac{8x}{2x}$

12. $\dfrac{6y^2}{8y^2}$ 13. $\dfrac{z^n}{z^3}$ 14. $\dfrac{w^n}{w}$ 15. $\dfrac{p^{3n}}{p^n}$ 16. $\dfrac{e^{2n}}{e^{5n}}$

17. $\dfrac{k^{n+1}}{k^n}$ 18. $\dfrac{s^{m-2}}{s}$ 19. $\dfrac{c^{t+4}}{c^{t+4}}$ 20. $\dfrac{x^{2m+1}}{x^{m+1}}$ 21. $\dfrac{y^{3n-1}}{y^{3n+1}}$

22. Is the quotient of two monomials always a monomial? Explain.

Written Exercises

Simplify each expression. Assume that no denominator has zero as a value. In Exercises 1–8 use your result to state a property of real numbers.

A 1. $\dfrac{15a^5b^3}{5a^2b}$ 2. $\dfrac{18cd^6}{2cd^4}$ 3. $\dfrac{-12e^2f}{4e^2}$ 4. $\dfrac{9h^5}{-9h^5}$

5. $\dfrac{8x^2y}{24x^4}$ 6. $\dfrac{6r^2s^3}{12rs^4}$ 7. $\dfrac{-21w^5z^2}{-7w^4z^4}$ 8. $\dfrac{36mn^4}{-18m^2n^3}$

9. $\dfrac{42x^3y^3}{6x^4y^3}$ 10. $\dfrac{27a^7b^4}{54a^6b}$ 11. $\dfrac{-12c^4d}{48cd}$ 12. $\dfrac{-14e^6f^{10}}{-28e^7f^{11}}$

13. $\dfrac{2.4xz^2}{0.4x^2z^2}$ 14. $\dfrac{3.9m^5n^2}{1.3mn^3}$ 15. $\dfrac{6x^5yz}{0.5x^4y^4z}$ 16. $\dfrac{0.4a^2bc^2}{8ab^2c^2}$

17. $\dfrac{8(x-y)^2}{2(x-y)^3}$ 18. $\dfrac{35(m+n)^4}{-5(m+n)^2}$ 19. $\dfrac{k(3x+2)^2}{k^2(3x+2)}$ 20. $\dfrac{m^3(p-3q)^4}{m^4(p-3q)^3}$

Find the solution set.

EXAMPLE $\dfrac{6^m}{6^3} = 6^7$

SOLUTION $\dfrac{6^m}{6^3} = 6^{m-3}$, so $6^{m-3} = 6^7$.

$\therefore\ m - 3 = 7$ or $m = 10$. The solution set is $\{10\}$.

B 21. $\dfrac{2^k}{2^5} = 2^8$ 22. $\dfrac{4^{3x}}{4^x} = 4^6$ 23. $\dfrac{5^{2y-1}}{5^{y+1}} = 5^5$ 24. $3^4 = 3^2 \cdot 3^{2x}$

25. $\dfrac{3^{m-2}}{9} = 3^7$ 26. $4^{2a} = 2^{a+6}$ 27. $\dfrac{4^{2x-3}}{8^{x-2}} = 2^{20}$ 28. $8 \cdot 2^{3n} \cdot 4^n = 2^{n+15}$

Simplify the given expression. Assume that no denominator has zero as a value.

29. $\dfrac{(4ab^3)(3a^2b)}{(2a^3b)(9ab^2)}$

30. $\dfrac{(4xy^2)^2}{-8x^2y^4}$

31. $\dfrac{(3r)(3r^2)^2}{(3r)^2(3r)^3}$

32. $\dfrac{(4m)^2(2m)^3}{(2m^2)^2(-m)}$

33. $\dfrac{(-2x^2)^3(7x)^2}{(4x)^2(7x)(x^2)}$

34. $\dfrac{(-4xy^2)^3}{(2xy^2)^4}$

35. $\dfrac{(x+1)^3(-2y^2)^3}{(-4y)^2(x+1)^3}$

36. $\dfrac{(-3ab^2)^2(-2a^2b)^3}{9a^2b(-4a^2b)^2}$

37. $\dfrac{(4m^2n^3)^2(-mn^2)^3}{-(2m^2n^2)^2(m^3n)^2}$

Use the basic property of quotients or the three results listed on page 324 to prove the theorems stated in Exercises 38–41.

C 38. If r and t are real numbers such that $r \neq 0$ and $t \neq 0$, then $\dfrac{r}{t} \cdot \dfrac{t}{r} = 1$.

39. If a and b are real numbers and $b \neq 0$, then $\left(\dfrac{a}{b}\right)^2 = \dfrac{a^2}{b^2}$.

(*Hint:* Recall that $\dfrac{a}{b} = a \cdot \dfrac{1}{b}$, and use the third law of exponents for multiplication (page 284).)

40. If a and b are real numbers and $b \neq 0$, then $\dfrac{-a}{b} = -\dfrac{a}{b}$.

41. If a and b are real numbers and $b \neq 0$, then $\dfrac{a}{-b} = -\dfrac{a}{b}$.

9-2 Dividing a Polynomial by a Monomial

Do you agree that $\dfrac{54+5}{7} = \dfrac{54}{7} + \dfrac{5}{7}$?

The relationship between multiplication and division (page 110), together with the distributive axiom (page 47), guarantees that the given equation must be a true statement. You have:

$$\frac{54+5}{7} = (54+5)\frac{1}{7} = 54 \cdot \frac{1}{7} + 5 \cdot \frac{1}{7} = \frac{54}{7} + \frac{5}{7}.$$

This example illustrates the following theorem, whose proof you have already given if you have solved Exercises 35 and 36 on page 114.

Theorem. For all real numbers a, b, and all nonzero real numbers c,

$$\frac{a+b}{c} = \frac{a}{c} + \frac{b}{c} \quad \text{and} \quad \frac{a-b}{c} = \frac{a}{c} - \frac{b}{c}.$$

You can use this theorem in the following examples.

EXAMPLE 1   Find a polynomial equivalent to

$$\frac{6y^3 + 9y^2 - 3y}{3y}$$

subject to the restriction $y \neq 0$.

SOLUTION $\dfrac{6y^3 + 9y^2 - 3y}{3y} = \dfrac{6y^3}{3y} + \dfrac{9y^2}{3y} - \dfrac{3y}{3y} = 2y^2 + 3y - 1$

We call $2y^2 + 3y - 1$ the *quotient* when the polynomial $6y^3 + 9y^2 - 3y$ is divided by the monomial $3y$. Example 1 illustrates the following rule.

> The quotient of a polynomial divided by a monomial is the sum of the quotients obtained by dividing each term of the polynomial by the monomial.

EXAMPLE 2   Express the quotient of $ax^2 + 2ax - a^2$ divided by ax as a sum, and use the result to state a fact about real numbers.

SOLUTION $\dfrac{ax^2 + 2ax - a^2}{ax} = \dfrac{ax^2}{ax} + \dfrac{2ax}{ax} - \dfrac{a^2}{ax} = x + 2 - \dfrac{a}{x}$

For all nonzero real numbers a and x,

$$\frac{ax^2 + 2ax - a^2}{ax} = x + 2 - \frac{a}{x}.$$

Oral Exercises

Express each quotient as a sum. Assume that no denominator has zero as a value.

1. $\dfrac{8a + 6b}{2}$

2. $\dfrac{6c - 12}{3}$

3. $\dfrac{15d^2 - 6d}{3d}$

4. $\dfrac{6e^3 - 17e^2}{e}$

5. $\dfrac{9f + 4}{f}$

6. $\dfrac{7g - 5}{g}$

7. $\dfrac{6x^2 - 3x + 12}{3}$

8. $\dfrac{m^2y - 2my^2}{my}$

9. $\dfrac{w^3z^2 - 5z}{w^2z}$

10. Is the quotient of a polynomial and a monomial always a polynomial? Explain.

Written Exercises

Express each quotient as a sum. Assume that no denominator has zero as a value. In Exercises 1–12 use your result to state a property of real numbers.

A   1. $\dfrac{25a - 15}{5}$

2. $\dfrac{24b + 16}{8}$

3. $\dfrac{12c^2 + 4c}{2c}$

4. $\dfrac{36d - 48d^2}{6d}$

5. $\dfrac{2e^2f + 4ef - f^2}{2f}$

6. $\dfrac{16x^3 + 12x^2 - 40x}{4x}$

7. $\dfrac{24y^4 - 16y^3 + 32y^2}{-8y^2}$

8. $\dfrac{30z^5 + 35z^3 - 45z}{-5z}$

9. $\dfrac{9w^2z^3 + 27wz^2 - 72wz}{9w^2z^2}$

10. $\dfrac{21s^2r - 7sr^2 + 63sr}{7sr^2}$

11. $\dfrac{8xy^3 - 6x^2y + 36xy}{6x^2y}$

12. $\dfrac{11c^3d^3 - 33cd + 66c^2d^2}{11c^2d^2}$

B   13. $\dfrac{18p^2q + 30pq^2 - 12p^2q^2}{-6p^2q^2}$

14. $\dfrac{16r^5 - 28r^4 + 35r^3}{-4r^3}$

15. $\dfrac{21c^2de - 20cd^2e + 24cde^2}{5c^2d^2e}$

16. $\dfrac{40ab^2c^2 - 24a^2bc^2 + 36a^2b^2c}{8a^2b^2c^2}$

17. $\dfrac{x^{2n} - x^n + x^{n-1}}{x^n}$

18. $\dfrac{2^{2n} - 2^n + 2^{n-1} - 2^{n-2}}{-2^n}$

For the functions P and Q defined in the exercises below, express $\dfrac{P(x)}{Q(x)}$ as a sum.

19. $P(x) = 12a^3 + 4a^2 - 8a; \quad Q(x) = 4a$

20. $P(x) = 20b^5 - 30b^3 + 60b^2; \quad Q(x) = -10b^2$

21. $P(x) = 36c^6 + 45c^5 - 81c^3; \quad Q(x) = 9c^3$

22. $P(x) = 72d^4 - 36d^2 + 54d^5; \quad Q(x) = -18d^2$

C   23. Evaluate $\dfrac{x}{y}$ when $\dfrac{x+y}{y} = 21.$

24. Evaluate $\dfrac{x}{y}$ when $\dfrac{x-y}{y} = 50.$

25. Find the solution set: $\dfrac{5r^2 + 10r}{5r} = r + 2.$

26. Are $\dfrac{2x^2 + 16x}{2x} = 8$ and $x + 8 = 8$ equivalent open sentences? Explain your answer.

27. Are $\dfrac{6m^2 - 10m}{2m} < 1$ and $3m - 5 < 1$ equivalent open sentences? Explain your answer.

28. Find the positive value of $\dfrac{a}{x}$ that is a solution of $\dfrac{a^2 - x^2}{x^2} = -0.84$.

9-3 Quotients of Polynomials: Rational Expressions

Do you recognize the division algorithm shown at the right? It is an expanded version of the familiar division algorithm of arithmetic. It shows that

$$770 - (32)(24) = 2.$$

$$
\begin{array}{r}
\text{Quotient} \\
\downarrow \\
\left.\begin{array}{r} 4 \\ 20 \end{array}\right\} 24 \\
\text{Divisor} \rightarrow 32\overline{)770} \leftarrow \text{Dividend} \\
\underline{640} \\
130 \\
\underline{128} \\
\text{Remainder} \rightarrow \quad 2
\end{array}
$$

To check this fact, you compute $(32)(24)$ and add this product to 2 to obtain 770. That is, you verify that

$$770 = (32)(24) + 2.$$

$$
\begin{array}{r}
32 \\
\times 24 \\
\hline
128 \\
64 \\
\hline
768 \\
+ 2 \\
\hline
770
\end{array}
$$

In general form, these two displayed statements are:

$$\text{Dividend} - \text{Divisor} \times \text{Quotient} = \text{Remainder}$$
$$\text{Dividend} = \text{Divisor} \times \text{Quotient} + \text{Remainder}$$

Each of these statements is equivalent to

$$\frac{\text{Dividend}}{\text{Divisor}} = \text{Quotient} + \frac{\text{Remainder}}{\text{Divisor}},$$

which, for the example shown, is

$$\frac{770}{32} = 24 + \frac{2}{32}.$$

Any number that is the quotient (ratio) of two integers is called a **rational number**, except, of course, that 0 cannot be a divisor.

In a similar way, any expression that is the quotient of two polynomials is called a **rational expression**. To see how to use the division algorithm to divide one polynomial by another, study the following example.

EXAMPLE 1   Divide $4x^2 - 12x + 11$ by $2x - 3$.

SOLUTION

1. Be sure the dividend and divisor are written with terms in descending degree in one variable.

$$2x - 3 \overline{)4x^2 - 12x + 11}$$

2. $4x^2 \div 2x = 2x$

$$\begin{array}{r} 2x \\ 2x - 3 \overline{)4x^2 - 12x + 11} \end{array}$$

3. $2x \cdot (2x - 3) = 4x^2 - 6x$

$$\begin{array}{r} 2x \\ 2x - 3 \overline{)4x^2 - 12x + 11} \\ 4x^2 - 6x \end{array}$$

4. $(4x^2 - 12x + 11)$
   $- (4x^2 - 6x) = -6x + 11$

$$\begin{array}{r} 2x \\ 2x - 3 \overline{)4x^2 - 12x + 11} \\ \underline{4x^2 - 6x} \\ - 6x + 11 \end{array}$$

5. $-6x \div 2x = -3$

$$\begin{array}{r} 2x - 3 \\ 2x - 3 \overline{)4x^2 - 12x + 11} \\ \underline{4x^2 - 6x} \\ - 6x + 11 \end{array}$$

6. $-3 \cdot (2x - 3) = -6x + 9$

$$\begin{array}{r} 2x - 3 \\ 2x - 3 \overline{)4x^2 - 12x + 11} \\ \underline{4x^2 - 6x} \\ - 6x + 11 \\ - 6x + 9 \end{array}$$

7. $(-6x + 11) - (-6x + 9) = 2$

$$\begin{array}{r} 2x - 3 \\ 2x - 3 \overline{)4x^2 - 12x + 11} \\ \underline{4x^2 - 6x} \\ - 6x + 11 \\ \underline{- 6x + 9} \\ 2 \end{array}$$

The result can be written as:

$$(4x^2 - 12x + 11) \div (2x - 3) = 2x - 3 + \frac{2}{2x - 3}, \ (2x - 3 \neq 0).$$

Check:
$(2x - 3)(2x - 3) + 2 = (4x^2 - 12x + 9) + 2 = 4x^2 - 12x + 11.$

The next example shows how to insert a "missing" term in a dividend by using 0 as a coefficient.

EXAMPLE 2   Divide $x^3 - 7x - 2$ by $x - 2$.

SOLUTION

$$\begin{array}{r} x^2 + 2x - 3 \\ x - 2 \overline{\smash{\big)}\, x^3 + 0x^2 - 7x - 2} \\ \underline{x^3 - 2x^2} \\ 2x^2 - 7x \\ \underline{2x^2 - 4x} \\ -3x - 2 \\ \underline{-3x + 6} \\ -8 \end{array}$$

$$\therefore (x^3 - 7x - 2) \div (x - 2) = x^2 + 2x - 3 - \frac{8}{x - 2}, \ (x - 2 \neq 0).$$

Note that in both Example 1 and Example 2, the variable is restricted to values for which the divisor is not 0.

Oral Exercises

In each exercise state how you would write the dividend before using the division algorithm.

1. $(2 + 3a + a^2) \div (a + 1)$
2. $(c^2 - 16) \div (c + 4)$
3. $(-3b - 18 + b^2) \div (b - 6)$
4. $(d^3 + 8) \div (d + 2)$
5. $(e^2 + 2e^3) \div (2e - 1)$
6. $(f^4 - 1) \div (f^2 + 1)$
7. Is the quotient of two polynomials always a polynomial? Explain.
8. Is the quotient of two polynomials, if the quotient exists, always a rational expression? Explain.

Written Exercises

Divide the first polynomial by the second.

A
1. $a^2 - 3a - 10, \ a + 2$
2. $b^2 - 5b + 6, \ b - 2$
3. $3c^2 + 13c + 12, \ 3c + 4$
4. $2d^2 - 7d + 6, \ d - 2$
5. $15e^2 - 11e - 14, \ 3e + 2$
6. $6f^2 + f - 2, \ 2f - 1$
7. $4 - 8g + 3g^2, \ 3g - 2$
8. $2h^2 + 11h - 18, \ 2h - 3$
9. $12j^2 - 18 + 4j, \ 2j + 3$
10. $k^2 - 25, \ k + 5$
11. $x^3 + 8, \ x + 2$
12. $y^4 - 16, \ y^2 + 4$

B
13. $3z^2 - 2w^2 - 5wz, \ z - 2w$
14. $10u^2 - 11uv - 6v^2, \ 5u + 2v$
15. $p^2 + 9, \ p + 3$
16. $25s^2 + 20s + 4, \ 5s + 2$
17. $6b^3 - 4b^2 - 18b + 12, \ 3b - 2$
18. $2c^3 - 7c^2 + 11c - 4, \ 2c - 1$
19. $d^4 - 2d^3 - d + 1, \ d^2 + 3$
20. $2e^4 - e^3 - e^2 + e - 1, \ e^2 - 1$

In Exercises 21–24 the factor set is the set of polynomials with integral coefficients.

21. One factor of $6f^4 - f^3 - 6f^2 + 5f - 2$ is $3f^2 - 2f + 1$. Find the other factor.

22. One factor of $6x^4 - 5x^3 - 5x^2 + 39x - 27$ is $x^2 - 2x + 3$. Find the other factor.

23. One factor of $2y^3 + 3y^2 - 11y - 6$ is $y - 2$. Find two other factors.

24. One factor of $3z^3 - 14z^2 - 7z + 10$ is $z + 1$. Find two other factors.

Determine a value for p so that the remainder in the indicated division is 0.

C **25.** $(6a^2 + 2a + p) \div (2a + 4)$ **26.** $(2c^3 - 9c^2 + c + p) \div (2c - 3)$

   27. $(6e^3 + 3e^2 + pe - 2) \div (2e + 1)$ **28.** $(f^3 - 4f^2 + pf - 6) \div (f - 2)$

   29. Determine p so that

$$(2h^2 + ph + 13) \div (h + 3) = 2h + 4 + \frac{1}{h + 3}.$$

30. Determine p such that the solution set of

$$\frac{3l^2 + pl + 7}{l - 2} = 3l - 5 - \frac{3}{l - 2} \quad \text{is} \quad \{l : l \neq 2\}.$$

31. A rectangle of width $(3x + 2y)$ cm has an area of $(6x^2 + 13xy + 6y^2)$ cm². Find the perimeter of the rectangle.

Self-Test 1

VOCABULARY   basic property of quotients (p. 324)
rational expression (p. 332)

Simplify.

1. $\dfrac{14a^2b}{-2ab}$

2. $\dfrac{-36c^2d^3}{-9c^3d}$

Obj. 1, p. 323

Express each quotient as a sum.

3. $\dfrac{15e^3 - 18e^2 + 9e}{3e}$

4. $\dfrac{8f^2g + 6fg^2 - 7fg}{2f^2g^2}$

Complete the indicated division.

5. $(2x^2 + 3x - 2) \div (x + 2)$ **6.** $y + 2\overline{)y^3 - 2y + 4}$

Obj. 2, p. 323

Check your answers with those at the back of the book.

Simplifying Rational Expressions

OBJECTIVES for Sections 9-4 and 9-5:
1. *Reduce fractions to lowest terms.*
2. *Simplify expressions involving powers with zero or negative integral exponents.*

9-4 Simplifying Rational Expressions

Any rational number can be represented by many different fractions. For example:

$$5 = \frac{5}{1} = \frac{10}{2} = \frac{-15}{-3} \qquad 0 = \frac{0}{7} = \frac{0}{-11}$$

$$-\frac{3}{11} = \frac{-3}{11} = \frac{3}{-11} \qquad 0.8 = \frac{8}{10} = \frac{4}{5}$$

You can use the basic property of quotients (page 324) to determine whether or not two fractions name the same number.

EXAMPLE 1 Show that $\dfrac{15}{20} = \dfrac{3}{4}$.   **SOLUTION** $\dfrac{15}{20} = \dfrac{3 \cdot 5}{4 \cdot 5} = \dfrac{3}{4} \cdot \dfrac{5}{5} = \dfrac{3}{4} \cdot 1 = \dfrac{3}{4}$.

Example 1 suggests how to prove the following result.

Theorem.   For all real numbers r, and all nonzero real numbers s and t,

$$\frac{rt}{st} = \frac{r}{s} \quad \text{and} \quad \frac{r}{s} = \frac{rt}{st}.$$

PROOF

$$\frac{rt}{st} = \frac{r}{s} \cdot \frac{t}{t} \qquad\qquad \text{Basic property of quotients}$$

$$= \frac{r}{s} \cdot 1 \qquad\qquad \text{First law of exponents for division}$$

$$= \frac{r}{s} \qquad\qquad \text{Identity axiom of multiplication}$$

$$\therefore \frac{rt}{st} = \frac{r}{s} \qquad\qquad \text{Transitive property of equality}$$

and $\qquad \dfrac{r}{s} = \dfrac{rt}{st}.$ $\qquad\qquad$ Reflexive property of equality

In the fraction $\dfrac{r}{s}$, you call r the **numerator** and s the **denominator**. Using these terms, you can express the preceding theorem in the following informal way.

> Dividing or multiplying the numerator and denominator of a fraction by the same nonzero number produces a fraction equivalent to the given fraction.

Thus, $-\dfrac{24}{18} = -\dfrac{24 \div 6}{18 \div 6} = -\dfrac{4}{3}$ and $\dfrac{7}{x} = \dfrac{7 \cdot x^3}{x \cdot x^3} = \dfrac{7x^3}{x^4}$, if $x \neq 0$.

We say the fractions

$$-\frac{4}{3} \quad \text{and} \quad \frac{7}{x}$$

are in *lowest terms* or in *simple form*. Any fraction whose numerator and denominator are integers or polynomials with integral coefficients is in **lowest terms** or **simple form** if the greatest common factor of the numerator and denominator is 1. Thus, to **simplify a fraction** or to **reduce it to lowest terms,** you divide the numerator and denominator by their greatest common factor.

EXAMPLE 2   Reduce $\dfrac{5n + 10}{n^2 - 4}$ to lowest terms.

SOLUTION   Factor the numerator and denominator. Then divide the numerator and denominator by their greatest common factor.

$$\frac{5n + 10}{n^2 - 4} = \frac{5(n + 2)}{(n - 2)(n + 2)} = \frac{5}{n - 2}, \text{ if } n \notin \{-2, 2\}.$$

Sometimes factors of the numerator and denominator are negatives of one another, as in the next example.

EXAMPLE 3   Simplify $\dfrac{8 - 2a}{a^2 - a - 12}$.

> Express the numerator as a product having -1 as a factor.

SOLUTION

$$\frac{8 - 2a}{a^2 - a - 12} = \frac{2(4 - a)}{(a + 3)(a - 4)} = \frac{2(-1)(a - 4)}{(a + 3)(a - 4)}$$

$$= \frac{-2}{a + 3}$$

$$= -\frac{2}{a + 3}, \text{ if } a \notin \{-3, 4\}$$

Notice that in Examples 2 and 3 we stated restrictions on the values of the variables. The restrictions are needed because in each case the original fraction and the fraction in lowest terms are equivalent only for those values of the variable for which neither denominator is zero.

Hereafter, in this book it will be assumed that the replacement sets of the variables in a fraction include no value for which the denominator is zero.

Oral Exercises

Name the greatest common factor of the numerator and denominator of the given fraction, and state all restrictions on the values of the variable.

1. $\dfrac{9}{27}$

2. $\dfrac{7x}{5x}$

3. $\dfrac{4y^2}{7y}$

4. $\dfrac{x+2}{x-2}$

5. $\dfrac{2w-3}{2w+1}$

6. $\dfrac{14(2a-3)}{13(2a-3)}$

7. $\dfrac{b+2}{(b+1)(b+2)}$

8. $\dfrac{c(c-1)}{c^2(c-1)}$

9. Do the open sentences $\dfrac{x}{x} = 1$ and $x = x$ have the same solution set over $\Re$? Explain.

Written Exercises

Reduce each fraction to lowest terms.

A

1. $\dfrac{8ab}{18ac}$

2. $\dfrac{22a^2b}{33ab^2}$

3. $\dfrac{-6ab}{12ab}$

4. $\dfrac{18c^2d}{-10cd^2}$

5. $\dfrac{8(c-1)}{13(c-1)^2}$

6. $\dfrac{6d(d+3)}{3(d+3)}$

7. $\dfrac{6e+4f}{16e+8f}$

8. $\dfrac{e^2-4e}{5(e-4)}$

9. $\dfrac{15f^2+24f}{3f^2}$

10. $\dfrac{16r}{12r^2-24r}$

11. $\dfrac{4s-16}{s^2-16}$

12. $\dfrac{r^2-36}{2r-12}$

13. $\dfrac{3p+2}{3p^2+5p+2}$

14. $\dfrac{q^2+4q-5}{q-1}$

15. $\dfrac{p^2-3p}{p^2-2p-3}$

16. $\dfrac{x^2-x}{x^2+x-2}$

B

17. $\dfrac{x^2-3xy-4y^2}{x^2-xy-12y^2}$

18. $\dfrac{2x^2-xy-3y^2}{3x^2+5xy+2y^2}$

19. $\dfrac{2w^2-w-3}{3w+3}$

20. $\dfrac{3w-2z}{9w^2-4z^2}$

21. $\dfrac{u^2+5u-14}{2-u}$

22. $\dfrac{2y^2+5y-3}{1-2y}$

23. $\dfrac{10b^3-15b^2-10b}{4b-16b^3}$

24. $\dfrac{2d^3-d^2-10d}{d^3-2d^2-8d}$

25. $\dfrac{5f^3+20f^2+20f}{25f^3-100f}$

Polynomials and Rational Expressions | **337**

Explain why each reduction is incorrect.

C **26.** $\dfrac{k+2}{3k} \stackrel{?}{=} \dfrac{1+2}{3} = 1$ **27.** $\dfrac{k-4}{k+4} \stackrel{?}{=} \dfrac{-4}{4} = -1$

28. $\dfrac{3x}{(2x)(5x)} \stackrel{?}{=} \dfrac{3}{2\cdot 5} = \dfrac{3}{10}$ **29.** $\dfrac{8+z}{4+z} \stackrel{?}{=} \dfrac{8}{4} = 2$

30. Prove: For any real number a and any nonzero real numbers b

and c: $\dfrac{a}{b} = \dfrac{\frac{a}{c}}{\frac{b}{c}}$.

31. Prove: If $b \neq c$ and $\dfrac{ab-ac}{b-c} < 0$, then a is a negative number.

ⅅIVERSION

You might be surprised to find that the "illegal" cancellation $\dfrac{2\cancel{6}}{\cancel{6}5}$ gives a correct answer, since $\dfrac{26}{65} = \dfrac{2}{5}$ is true!

Can you find any other fractions where this kind of illegal cancellation produces right answers?

9-5 Zero and Negative Exponents

You are familiar with positive exponents, such as those in the powers 3^5, 2^3, and $(-5)^2$. Can you assign a meaning to such expressions as 2^0, 2^{-5}, or, more generally, to a^{-n}, where $n \in \{0, 1, 2, 3, \ldots\}$?

Before a formal analysis, consider the number pattern

n	3	2	1	0	−1	−2	−3
2^n	8	4	2				

Do you agree that each time n decreases by 1 the value of 2^n is half the preceding value? For the pattern to continue, we would need to fill in each blank with half the preceding entry. Then the table would look like this:

n	3	2	1	0	−1	−2	−3
2^n	8	4	2	1	$\frac{1}{2}$	$\frac{1}{4}$	$\frac{1}{8}$

In extending the meaning of a power to include *any integer* as an exponent, we insist that operations with powers must continue to obey the laws already established for positive integral exponents.

The first law of exponents for division (page 325) implies that

$$\frac{2^7}{2^7} = 1.$$

If the second law were also to hold for this quotient, you would find:

$$\frac{2^7}{2^7} = 2^{7-7} = 2^0.$$

This suggests that it is appropriate to define 2^0 to be 1, and, in general, to make the following definition.

> For every nonzero real number a,
> $$a^0 = 1.$$

No meaning is assigned to the expression 0^0.

Now consider the quotient

$$\frac{2^3}{2^8} = \frac{1}{2^{8-3}} = \frac{1}{2^5}.$$

If you were to apply the second law of exponents for division, ignoring the restriction that $m > n$, you would find

$$\frac{2^3}{2^8} = 2^{3-8} = 2^{-5}.$$

This suggests that you define 2^{-5} to be $\frac{1}{2^5}$, and that you make the following definition.

> For every nonzero real number a and every integer n,
> $$a^{-n} = \frac{1}{a^n};$$
> that is, a^{-n} is the reciprocal of a^n.

Thus:

$$5^{-3} = \frac{1}{5^3} = \frac{1}{125}$$

With these definitions, the laws of exponents that you have learned continue to apply. For example, if $a \neq 0$:

$$a^{-4} \cdot a^{-3} = a^{-4+(-3)} = a^{-7}$$

To justify this chain of equalities, you may proceed in this way:

$$a^{-4} \cdot a^{-3} = \frac{1}{a^4} \cdot \frac{1}{a^3} = \frac{1 \cdot 1}{a^4 \cdot a^3} = \frac{1}{a^{4+3}} = \frac{1}{a^7} = a^{-7}.$$

You can also justify each of the following statements for every nonzero value of each variable.

$$\frac{a^{-3}}{a^{-5}} = a^{-3-(-5)} = a^2$$

$$a^2 \cdot a^{-2} = a^{2+(-2)} = a^0 = 1$$

$$\frac{1}{a^{-2}} = a^2$$

$$(3a^{-2})^4 = 3^4 \cdot a^{-2 \cdot 4} = 81a^{-8}$$

$$\left(\frac{1}{b}\right)^{-1} = b$$

$$\left(\frac{a}{b}\right)^{-1} = \frac{b}{a}$$

$$\frac{(a^0 b^{-2})^5}{2a^{-1}} = \frac{b^{-10}}{2a^{-1}} = \frac{a}{2b^{10}}$$

Oral Exercises

In Exercises 1–8 state the most common name of the given number.

EXAMPLE $\dfrac{1}{2^{-3}}$ **SOLUTION** $\dfrac{1}{2^{-3}} = 2^3 = 8$

1. 3^{-3}

2. 4^{-1}

3. 6^0

4. $(-4)^{-2}$

5. $(3-1)^{-1}$

6. 1^{-11}

7. $3^2 \cdot 3^{-3}$

8. $\dfrac{1}{2^{-1}} + (-4)^0$

Express each of the following as a number or as a power of a variable (with a positive exponent). Assume that the replacement set of all variables is {the nonzero integers}.

9. $a^{-2} \cdot a^4$

10. $3(-b)^0$

11. $c^3 \div c^{-4}$

12. $\dfrac{2d^6}{2d^2}$

13. $\dfrac{e^5}{e^{-2}}$

14. $\dfrac{f^{-5}}{f^2}$

15. $(x^0)^{-3}$

16. $\dfrac{y^0 + z^0}{y^{-1}}$

17. 0^4

18. $\dfrac{1}{3^0}$

Written Exercises

In all these exercises assume that no variable has 0 as a value. Simplify each expression. Express your result using only positive exponents.

A   1. $\dfrac{3}{2^{-1}}$

2. $\dfrac{6}{3^{-2}}$

3. $\dfrac{4^{-2}}{-4}$

4. $\dfrac{5^{-1}}{-5}$

5. $3a^{-2}$

6. $(3a)^{-2}$ **7.** $8a^{-3}b$ **8.** $-2x^{-2}y$ **9.** $\dfrac{c^{-2}}{d^{-1}}$ **10.** $\dfrac{2e^2}{4e^{-3}}$

11. $(x^2y)^{-3}$ **12.** $(xy^{-2})^3$ **13.** $(x^{-2}z)^0$ **14.** $(x^0z)^{-2}$ **15.** $(m+n)^{-1}$

16. $a^{-1}+b^{-1}$ **17.** $\dfrac{2^{-2}r^2s}{4^{-1}r^{-2}s}$ **18.** $\dfrac{(-3)^{-3}x^{-2}y}{9^{-1}xy^{-1}}$ **19.** $\dfrac{d^{-2}e^0f}{d^0f^{-1}e^2}$ **20.** $\dfrac{x^3y^{-2}z^0}{x^{-1}yz^{-1}}$

B **21.** $\left[\dfrac{(-4)^2}{2^4}\right]^{-1}$ **22.** $\left[\dfrac{7^{-1}}{2^{-2}}\right]^{-2}$ **23.** $\left[\dfrac{4^{-1}a}{4b^{-1}}\right]^2$ **24.** $\left[\dfrac{3^{-2}x^2}{2^{-2}x^{-2}}\right]^3$

25. $9y^{-1}(3^{-1}y+3^{-2}y^2)$ **26.** $4z^2(4^{-2}z+4z^{-2})$ **27.** $3^{-2}w^{-1}(3w^2+3^3w^{-3})$

Rewrite each expression so that no denominator contains a variable and no variable occurs as a base more than once.

28. $\dfrac{(4^2)^{-1}a^0b^{-3}}{2^{-4}a^2b}$ **29.** $\dfrac{c^{-3}d^{-2}e^{-5}}{c^0d^{-4}e^{-2}}$ **30.** $\dfrac{(2^{-2})^{-1}r^{-3}s^{-2}t^4}{(2^{-1})^2rs^4t^{-2}}$

31. $\dfrac{6^{-1}u^{-2}v^4w^{-4}}{3^0uv^{-1}w^{-3}}$ **32.** $\dfrac{-18a^2b^{-3}c}{8ab^{-4}c^{-1}}$ **33.** $\dfrac{22xyz^{-2}}{-24x^{-2}yz}$

Evaluate each expression given that the value of x is 3 and of y is -2.

C **34.** $\dfrac{6}{x^{-1}+y^{-1}}$ **35.** $\dfrac{(x+y)^{-1}}{x^{-1}-y^{-1}}$ **36.** $\dfrac{x^{-2}+y^{-2}}{(x^{-1}y)^{-2}}$ **37.** $\left(\dfrac{1}{x^{-2}}+\dfrac{2}{y^{-2}}\right)^{-1}$

38. Simplify $\dfrac{8^{2x+1}\cdot 2^x\cdot 16^{2x}}{4^{3x}}$. (*Hint:* Express each factor as a power of 2.)

39. If $a=2^k$ and $b=2^{k+3}$, what is the value of $\dfrac{2a}{b}$?

Self-Test 2

VOCABULARY   numerator (p. 336)   lowest terms of a
                 denominator (p. 336)   fraction (p. 336)

Reduce to lowest terms.

1. $\dfrac{-36ay}{54yz}$ **2.** $\dfrac{2a-4b}{3a-6b}$ **3.** $\dfrac{6x^2-x-2}{3x-2}$ *Obj. 1, p. 335*

Write in a form involving no zero or negative exponents.

4. $p^3\cdot p^{-5}$ **5.** $\dfrac{x^3y^{-2}z^0}{x^{-2}yz^{-2}}$ *Obj. 2, p. 335*

Check your answers with those at the back of the book.

Operations with Rational Expressions

9-6 Products and Quotients of Rational Expressions

If you state the conclusion of the basic property of quotients (page 324) as

$$\frac{r}{t} \cdot \frac{s}{u} = \frac{rs}{tu} \qquad (t \neq 0,\ u \neq 0),$$

you obtain the following rule.

Rule for Naming Products of Real Numbers

The product of fractions is the fraction whose numerator is the product of the numerators and whose denominator is the product of the denominators of the given fractions.

Thus:

$$\frac{24}{35} \cdot \frac{6}{15} = \frac{24 \cdot 6}{35 \cdot 15} = \frac{144}{525}$$

The preceding rule for naming products of real numbers is also used in multiplying rational expressions. For example:

$$\frac{a-2}{a} \cdot \frac{a+3}{a-3} = \frac{(a-2)(a+3)}{a(a-3)} = \frac{a^2 + a - 6}{a^2 - 3a}$$

$$4t \cdot \frac{5t}{s} = \frac{4t}{1} \cdot \frac{5t}{s} = \frac{20t^2}{s}$$

$$\frac{x}{y} \cdot \frac{y}{x} = \frac{xy}{yx} = \frac{xy}{xy} = 1$$

$$\left(\frac{u}{v}\right)^3 = \left(\frac{u}{v}\right)\left(\frac{u}{v}\right)\left(\frac{u}{v}\right) = \frac{u \cdot u \cdot u}{v \cdot v \cdot v} = \frac{u^3}{v^3}$$

The last example illustrates the first part of the following theorem:

Theorem. Let a and b denote real numbers, and let n denote an integer.

1. If $b \neq 0$, then $\left(\dfrac{a}{b}\right)^n = \dfrac{a^n}{b^n}$.

2. If $a \neq 0$ and $b \neq 0$, then $\left(\dfrac{a}{b}\right)^{-n} = \dfrac{a^{-n}}{b^{-n}} = \dfrac{b^n}{a^n}$.

3. If $a \neq 0$ and $b \neq 0$, then $\left(\dfrac{a}{b}\right)^0 = \dfrac{a^0}{b^0} = 1$.

You can often simplify products of rational expressions by factoring numerators and denominators.

EXAMPLE 1 Simplify $\dfrac{3y - 6}{5y} \cdot \dfrac{y^2 - y - 6}{y^2 - 4}$.

SOLUTION
$$\dfrac{3y - 6}{5y} \cdot \dfrac{y^2 - y - 6}{y^2 - 4} = \dfrac{3(y - 2)}{5y} \cdot \dfrac{(y - 3)(y + 2)}{(y - 2)(y + 2)}$$

$$= \dfrac{3(y - 3)(y - 2)(y + 2)}{5y(y - 2)(y + 2)}$$

$$= \dfrac{3(y - 3)}{5y} = \dfrac{3y - 9}{5y}$$

The relationship between multiplication and division asserts that a quotient equals the product of the dividend and the reciprocal of the divisor. This fact, together with the fact that the reciprocal of $\dfrac{s}{u}$ is $\dfrac{u}{s}$ if $s \neq 0$ and $u \neq 0$, leads to the following theorem.

Theorem. For all real numbers r, s, t, and u such that $t \neq 0$, $s \neq 0$, and $u \neq 0$,

$$\dfrac{r}{t} \div \dfrac{s}{u} = \dfrac{r}{t} \cdot \dfrac{u}{s}.$$

Using this theorem and the rule for naming products, you find that if $t \neq 0$, $s \neq 0$, and $u \neq 0$, then

$$\dfrac{r}{t} \div \dfrac{s}{u} = \dfrac{ru}{ts}.$$

For example: $\dfrac{3x}{4y} \div \dfrac{x}{2y} = \dfrac{3x}{4y} \cdot \dfrac{2y}{x} = \dfrac{3 \cdot 2 \cdot x \cdot y}{2 \cdot 2 \cdot x \cdot y} = \dfrac{3}{2}$

EXAMPLE 2 Simplify $\dfrac{9a^2 - 25}{2a - 2} \div \dfrac{6a - 10}{a^2 - 1}$.

SOLUTION
$$\dfrac{9a^2 - 25}{2a - 2} \div \dfrac{6a - 10}{a^2 - 1} = \dfrac{9a^2 - 25}{2a - 2} \cdot \dfrac{a^2 - 1}{6a - 10}$$

$$= \dfrac{(3a - 5)(3a + 5)}{2(a - 1)} \cdot \dfrac{(a - 1)(a + 1)}{2(3a - 5)}$$

$$= \dfrac{(3a - 5)(3a + 5)(a - 1)(a + 1)}{4(a - 1)(3a - 5)}$$

$$= \dfrac{(3a + 5)(a + 1)}{4}$$

$$= \dfrac{3a^2 + 8a + 5}{4}$$

Oral Exercises

Simplify.

1. $\dfrac{1}{4} \cdot \dfrac{4}{7}$ **2.** $\dfrac{5}{9} \cdot \dfrac{2}{3}$ **3.** $\left(-\dfrac{1}{4}\right)^2 \cdot \dfrac{1}{4}$ **4.** $-\dfrac{2}{x} \cdot \dfrac{x}{3}$ **5.** $\dfrac{z}{13} \cdot \dfrac{3}{z}$ **6.** $\dfrac{z^2}{5} \cdot \dfrac{5}{z}$

7. $\left(\dfrac{k}{2}\right)^{-2}$ **8.** $\left(-\dfrac{3}{4}\right)^{-3}$ **9.** $\dfrac{a + b}{c} \cdot \dfrac{c}{a - b}$ **10.** $\dfrac{3x + 1}{3} \cdot \dfrac{x}{3x - 1} \cdot \dfrac{3x - 1}{x}$

Express each quotient as a product.

11. $\dfrac{13}{4} \div \dfrac{4}{7}$ **12.** $\dfrac{k}{x} \div \left(-\dfrac{h}{y}\right)$ **13.** $\dfrac{c}{c + 1} \div \dfrac{c + 1}{c}$

14. $\dfrac{2l}{l - 2} \div \dfrac{5}{l}$ **15.** $-\dfrac{x}{2y} \div \dfrac{y + 1}{2x}$ **16.** $\dfrac{k - 1}{k + 1} \div \left[-\left(\dfrac{k - 1}{k + 1}\right)\right]$

17. Is the product of any two rational expressions a rational expression?

18. Is the quotient of any two nonzero rational expressions a rational expression?

19. Explain why the set of rational numbers is closed with respect to (a) multiplication, (b) division, excluding division by 0.

Written Exercises

Simplify.

A   **1.** $\dfrac{12}{35} \cdot \dfrac{25}{9}$ **2.** $\dfrac{32}{21} \cdot \dfrac{27}{64}$ **3.** $\left(-\dfrac{63}{34}\right) \div \left(\dfrac{21}{22}\right)$ **4.** $\left(-\dfrac{57}{50}\right) \div \left(-\dfrac{38}{24}\right)$

5. $\dfrac{6a}{1} \cdot \dfrac{1}{6}$ **6.** $\dfrac{9}{4} \cdot \dfrac{2b^2}{3}$ **7.** $\dfrac{6c^2}{5} \div \left(-\dfrac{3c}{2}\right)$ **8.** $\dfrac{8d}{-5} \div d^3$

9. $\dfrac{6ef}{7} \cdot \dfrac{14e}{3ef^2}$ **10.** $\dfrac{-24rs^2}{7r} \cdot \dfrac{21r^2s}{16s}$ **11.** $\dfrac{wz}{3w - 3z} \cdot \dfrac{w^2 - wz}{wz}$

12. $\dfrac{4p - 4q}{pq} \cdot \dfrac{pq}{2p - 2q}$

13. $\dfrac{3k - 9}{5k - 15} \div \dfrac{8k - 4}{10k - 5}$

14. $\dfrac{8g + 24h}{2g - 4h} \div \dfrac{2g + 6h}{4g - 8h}$

15. $\dfrac{2a + 1}{a^2 - 16} \cdot \dfrac{a^2 - 4a}{4a^2 - 1}$

16. $\dfrac{9b^2 - 25}{b^2 - 1} \cdot \dfrac{3b - 3}{9b - 15}$

17. $\dfrac{4c - 2}{c^2 - 1} \div \dfrac{12}{c^2 - c}$

18. $\dfrac{d^2 - e^2}{d} \div \dfrac{d - e}{5d}$

19. $\dfrac{4x - 16}{3x} \div (3x - 12)$

20. $\dfrac{6y - 27}{4y} \div (4y - 18)$

21. $(4z - 4) \div \dfrac{z^2 - 1}{z}$

22. $(6w - 24) \div \dfrac{w^2 - 16}{6w}$

23. $\dfrac{1}{5x - 10} \div \dfrac{x^2}{25x^2 - 100}$

For the functions P and Q defined in the exercises below, express
$P(x) \cdot Q(x)$ as a rational expression in simple form.

24. $P(x) = \dfrac{4x + 16}{2x}; \quad Q(x) = \dfrac{6x}{x + 4}$

25. $P(x) = \dfrac{x^2 - x - 6}{x^2}; \quad Q(x) = \dfrac{x^3 + 6x^2}{x^2 + 8x + 12}$

26. $P(x) = \dfrac{x^2 - 4}{x^2 - 9}; \quad Q(x) = \dfrac{2x^2 + 6x}{2 - x}$

27. $P(x) = \dfrac{x + 5}{x^2 - 1}; \quad Q(x) = \dfrac{x^2 + 1}{x^2 - 25}$

Simplify.

B

28. $\dfrac{n^2 - 5n + 6}{2n + 4} \cdot \dfrac{n + 2}{2n - 6}$

29. $\dfrac{x^2 + 9x + 14}{x^2 - 3x - 10} \cdot \dfrac{x^2 + 2x - 35}{x^2 + 4x - 21}$

30. $\dfrac{y^2 + 12y + 36}{36 - y^2} \cdot \dfrac{y^2 - 5y - 6}{2y^2 - 2y - 84}$

31. $\dfrac{k^2 - k - 6}{k^2 + 2k - 15} \cdot \dfrac{k^2 - 25}{k^2 - 4k - 5}$

32. $\dfrac{2z - 8}{z^2 - 4} \div \dfrac{4 - z}{3z^2 + 18z + 24}$

33. $\dfrac{w^2 - w - 20}{w^2 + 7w + 12} \div \dfrac{w^2 - 7w + 10}{w^2 + 9w + 18}$

34. $\dfrac{m^2 - 7m}{3m^2 - 48} \div \dfrac{2m^2 - 18}{m^2 - 7m + 12}$

35. $\dfrac{v^2 - 5v + 6}{v^2 - 4} \div \dfrac{v^2 - 2v - 3}{v^2 + 3v + 2}$

36. $\dfrac{6u^2 - u - 2}{12u^2 + 5u - 2} \div \dfrac{4u^2 - 1}{8u^2 - 6u + 1}$

37. $\dfrac{3n^2 + 2n - 1}{5n^2 - 9n - 2} \cdot \dfrac{10n^2 - 13n - 3}{2n^2 - n - 3}$

C

38. $\dfrac{4}{3p - 9} \cdot \dfrac{p^3 - 9p}{8} \div \dfrac{p + 3}{4}$

39. $\dfrac{2k^2 - k}{5k^2 - k} \cdot \dfrac{3k + 2}{2k^2 - k - 1} \cdot \dfrac{10k^2 + 3k - 1}{6k^2 + k - 2}$

40. $\dfrac{2l + 3}{4l^2 - 9} \cdot \dfrac{5 - 11l + 2l^2}{1 - 2l} \cdot \dfrac{2l - 1}{5 - l}$

41. $\dfrac{c^2 + 2c + 1}{c^2 + 4c} \cdot \dfrac{c^2 - c}{c^2 - 2c - 3} \cdot \dfrac{c^2 - 16}{c^2 - 3c - 4}$

42. $\left(\dfrac{d^2 + 8d + 12}{d^2 - 6d - 7}\right)^{-1} \cdot \dfrac{d^2 + 11d + 18}{d^2 + 4d - 5} \cdot \dfrac{d^2 + 2d - 15}{d^2 - 7d - 8}$

43. $\dfrac{e^2 - 4e + 3}{e^4 - 6e^3 - 7e^2} \cdot \dfrac{1}{(e^2 + e)^{-1}} \cdot \left(\dfrac{e^2 - 2e - 3}{e^2 - e - 6}\right)^{-1}$

44. $\dfrac{f^2 + 5f - 6}{f^2 + 7f + 12} \cdot \dfrac{f^2 - f - 20}{f^2 - 6f + 5} \div \dfrac{f^2 - 36}{f^2 - 9}$

45. Show that the reciprocal of $a \div \dfrac{b}{c}$ is $\dfrac{b}{ac}$ $(a \neq 0,\ b \neq 0,\ c \neq 0)$.

46. Show that $\dfrac{\frac{a}{b}}{\frac{c}{d}} = \dfrac{ad}{bc}$ $(b \neq 0,\ c \neq 0,\ d \neq 0)$.

Exercises 47–49 refer to the three parts of the theorem stated on page 343.

47. Use the meaning of a power (page 7) and the rule for naming products (page 342) to prove part 1 of the theorem.

48. Use the meaning of negative exponents and the first part of the theorem to prove the second part.

49. Prove the third part of the theorem.

9-7 Sums and Differences of Rational Expressions (1)

When you write the conclusion of the theorem stated on page 328 as

$$\frac{a}{c} + \frac{b}{c} = \frac{a+b}{c} \quad \text{and} \quad \frac{a}{c} - \frac{b}{c} = \frac{a-b}{c} \quad (c \neq 0),$$

you see how to find the sum or difference of fractions with equal denominators.

> ### Rule for Naming Sums and Differences of Real Numbers
>
> The sum (difference) of fractions with equal denominators is the fraction whose numerator is the sum (difference) of the numerators and whose denominator is the common denominator of the given fractions.

Thus, $\qquad\qquad \dfrac{12}{17} + \dfrac{23}{17} = \dfrac{12+23}{17} = \dfrac{35}{17}$

and $\qquad\qquad \dfrac{5}{11} - \dfrac{8}{11} = \dfrac{5-8}{11} = \dfrac{-3}{11} = -\dfrac{3}{11}.$

The rule for naming sums and differences of real numbers is also used in adding and subtracting rational expressions.

EXAMPLE State a rational expression in lowest terms equivalent to each expression.

a. $\dfrac{12}{5x} + \dfrac{x-2}{5x}$

b. $\dfrac{3r}{2r-s} + \dfrac{5r-2s}{2r-s} - \dfrac{r-3s}{2r-s}$

c. $\dfrac{x^2+3x}{x^2+2x-15} - \dfrac{2x+12}{x^2+2x-15}$

SOLUTION a. $\dfrac{12}{5x} + \dfrac{x-2}{5x} = \dfrac{12+(x-2)}{5x} = \dfrac{x+10}{5x}$

b. $\dfrac{3r}{2r-s} + \dfrac{5r-2s}{2r-s} - \dfrac{r-3s}{2r-s} = \dfrac{3r+(5r-2s)-(r-3s)}{2r-s}$

$= \dfrac{3r+5r-2s-r+3s}{2r-s}$

$= \dfrac{7r+s}{2r-s}$

c. $\dfrac{x^2+3x}{x^2+2x-15} - \dfrac{2x+12}{x^2+2x-15} = \dfrac{(x^2+3x)-(2x+12)}{x^2+2x-15}$

$= \dfrac{x^2+3x-2x-12}{x^2+2x-15}$

$= \dfrac{x^2+x-12}{x^2+2x-15}$

$= \dfrac{(x+4)(x-3)}{(x+5)(x-3)} = \dfrac{x+4}{x+5}$

Oral Exercises

Express each sum as a rational expression in lowest terms.

1. $\dfrac{13}{19} + \dfrac{4}{19} - \dfrac{7}{19}$

2. $\dfrac{4}{21} + \dfrac{13}{21} - \dfrac{17}{21}$

3. $\dfrac{3}{7a} + \dfrac{2}{7a} - \dfrac{4}{7a}$

4. $\dfrac{11}{10b} - \dfrac{7}{10b} + \dfrac{3}{10b}$

5. $\dfrac{c+1}{3} + \dfrac{2}{3}$

6. $\dfrac{d+1}{3e} + \dfrac{d-1}{3e}$

7. $\dfrac{5}{r+s} - \dfrac{r+5}{r+s}$

8. $\dfrac{2m}{m-n} - \dfrac{m-3n}{m-n}$

9. $\dfrac{2x+3y}{4x-2y} - \dfrac{y-2x}{4x-2y}$

10. $\dfrac{k+9}{3k-6h} + \dfrac{2k+3}{3k-6h}$

11. $\dfrac{x}{y-3z} - \dfrac{-x-2}{y-3z}$

12. $-\dfrac{p+3}{p+5q} - \dfrac{p-1}{p+5q} + \dfrac{2}{p+5q}$

13. $\dfrac{w}{4w+8} - \dfrac{4-w}{4w+8}$

14. $\dfrac{-t}{2(3-t)} + \dfrac{6-t}{2(3-t)}$

15. $\dfrac{x+y}{4(x-y)} + \dfrac{x-3y}{4(x-y)} + \dfrac{4x-4y}{4(x-y)}$

Written Exercises

For the functions P and Q defined in the exercises below, express (a) $P(x) + Q(x)$ and (b) $P(x) - Q(x)$ as a rational expression in lowest terms.

A 1. $P(x) = \dfrac{2x + 7}{5}$; $Q(x) = \dfrac{x - 1}{5}$

2. $P(x) = \dfrac{9}{x - 3}$; $Q(x) = \dfrac{3x}{x - 3}$

3. $P(x) = \dfrac{2x}{x^2 + 1}$; $Q(x) = \dfrac{2 - x}{x^2 + 1}$

4. $P(x) = \dfrac{3x}{x^2 - 16}$; $Q(x) = \dfrac{2x - 1}{x^2 - 16}$

Express each of the following as a rational expression in lowest terms.

B 5. $\dfrac{x + 2}{2x^2 - 5x - 3} - \dfrac{1 - x}{2x^2 - 5x - 3}$

6. $\dfrac{3k - 5}{k^2 - 2k + 1} + \dfrac{k + 1}{k^2 - 2k + 1}$

7. $\dfrac{1 - y}{y^2 + 5y + 6} + \dfrac{2y + 1}{y^2 + 5y + 6}$

8. $\dfrac{3w - z}{2z^2 - 3wz + w^2} - \dfrac{w + z}{2z^2 - 3wz + w^2}$

9. $\dfrac{2m - 3}{m^2 - m + 2} + \dfrac{m + 4}{m^2 - m + 2}$

10. $\dfrac{7 - a}{a^2 + 3a - 4} + \dfrac{2a - 3}{a^2 + 3a - 4}$

11. $\dfrac{3}{d - 2} - \left(\dfrac{d + 3}{d - 2} - \dfrac{d}{d - 2} \right)$

12. $\dfrac{u^3}{(u + 1)^2} - \left(\dfrac{3u^2 - 1}{(u + 1)^2} - \dfrac{2u^2 - u}{(u + 1)^2} \right)$

13. $\dfrac{3r^2 - 4}{r^2 + 4} - \dfrac{r^2}{r^2 + 4} + \dfrac{4}{r^2 + 4}$

14. $\dfrac{6x}{x^2 - y^2} - \dfrac{5x + 4y}{x^2 - y^2} + \dfrac{3y}{x^2 - y^2}$

C 15. Show that for all real numbers x and y where $x \neq y$,

$$\frac{1}{x - y} + \frac{1}{y - x} = 0.$$

16. Find a rational expression which when added to $\dfrac{2x - 7}{7 - 3x}$ gives a sum of $\dfrac{x - 2}{7 - 3x}$.

17. Find a rational expression which when added to $-\dfrac{7x + 3}{x^2 - 9}$ gives a sum of $\dfrac{x^2 + x}{x^3 - 9x}$.

9-8 Sums and Differences of Rational Expressions (2)

The first step in finding the sum

$$\frac{3}{8} + \frac{5}{14}$$

is to replace $\frac{3}{8}$ and $\frac{5}{14}$ with equivalent fractions having a common denominator. Any positive integer that has both 8 and 14 as factors will serve as the common denominator. But, for convenience, we shall use the least such integer, that is, the *least common denominator* (L.C.D.).

To find the L.C.D. systematically, factor 8 and 14 over the set of primes:

$$8 = 2^3 \text{ and } 14 = 2 \cdot 7$$

$$\text{L.C.D.} = 2^3 \cdot 7 = 56$$

Notice that:

> The L.C.D. of two or more given fractions is the product of all the different prime factors of the given denominators, each prime occurring in the product the greatest number of times that it appears in any one of the denominators.

EXAMPLE 1   What is the L.C.D. of $\dfrac{1}{4x}$ and $\dfrac{1}{30x^2y}$?

SOLUTION
$$4x = 2^2 \quad \cdot \quad x$$
$$30x^2y = 2 \cdot 3 \cdot 5 \cdot x^2 \cdot y$$
$$\text{L.C.D.} = 2^2 \cdot 3 \cdot 5 \cdot x^2 \cdot y = 60x^2y$$

If you know that the L.C.D. of the fractions $\frac{3}{8}$ and $\frac{5}{14}$ is 56, as we just found, you can use this denominator to find fractions naming $\frac{3}{8}$ and $\frac{5}{14}$ but having 56 as denominator. Thus:

$$56 = 8 \cdot 7, \text{ so that } \frac{3}{8} = \frac{3 \cdot 7}{8 \cdot 7} = \frac{21}{56},$$

and

$$56 = 14 \cdot 4, \text{ so that } \frac{5}{14} = \frac{5 \cdot 4}{14 \cdot 4} = \frac{20}{56}.$$

Hence, to find a single fraction naming $\frac{3}{8} + \frac{5}{14}$, you write

$$\frac{3}{8} + \frac{5}{14} = \frac{21}{56} + \frac{20}{56} = \frac{21 + 20}{56} = \frac{41}{56}.$$

In the following examples, a procedure like the one just used to find a fraction naming the sum of two rational numbers is employed to find fractions naming a sum and a difference of rational expressions whose denominators are not equivalent polynomials.

Polynomials and Rational Expressions | **349**

EXAMPLE 2 Simplify $3 + \dfrac{2}{3y + 6} + \dfrac{y - 3}{y^2 - 4}$.

SOLUTION   1. Find the L.C.D. by factoring each denominator:

$$3y + 6 = 3(y + 2) \qquad y^2 - 4 = (y - 2)(y + 2)$$

$$\text{L.C.D.} = 3(y + 2)(y - 2)$$

2. Write the fractions in the given expression with factored denominators:

$$\frac{3}{1} + \frac{2}{3(y + 2)} + \frac{y - 3}{(y - 2)(y + 2)}$$

3. Replace each fraction with an equivalent one having $3(y - 2)(y + 2)$ as a denominator:

$$\frac{3 \cdot 3(y - 2)(y + 2)}{1 \cdot 3(y - 2)(y + 2)} + \frac{2(y - 2)}{3(y + 2)(y - 2)} + \frac{(y - 3) \cdot 3}{(y - 2)(y + 2) \cdot 3}$$

4. Simplify the sum:

$$\frac{9(y - 2)(y + 2) + 2(y - 2) + 3(y - 3)}{3(y - 2)(y + 2)}$$

$$= \frac{9y^2 - 36 + 2y - 4 + 3y - 9}{3(y - 2)(y + 2)}$$

$$= \frac{9y^2 + 5y - 49}{3y^2 - 12}$$

Unless instructed otherwise, you need not number and describe steps as in Example 2. The way that you usually give the solution is shown in Example 3.

EXAMPLE 3 Simplify $\dfrac{3}{x^2 - x} - \dfrac{2}{x^2 + x - 2}$.

SOLUTION $\dfrac{3}{x^2 - x} - \dfrac{2}{x^2 + x - 2} = \dfrac{3}{x(x - 1)} - \dfrac{2}{(x - 1)(x + 2)}$

$$\text{L.C.D.} = x(x - 1)(x + 2)$$

$$\frac{3}{x^2 - x} - \frac{2}{x^2 + x - 2} = \frac{3(x + 2)}{x(x - 1)(x + 2)} - \frac{2 \cdot x}{x(x - 1)(x + 2)}$$

$$= \frac{3x + 6 - 2x}{x(x - 1)(x + 2)} = \frac{x + 6}{x(x - 1)(x + 2)}$$

$$= \frac{x + 6}{x^3 + x^2 - 2x}$$

Oral Exercises

State the L.C.D. of the fractions in each expression.

1. $\dfrac{1}{3} + \dfrac{5}{6}$

2. $\dfrac{4}{3x} - \dfrac{2}{5xy}$

3. $\dfrac{2}{(x-3)(x+2)} + \dfrac{3}{5(x+2)}$

Give the expression by which the question mark must be replaced in order that the given pair of fractions be equivalent.

4. $\dfrac{2}{x-5} = \dfrac{?}{x(x-5)}$

5. $\dfrac{7y^2}{4y-1} = \dfrac{21y^3}{?}$

6. $\dfrac{7}{4xy} = \dfrac{?}{4x^2y + 4xy^3}$

7. Is the sum of any two rational expressions a rational expression?

8. Is the difference of any two rational expressions a rational expression?

9. Explain why the set of rational numbers is closed with respect to addition and subtraction.

Written Exercises

Write each expression as a fraction in lowest terms.

A 1. $2 + \dfrac{5}{6}$

2. $5 - \dfrac{3}{8}$

3. $\dfrac{7}{2} + \dfrac{3}{10}$

4. $\dfrac{5}{9} - \dfrac{6}{7}$

5. $\dfrac{3}{a} + \dfrac{4}{b}$

6. $\dfrac{5}{8} - \dfrac{3}{b}$

7. $\dfrac{2}{cd} - \dfrac{7}{c^2d^2}$

8. $\dfrac{5}{ef} + \dfrac{9}{fe}$

9. $\dfrac{r+s}{r} + \dfrac{2r-s}{2s}$

10. $\dfrac{m-n}{mn} - \dfrac{n-m}{nm}$

11. $\dfrac{3w}{2w+6} - \dfrac{w-1}{w+3}$

12. $\dfrac{2}{6-2z} - \dfrac{3}{z-3}$

13. $\dfrac{9}{5x-10} + \dfrac{4}{4x-8}$

14. $\dfrac{5}{4a+2b} - \dfrac{4}{2a+b}$

15. $\dfrac{c}{c-5} + \dfrac{c-1}{c+5}$

16. $\dfrac{6}{d+3} - \dfrac{3}{d^2-9}$

17. $\dfrac{1}{k-h} + \dfrac{1}{h-k}$

18. $\dfrac{6}{6p-8} - \dfrac{10}{10p+12}$

19. $\dfrac{s}{r-4} - \dfrac{3s}{4-r}$

20. $\dfrac{2}{a^2+2ab+b^2} - \dfrac{3}{a+b}$

For the functions P and Q defined in the exercises below, express (a) $P(x) + Q(x)$ and (b) $P(x) - Q(x)$ as rational expressions in lowest terms.

21. $P(x) = \dfrac{3}{3x-1}$; $Q(x) = \dfrac{5}{3x}$

22. $P(x) = \dfrac{4}{x-4}$; $Q(x) = \dfrac{4}{x+4}$

23. $P(x) = \dfrac{x+1}{3x^2-3x}$; $Q(x) = \dfrac{2}{4x-4}$

24. $P(x) = \dfrac{-1}{x^2+4x}$; $Q(x) = \dfrac{7}{x^2-16}$

Write each expression as a fraction in lowest terms.

B 25. $\dfrac{5x - y}{3x + y} - \dfrac{6x - 5y}{2x - y}$

26. $\dfrac{x + 2y}{2x - y} - \dfrac{2x + y}{x - 2y}$

27. $\dfrac{1}{k^2 - 2k + 1} + \dfrac{1}{k^2 - 4k + 3}$

28. $\dfrac{6 - 3s}{s^2 + 4s + 4} + \dfrac{2s}{s^2 + 3s + 2}$

29. $\dfrac{r + 1}{r^2 - 5r + 4} - \dfrac{r}{16 - r^2}$

30. $p - \dfrac{3p - 6}{p^2 + 6p + 9}$

31. $\dfrac{m}{m^2 - 25} + \dfrac{m - 1}{25 - m^2}$

32. $\dfrac{4}{n^2 - 5n + 6} + \dfrac{6}{n^2 - 4}$

C 33. $\dfrac{3h - 9}{2h^2 + 5h - 12} - 2h$

34. $\dfrac{9}{2 + u} + \dfrac{-7}{u - 2} - \dfrac{3u - 1}{4 - u^2}$

35. $\dfrac{c + d}{c} - \dfrac{d + 1}{d} + \dfrac{2c - 1}{c}$

36. $e + \dfrac{e - 2}{5 - e} + \dfrac{4 - e^2}{e^2 - 25}$

37. $\dfrac{3}{p + 3} + p - \dfrac{2p}{p^2 - 9}$

38. $q + \dfrac{1}{2q - 1} - \dfrac{2}{(2q - 1)^2}$

39. $\dfrac{1}{6l^2 + 7l - 20} + \dfrac{2}{2l^2 + l - 10} - \dfrac{3}{3l^2 - 10l + 8}$

40. $\dfrac{m - n}{(m + n)^2} - \dfrac{1}{m - n} - \dfrac{n}{m^2 - n^2}$

41. $\dfrac{r - s}{r^2 + 2rs - 3s^2} + \dfrac{2r - s}{r^2 + 4rs + 3s^2} - \dfrac{r + 2s}{r^2 - s^2}$

Write each expression as a single fraction in lowest terms.

42. $6 \div \left(\dfrac{4}{z} - 2 \right)$

43. $(w + 2) \div \left(w - \dfrac{4}{w} \right)$

44. $\left(k - \dfrac{6}{k} \right) \div (6 - k^2)$

45. $\dfrac{\dfrac{x}{y} + 1}{\dfrac{x}{y} - 1}$

46. $\dfrac{\dfrac{x^2 + y^2}{xy} + 2}{\dfrac{x^2 - y^2}{2xy}}$

47. $\dfrac{\dfrac{p - q}{p^2 + q^2} - \dfrac{1}{p}}{\dfrac{p - q}{p^2 + q^2} - \dfrac{1}{q}}$

Self-Test 3

VOCABULARY   least common denominator (p. 349)

Simplify.

1. $\dfrac{10h^3}{7} \div 5h$

2. $\dfrac{a^2 - 4}{2a} \div \dfrac{2 - a}{a}$ *Obj. 1, p. 342*

3. $\dfrac{4x - 3}{x^2 + 2x + 2} - \dfrac{x - 3}{x^2 + 2x + 2}$

4. $\dfrac{3}{p} + \dfrac{2}{5p^3}$ *Obj. 2, p. 342*

Check your answers with those at the back of the book.

programming in BASIC

Another way to find an L.C.D. is to list multiples of each denominator in order and then find their least common multiple (L.C.M.) by inspection.
Consider $\frac{5}{24}$ and $\frac{7}{60}$. Multiples of 24 are:

$$24, 48, 72, 96, \mathbf{120}, 144, 168, 192, 216, \mathbf{240}, \ldots$$

Multiples of 60 are:

$$\mathbf{60}, \mathbf{120}, 180, \mathbf{240}, \ldots$$

You can see that 120 and 240 are common multiples, and that 120 is the L.C.M.

Exercises

1. Write a program that will find the L.C.M. of M and N, with $M > N$. Find successive multiples of M (test 1 to N) and try N as a factor. When you come to a multiple of M that is divisible by N, that is the L.C.M.

2. When the L.C.M. of M and N is MN, what does that mean?

Chapter Summary

1. **Basic Property of Quotients:** For all real numbers r and s, and all nonzero real numbers t and u, $\dfrac{rs}{tu} = \dfrac{r}{t} \cdot \dfrac{s}{u}$.

2. **Laws of Exponents for Division:** For all positive integers m and n, and every nonzero real number a:

   a. If $m = n$, then $\dfrac{a^m}{a^n} = 1$ b. If $m > n$, then $\dfrac{a^m}{a^n} = a^{m-n}$

   c. If $m < n$, then $\dfrac{a^m}{a^n} = \dfrac{1}{a^{n-m}}$

3. For every nonzero real number a, $a^0 = 1$.

4. For every nonzero real number a and every positive integer n, $a^{-n} = \dfrac{1}{a^n}$.

Chapter Review

1. Simplify $\dfrac{8ab^3}{14a^4b^2}$ 9-1

    a. $\dfrac{4a^3b}{7}$ b. $\dfrac{4b}{7a^3}$ c. $\dfrac{4a^3}{7b}$ d. $\dfrac{4b^5}{7a^5}$

2. Express the quotient $\dfrac{12m^2 + 4m}{4m}$ as a sum. 9-2

   a. $3m^2 + 4m$ **b.** $3m + 4$ **c.** $3m^2 + 1$ **d.** $3m + 1$

3. Divide $2y^2 - y - 15$ by $2y + 5$. 9-3

   a. $y - 3$ **b.** $y + 3$ **c.** $y^2 - 3$ **d.** $y^2 + 3$

4. Simplify $\dfrac{x^2 - 7x + 12}{x^2 - 16}$. 9-4

   a. $\dfrac{x - 3}{x - 4}$ **b.** $\dfrac{x - 3}{x + 4}$ **c.** $\dfrac{x + 3}{x - 4}$ **d.** $\dfrac{x - 4}{x + 4}$

5. Simplify $\dfrac{6m^2}{3m^{-1}}$. 9-5

   a. $2m$ **b.** $\dfrac{2}{m}$ **c.** $\dfrac{m^3}{2}$ **d.** $2m^3$

6. Simplify $\dfrac{2p^3}{5} \div \dfrac{p^2}{15}$. 9-6

   a. $6p$ **b.** $\dfrac{6}{p}$ **c.** $10p$ **d.** $6p^2$

7. Express as a rational expression in lowest terms: 9-7

$$\frac{5}{x - 3y} - \frac{2x}{x - 3y} + \frac{x + y}{x - 3y}.$$

   a. $\dfrac{5 - 3x + y}{x - 3y}$ **b.** $\dfrac{5 + 3x - y}{x - 3y}$

   c. $\dfrac{5 + x - y}{x - 3y}$ **d.** $\dfrac{5 - x + y}{x - 3y}$

8. Simplify $\dfrac{1}{3a - b} - \dfrac{1}{3a + b}$. 9-8

   a. 0 **b.** $\dfrac{2}{3a - b}$

   c. $\dfrac{2b}{(3a - b)(3a + b)}$ **d.** $\dfrac{3a - 2b}{(3a - b)(3a + b)}$

Chapter Test

1. Simplify $\dfrac{15x^2y}{25xy}$. 9-1

2. Express the quotient $\dfrac{6a^3b^2 - 9a^2b + 3a}{3a}$ as a sum. 9-2

3. Divide $5t^2 + 10t - 15$ by $t - 1$. 9-3

4. Simplify $\dfrac{2x^2 - 9x - 5}{x^2 - 4x - 5}$. 9-4

5. Simplify $(m^2n^{-3})^{-1}$. 9-5

6. Simplify $\dfrac{a^2 - 3a}{a^2 - 1} \div \dfrac{2a^3 - 6a^2}{3a - 3}$ 9-6

7. Express as a rational expression in lowest terms: 9-7

$$\dfrac{x + 5y}{3x + 2y} - \dfrac{x - 6y}{3x + 2y} + \dfrac{3x - 9y}{3x + 2y}$$

8. Simplify $\dfrac{2m + 6}{m^2 - 9} + \dfrac{3m + 2}{m - 3}$. 9-8

Margaret Mead
1901–1978

Margaret Mead, one of the world's foremost anthropologists, first gained international acclaim with the publication, in 1928, of her now famous book, *Coming of Age in Samoa,* describing the primitive tribal customs and way of life of Polynesian islanders. She was curator of ethnology at the American Museum of Natural History and occupied chairs of anthropology at Columbia and Fordham Universities. Throughout her life she continued doing field studies of primitive Pacific island cultures, and was responsible in large part, for popularizing anthropology.

Careers

in Commercial Art

Many commercial artists work as staff artists in advertising departments of large businesses and department stores, advertising agencies, television and motion picture studios, and in printing or publishing firms. Some commercial artists are free-lancers, working in small design studios or out of their own homes. Commercial artists may design posters or illustrate displays and advertising brochures, or they may work in the field of textile or wallpaper design. Most commercial artists are graduates of art or trade schools, admission to which is based upon high school grades, a portfolio of art work, and an interview.

Moshe Safdie's award-winning Habitat '67, in Montreal, creates outdoor living space in an urban environment by stacking blocks of apartments in pyramid-shaped units. The roofs of the dwellings below serve as gardens and decks for the ones above.

10 Rational Expressions in Open Sentences

Open Sentences Involving Rational Coefficients

OBJECTIVES for Sections 10-1 and 10-2:
1. *Solve open sentences with coefficients named by fractions.*
2. *Apply open sentences with coefficients named by fractions, decimals, and percents.*

10-1 Open Sentences with Coefficients Named by Fractions

In many open sentences, one or more of the numerical coefficients may be named by a fraction. For example, when the equation

$$\frac{4n}{3} - \frac{6n}{5} = 8$$

is written equivalently as

$$\tfrac{4}{3}n - \tfrac{6}{5}n = 8,$$

the coefficients of n are seen to be $\tfrac{4}{3}$ and $-\tfrac{6}{5}$. To solve such an equation, you find the L.C.D. and apply the multiplicative property of equality.

EXAMPLE 1   Find the solution set of $\dfrac{2x}{3} - \dfrac{2x + 5}{6} = \dfrac{1}{2}$.

SOLUTION $\dfrac{2x}{3} - \dfrac{2x + 5}{6} = \dfrac{1}{2}$ The L.C.D. is 6.

$6\left(\dfrac{2x}{3} - \dfrac{2x + 5}{6}\right) = 6 \cdot \dfrac{1}{2}$ Here the multiplicative property of equality is used: If $a = b$, then $ac = bc$.

$6\left(\dfrac{2x}{3}\right) - 6\left(\dfrac{2x + 5}{6}\right) = 6 \cdot \dfrac{1}{2}$

$2(2x) - (2x + 5) = 3$

$4x - 2x - 5 = 3$

$2x - 5 = 3$

$2x = 8$

$x = 4$

Check: $\dfrac{2(4)}{3} - \dfrac{2(4) + 5}{6} \overset{?}{=} \dfrac{1}{2}$

$\dfrac{8}{3} - \dfrac{13}{6} \overset{?}{=} \dfrac{1}{2}$

$\dfrac{16}{6} - \dfrac{13}{6} = \dfrac{3}{6} = \dfrac{1}{2}$ ✓

$\therefore$ the solution set is $\{4\}$.

EXAMPLE 2   Solve $\dfrac{z + 2}{5} + \dfrac{z - 3}{2} < 1$.

SOLUTION $\dfrac{z + 2}{5} + \dfrac{z - 3}{2} < 1$ The L.C.D. is 10.

$10\left(\dfrac{z + 2}{5} + \dfrac{z - 3}{2}\right) < 10 \cdot 1$ Here the multiplicative axiom of inequality is used: If $a < b$ and $c > 0$, then $ac < bc$.

$10\left(\dfrac{z + 2}{5}\right) + 10\left(\dfrac{z - 3}{2}\right) < 10 \cdot 1$

$2(z + 2) + 5(z - 3) < 10$

$2z + 4 + 5z - 15 < 10$

$7z - 11 < 10$

$7z < 21$

$z < 3$

$\therefore$ the solution set is $\{z\colon z < 3\}$.

Oral Exercises

State the L.C.D. of the terms of each open sentence; then give the equivalent sentence formed by multiplying both members by the L.C.D.

EXAMPLE $\dfrac{x}{4} + \dfrac{x}{6} = 3$

SOLUTION The L.C.D. is 12. The equivalent sentence is $3x + 2x = 36$.

1. $\dfrac{a}{2} + \dfrac{a}{4} = 3$

2. $\dfrac{b}{2} + \dfrac{3b}{7} < \dfrac{13}{14}$

3. $\dfrac{2c}{7} - 2 \geq \dfrac{2}{3}$

4. $d - \dfrac{3d}{5} = 4$

5. $3e - \dfrac{e}{6} = \dfrac{1}{9}$

6. $\dfrac{3x}{5} - \dfrac{x}{5} > 2$

7. $\dfrac{f - 2}{8} + \dfrac{f}{6} > 5$

8. $\dfrac{y + 6}{10} < -\dfrac{y}{20}$

9. How are the least common denominator and the least common multiple related?

Written Exercises

Find the solution set of each open sentence.

A 1–8. Oral Exercises 1–8 above.

9. $\dfrac{a}{6} - \dfrac{a}{3} = -2$

10. $\dfrac{2b}{3} + \dfrac{b}{10} = \dfrac{1}{5}$

11. $\dfrac{3c - 4}{6} < 2$

12. $\dfrac{5d + 2}{6} > 7$

13. $\dfrac{e}{3} + 1 \leq \dfrac{e}{2} - e$

14. $\dfrac{f + 2}{3} \geq \dfrac{f}{2}$

15. $\dfrac{x + 1}{2} = \dfrac{2x - 1}{5}$

16. $\dfrac{y + 15}{3} = \dfrac{y}{5}$

17. $2z - \dfrac{1}{3} - \dfrac{5z}{6} = 1 + \dfrac{z}{2}$

18. $\dfrac{5}{6} - \dfrac{31w}{9} = \dfrac{2}{3} - \dfrac{7w}{2}$

19. $\dfrac{3}{8}k - \dfrac{1}{4}k < \dfrac{3}{2}$

20. $\dfrac{1}{6}m - \dfrac{2}{3} \leq \dfrac{1}{3}m$

B 21. $\dfrac{3x}{4} + \dfrac{4x + 1}{6} = \dfrac{4 - 2x}{-3}$

22. $\dfrac{5z - 29}{5} = \dfrac{5 - 3z}{2} - \dfrac{4z}{15}$

23. $\dfrac{3w}{8} - \dfrac{6w - 5}{3} \geq -\dfrac{6w + 5}{6}$

24. $\dfrac{y + 5}{9} > \dfrac{2(y + 1)}{3} + y + 3$

25. $0.12(1000 - k) + 0.08k = 96$

26. $145 - 0.06n = 0.04(n + 1500)$

27. $\dfrac{5}{8}(3x - 1) - \dfrac{3}{4}(2x + 5) + 1 = 0$

28. $\dfrac{2z + 3}{6} + \dfrac{6z + 13}{4} = \dfrac{1 - 6z}{10}$

29. $0.03(2p - 3) - 0.08(4p + 5) = 0.36$

30. $\dfrac{3x + 5}{3} + \dfrac{3x - 4}{6} = 3x + 2$

Rational Expressions in Open Sentences | **359**

In Exercises 31–36 find the slope of the graph of the given equation.

EXAMPLE $\dfrac{x-2}{3} - \dfrac{2y+3}{2} = 8$

SOLUTION $\dfrac{x-2}{3} - \dfrac{2y+3}{2} = 8$

$$6\left(\dfrac{x-2}{3}\right) - 6\left(\dfrac{2y+3}{2}\right) = 6 \cdot 8$$

$$2(x-2) - 3(2y+3) = 48 \ \bullet$$

$$2x - 4 - 6y - 9 = 48$$

$$2x - 6y - 13 = 48$$

$$-6y = -2x + 61$$

$$y = \dfrac{1}{3}x - \dfrac{61}{6}$$

$\therefore$ the slope is $\tfrac{1}{3}$. Answer.

31. $\dfrac{x}{5} - \dfrac{y}{3} = \dfrac{2}{3}$

32. $-\dfrac{1}{3} = \dfrac{y}{12} - \dfrac{x}{8}$

33. $\dfrac{3y+5}{2} = \dfrac{4-x}{4} + 12$

34. $\dfrac{y+1}{6} - \dfrac{x+2}{3} + 1 = 0$

35. $\dfrac{7}{2} = \dfrac{5y+3}{4} + \dfrac{x+2}{2}$

36. $\dfrac{2x-5}{-4} - \dfrac{8-6y}{2} = x - 2$

Solve each system of equations.

37. $y = \dfrac{2x}{3} - 4$

$x - \dfrac{3}{4}y = 6$

38. $x = 3 - \dfrac{1}{3}y$

$\dfrac{5x}{4} + y = \dfrac{11}{2}$

39. $\dfrac{2x}{5} + y = -2$

$\dfrac{2x}{5} - \dfrac{y}{2} = 1$

40. $\dfrac{y}{8} - \dfrac{2x}{3} = -3$

$\dfrac{5}{6}x + \dfrac{1}{4}y = 7$

41. $\dfrac{1}{5} - \dfrac{3y+2}{5} = \dfrac{2x+1}{7}$

$\dfrac{y+4}{8} + \dfrac{3x-2}{4} = 2$

42. $\dfrac{7y}{6} - \dfrac{x}{2} = -3\dfrac{1}{6}$

$\dfrac{x+3}{7} - \dfrac{y+4}{3} = 0$

C 43. $\dfrac{1}{x} - \dfrac{1}{y} = 4$

$\dfrac{2}{x} - \dfrac{1}{2y} = 11$

44. $\dfrac{1}{x} + \dfrac{1}{y} = 7$

$\dfrac{2}{x} + \dfrac{3}{y} = 16$

45. $\dfrac{6}{y} - \dfrac{5}{x} = 3$

$\dfrac{10}{x} + \dfrac{9}{y} = 1$

46. $\dfrac{a}{x} + \dfrac{2a}{y} = 11$

$\dfrac{a}{x} - \dfrac{2a}{y} = -1$

47. Prove the theorem· If $\dfrac{a}{b} = 0$, then $a = 0$.

48. Prove the theorem: For all real numbers $a, b, c, c \neq 0$, if $ac = bc$, then $a = b$.

10-2 Percent Problems

The word *percent* (%) stands for *divided by 100* or *hundredths*. Thus, 7% is another way of writing $\frac{7}{100}$ or 0.07. Similarly,

$$100\% = \frac{100}{100} = 1;$$

and

$$\tfrac{3}{4} = (\tfrac{3}{4} \cdot 100)\tfrac{1}{100} = 75 \cdot \tfrac{1}{100} = \tfrac{75}{100} = 75\%.$$

When you multiply a number called the *base* (b) by a *percent* or *rate* (r), the product is called the *percentage* (p); that is,

$$p = br.$$

EXAMPLE 1   **a.** What is 12% of 54?
                b. What percent of 105 is 63?
                c. 240 is 125% of what number?

SOLUTION   Use $p = br$.

              a. $p = (0.12) \cdot 54 = 6.48.$ Answer.

              b. $63 = 105r;\ r = \frac{63}{105} = 0.6 = 60\%.$ Answer.

              c. $240 = b(1.25);\ b = \dfrac{240}{1.25} = 192.$ Answer.

When you invest money, the rate of interest is expressed as a percent. The (simple) interest i on p dollars invested at r percent per year for t years is found by using the relationship

$$i = prt.$$

Thus, the interest (income) from $2000 invested at 5% for two years is

$$i = (\$2000)(0.05)(2) = \$200.$$

EXAMPLE 2   Mr. Charles has $8000 invested, part at an interest rate of 4% and the remainder at 5%. How much does he have invested at each rate if his annual interest from both investments amounts to $380?

SOLUTION   1. The problem asks for the number of dollars invested at 4% and the number of dollars invested at 5%.

         2.   Let x = number of dollars invested at 4%.
           Then $8000 - x$ = number of dollars invested at 5%, and
                $0.04x$ = interest received at 4%,
           $0.05(8000 - x)$ = interest received at 5%.

         3. interest on 4%   interest on 5%   total interest
            investment   investment

$$0.04x \quad + \quad 0.05(8000 - x) = \quad 380$$

(Solution continued on next page.)

4. $100[0.04x + 0.05(8000 - x)] = 100(380)$
$$4x + 5(8000 - x) = 38{,}000$$
$$4x + 40{,}000 - 5x = 38{,}000$$
$$-x = -2000$$
$$x = 2000$$

Then: $8000 - x = 8000 - 2000 = 6000$

5. Now you should check that the sum of 4% of $2000 and 5% of $6000 is equal to $380. Thus:

Mr. Charles has $2000 invested at 4% and $6000 invested at 5%.
Answer.

Chemists, druggists, and others frequently are confronted with situations where they find it necessary to mix ingredients. In such problems percent is commonly used to describe the composition of the mixture.

EXAMPLE 3   Miss Hall, a chemist, wants to make a 12% alcohol solution. How many milliliters of an 8% alcohol solution must be added to 10 mL of a 20% alcohol solution to obtain the 12% solution?

SOLUTION   1. The problem asks for the number of milliliters of an 8% solution needed to make the required strength.

2.   Let x = number of mL of 8% alcohol needed.
Then $x + 10$ = number of mL of the final 12% solution,
           $0.08x$ = number of mL of alcohol in the 8% solution,
$0.12(x + 10)$ = number of mL of alcohol in the final solution.

3.

mL of alcohol in 8% solution		mL of alcohol in 20% solution		mL of alcohol in final 12% solution
$0.08x$	$+$	$0.20(10)$	$=$	$0.12(x + 10)$

4. $100[0.08x + 0.20(10)] = 100[0.12(x + 10)]$
$$8x + 20(10) = 12(x + 10)$$
$$8x + 200 = 12x + 120$$
$$80 = 4x$$
$$20 = x$$

5. *Check:* $0.08(20) + 0.20(10) \overset{?}{=} 0.12(20 + 10)$
$$1.6 + 2 \overset{?}{=} 3.6$$
$$3.6 = 3.6 \ \checkmark$$

∴. Miss Hall needs to add 20 mL of the 8% solution.   **Answer.**

Written Exercises

Replace each question mark to make a true statement. Give all results to the nearest tenth or tenth of a percent.

A
1. 22% of 684 = _?_
2. 72% of 1313 = _?_
3. _?_% of 215 = 145
4. _?_% of 19.4 = 14.6
5. 18% of _?_ = 56
6. 64% of _?_ = 183.6
7. 128% of 156 = _?_
8. 214% of 36.4 = _?_
9. _?_% of 118 = 138
10. _?_% of 342 = 530
11. 130% of _?_ = 44.8
12. 235% of _?_ = 51.6
13. 0.8% of 114 = _?_
14. $\frac{2}{5}$% of 2518 = _?_
15. _?_% of 6420 = 14
16. _?_% of 14,375 = 75
17. 0.75% of _?_ = 36
18. $\frac{2}{3}$% of _?_ = 39

Problems

A
1. If an ore contains 8% copper, how many metric tons of ore are needed to obtain 25 t of copper?

2. Lisa received 448 votes for president of Clarke High School. If this represented 56% of the votes cast for president, how many votes were cast?

3. Glenn Nelson paid $275 in sales tax on his new car. If this represents 5% of the price of the car, what was the price of the car?

4. Norma Croteau invested $8000, part at 6% and the rest at 9% per year. How much did she invest at each rate if her total return per year on the investment was $660?

5. How many milliliters of water must be added to 16 mL of a 25% salt solution to obtain a 10% solution?

6. How many liters of pure acid must be added to 20 L of a 20% acid solution to make a 50% solution?

7. How many grams of a 35% silver alloy must be melted with how many grams of a 65% silver alloy to obtain 20 g of a 41% silver alloy?

8. Raypost Corporation stock lost $12\frac{1}{2}$% of its value in one day's trading in the stock market. If the stock sold for $64.75 a share at the end of the day, what was its price at the start of the day?

9. A survey established that 1275 people out of 2900 polled walked at least one kilometer per day. What percent of the people polled did not walk at least one kilometer per day?

10. How much water must be evaporated from 150 L of a 4% salt solution to leave a solution that is 6% salt?

B
11. A buyer for a store paid $38 each for some calculators. What is the least price at which he can sell each calculator if the profit is to be at least 25% of the selling price?

12. Marilyn invested a total of $12,000 in two local businesses and received yearly dividends of 5% and 8%. If her dividends totaled at least $840, what is the smallest sum she could have invested in the business paying 8%?

13. A 40 L solution of salt water contains 15% salt. What is the least amount of water that can be added to form a solution that is no more than 5% salt?

14. After investing $1800 in bonds paying 7% annually, Hank Sanders deposited a second sum in a bank paying $5\frac{1}{2}$% annually. His yearly return on the two investments was the same as if both sums had been invested at 6%. Find the amount of his bank deposit.

15. What is the least amount of money that can be invested at 7% if a total of $10,000 is to be invested, part at 5% and part at 7%, and the return must be at least $616 per year?

16. The Cheyenne City basketball team has a record of 28 wins and 19 losses. What is the least number of the remaining 28 games it must win to finish the season winning at least 51% of all games played?

17. A dealer bought a tape cassette recorder for $98. She wants to make a profit of at least 25% of the amount to be paid by the purchaser after allowing a discount of 8%. What should the marked price of the recorder be before the discount?

18. Mrs. Garcia bought several major appliances and a new car. She paid a sales tax of 5% on the appliances, and an excise tax of 6% on the car. The total cost of the purchases was $5600, and she paid a total of $322 in taxes. How much did the car cost?

C 19. A car radiator contains 19 L of a 25% antifreeze solution. At least how many liters should be drawn off and replaced with pure antifreeze to fill the radiator with at least a 40% antifreeze solution?

20. The Thibodeau Scholarship Fund invested part of $150,000 at 6% and the remainder at 8%. If it had invested twice as much at 8% and the rest at 6%, it would have increased its income by $800. How much was invested at 6%?

21. Kathy Keefe invests some money at 7% per year, $\frac{3}{4}$ as much at 6% as she does at 7%, and $150 less at 5% than she does at 6%. If her yearly income from these investments is $633, how much has she invested at each rate?

22. When a "super" rubber ball is dropped it rebounds to a height that is 75% of that from which it is dropped. From what height was it dropped if on one occasion it had traveled 286 m at the time it struck the ground for the fourth time?

23. Alice Ponti, an industrial chemist, has a solution that is 60% alcohol. After she draws off 5 L and replaces it with 5 L of alcohol the resulting solution is 85% alcohol. How many liters does she have?

24. The capacity of a cooling system is 16 L. If it is full of a 15% anti-freeze solution, how many liters must be replaced by a 95% solution to give 16 L of a 65% solution?

Self-Test 1

VOCABULARY   percent (p. 361)   percentage (p. 361)

Solve.

1. $\dfrac{x}{9} = \dfrac{1}{3} + \dfrac{x}{12}$ **2.** $\dfrac{y + 1}{3} = \dfrac{y + 2}{2}$ *Obj. 1, p. 357*

3. $\dfrac{4z - 1}{3} \le 13$ **4.** $7 - \dfrac{5w}{4} > -3$

5. What is 46% of 95?   **6.** What percent of 48 is 21?   *Obj. 2, p. 357*

7. A furniture dealer bought a table for $32 and sold it for $44. What was the percent of markup?

8. How many liters of pure acid must be added to 5 L of a 40% acid solution to produce a 50% solution?

Check your answers with those at the back of the book.

Rational Expressions in Equations and Problems

OBJECTIVES *for Sections 10-3 through 10-6:*
1. Solve equations involving rational expressions.
2. Apply equations involving rational expressions to the solution of problems.

10-3 Fractional Equations

An equation such as

$$3 + \frac{10}{x^2 - 1} = \frac{5}{x - 1}$$

in which a variable appears in the denominator of one or more terms is called a **fractional equation.** The method used to solve a fractional equation is similar to the one used in Section 10-1 to solve an equation in which fractions appear only in numerical coefficients. But when you multiply each member of a fractional equation by the L.C.D. of the terms, you may not obtain an equivalent equation.

EXAMPLE 1   Solve $3 + \dfrac{10}{x^2 - 1} = \dfrac{5}{x - 1}$ over $\Re$.

SOLUTION $3 + \dfrac{10}{x^2 - 1} = \dfrac{5}{x - 1}$ L.C.D. $= x^2 - 1$

$$(x^2 - 1)\left[3 + \dfrac{10}{(x + 1)(x - 1)}\right] = (x^2 - 1)\left[\dfrac{5}{x - 1}\right]$$

$$3(x^2 - 1) + 10 = 5(x + 1)$$

$$3x^2 - 3 + 10 = 5x + 5$$

$$3x^2 - 5x + 2 = 0$$

$$(3x - 2)(x - 1) = 0$$

$$x = \tfrac{2}{3} \quad \text{or} \quad x = 1$$

When you test $\tfrac{2}{3}$ and 1 in the original equation, notice what happens:

$$3 + \dfrac{10}{(\frac{2}{3})^2 - 1} \overset{?}{=} \dfrac{5}{\frac{2}{3} - 1}$$

$$3 + \dfrac{10}{\frac{4}{9} - 1} \overset{?}{=} \dfrac{5}{\frac{2}{3} - 1}$$

$$3 + \dfrac{10}{-\frac{5}{9}} \overset{?}{=} \dfrac{5}{-\frac{1}{3}}$$

$$3 - \dfrac{90}{5} \overset{?}{=} -15$$

$$3 - 18 \overset{?}{=} -15$$

$$-15 = -15 \ \checkmark$$

$$3 + \dfrac{10}{(1)^2 - 1} \overset{?}{=} \dfrac{5}{1 - 1}$$

$$3 + \dfrac{10}{0} \overset{?}{=} \dfrac{5}{0}$$

The fractions $\tfrac{10}{0}$ and $\tfrac{5}{0}$ do not name real numbers, and 1 is not a root of the original equation.

$\therefore$ the solution set over $\Re$ is $\{\tfrac{2}{3}\}$.   **Answer.**

The equation obtained by multiplying each member of the original equation by $x^2 - 1$ has the "extra" root 1, a number for which the multiplier $x^2 - 1$ represents 0. Whenever you multiply an equation in a variable by a polynomial in that variable, the solution set of the resulting equation always contains all the roots of the original equation. But it sometimes also contains numbers that are *not* roots of the original equation. Therefore, you should observe the following rule.

When you multiply both members of an equation by a polynomial, always test each root of the resulting equation in the original equation. Those values and only those values producing true statements are the members of the solution set of the original equation.

The fact that by multiplying the members of an equation by a polynomial you may gain "extra" roots, but cannot lose "true" roots, follows from the fact that for every value of the variable the polynomial denotes some real number. (See Exercises 34–37, page 368)

EXAMPLE 2   Solve $\dfrac{3}{x+2} - \dfrac{2}{3} = \dfrac{x}{x+2}$ over $\mathcal{R}$.

SOLUTION

$$\dfrac{3}{x+2} - \dfrac{2}{3} = \dfrac{x}{x+2} \qquad \text{L.C.D.} = 3(x+2)$$

$$3(x+2)\left[\dfrac{3}{x+2} - \dfrac{2}{3}\right] = 3(x+2)\left[\dfrac{x}{x+2}\right]$$

$$9 - 2(x+2) = 3x$$

$$9 - 2x - 4 = 3x$$

$$5 = 5x$$

$$1 = x$$

Check: $\quad \dfrac{3}{1+2} - \dfrac{2}{3} \overset{?}{=} \dfrac{1}{1+2}$

$$\dfrac{3}{3} - \dfrac{2}{3} \overset{?}{=} \dfrac{1}{3}$$

$$\dfrac{1}{3} = \dfrac{1}{3} \checkmark$$

$\therefore$ the solution set over $\mathcal{R}$ is $\{1\}$.   **Answer.**

Oral Exercises

Is the given equation a fractional equation?

1. $\dfrac{2x}{3} - \dfrac{1}{4} = \dfrac{5}{9}$

2. $\dfrac{3(x-1)}{5} - \dfrac{x}{4} = x$

3. $\dfrac{5}{3x} - 7 = \dfrac{1}{4}$

4–6. Name the L.C.D. of the terms in each of the equations above.

State all real numbers which cannot be used as replacements for x in the given sentence.

7. $\dfrac{3}{x-2} = \dfrac{4}{5}$

8. $\dfrac{9}{x+3} = \dfrac{1}{6}$

9. $\dfrac{5}{x} + \dfrac{3}{x-1} = x$

10. $\dfrac{2}{x^2} + \dfrac{4}{x+4} = x^2$

11. $\dfrac{x+1}{x-2} + \dfrac{x-1}{x+2} = 1$

12. $\dfrac{9}{x^2-4} = \dfrac{7}{x}$

13. Are $\dfrac{x^2+1}{x^2+1} = 1$ and $x^2 + 1 = x^2 + 1$ equivalent equations over $\mathcal{R}$? Explain.

14. Are $\dfrac{x}{x} = 1$ and $x = x$ equivalent equations over $\mathcal{R}$? Explain.

Written Exercises

Solve each equation over $\mathcal{R}$.

A 1. $\dfrac{x}{x+2} = \dfrac{3}{5}$

2. $\dfrac{1}{2} = \dfrac{6}{y} - \dfrac{5}{y}$

3. $\dfrac{6}{x-2} = 6$

4. $\dfrac{11}{w} = 2 + \dfrac{5}{w}$

5. $\dfrac{5}{2t} - \dfrac{3}{2} = \dfrac{3}{t} - 2$

6. $3 - \dfrac{1}{k} = \dfrac{7}{5k} - \dfrac{9}{5}$

7. $\dfrac{a-2}{5a} = \dfrac{1}{6} - \dfrac{4}{15a}$

8. $\dfrac{5}{4b} - 1 = \dfrac{3}{4b} - \dfrac{b-3}{4b}$

9. $\dfrac{4}{c} - 3 = \dfrac{5}{2c+3}$

10. $\dfrac{15}{d} - \dfrac{11}{4} = \dfrac{2d-5}{4d}$

11. $\dfrac{e}{e-2} - 7 = \dfrac{2}{e-2}$

12. $\dfrac{2}{f-3} + 1 = \dfrac{f-1}{f-3}$

B 13. $g + 6 = \dfrac{g+6}{g-1}$

14. $\dfrac{6-h}{6h} = \dfrac{1}{h+1}$

15. $\dfrac{1}{x^2-x} - \dfrac{3}{x} = -1$

16. $\dfrac{2y}{y+1} = \dfrac{2y-1}{y-1}$

17. $\dfrac{1}{z+4} + \dfrac{4}{z-1} = \dfrac{5}{z^2+3z-4}$

18. $\dfrac{3}{w-2} = \dfrac{w^2+2}{w^2-4}$

19. $\dfrac{4}{t-2} - \dfrac{7}{t-3} = \dfrac{2}{15}$

20. $\dfrac{r+3}{r-1} + \dfrac{r+1}{r-3} - 2 = 0$

21. $\dfrac{4}{s-3} - \dfrac{2s}{9-s^2} = \dfrac{1}{s+3}$

Solve for x.

22. $y = \dfrac{kz}{x+k}$

23. $h = \dfrac{x-2\pi r^2}{2\pi r}$

24. $l = \dfrac{nE}{R+nx}$

25. $P = \dfrac{A}{1+rx}$

Solve over $\mathcal{R}$.

C 26. $\dfrac{3}{3a-2} + \dfrac{3}{2a+2} + \dfrac{5}{1-2a} = 0$

27. $\dfrac{2}{b-3} + \dfrac{1}{1-b} = \dfrac{2}{2b+3}$

28. $\dfrac{1}{c+2} - \dfrac{2}{c-1} = \dfrac{1}{2-c}$

29. $\dfrac{d+3}{d^2-1} + \dfrac{d-3}{d^2-d} = \dfrac{2d}{d^2+d}$

30. $\dfrac{6e}{e+3} - \dfrac{e-6}{e-3} = \dfrac{5e^2-12}{e^2-9}$

31. $\dfrac{f}{f-3} + \dfrac{f^2}{f^2-8f+15} = \dfrac{2f+1}{f-5}$

32. For what value of k will the solution set of $\dfrac{2x-3}{k} = x + 2$ be $\{-1\}$?

33. For what value of k will the solution set of $\dfrac{4x-k}{x-3} = 3$ be the empty set?

In Exercises 34–37 let b and c be the values of the left and right members of an equation in x for some given value of x. Let a be the value of a given polynomial in x for the given value of x. Prove that each assertion is true.

34. If $b = c$ and $a \neq 0$, then $ab = ac$.

35. If $b = c$ and $a = 0$, then $ab = ac$.

36. If $b \neq c$ and $a \neq 0$, then $ab \neq ac$.

37. If $b \neq c$ and $a = 0$, then $ab = ac$.

10-4 Number Problems

Fractional equations are often used to solve number problems.

EXAMPLE 1 The numerator of a fraction is 16 less than the denominator, and the fraction is equal to $\frac{3}{5}$. Find the fraction.

SOLUTION
1. The problem asks for a fraction.

2. Let x = the denominator of the fraction, and $x - 16$ = the numerator of the fraction.

3. The fraction is equal to $\frac{3}{5}$.
$$\frac{x - 16}{x} = \frac{3}{5}$$

4. $5x\left(\dfrac{x - 16}{x}\right) = 5x\left(\dfrac{3}{5}\right)$

$$5(x - 16) = 3x$$
$$5x - 80 = 3x$$
$$2x = 80$$
$$x = 40 \qquad \text{Then } x - 16 = 40 - 16 = 24$$

5. Does $\dfrac{24}{40} = \dfrac{3}{5}$?

$$\frac{3}{5} = \frac{3}{5} \checkmark$$

$\therefore$ the fraction is $\frac{24}{40}$. **Answer.**

EXAMPLE 2 The difference of the reciprocals of two positive numbers is $\frac{2}{5}$ and one of the numbers is 5 times the other. Find the numbers.

SOLUTION
1. The problem asks for two numbers.

2. Let x = one number and $5x$ = the other number.

3. $\dfrac{1}{x} - \dfrac{1}{5x} = \dfrac{2}{5}$

4. $5x\left[\dfrac{1}{x} - \dfrac{1}{5x}\right] = 5x\left(\dfrac{2}{5}\right)$

$$5 - 1 = 2x$$
$$4 = 2x$$
$$x = 2 \qquad 5x = 10$$

5. The check is left for you.

The numbers are 2 and 10. **Answer.**

Problems

A 1. The sum of two numbers is 33 and their quotient is $\frac{5}{6}$. Find the numbers.

2. The difference of two numbers is 36 and their quotient is $\frac{2}{5}$. Find the two numbers.

3. Five sixths of a number is 6 more than half of the number. Find the number.

4. What number added to both the numerator and denominator of the fraction $\frac{4}{7}$ results in a fraction equal to $\frac{4}{5}$?

5. The sum of the reciprocals of two consecutive positive integers is $\frac{11}{30}$. Find the integers.

6. Find two consecutive even integers such that 5 times the reciprocal of the lesser is equal to 6 times the reciprocal of the greater.

7. One number is 16 more than another number. Four ninths of the greater number is equal to four fifths of the lesser number. Find the numbers.

8. The numerator of a fraction is 6 greater than 3 times the denominator, and the fraction is equal to $\frac{9}{2}$. Find the fraction.

9. The denominator of a fraction is 4 less than 5 times the numerator, and the fraction is equal to $\frac{1}{4}$. Find the fraction.

10. When a number is added to the numerator of $\frac{4}{7}$ and twice the number is subtracted from the denominator, the result is a fraction equal to 7. Find the number.

B 11. Find two numbers whose sum is 24, and the sum of whose reciprocals is $\frac{24}{119}$.

12. Find two numbers whose difference is 5, and the difference of whose reciprocals is $\frac{5}{84}$.

13. The sum of a number and its reciprocal is $\frac{13}{6}$. Find the number.

14. Find the least positive integer for which the sum of its reciprocal and $\frac{3}{7}$ is greater than 3 times the reciprocal.

15. One positive number is 3 times another. If 54 is divided by each number, the greater quotient exceeds the lesser by 3. Find the two numbers.

16. The difference of a number and its reciprocal is $\frac{39}{40}$. Find the number.

17. Find two consecutive integers such that the reciprocal of the lesser increased by 3 times the reciprocal of the greater is equal to 9 times the reciprocal of the product of the integers.

18. The sum of two numbers is 90. When the greater number is divided by the lesser, the partial quotient is 3 and the remainder is 10. Find the numbers.

C 19. An integer is to be subtracted from both the numerator and the denominator of $\frac{7}{15}$ to yield a fraction whose value is greater than $\frac{1}{5}$. Find the greatest value of such an integer less than 10 and the least value greater than 10.

10-5 Work Problems

It is often useful to be able to solve problems which involve finding how long it takes to accomplish a task when a steady rate of work is assumed.

EXAMPLE 1 To reduce the water level behind an earthquake-damaged dam, two pumps are used to pump water from the lake behind the dam. If one pump can lower the water level 1 meter in 12 hours and the other can lower the water level 1 m in 18 h, how long must both pumps operate together to lower the surface 15 m to a safe level?

SOLUTION
1. The problem asks for the number of hours the pumps will take to lower the water level 15 m.

2. Let x = the number of hours the pumps operate. Then, since the pumps lower the level of the lake $\frac{1}{12}$ and $\frac{1}{18}$ m/h, respectively, $\dfrac{x}{12}$ represents the number of meters the first pump lowers the surface and $\dfrac{x}{18}$ represents the number of meters the second pump lowers the surface.

3. $\underbrace{\text{Meters lowered} \atop \text{by first pump}} + \underbrace{\text{meters lowered} \atop \text{by second pump}} = \underbrace{\text{total number of} \atop \text{meters lowered}}$

$$\frac{x}{12} \quad + \quad \frac{x}{18} \quad = \quad 15$$

4. $36\left[\dfrac{x}{12} + \dfrac{x}{18}\right] = 36(15)$

$$3x + 2x = 540$$
$$5x = 540$$
$$x = 108$$

5. *Check:* In 108 h, the first pump lowers the water level $\frac{108}{12} = 9$ m, and the second pump lowers it $\frac{108}{18} = 6$ m. $9 + 6 = 15$.

∴ it takes the two pumps 108 h to lower the water level in the lake by 15 m. **Answer.**

EXAMPLE 2   The Bridgeport Municipal Swimming Pool has two pumps it uses to fill the pool. If one pump alone takes 10 h to fill the pool and the other pump alone takes 15 h to fill the pool, how long will it take both pumps working together to fill the pool?

SOLUTION   1. The problem asks for the number of hours it will take the pumps to fill the pool.

2. Let $x =$ the number of hours it takes both pumps to fill the pool. Then, since the pumps fill $\frac{1}{10}$ and $\frac{1}{15}$ of the pool per hour, respectively, $\frac{x}{10}$ represents the part of the pool filled by one pump and $\frac{x}{15}$ the part filled by the other.

3. $\underbrace{\text{Part of pool filled by}}_{} + \underbrace{\text{Part of pool filled by}}_{} = \underbrace{\text{Completely filled}}_{}$
   one pump   the second pump   pool

$$\frac{x}{10} \quad + \quad \frac{x}{15} \quad = \quad 1$$

4 and 5. Solving the equation and checking to see that it takes both pumps 6 h to fill the pool is left to you.

Problems

A   **1.** Ralph can paint his house in 3 days (d), but it takes his son 5 d to paint it. How long will it take them to paint the house if they work together?

2. Mary Beth can keypunch a certain number of computer cards in 5 h, while Sue requires 6 h to punch the same number. How long will it take both girls to punch that number of cards if they work together?

3. It takes one crew of cleaners 6 h to wash and wax the floors in the Acme Mall. Another crew does the same job in 9 h. How long would it take both crews working together to do the job?

4. It takes Linda 9 min to fill one container of raspberries and Julie 7 min to fill the same size container. How long would it take both of them working together to fill 12 containers?

5. One outlet of a storage tank will empty the tank in 4.25 h and a second outlet will empty it in 2 h. How long will it take both outlets working together to empty the tank?

6. A computer can process a town's monthly water billings in 10 h. A newer computer can process the same billings in 6 h. How long would it take to complete the billings using both computers?

B   **7.** One electronic reader can read a deck of computer cards in $\frac{1}{2}$ the time of another reader. Together they can read the deck in 8 min. How long would it take each reader alone to read the deck?

8. Faith can type the weekly report for Blossomville in 4 h. Working together, she and Jack can do the same job in 1 h 45 min. How long would it take Jack alone to do the job?

9. Carol Lee can complete her paper route in 2 h. When her brother Chet helps her, it takes them 75 min to complete the route. How long would it take Chet alone to complete the route?

10. It takes 8 min to fill a certain tank and 12 min to drain it when it is full. With the drain open and the tank empty, how long would it take to fill the tank?

11. A tank can be filled by two inlets in 0.4 h and 0.5 h, respectively, and emptied by an outlet in 0.25 h. How long would it take to fill the empty tank if the inlets and outlet were working together?

12. Jake takes 4 h to do a job that takes Frank 3 h. One day they started out working on the job together, but after 1.5 h Jake left. How long did it take Frank to finish the job?

C 13. Andy can paint the fence around his house in 15 h, while his brother Franklin can do it in 18 h. If Andy paints for 7 h alone and then turns the rest of the job over to Franklin, how long will it take Franklin to finish painting the fence?

14. Mike and Michelle hand-printed the tickets to a play in 20 h. Michelle completed 3 tickets in the time it took Mike to do 2 tickets. How long would it take Mike to do all the tickets?

15. Clarissa punched $\frac{1}{3}$ of a set of punch cards in 2 hours. She was then joined by Dave, and together they finished the set in 2 more hours. How long would it have taken Dave to do the job alone?

16. Professor Yamoto can grade a set of examination papers in an hour and a half, but her teaching assistant requires 2 hours to grade the same set. After working together for 30 minutes, Professor Yamoto left to teach a class and her assistant completed the grading. How long did it take him?

10-6 Motion Problems

As mentioned on page 168, you can use the basic motion equation

$$\text{distance} = \text{rate} \times \text{time}$$
$$d = r \times t$$

to solve problems involving uniform motion.

EXAMPLE Homer can ride 15 km on his bicycle in the same time it takes him to walk 7 km. If his rate riding is 5 km/h faster than his rate walking, how fast does he walk?

SOLUTION 1. The problem asks for his rate walking.

2 and 3. Let r = his rate walking.

	r	d	$t = \dfrac{d}{r}$
Riding	$r + 5$	15	$\dfrac{15}{r + 5}$
Walking	r	7	$\dfrac{7}{r}$

$$\frac{\text{Time}}{\text{riding}} = \frac{\text{Time}}{\text{walking}}$$

$$\frac{15}{r + 5} = \frac{7}{r}$$

4. $r(r + 5)\left[\dfrac{15}{r + 5}\right] = r(r + 5)\left[\dfrac{7}{r}\right]$

$\begin{aligned} 15r &= 7(r + 5) \\ 15r &= 7r + 35 \end{aligned}$ $\qquad \begin{aligned} 8r &= 35 \\ r &= \tfrac{35}{8} = 4\tfrac{3}{8} \end{aligned}$

5. Checking that his rate walking is $4\tfrac{3}{8}$ km/h is left to you.

Problems

A **1.** A boat that sails at the rate of 28 km/h in still water can go 100 km down the river in the same time it takes to go 72 km up the river. What is the speed of the current in the river to the nearest km/h?

2. An airplane whose cruising speed in still air is 350 km/h can travel 830 km with the wind in the same time it travels 570 km against the wind. Find the speed of the wind.

3. A freight train travels 160 km in the same time an express train travels 200 km. If the rate of the express is 20 km/h greater than that of the freight, find the rate of each.

4. Pauline can row in still water at a rate twice that of the current in a certain river. What is the rate of the current in the river if it takes Pauline 2 h less to row 4 km up the river than it does to row 15 km down the river?

5. The Korowski family made a 30-km trip in $2\tfrac{3}{4}$ h, partly across a lake by boat at 12 km/h and partly walking along a nature trail at 6 km/h. How far did they travel by boat?

6. In still water Phil and Joyce can paddle a canoe 14 km/h. If it takes them 4 h to paddle 21 km up a river and then return, what is the speed of the current in the river?

7. An airplane flies from Keene to Marlborough at a speed of 288 km/h and returns at a speed of 240 km/h. If the first trip requires one hour less than the second, find the distance frome Keene to Marlborough.

B   8. A bus trip of 288 km would have taken four fifths as long if the average speed had been increased by 15 km/h. Find the rate at which the bus traveled.

9. Colleen drove 120 km to visit her family. She averaged 20 km/h more on the return trip than she did on the trip going. If her total travel time was $3\frac{1}{2}$ h, what was her average rate on the return trip?

10. Mario had driven 32 km at a constant speed, but found that he had to increase his speed by 32 km/h in order to cover the last 48 km in time to make an appointment. If his total traveling time was one hour, what was his original rate?

11. Sondra drove from her home to the airport at an average rate of 72 km/h. After waiting 40 min at the airport, she took off in a plane on a flight which averaged 192 km/h. If her total trip took $3\frac{1}{2}$ h, and if she traveled 384 km, how far is it from her home to the airport?

12. An express train required 3 h longer to travel 480 km than a plane required to travel 1920 km. If the plane travels 8 times as fast as the train, find the speed of the train.

13. On a 6400 km rocket-test range, one rocket takes 8 min longer than a second rocket, which travels 40 km a minute faster than the first rocket. Find the speed of the second rocket.

C   14. Jane gave Karen a 5 m head start in a 100 m race and Jane was beaten by 0.25 m. In how many meters more would Jane have overtaken Karen?

15. Two cars race on a 6.4 km track. The sum of the rates at which they travel is 320 km/h. If the faster car gains one lap in 40 min, find the rate of each car.

16. Marjorie rows 2 km up a river in order to board a motorboat which takes her 16 km down the river. In still water she rows at 3.3 km/h, and the motorboat travels at 19.2 km/h. If her trip upstream takes just as long as the boat's trip downstream, find the speed of the current in the river.

17. A state trooper clocks a speeding motorist going 130 km/h. The trooper gives chase; within 1.5 min the trooper has reached a speed of 150 km/h and has traveled 1.25 km. If the trooper continues at this speed, how long does it take to overtake the speeder?

Self-Test 2

VOCABULARY   fractional equation (p. 365)

Solve.

1. $3 + \dfrac{2}{x} = 4$

2. $\dfrac{y+1}{y-2} = \dfrac{5}{4}$

Obj. 1, p. 365

3. $\dfrac{6}{z+2} = \dfrac{8}{z}$

4. $\dfrac{6}{2w} + \dfrac{3}{2w+1} = -1.6$

5. Find two numbers whose sum is 117 and whose quotient is $\frac{5}{8}$.

Obj. 2, p. 365

6. One pipe can fill a tank in 5 h. A second pipe can fill it in 3 h. How long will it take both pipes together to fill the tank?

7. Robert drove 128 km from Marlowe to Pelham. He drove home from Pelham at twice the speed he averaged going to Pelham. If the round trip took 6 h, how fast did he travel each way?

Check your answers with those at the back of the book.

Rational Expressions in Types of Variations

***OBJECTIVES** for Sections 10-7 and 10-8:*
1. Find a selected member of an inverse, joint, or combined variation.
2. Solve problems involving inverse, joint, or combined variation.

10-7 Inverse Variation

During a test of a new paint, rectangular pieces of wood covered by the same amount of paint were exposed to varying climatic conditions. The lengths and widths of the test rectangles are shown in the table. You can verify that for each rectangle

$$l = \frac{100}{w},$$

which agrees with the fact that the area of a rectangle is equal to the product of its length and its width. This relationship is an example of an *inverse variation*.

Test Block	Length l (cm)	Width w (cm)
A	10	10
B	20	5
C	25	4
D	40	2.5
E	50	2

An **inverse variation** is any function specified by an equation of the form

$$y = \frac{k}{x}, \quad k \text{ a nonzero constant,}$$

where it is understood that the domain of the function excludes 0. Since

$$y = k\left(\frac{1}{x}\right),$$

you can say (page 211) that y is *directly proportional* to $\frac{1}{x}$, which is the multiplicative inverse of x. For that reason, it is customary to say that y *is inversely proportional to x*, or that y *varies inversely as x*. The nonzero number k is called the **constant of proportionality** or the **constant of variation,** and is ordinarily a positive number.

The graph of an inverse variation

$$y = \frac{k}{x}, \quad k > 0,$$

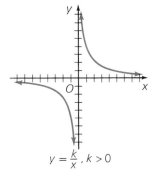

when the domain is $\Re$ (excluding 0) is pictured in Figure 1. Note that the graph does not intersect the x-axis or the y-axis. The graph of an inverse variation is called a **hyperbola.** In most practical applications, only the portion of the hyperbola in the first quadrant is used.

When you are given that a function is an inverse variation and one ordered pair in the function is known, you can find the equation that specifies the function.

$$y = \frac{k}{x}, k > 0$$

Figure 1

EXAMPLE   Given that y varies inversely as x, and $y = 12$ when $x = 27$, find **(a)** the equation specifying the variation and **(b)** the value of y when x has the value 3.

SOLUTION   **a.** You have $y = \frac{k}{x}$ for any inverse variation. Then, substituting 12 for y and 27 for x, you have

$$12 = \frac{k}{27},$$

from which $k = 12(27) = 324.$ $\therefore y = \frac{324}{x}.$ Answer.

b. Since $y = \frac{324}{x}$, if x has the value 3, you have

$$y = \frac{324}{3} = 108.$$

$\therefore$ the required value of y is 108.   Answer.

Oral Exercises

State the relationship between the given variables as an equation, using k for the constant of variation.

1. The length l and width w of a rectangle of given area vary inversely as each other.

2. The volume V of a gas at a fixed temperature varies inversely as the pressure P.

3. The current I in an electrical circuit of fixed voltage varies inversely as the resistance R.

4. The height h of a right circular cylinder of fixed volume varies inversely as the area A of the base.

5. The frequency f of an electromagnetic wave is inversely proportional to the length l of the wave.

6. Give examples of relationships that depict inverse variation and direct variation. How does direct variation differ from inverse variation?

Written Exercises

A

1. If y varies inversely as x, and $y = 6$ when $x = 4$, find y when $x = 10$.

2. If m varies inversely as n, and $m = 10$ when $n = 7$, find m when $n = 5$.

3. If r varies inversely as s, and $r = 20$ when $s = 70$, find s when $r = 30$.

4. If a varies inversely as b, and $a = 60$ when $b = 80$, find b when $a = 20$.

5. If w varies inversely as t, and $w = \frac{3}{4}$ when $t = 12$, find t when $w = \frac{3}{8}$.

6. If c varies inversely as d, and $c = 14$ when $d = 8$, find c when $d = 9$.

In Exercises 7–14 find the number c so that the given ordered pairs belong to the same inverse variation.

7. $(7, 6)$, $(c, 7)$ 8. $(16, 40)$, $(10, c)$ 9. $(18, 15)$, $(12, c)$

10. $(c, 25)$, $(25, 75)$ 11. $(\frac{2}{9}, c)$, $(\frac{1}{27}, \frac{1}{5})$ 12. $(1.6, 1.2)$, $(c, 0.8)$

B

13. (e, c), $(2e, 3e)$ 14. $(\frac{3}{2}k, \frac{1}{2}k)$, $(c, \frac{3}{4}k)$

15. If x is inversely proportional to y, how does x change when y is doubled?

16. If x is inversely proportional to y, how does x change when y is quadrupled?

17. If x is inversely proportional to y, does it follow that y is inversely proportional to x?

18. If a is inversely proportional to b, and if b is inversely proportional to c, what is the relationship of a to c?

Problems

A 1. The frequency of a radio wave is inversely proportional to the length of the wave. If a wave of length 300 m has a frequency of 1000 kHz, find the frequency of a wave whose length is 400 m.

2. The interest rate required to yield a given income is inversely proportional to the invested capital. Abraham has an income from an invested capital of $25,000 at a rate of 6%. How much money must he have invested to receive the same income if the return rate is 8%?

3. The height of a circular cylinder of fixed volume varies inversely as the area of the base. Jill wants to replace a cylindrical cistern 4 m deep, whose base area is 30 m², with a new cistern of the same volume whose base area is 20 m². How deep will the new cistern be?

4. A pulley of diameter 30 cm is belted to a pulley of diameter 20 cm, which is spinning at 240 revolutions per minute (r/min). How fast is the larger pulley revolving if the speed is inversely proportional to the diameter?

5. If the rotational speed of a gear wheel is inversely proportional to the number of teeth on the wheel, how fast is a gear wheel with 30 teeth revolving if it is meshed with a gear wheel with 50 teeth which is revolving at 1500 r/min?

6. James Mooney decided to make monthly purchases of shares in Greenfield Consolidated. He used the dollar-cost averaging system according to which the number of shares purchased varies inversely as the price per share. On March 6, he purchased 175 shares at a price of $9 per share. How many shares would he purchase on April 8 if the price then is $15 per share?

B 7. The force of gravitational attraction between two objects is inversely proportional to their distance apart. If two objects have a gravitational force of 550 N when they are 2200 m apart, how far apart are they when their gravitational force is 665.5 N?

10-8 Joint and Combined Variation

The terms "direct variation" and "inverse variation" are used in other ways than those previously mentioned. For example, because the force of attraction of an object at or above Earth's surface is related to the distance d from the center of Earth to the object by an equation of the form

$$F = \frac{k}{d^2},$$

we say that the force varies inversely as the *square* of the distance.

Similarly because the volume V of a sphere is related to its radius r by

$$V = \frac{4\pi}{3}r^3,$$

we say that the volume varies directly as the *cube* of the radius.

EXAMPLE 1 The distance a particle falls in a certain medium is directly proportional to the square of the length of time it falls. If a particle falls 16 m in 2 s in this medium, how far will it fall in 12 s?

SOLUTION 1. The problem asks for the distance fallen in 12 s.

2–4. Using $d = kt^2$, with $d = 16$ when $t = 2$, we have

$$16 = k(2)^2, \quad \text{or} \quad k = 4.$$

Thus, for this medium, $d = 4t^2$. Substituting 12 for t, we find that

$$d = 4(12)^2 = 4(144) = 576.$$

5. The check is left for you. The particle falls 576 m.   Answer.

If one variable varies directly as the product of two or more other variables, we call the resulting relationship a **joint variation.**

EXAMPLE 2 If y varies jointly as x and the square of z, and $y = 180$ when $x = 4$ and $z = 3$, find y when $x = 3$ and $z = 4$.

SOLUTION *Method 1.* You know that $y = kxz^2$. To solve for k, replace y with 180, x with 4, and z with 3:

$$180 = k(4)(3)^2$$

$$k = \frac{180}{4 \cdot 9} = \frac{180}{36} = 5$$

To find y when $x = 3$ and $z = 4$, replace x with 3 and z with 4 in $y = 5xz^2$.

$$y = 5(3)(4)^2 = 5 \cdot 3 \cdot 16 = 240$$

The required value of y is 240.   Answer.

Method 2. Using proportions: If $y = kxz^2$, then $k = \dfrac{y}{xz^2}$, and, using $x_1, y_1,$ z_1 and x_2, y_2, z_2 in the variation, you have the proportion:

$$\frac{y_1}{x_1 z_1^{\,2}} = \frac{y_2}{x_2 z_2^{\,2}}$$

Setting $y_2 = 180$, $x_2 = 4$, $z_2 = 3$, $x_1 = 3$, and $z_1 = 4$, you can solve for y_1:

$$\frac{y_1}{3(4)^2} = \frac{180}{4(3)^2}$$

$$y_1 = \frac{8640}{36} = 240$$

$\therefore$ the required value of y is 240.   Answer.

If z varies jointly as x and $\frac{1}{y}$, you have

$$z = kx\left(\frac{1}{y}\right) = k\left(\frac{x}{y}\right), \quad k \neq 0,$$

and you say that z varies directly as x and inversely as y. Any such combination of direct and inverse variation is called a **combined varia-**
tion. It is easy to prove that if $z = k\dfrac{x}{y}$, then $k = \dfrac{zy}{x}$, and

$$\frac{z_1 y_1}{x_1} = \frac{z_2 y_2}{x_2}.$$

This equation may be used for solving or for checking.

EXAMPLE 3   The pressure p required to force water through a pipe varies directly as the square of the speed s of the water and inversely as the diameter d of the pipe. If it requires 80 newtons per square meter pressure to drive water at 40 km/h through a pipe with a 2 cm diameter, what would be the pressure required to drive water at 35 km/h through a pipe with a diameter of 1.75 cm?

SOLUTION   1. The problem asks for the pressure needed to drive water at 35 km/h through a 1.75-cm pipe.

2–4. Since $p = \dfrac{ks^2}{d}$, you can replace p with 80, s with 40, and d with 2 to obtain

$$80 = \frac{k(40)^2}{2},$$

from which you have $k = \frac{1}{10}$. Then in $p = \dfrac{s^2}{10d}$, you can replace s with 35 and d with 1.75 to obtain

$$p = \frac{(35)^2}{10(1.75)} = \frac{1225}{17.5} = 70.$$

5. You can check using

$$\frac{p_1 d_1}{s_1^{\,2}} = \frac{p_2 d_2}{s_2^{\,2}}$$

$$\frac{80(2)}{(40)^2} \overset{?}{=} \frac{70(1.75)}{(35)^2}$$

$$\tfrac{1}{10} = \tfrac{1}{10} \ \checkmark$$

∴ the required pressure is 70 newtons per square meter.   Answer.

Oral Exercises

Express the relationship in words assuming that k is the constant of variation.

1. $a = kbc$
2. $d = ke^2f$
3. $x = \dfrac{ky}{d}$
4. $r = \dfrac{ks^2}{p^3}$

5. $t = \dfrac{kmn^2}{u}$
6. $v = \dfrac{k}{pq}$
7. $R = \dfrac{kl}{vu^2}$
8. $A = \dfrac{kmn^2t}{j^3}$

9. Translate into a formula: The centrifugal force of an object moving in a circular path is directly proportional to the square of the velocity of the object and inversely proportional to the radius of its path.

10. How does joint variation differ from direct variation? from inverse variation?

Written Exercises

A

1. If y varies directly as x^2, and $y = 50$ when $x = 4$, find y when $x = 7$.
2. If x varies directly as t^3, and $x = 64$ when $t = 4$, find x when $t = 2$.
3. If a varies inversely as b^2, and $a = \frac{1}{2}$ when $b = 2$, find a when $b = 8$.
4. If c varies inversely as d^3, and $c = \frac{1}{3}$ when $d = 3$, find c when $d = 2$.
5. If y varies jointly as x and z, and $y = 10$ when $x = 2$ and $z = 4$, find y when $x = 5$ and $z = 7$.
6. If p varies jointly as r and s, and $p = 4$ when $r = 6$ and $s = 2$, find p when $r = 12$ and $s = 10$.
7. If z varies directly as x and inversely as y, and $z = 10$ when $x = 38$ and $y = 4$, find z when $x = 48$ and $y = 8$.
8. If A varies directly as B and inversely as C, and $A = 5$ when $B = 80$ and $C = 5$, find A when $B = 100$ and $C = 10$.
9. If W varies jointly as m and r^2, and $W = 10$ when $m = 14$ and $r = 4$, find W when $m = 6$ and $r = 2$.
10. If x varies jointly as y^2 and z^3, and $x = 243$ when $y = 2$ and $z = 3$, find x when $y = \frac{1}{2}$ and $z = 2$.

B

11. If x varies inversely as y^2, how does x change when y is doubled?
12. If x varies directly as z^2, how does x change when z is doubled?
13. If x is directly proportional to y^2, does it follow that y is directly proportional to x^2? Explain.
14. If x is inversely proportional to y^2, does it follow that y is inversely proportional to x^2? Explain.
15. In the formula $x = \dfrac{y^2zw}{4}$, z remains constant. If y is doubled and w is tripled, how is x changed?

In Exercises 16–19 assume that Q varies directly as x and inversely as y.

C   16. If x is doubled and y is doubled, what happens to Q?

17. If x is tripled and y is doubled, what happens to Q?

18. If x is halved and y is doubled, what happens to Q?

19. If x is doubled and y is halved, what happens to Q?

20. In the formula $F = \dfrac{m\pi l^2}{p}$, m and π remain constant, l is halved, and p is doubled. How does F change?

Problems

A   1. The volume of a right circular cylinder varies jointly as its height and the square of its radius. When the height is 5 cm and the radius is 3 cm, the volume is 45π cm³. Find the volume (in terms of π) when the height is 2 cm and the radius is 7 cm.

2. The number of persons needed to do a job varies directly as the amount of work to be done and inversely as the time in which the job must be done. If 2 people can erect 150 m of fence in 9 h, how long will it take 5 people to erect 260 m of fence?

3. If a wire carries an electric current for a given time, then the heat developed varies jointly as the resistance and the square of the current. If a current of 6 amperes (A) produces 130 joules (J) of heat in a wire having a resistance of 15 ohms (Ω), find the heat produced by a current of 9 A in a wire having a resistance of 10 Ω.

B   4. The heat loss through a windowpane varies jointly as the difference of the inside and outside temperatures and the window area, and inversely as the thickness of the pane. If 198 J are lost through a pane 40 cm × 28 cm in area, and 0.8 cm thick, in 1 h when the temperature difference is 11°C, how many joules are lost in 1 h through a pane 0.5 cm thick having $\frac{1}{4}$ the area, when the temperature difference is 5°C?

5. The crushing load of a square wooden post varies directly as the fourth power of its thickness and inversely as the square of its length. If a post 10 cm thick and 2 m high is crushed by a load of 90 t, what is the crushing load of a post 8 cm thick and 3 m high?

6. The volume of a pyramid varies jointly as its altitude and the area of its base. A pyramid whose base is a square 4 cm on each side and whose altitude is 6 cm has a volume equal to 32 cm³. Find the volume of a pyramid 8 cm high with a base that is a square 3 cm on each side.

Self-Test 3

Obj. 1. p. 376

VOCABULARY inverse variation (p. 377)   combined variation (p. 381)
joint variation (p. 380)

1. If x varies inversely as y, and $x = 10$ when $y = 12$, find x when $y = 8$.

2. If x varies directly as y and inversely as z, and $x = 6$ when $y = 18$ and $z = 9$, find x when $y = 12$ and $z = 3$.

3. The densities of spherical objects of equal mass are inversely proportional to the cubes of the spheres' radii. If a sphere with a radius of 3 cm has a density of 38 kg/m³, what is the density of a sphere with the same mass and a radius of 5 cm?

Obj. 2, p. 376

Check your answers with those at the back of the book.

Chapter Summary

1. If an open sentence has fractional coefficients, the fractions may be eliminated by multiplying both members by their L.C.D.

2. An equation with variables in the denominator of one or more terms is called a *fractional equation.*

3. Fractions can be eliminated in a fractional equation by multiplying both members by the L.C.D. of the terms. Careful checking is necessary since this may introduce extra roots.

4. A function defined by an equation of the form $y = \dfrac{k}{x}$, with k a nonzero constant, is called an *inverse variation.* y is said to vary inversely as x, or to be inversely proportional to x.

5. Equations such as $y = kxz$ and $y = k\left(\dfrac{x}{z}\right)$ define *joint and combined variations*, respectively. In $y = kxz$, y is said to vary jointly as x and z. In $y = k\left(\dfrac{x}{z}\right)$, y is said to vary directly as x and inversely as z.

Chapter Review

10-1

1. $\dfrac{2x - 3}{3} + \dfrac{x}{4} = 2$ is equivalent to:

   a. $2x - 3 + x = 2$ **b.** $8x - 3 + 3x = 24$

   c. $8x - 12 + 3x = 24$ **d.** $8x - 12 + 3x = 2$

2. Find the solution set of $\dfrac{a - 3}{5} - \dfrac{a}{2} < 3$.

   a. $\{a: a > -12\}$ **b.** $\{a: a < 12\}$

   c. $\{a: a > -11\}$ **d.** $\{a: a > -3\}$

10-2

3. Which of these is *not* equal to 8%?

   a. 0.08 **b.** $\dfrac{8}{100}$ **c.** $\dfrac{2}{25}$ **d.** 0.8

4. If x dollars are invested at 6% interest, then the amount of interest received is:

   a. $6x$ **b.** $\dfrac{x}{6}$ **c.** $0.06x$ **d.** $\dfrac{x}{100}$

10-3

5. If A is the solution set of $\dfrac{1}{x^2 - 1} + 2 = \dfrac{x}{x + 1}$, and B is the solution set of $(x^2 - 1)\left(\dfrac{1}{x^2 - 1} + 2\right) = (x^2 - 1)\dfrac{x}{x + 1}$, which statement *must* be true?

   a. $A = B$ **b.** $A \subset B$ **c.** $B \subset A$

6. The solution set of $\dfrac{30}{t^2 - 9} + 2 = \dfrac{5}{t - 3}$ is:

   a. $\{-\frac{1}{2}\}$ **b.** $\{3\}$ **c.** $\{-\frac{1}{2}, 3\}$ **d.** $\varnothing$

10-4

7. The sum of two numbers is 3, and their quotient is $-1\frac{1}{2}$. Find the numbers.

   a. -3 and 0 **b.** 6 and -9 **c.** 2 and -5 **d.** 9 and -6

10-5

8. Which equation correctly expresses the question: If one gardener can prepare a flat of seedlings in 9 min, and another can prepare a flat in 12 min, how long will it take them working together to prepare 7 flats?

   a. $\dfrac{x}{9} + \dfrac{x}{12} = 1$ **b.** $9x + 12x = 7$

   c. $\dfrac{9}{x} + \dfrac{12}{x} = 7$ **d.** $\dfrac{x}{9} + \dfrac{x}{12} = 7$

10-6

9. One hydrofoil travels 5 km in the same time another hydrofoil travels 6 km. The faster hydrofoil travels 15 km/h faster than the slower one. Find the speed of the slower hydrofoil.

   a. 65 km/h   **b.** 70 km/h   **c.** 75 km/h   **d.** 80 km/h

10. Which of these equations does *not* express an inverse variation? 10-7

    a. $ab = 6$ **b.** $m = 5s$ **c.** $h = \dfrac{12}{k}$ **d.** $y = \dfrac{3x}{z}$

11. If P varies inversely as V, and $P = 18$ when $V = 24$, find V when $P = 12$.

    a. 16   **b.** 5184   **c.** 36   **d.** 288

12. "A varies jointly as c and the cube of d." Which equation correctly expresses this relationship? 10-8

    a. $A = \dfrac{kc}{d^3}$ **b.** $A = kcd^3$ **c.** $A = \dfrac{k}{cd^3}$ **d.** $A = k(cd)^3$

13. If V varies directly as w and inversely as u^2, and $V = 12$ when $w = 2$ and $u = 3$, find V when $w = 6$ and $u = 6$.

    a. $V = 9$ **b.** $V = 54$ **c.** $V = 324$ **d.** $V = 3$

Chapter Test

Find the solution set:

1. $\dfrac{x}{5} = \dfrac{x-3}{4} - 1$ **2.** $\dfrac{3a-4}{2} < \dfrac{2a+9}{3}$ 10-1

3. An investor has $10,000 invested, part at 6% and part at 7%. If the annual income from these investments is $665, how much is invested at each rate? 10-2

Find the solution set:

4. $\dfrac{2}{s} - \dfrac{3}{2s} = 1$ **5.** $\dfrac{y}{y-2} = \dfrac{2}{y} + \dfrac{3}{2}$ 10-3

6. A negative number is 9 times its reciprocal. Find the number. 10-4

7. Working together, Collins and Yung can do a job in 2 h. Alone, Yung can do the job in 3 h. How much time would Collins need to do the job alone? 10-5

8. One hiker walks 4 km in the time a second hiker covers 6 km. If the faster hiker's speed is 1 km/h more than the other's speed, find the speed of each hiker. 10-6

9. The space per shrub in a nursery plot is inversely proportional to the number of shrubs planted. When 16 shrubs are planted, the average space for each is 6 m². How many can be planted if only 4 m² is required per plant? 10-7

10. The time required to paint a room varies directly as the surface area of the walls and inversely as the number of painters. If it takes 6 h for 2 painters to paint a room with 80 m² of wall space, how long will it take 4 painters to paint a room with 240 m² of walls? 10-8

Lise Meitner
1878–1968

Lise Meitner, born in Vienna, received her doctorate in physics from the University of Vienna in 1905. She began investigating radioactivity, a new and exciting field of study pioneered by Marie and Pierre Curie. Working in Berlin with the German chemist, Otto Hahn, whose chief area of interest was the chemical properties of radioactive substances, Meitner concentrated her attention on the rays emitted by radioactive materials, especially the beta ray. Except for a period of time during World War I, this partnership lasted for some 30 years. This covered a time span which was extremely fruitful in the research into the structure of the atom. Hahn and Meitner's pioneering work in the physics of the nucleus led to the discovery of nuclear fission.

In 1935, they were joined in their work by a younger chemist, Fritz Strassman; however it wasn't long before interference by the Nazis made their work difficult. In 1938 when the German army occupied Austria, Meitner's position as an Austrian Jew in Berlin became precarious. She was smuggled into Holland by a Dutch colleague, Dirk Coster, and subsequently moved to Sweden where she continued her research at the Nobel Institute of Physics in Stockholm. In 1965 she shared the Enrico Fermi prize with Hahn and Strassman.

Careers

in Teaching

Although school enrollments are not expected to rise rapidly, teachers with superior educational backgrounds and the ability to understand students and to communicate knowledge skillfully will still be in demand. Teaching requires organizational and intellectual skills, attention to clerical detail. A bachelor's degree is the minimum requirement for certification. A master's degree is preferred in many positions.

Rational Expressions in Open Sentences | **387**

The 300-m diameter radio/radar antenna enables scientists at the Arecibo Observatory in Puerto Rico to use radar to study solar-system objects as far out as Saturn.

11

Irrational Numbers and Radicals

Irrational Numbers

OBJECTIVES for Sections 11-1 through 11-3:
1. *Simplify square-root radical expressions and express given real numbers using radical signs.*
2. *Convert decimals which represent rational numbers into common-fraction form.*
3. *Find rational approximations for irrational square roots using a square-root table or the method of successive approximations.*

11-1 Square Roots

Just as the inverse of addition is subtraction, and of multiplication is division, the inverse operation of squaring a number is finding a *square root*. Thus, a number b is called a **square root** of a positive real number a if

$$b^2 = a.$$

Since $b^2 = (-b)^2$, a positive real number a has two square roots. The *positive*, or *principal*, square root of a is represented by the symbol

$$\sqrt{a}$$

and the negative square root by

$$-\sqrt{a}.$$

Of course, if $a = 0$, then both $\sqrt{a}$ and $-\sqrt{a}$ represent 0. For example,

$$\sqrt{9} = 3, \quad -\sqrt{9} = -3, \quad \sqrt{16} = 4, \quad \text{and} \quad \sqrt{0} = 0.$$

The symbol $\sqrt{}$ is called a **radical sign,** and any expression occurring beneath the symbol represents the *radicand.*

Because the square of each real number is a nonnegative real number, a negative real number has no real square root. Thus, such expressions as $\sqrt{-2}$, $\sqrt{-3}$, and so on, do not name real numbers.

By our definition, the expression $\sqrt{x^2}$ always denotes a nonnegative number regardless of whether x represents a nonnegative or a negative number. Thus,

$$\sqrt{(-3)^2} = 3, \quad \sqrt{(-1)^2} = 1, \quad \text{and} \quad \sqrt{(-5)^2} = 5.$$

Since $|a| \geq 0$ for all real numbers a, you have:

For any real number a:

$$\sqrt{a^2} = |a|$$

Thus you can write, for example,

$$\sqrt{r^2} = |r|,$$
$$\sqrt{(z + 1)^2} = |z + 1|,$$
$$\sqrt{x^4} = |x^2| = x^2 \quad (|x^2| = x^2 \text{ for all } x \in \mathcal{R}).$$

Oral Exercises

Simplify each expression.

EXAMPLE $-\sqrt{16}$ **SOLUTION** $-\sqrt{16} = -4$

1. $\sqrt{49}$
2. $-\sqrt{1}$
3. $\sqrt{100}$
4. $-\sqrt{\frac{1}{9}}$
5. $\sqrt{25}$
6. $\sqrt{(-4)^2}$
7. $\sqrt{8^2}$
8. $-\sqrt{\frac{25}{16}}$

9. Is the sentence $\sqrt{a^2} = a$ true for every nonnegative value of a? for every negative value of a?

10. Is the sentence $\sqrt{a^2} = |a|$ true for every nonnegative value of a? for every negative value of a?

11. Does $\sqrt{a^2 - 1}$ represent a real number for all real values of a?

12. Does $\sqrt{a^2 + 1}$ represent a real number for all real values of a?

Written Exercises

Simplify each expression.

A   1. $\sqrt{64}$ 2. $\sqrt{100}$ 3. $-\sqrt{81}$ 4. $-\sqrt{36}$

5. $\sqrt{(14)^2}$ 6. $-\sqrt{(11)^2}$ 7. $\sqrt{(-5)^2}$ 8. $-\sqrt{(-12)^2}$

9. $\sqrt{\frac{9}{64}}$ 10. $-\sqrt{\frac{81}{16}}$ 11. $-\sqrt{\frac{36}{49}}$ 12. $\sqrt{\frac{9}{100}}$

In Exercises 13–18 express the given number using a radical symbol.

EXAMPLE $-\frac{4}{7}$ **SOLUTION** $-\frac{4}{7} = -\sqrt{(\frac{4}{7})^2} = -\sqrt{\frac{16}{49}}$

13. 2   14. 13   15. -12 16. -1 17. $\frac{3}{5}$ 18. $-\frac{5}{8}$

Give an equivalent expression that does not contain a radical symbol. Use absolute-value notation as needed and assume that each variable denotes a real number.

B   19. $\sqrt{16a^2}$ 20. $\sqrt{36a^2b^2}$ 21. $\sqrt{49a^4}$ 22. $-\sqrt{81b^4}$ 23. $\sqrt{\frac{a^2}{b^2}}$

24. $-\sqrt{25a^4b^4}$ 25. $\sqrt{(3ab)^2}$ 26. $-\sqrt{(-2ab)^2}$ 27. $\sqrt{(a + b)^2}$ 28. $\sqrt{(a + b)^4}$

29. Show that $\sqrt{9x^2 + 12xy + 4y^2} = |3x + 2y|$ is true for all values of x and y.

C   30. Does $\{(x, y): y = \sqrt{x}\}$ define a function? If so, what is the domain? the range?

31. Does $\{(x, y): y = -\sqrt{x}\}$ define a function? If so, what is the domain? the range?

32. Does $\{(x, y): y^2 = x\}$ define a function? If so, what is the domain? the range?

33. Find the fallacy in the following argument that every real number is equal to its opposite.

$$\text{If } a \in \mathfrak{R}, \text{ then } a = \sqrt{a^2} = \sqrt{(-a)^2} = -a.$$

34. Prove that each positive real number has at most one positive square root. That is, if a, b, and c are any positive real numbers such that $b^2 = a$ and $c^2 = a$, then $b = c$.

11-2 Decimals and Fractions

In Chapter 1 you learned that every rational number can be named by either a terminating or a repeating decimal. For example,

$$\tfrac{3}{25} = 0.12$$
$$\tfrac{5}{6} = 0.833 \ldots$$

You also learned that every terminating or repeating decimal names a unique rational number. To find a common fraction equivalent to a given *terminating decimal*, you simply write the decimal in common-fraction form. Thus,

$$2.13 = \frac{213}{100}, \quad 0.029 = \frac{29}{1000}, \quad \text{and} \quad 1.0003 = \frac{10{,}003}{10{,}000}.$$

On the other hand, to find a common fraction equivalent to a given *repeating decimal*, you can proceed as shown below.

For $3.\overline{21}$:

$$\text{Let } N = 3.\overline{21}$$

Multiply: $\quad 100N = 321.\overline{21}$
Subtract: $\quad\ \ \underline{N = \ \ \ 3.\overline{21}}$
$$\qquad\quad 99N = 318.00$$

$$\therefore\ 3.\overline{21} = \frac{318}{99} = \frac{106}{33} = 3\tfrac{7}{33}$$

For $0.34\overline{3}$:

$$\text{Let } N = 0.34\overline{3}$$

Multiply: $\quad 10N = 3.4\overline{3}$
Subtract: $\quad\ \underline{N = 0.34\overline{3}}$
$$\qquad\quad 9N = 3.09$$

$$\therefore\ 0.34\overline{3} = \frac{3.09}{9} = \frac{1.03}{3} = \frac{103}{300}$$

Notice that multiplying by 100 and by 10, respectively, shifts the repeating block of digits 2 and 1 units, respectively, to the left. The subtraction then produces a terminating decimal numeral. In general, if a repeating block of digits contains n digits, you multiply by 10^n.

You also learned in Chapter 1 that a decimal which does not terminate or repeat names an irrational number, and that you can construct such a decimal as follows:

$$0.37337333733337 \ldots,$$
$$0.24681012141618 \ldots$$

Ordinarily, the decimal numerals for irrational numbers do not have systematic patterns for their digits as do the examples given above. You already know one such irrational number,

$$\pi = 3.14159 \ldots,$$

and you have used *rational approximations* for it, such as

$$\pi \approx 3.14 \quad \text{or} \quad \pi \approx 3.1416,$$

in computations of, for instance, the circumference ($C = 2\pi r$) or the area ($A = \pi r^2$) of a circle.

In the preceding section you found square roots of numbers such as 16, 36, and $\frac{9}{25}$ that can be factored into two equal factors:

$$16 = 4 \times 4, \; 4 = \sqrt{16}; \quad 36 = 6 \times 6, \; 6 = \sqrt{36};$$
$$\tfrac{9}{25} = \tfrac{3}{5} \times \tfrac{3}{5}, \; \tfrac{3}{5} = \sqrt{\tfrac{9}{25}}$$

Such rational numbers are called **perfect squares** because they are squares of rational numbers. It can be shown that the square roots of a positive integer that is *not* the square of an integer *cannot* be rational numbers. You will learn how to find rational approximations to such irrational numbers in the next section.

Oral Exercises

Tell whether the given symbol represents a real number. If the number represented is real, tell whether it is rational or irrational.

1. $\sqrt{3}$ 2. $\sqrt{-25}$ 3. $-\sqrt{25}$ 4. $\sqrt{(-4)^2}$

5. $\sqrt{12}$.   6. 3π 7. $\sqrt{\tfrac{2}{3}}$ 8. $\sqrt{\tfrac{25}{36}}$

9. $\dfrac{1}{\sqrt{5}}$ 10. 0.213 11. $1.2\overline{72}$ 12. $\sqrt{0.64}$

13. $\sqrt{2.2}$ 14. $3.14159 \ldots$ 15. $3.101001000 \ldots$

16. If x and y are positive numbers such that $x^2 > y^2$, then $x > y$. Is this true or false? Explain.

17. If $0 < x < 1$, then $\sqrt{x} < x$. Is this true or false? Explain.

Written Exercises

Express in the form $\dfrac{a}{b}$, where a and b are integers.

A
1. 0.73 2. 2.78 3. $0.\overline{23}$ 4. $1.\overline{25}$

5. $7.\overline{214}$ 6. $6.1\overline{14}$ 7. $0.\overline{75}$ 8. $0.2\overline{34}$

Solve each equation over $\mathcal{R}$.

EXAMPLE $x^2 + 11 = 40$

SOLUTION $x^2 + 11 = 40$
$$x^2 = 40 - 11 = 29$$
$\therefore x$ must name one of the square roots of 29, and the solution set is $\{\sqrt{29}, -\sqrt{29}\}$.

9. $x^2 - 5 = 0$ 10. $y^2 - 13 = 0$ 11. $z^2 + 7 = 21$

12. $x^2 - 9 = 16$ 13. $2y^2 + 5 = 15$ 14. $4z^2 - 13 = 23$

Irrational Numbers and Radicals | **393**

Find the sum. Express the result in the form $\frac{a}{b}$, where a and b are integers. (*Hint:* First convert to common-fraction form.)

B **15.** 0.3 and $0.\overline{3}$ **16.** 0.5 and $0.\overline{5}$ **17.** 0.125 and $\frac{5}{12}$

    18. 0.375 and $\frac{1}{3}$ **19.** $0.\overline{9}$ and 0.1 **20.** $0.\overline{24}$ and $\frac{1}{11}$

    21. $0.\overline{3}$ and $0.\overline{6}$ **22.** $0.\overline{3}$ and $0.\overline{2}$ **23.** 0.8 and $0.\overline{8}$

    24. 0.7 and $\frac{1}{9}$ **25.** $0.\overline{2}$ and $0.\overline{8}$ **26.** 0.9 and $0.\overline{9}$

Find the product. Express the result in the form $\frac{a}{b}$, where a and b are integers. (*Hint:* First convert to common-fraction form.)

    27. 0.3 and $0.\overline{3}$ **28.** $0.\overline{2}$ and 3 **29.** $3.\overline{6}$ and $\frac{1}{5}$

    30. 0.9 and $0.\overline{8}$ **31.** $1.\overline{3}$ and 0.25 **32.** $1.\overline{4}$ and 0.75

    33. $0.\overline{45}$ and 2.2 **34.** $6.\overline{12}$ and 3.2 **35.** $5.\overline{38}$ and $1.\overline{92}$

C **36.** Express $3.\overline{9}$, $7.\overline{9}$, and $15.\overline{9}$ in common-fraction form.

    37. Use the result of Exercise 36 to guess the value of any repeating decimal of the form $x.\overline{9}$, where x is a whole number.

    38. Show that if a fraction in lowest terms can be represented by a terminating decimal, its denominator can have only powers of 2 and 5 as factors.

◻IVERSION

One, and only one, of the following numbers is a perfect square. Which one is it?

4,444,355,556   9,688,743,122   7,112,594,838

programming in BASIC
Exercise

Write a program that will print out the decimal numeral for a rational number, digit by digit. To make the problem simpler, take the numerator N less than the denominator D. (*Hint:* Find 10 × N, divide by D, and PRINT the integral part of the quotient. Find the remainder, multiply it by 10 and so on. End the print-out if the remainder is 0. Otherwise end the print-out when you have D + 1 digits.) Sample RUN:

```
INPUT NUMERATOR N:?3
INPUT DENOMINATOR D (>N):?14
  3 / 14 = . 2 1 4 2 8 5 7 1 4 2 8 5 7 1 4
END
```

Note the repeating blocks of digits in this case.

11-3 Finding Rational Approximations to Square Roots

It is often useful to be able to find a rational number that is a good approximation to a square root that is an irrational number. Before looking at a way of doing this, first note that you can always locate an irrational square root between two consecutive integers. For example, because

$$\sqrt{4} = 2 \quad \text{and} \quad \sqrt{9} = 3,$$

and because the greater a number the greater its positive square root, you can see that $\sqrt{5}$, $\sqrt{6}$, $\sqrt{7}$, and $\sqrt{8}$ are all greater than 2 and less than 3.

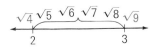

Figure 1

In general, whenever

$$a^2 < b^2 < c^2,$$

with $a > 0$, $b > 0$, $c > 0$, you have

$$a < b < c.$$

Let us now find a rational approximation to the positive square root of a positive number that is not a perfect square.

EXAMPLE 1   Find $\sqrt{21}$ correct to three digits.

SOLUTION   Since $\sqrt{16} = 4$ and $\sqrt{25} = 5$, you know that
$$4 < \sqrt{21} < 5.$$

1. Since $4 < \sqrt{21} < 5$, you take as a first approximation to $\sqrt{21}$ the *average* of 4 and 5.
$$\frac{4 + 5}{2} = \frac{9}{2} = 4.5$$

2. Divide 21 by 4.5, carrying out the operation to twice as many digits as the first approximation.
$$21 \div 4.5 \approx 4.667. \quad \text{Thus } 4.5 < \sqrt{21} < 4.667.$$

3. Take the average of the two approximate factors (the quotient and the divisor) as the second approximation to $\sqrt{21}$.
$$\frac{4.5 + 4.667}{2} = \frac{9.167}{2} = 4.584$$

4. Divide: $21 \div 4.584 \approx 4.5811518$. You can now be sure that $4.5811518 < \sqrt{21} < 4.584$, so that, correct to three digits,
$$\sqrt{21} \approx 4.58. \quad \text{Answer.}$$

You can verify the result of Example 1 by looking up $\sqrt{21}$ in the table of square roots at the back of the book. Unless instructed otherwise, use the square-root table to find the square root of any number listed in it.

When you need to find a rational approximation to the square root of a number not listed in the table, you may use the method of Example 1, which is sometimes called the **method of successive approximations.**

EXAMPLE 2   Find $\sqrt{45.8}$ correct to four digits.

SOLUTION   Since $\sqrt{36} = 6$ and $\sqrt{49} = 7$, $6 < \sqrt{45.8} < 7$.

Average: $\dfrac{6 + 7}{2} = \dfrac{13}{2} = 6.5$

Divide: $45.8 \div 6.5 \approx 7.046$

Compare: 6.5 and 7.046 do not agree to four digits.

Average: $\dfrac{6.5 + 7.046}{2} = \dfrac{13.546}{2} = 6.773$

Divide: $45.8 \div 6.773 \approx 6.7621438$

Compare: 6.773 and 6.7621438 do not agree to four digits.

Average: $\dfrac{6.773 + 6.7621438}{2} = \dfrac{13.5351438}{2} = 6.7675719$

Divide: $45.8 \div 6.7675719 \approx 6.7675671$

Compare: 6.7675719 and 6.7675671 agree to four digits.

$\sqrt{45.8} \approx 6.768$, correct to four digits (rounded).   **Answer.**

Oral Exercises

State two consecutive integers between which each square root lies.

1. $\sqrt{14}$ **2.** $\sqrt{32}$ **3.** $\sqrt{2.8}$ **4.** $\sqrt{12.5}$ **5.** $\sqrt{45}$ **6.** $\sqrt{59}$ **7.** $\sqrt{5.07}$ **8.** $\sqrt{55.1}$

Notice that:

$(0.1)^2 = 0.01$	$(0.4)^2 = 0.16$	$(0.7)^2 = 0.49$
$(0.2)^2 = 0.04$	$(0.5)^2 = 0.25$	$(0.8)^2 = 0.64$
$(0.3)^2 = 0.09$	$(0.6)^2 = 0.36$	$(0.9)^2 = 0.81$

In Exercises 9–14 use this table to place the given number between two numbers, correct to tenths.

EXAMPLE $\sqrt{0.19}$ **SOLUTION** $0.4 < \sqrt{0.19} < 0.5$

9. $\sqrt{0.40}$ **10.** $\sqrt{0.33}$ **11.** $\sqrt{0.726}$ **12.** $\sqrt{0.479}$ **13.** $\sqrt{0.039}$ **14.** $\sqrt{0.803}$

15. Tell how you would evaluate $\sqrt{5} - 7$ correct to hundredths.

16. Tell how you would find $\sqrt{68.5}$ correct to three digits.

Written Exercises

Use the method of successive approximations to find each square root correct to three digits. Show your work.

A 1. $\sqrt{17}$ 2. $\sqrt{28}$ 3. $\sqrt{53}$ 4. $\sqrt{5.2}$ 5. $\sqrt{65.6}$ 6. $\sqrt{452}$

Evaluate each expression correct to hundredths. Use the table of square roots as necessary.

EXAMPLE $2(8 - \sqrt{19})$

SOLUTION   From the table, $\sqrt{19} \approx 4.359$.

$$2(8 - \sqrt{19}) \approx 2(8 - 4.359) = 2(3.641) = 7.282$$

Correct to hundredths, $2(8 - \sqrt{19}) \approx 7.28$

7. $13 - \sqrt{13}$ 8. $13 + 2\sqrt{18}$ 9. $3\sqrt{38} - 12$ 10. $\sqrt{14} + 2\sqrt{17}$ 11. $\dfrac{\sqrt{18} - 4}{2}$

Use the method of successive approximations to find each square root correct to hundredths.

B 12. $\sqrt{0.5}$ 13. $\sqrt{0.31}$ 14. $\sqrt{0.46}$ 15. $\sqrt{0.888}$ 16. $\sqrt{0.222}$

What is the distance (correct to tenths) between the two points on the number line whose coordinates are given?

17. $6 + \sqrt{2}, 12 - 2\sqrt{3}$ 18. $10 - 2\sqrt{6}, 2 + 3\sqrt{5}$ 19. $\sqrt{18} - \sqrt{7}, \sqrt{41} - \sqrt{28}$

20. Which is the greater number, $(5 + \sqrt{3})$ or $(8 - \sqrt{2})$?

21. If $2x + \sqrt{3} = 7$, find x correct to two decimal places.

22. Show that $(2 - \sqrt{3})^2 < 2 - \sqrt{3}$.

23. Find two nonzero rational numbers x and y such that $x\sqrt{2} + y\sqrt{2}$ is rational.

C 24. On a number line, graph the solution set of $\sqrt{x} > x$.

25. Compute $(\sqrt{a} + \sqrt{b})(\sqrt{a} - \sqrt{b})$ for these values of (a, b): $(4, 9)$, $(9, 4)$, $(16, 25)$, $(36, 4)$, and $(100, 100)$. On the basis of your results, suggest a simple expression that appears to be equivalent to the expression $(\sqrt{a} + \sqrt{b})(\sqrt{a} - \sqrt{b})$ when a and b are positive numbers.

26. If p is a rational number and $\sqrt{q}$ is an irrational number, prove that $(p + \sqrt{q})$ is an irrational number.

Self-Test 1

Simplify each expression.

1. $\sqrt{36}$ 2. $-\sqrt{121}$ 3. $\sqrt{(-\frac{1}{5})^2}$ 4. $\sqrt{\frac{64}{81}}$ *Obj. 1, p. 389*

Express using a radical sign.

5. 12 6. -5 7. $\frac{1}{3}$ 8. $-\frac{5}{6}$

Express in the form $\frac{a}{b}$ where a and b are integers and $b \neq 0$.

9. 0.83 10. 1.24 11. $0.\overline{17}$ 12. $6.\overline{85}$ *Obj. 2, p. 389*

Find, correct to three digits.

13. $\sqrt{15}$ 14. $\sqrt{0.63}$ 15. $-\sqrt{8.24}$ 16. $\sqrt{50.5}$ *Obj. 3, p. 389*

Check your answers with those at the back of the book.

DIVERSION

Find as many ordered pairs of positive integers (x, y) as you can such that $\sqrt{x} - \sqrt{y} = y$.

programming in BASIC
Exercise

Write a program that will find a square root by the method of successive approximations. Print out the test factor, the quotient, and their average at each stage. Stop the program when the difference between the factor and the quotient is less than 0.00001.

Using the Pythagorean Theorem

OBJECTIVES for Sections 11-4 and 11-5:
1. *Use the Pythagorean Theorem and its converse.*
2. *Use the distance formula to find the distance between two points in a coordinate plane.*

11-4 The Pythagorean Theorem

Look at the tile pattern shown in Figure 2. You can verify by counting the small triangles that the area of the large red square is equal to the sum of the areas of the two small red squares. This illustrates a special case of a theorem about a right triangle (in this case, the one outlined by the heavy black line) that is credited to Pythagoras, a famous Greek philosopher and mathematician (about 580–500 B.C.). It is believed that his method of proof was based on comparison of areas.

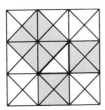

Figure 2

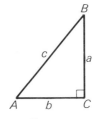

Figure 3

In stating the general theorem, we shall use the customary labeling of a right triangle, as shown in Figure 3. The vertex of the right angle is labeled C. The length of the side opposite C, called the *hypotenuse*, is represented by c. The lengths of the other two sides are represented by a and b, respectively. Recall that the area of a square is the square of the length of one of its sides.

Pythagorean Theorem

In any right triangle, the square of the length c of the hypotenuse is equal to the sum of the squares of the lengths a and b of the other two sides; that is:

$$c^2 = a^2 + b^2$$

The diagrams in Figure 4 illustrate the Pythagorean Theorem in terms of areas.

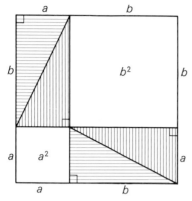

Figure 4

Notice how the right triangles in the left-hand diagram are rearranged in the right-hand diagram. It can be proved that the quadrilateral marked c^2 is actually a square.

Some common right triangles are shown in Figure 5. Verify that $c^2 = a^2 + b^2$ in each case.

Figure 5

EXAMPLE 1   A support rope 5 m long is attached to the top of a pole 2 m tall and is then stretched taut. How far from the base of the pole will the line touch the ground?

SOLUTION   1. The problem asks for the distance from the base of the pole to the end of the rope where it touches the ground.

2. Let $x =$ the distance in meters from the base of the pole to the end of the rope.

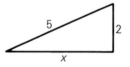

3. From the Pythagorean Theorem:
$$x^2 + 2^2 = 5^2$$

4.
$$x^2 + 4 = 25$$
$$x^2 = 21$$

Hence x denotes a square root of 21, and we select the positive square root since the required distance is positive; thus
$$x = \sqrt{21}.$$

5. The check is left to you.

∴ the distance from the base of the pole to the end of the rope is $\sqrt{21}$ m, or approximately 4.6 m.   **Answer.**

The converse of the Pythagorean Theorem is also true:

Converse of the Pythagorean Theorem

If the lengths of the sides of a triangle are such that the sum of the squares of the lengths of the two shorter sides is equal to the square of the length of the longest side, the triangle is a right triangle, with the right angle opposite the longest side.

Any set of positive integers satisfying the equation

$$c^2 = a^2 + b^2$$

is called a set of *Pythagorean numbers* or a *Pythagorean triple*.

EXAMPLE 2   Show that {9, 12, 15} is a set of Pythagorean numbers.

SOLUTION $(9)^2 + (12)^2 \overset{?}{=} (15)^2,$ $81 + 144 \overset{?}{=} 225,$ $225 = 225$ ✓

Written Exercises

Exercises 1–9 refer to the right triangle pictured at the right. Find the missing length. Use the table of square roots as necessary and give each length correct to hundredths.

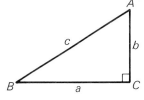

A   1. $a = 3, b = 6, c = \underline{?}$ 2. $a = 7, b = 5, c = \underline{?}$ 3. $b = 4, c = 9, a = \underline{?}$
4. $b = 10, c = 12, a = \underline{?}$ 5. $a = 2, c = 8, b = \underline{?}$ 6. $a = \sqrt{3}, b = \sqrt{5}, c = \underline{?}$
7. $a = \sqrt{8}, b = 3, c = \underline{?}$ 8. $a = \sqrt{37}, b = \sqrt{11}, c = \underline{?}$ 9. $c = \sqrt{30}, a = \sqrt{3}, b = \underline{?}$

State whether or not each set is a Pythagorean triple.

10. {5, 12, 13}   11. {4, 7, 8}   12. {9, 12, 15}   13. {3, 7, 8}
14. {6, 8, 10}   15. {3, 9, 11}   16. {8, 15, 17}   17. {9, 40, 41}
18. {5, 15, 17}   19. {20, 48, 52}   20. $\{\frac{3}{4}, \frac{1}{2}, 1\}$ 21. $\{2, \sqrt{3}, \sqrt{7}\}$

Problems

Make a sketch for each problem. Approximate each square root to the nearest hundredth.

A   1. Find the length of a diagonal of a rectangle whose dimensions are 6 m by 8 m.

2. Find the length of a diagonal of a square whose sides are each 6 cm.

3. A wire from the top of a telephone pole to a point on the ground 5 m from the pole is 9 m long. How high is the pole?

4. Two sides of a template in the shape of a right triangle are 16 cm long each. How long is the third side?

5. Melanie took a homing pigeon 9 km east of her home and then 12 km south, where she released the pigeon. If the pigeon flew directly home, how far did it fly?

6. If the foot of a 3 m ramp is 2 m from a loading platform, how high is the platform?

7. A cable from the top of a mast on a ship attached to a point 7 m from the base of the mast is 25 m long. How high is the mast?

8. A ship sailed in a northeasterly direction for 11 km and then sailed due south for 5 km. If the ship was then due east of its departure point, how far was it from its departure point?

B   9. A diagonal of a square is 8 cm long. Find the length of a side of the square.

10. The hypotenuse of a certain right triangle is twice as long as the shortest side. If the third side is 9 m, what is the length of the shortest side?

11. Find the dimensions of a rectangle which is twice as long as it is wide if one of its diagonals measures 20 mm.

12. Two hikers walk 6 km due west from their base camp, then north for 2 km, then due east. They spend the night at a point 4 km northwest of base camp. How far east did they walk?

C   13. Find the altitude of an equilateral triangle if each side measures 2 cm.

14. If two sides of a right triangle measure 8 cm and 12 cm, find two possible lengths for the third side.

15. Find the length of a diagonal of a rectangular box of length 8 cm, width 6 cm, and depth 4 cm.

16. Prove that a triangle with sides $(a^2 - b^2)$, $2ab$, and $(a^2 + b^2)$ is a right triangle. (Assume that $a > b > 0$.)

17. Develop a formula for finding the length d of the diagonal of a cube whose edge has a length of k units.

11-5 The Distance Formula

The distance between two distinct points is defined to be a positive number. For example, if P and Q are points on the number line, you find

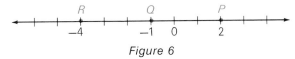

Figure 6

the distance between them, denoted by $d(P, Q)$, by computing the absolute value of the difference between the coordinates of the points. In Figure 6:

$$d(Q, P) = |2 - (-1)| = 3 \quad \text{and} \quad d(P, Q) = |-1 - 2| = 3$$

Therefore: $\qquad\qquad\qquad d(Q, P) = d(P, Q)$

Similarly:

$$d(Q, R) = |-4 - (-1)| = 3 \quad \text{and} \quad d(R, P) = |2 - (-4)| = 6$$

You can use the same method to find the length of any segment parallel to either coordinate axis in a coordinate plane.

Suppose that you wish to find the distance between the points $A(2, 2)$ and $B(-3, 6)$ shown in Figure 7. By drawing the horizontal and vertical segments intersecting at C, as shown, you can form a right triangle having $\overline{AB}$ as hypotenuse. You can find the length of each horizontal and vertical side and then use the Pythagorean Theorem.

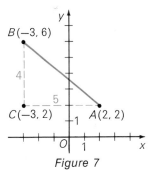

Figure 7

$$d(A, C) = |-3 - 2| = 5$$
$$d(C, B) = |6 - 2| = 4$$
$$[d(A, B)]^2 = 5^2 + 4^2 = 25 + 16 = 41$$
$$d(A, B) = \sqrt{41}$$

An answer left in a form such as $\sqrt{41}$ is said to be in *radical form*.

A formula for the distance between any two points $P_1(x_1, y_1)$ and $P_2(x_2, y_2)$ in a coordinate plane can be derived in a similar way. In Figure 8 you have: $d(P_1, C) = |x_2 - x_1|$ and $d(C, P_2) = |y_2 - y_1|$.

Since triangle P_1P_2C is a right triangle,

$$[d(P_1, P_2)]^2 = [d(P_1, C)]^2 + [d(C, P_2)]^2$$
$$= |x_2 - x_1|^2 + |y_2 - y_1|^2.$$

But $\qquad\qquad |x_2 - x_1|^2 = (x_2 - x_1)^2$

and $\qquad\qquad |y_2 - y_1|^2 = (y_2 - y_1)^2,$

and so $\qquad [d(P_1, P_2)]^2 = (x_2 - x_1)^2 + (y_2 - y_1)^2,$

$$d(P_1, P_2) = \sqrt{(x_2 - x_1)^2 + (y_2 - y_1)^2}.$$

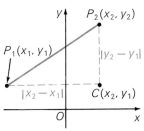

Figure 8

Irrational Numbers and Radicals | **403**

Distance Formula

For points $P_1(x_1, y_1)$ and $P_2(x_2, y_2)$:

$$d(P_1, P_2) = \sqrt{(x_2 - x_1)^2 + (y_2 - y_1)^2}$$

EXAMPLE   Find the distance between $P(-3, 2)$ and $Q(5, 1)$.

SOLUTION 1 $d(P, Q) = \sqrt{[5 - (-3)]^2 + (1 - 2)^2}$
                             $= \sqrt{8^2 + (-1)^2} = \sqrt{65}.$

SOLUTION 2 $d(Q, P) = \sqrt{[(-3) - 5]^2 + [2 - 1]^2}$
                             $= \sqrt{(-8)^2 + (1)^2} = \sqrt{65}.$

The two solutions show that it does not matter which point is considered (x_1, y_1) and which point (x_2, y_2).

Written Exercises

Find the distance between the two points having the given coordinates. Leave irrational answers in radical form.

A   **1.** $(0, 0)$, $(4, 3)$ **2.** $(0, 0)$, $(-5, 2)$ **3.** $(2, 3)$, $(-3, 4)$

        4. $(4, 2)$, $(6, -1)$ **5.** $(3, 3)$, $(8, 4)$ **6.** $(-3, -1)$, $(0, 1)$

        7. $(6, 2)$, $(-3, -5)$ **8.** $(-5, 4)$, $(4, -5)$ **9.** $(-3, 0)$, $(0, -3)$

Show by using the converse of the Pythagorean Theorem that the triangle whose vertices have the given coordinates is a right triangle.

B   **10.** $(0, 0)$, $(3, 4)$, $(3, 0)$ **11.** $(-6, -6)$, $(-4, 2)$, $(-1, -1)$

        12. $(-1, -2)$, $(-4, -2)$, $(-4, -6)$ **13.** $(3, 2)$, $(10, 3)$, $(6, -1)$

        14. $(5, 1)$, $(2, 2)$, $(4, 8)$ **15.** $(6, 4)$, $(-3, 1)$, $(9, -5)$

        16. $(0, 6)$, $(9, -6)$, $(-3, 0)$ **17.** $(0, 0)$, $(6, 0)$, $(3, 3)$

        18. Given the vertices of a right triangle $(3, 2)$, $(10, 3)$, and $(6, -1)$, write an equation of the line that passes through the endpoints of the hypotenuse.

C   **19.** Find two values of x such that the point $(x, 8)$ is 5 units from the point $(2, 4)$.

        20. Find two values of b such that the point $(6, b)$ is 13 units from the point $(11, 2)$.

        21. Show that the distance d between the origin and the point (x, y) is given by the formula $d = \sqrt{x^2 + y^2}$.

Self-Test 2

VOCABULARY   Pythagorean Theorem (p. 399)   distance formula (p. 404)

1. Find the value of x in the triangle at the right.

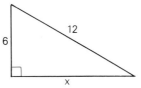

    Obj. 1, p. 399

2. Find the length of a diagonal of a rectangle whose dimensions are 3 m by 9 m. Use the table of square roots.
3. Find the distance between $A(2, -3)$ and $B(-1, 4)$. Leave your answer in radical form.

    Obj. 2, p. 399

Check your answers with those at the back of the book.

Working with Radicals

OBJECTIVES for Sections 11-6 through 11-8:
1. *Simplify radicals.*
2. *Simplify sums and products of radical expressions.*

11-6 Simplifying Radicals (1)

Expressions such as $\sqrt{a}$ are called **radicals.** A radical with a monomial radicand is said to be in **simplified form** provided the radicand contains no factor that is a perfect square. Thus, $\sqrt{15}$ is in simple form because when the radicand is written in factored form, $\sqrt{3 \cdot 5}$, no factor is a perfect square. On the other hand, $\sqrt{12}$ is not in simple form because $\sqrt{12} = \sqrt{3 \cdot 4}$, and the factored form contains the perfect square 4.

To find a means of simplifying radicals whose radicands contain perfect squares, first notice that

$$\sqrt{4 \cdot 9} = \sqrt{36} = 6$$

and

$$\sqrt{4} \cdot \sqrt{9} = 2 \cdot 3 = 6.$$

Irrational Numbers and Radicals | **405**

This suggests the following fact about square roots (see Exercises 35–38, page 407).

Product Property of Square Roots

For any nonnegative real numbers a and b:
$$\sqrt{a} \cdot \sqrt{b} = \sqrt{ab}$$

You can use the product property of square roots to simplify radicals whose radicands contain factors that are perfect squares.

EXAMPLE   Simplify $\sqrt{12}$.

SOLUTION $\sqrt{12} = \sqrt{4 \cdot 3} = \sqrt{4} \cdot \sqrt{3} = 2\sqrt{3}$.

Oral Exercises

Simplify. Assume that all variables denote positive numbers.

1. $\sqrt{2} \cdot \sqrt{3}$ 2. $\sqrt{5} \cdot \sqrt{7}$ 3. $\sqrt{8}$ 4. $\sqrt{24}$ 5. $\sqrt{3x^2}$ 6. $\sqrt{16y}$

7. $\sqrt{25z^2}$ 8. $\sqrt{x^2y}$ 9. $\sqrt{4ab^2}$ 10. $\sqrt{7a^2b}$ 11. $\sqrt{9c^2d^2}$ 12. $\sqrt{36cd^4}$

13. Is $\sqrt{2x} \cdot \sqrt{3x} = \sqrt{6x}$ true for every positive number x? Explain.

14. Is $b\sqrt{3a} \cdot b\sqrt{4a} = 2ab^2\sqrt{3}$ true for all positive numbers a and b? Explain.

Written Exercises

Assume that each variable denotes a positive real number and that the value of any expression under a radical sign is nonnegative. Simplify the given expression.

A   1. $\sqrt{48}$ 2. $\sqrt{250}$ 3. $\sqrt{24}$ 4. $\sqrt{18}$ 5. $\sqrt{32}$

6. $\sqrt{50}$ 7. $\sqrt{7a^2}$ 8. $\sqrt{13b^2}$ 9. $\sqrt{4} \cdot \sqrt{7}$ 10. $\sqrt{8} \cdot \sqrt{15}$

11. $\sqrt{c^3}$ 12. $\sqrt{d^4}$ 13. $\sqrt{9e^3}$ 14. $\sqrt{4f^5}$ 15. $\sqrt{49g}$

16. $\sqrt{64h}$ 17. $\sqrt{4s} \cdot \sqrt{8s^2}$ 18. $\sqrt{5r} \cdot \sqrt{12r^3}$ 19. $\sqrt{54m^2n}$ 20. $\sqrt{44tu^3}$

B   21. $\sqrt{(a + b)^2}$ 22. $\sqrt{(a^2 + b)^2}$ 23. $\sqrt{4(x^2 + y^2)^2}$ 24. $\sqrt{3(x + y)^3}$

25. $\sqrt{16(x + 2y)^4}$ 26. $\sqrt{2(r^2 - s^2)^2}$ 27. $\sqrt{24(r - s)^6}$ 28. $2\sqrt{48(r - s)^3}$

C   29. $\sqrt{a^2 + 2ab + b^2}$ 30. $\sqrt{4a^2 + 12ab + 9b^2}$

31. $\sqrt{9 - 24x + 16x^2}$ 32. $\sqrt{(x + y)^2 \cdot (x - y)^2}$

33. $r\sqrt{(r^2 - s^2) \cdot (r^2 + s^2)^2}$ 34. $(r + s)\sqrt{(r - s)^2 \cdot (r^2 - s^2)}$

In Exercises 35–38 provide reasons for the given step in the proof of the product property of square roots on page 406.

35. $(\sqrt{a} \cdot \sqrt{b})^2 = (\sqrt{a})^2 \cdot (\sqrt{b})^2$

36. $(\sqrt{a})^2 = a, (\sqrt{b})^2 = b$

37. $(\sqrt{a} \cdot \sqrt{b})^2 = ab$

38. $\sqrt{a} \cdot \sqrt{b} = \sqrt{ab}$

39. Find the distance between $P(\sqrt{a}, \sqrt{b})$ and $Q(a\sqrt{a}, a\sqrt{b})$. Simplify your answer.

11-7 Simplifying Radicals (2)

As the examples

$$\sqrt{\frac{100}{4}} = \sqrt{25} = 5 \qquad \frac{\sqrt{100}}{\sqrt{4}} = \frac{10}{2} = 5$$

suggest, another property of square roots is the following (see Exercise 36, page 409).

Quotient Property of Square Roots

For any nonnegative real number a and positive real number c:

$$\frac{\sqrt{a}}{\sqrt{c}} = \sqrt{\frac{a}{c}}$$

This property, together with the basic property of quotients, provides a means of converting a fraction with an irrational denominator, $\dfrac{\sqrt{a}}{\sqrt{b}}$, or a radical containing a fraction, $\sqrt{\dfrac{a}{b}}$, $b > 0$, into an equivalent fraction $\dfrac{\sqrt{ab}}{b}$ with a rational denominator. For example,

$$\sqrt{\frac{2}{3}} = \frac{\sqrt{2}}{\sqrt{3}} = \frac{\sqrt{2} \cdot \sqrt{3}}{\sqrt{3} \cdot \sqrt{3}} = \frac{\sqrt{6}}{3}, \text{ or } \tfrac{1}{3}\sqrt{6}$$

$$\frac{2}{\sqrt{8}} = \frac{2 \cdot \sqrt{2}}{\sqrt{8} \cdot \sqrt{2}} = \frac{2\sqrt{2}}{\sqrt{16}} = \frac{2\sqrt{2}}{4} = \frac{\sqrt{2}}{2}, \text{ or } \tfrac{1}{2}\sqrt{2}$$

$$\sqrt{\frac{2}{6x}} = \sqrt{\frac{1}{3x}} = \frac{1}{\sqrt{3x}} = \frac{1 \cdot \sqrt{3x}}{\sqrt{3x} \cdot \sqrt{3x}} = \frac{\sqrt{3x}}{3x}, \text{ or } \frac{1}{3x}\sqrt{3x}$$

The process illustrated in the examples above is called **rationalizing the denominator**, and provides another way to simplify a radical.

Including the method of simplification on pages 405–406, we have the following definition:

> A term containing such forms as $\sqrt{a}$ is in *simplified form* when:
> 1. The radicand has no factor that is a perfect square.
> 2. The radicand does not contain a fraction.
> 3. No radical appears in a denominator.

You may apply this rule to expressions involving radicals that have polynomials as radicands when it can be assumed that such polynomials represent nonnegative real numbers (positive if in a denominator).

Rationalizing the denominator may simplify the calculations involved in finding an approximation to a radical expression. For example, contrast these computations:

$$\frac{1}{\sqrt{2}} \approx \frac{1}{1.414} \approx 0.707 \quad \text{and} \quad \frac{1}{\sqrt{2}} = \frac{\sqrt{2}}{2} \approx \frac{1.414}{2} = 0.707$$

Oral Exercises

Simplify. Assume that all variables denote positive real numbers.

1. $\dfrac{1}{\sqrt{3}}$ **2.** $\dfrac{1}{\sqrt{6}}$ **3.** $\dfrac{3}{\sqrt{5}}$ **4.** $\dfrac{2}{\sqrt{10}}$ **5.** $\sqrt{\dfrac{2}{5}}$

6. $\sqrt{\dfrac{5}{12}}$ **7.** $\sqrt{\dfrac{1}{a}}$ **8.** $\sqrt{\dfrac{2}{b}}$ **9.** $\dfrac{1}{\sqrt{c^3}}$ **10.** $\dfrac{1}{\sqrt{d^5}}$

11. Is $\dfrac{x}{\sqrt{y}} = \dfrac{x\sqrt{y}}{y}$ true for all positive numbers x and y? Explain.

12. Is $\dfrac{y\sqrt{x}}{\sqrt{y}} = \sqrt{xy}$ true for all positive numbers x and y? Explain.

Written Exercises

Simplify. Assume that all variables denote positive real numbers.

A **1.** $\dfrac{3}{\sqrt{2}}$ **2.** $\dfrac{5}{\sqrt{3}}$ **3.** $\dfrac{8}{\sqrt{a}}$ **4.** $\dfrac{8}{\sqrt{2b}}$ **5.** $\sqrt{\dfrac{2}{c}}$ **6.** $\sqrt{\dfrac{3c}{d}}$

7. $\dfrac{\sqrt{27}}{\sqrt{3}}$ **8.** $\dfrac{\sqrt{32}}{\sqrt{2}}$ **9.** $\sqrt{\dfrac{12}{3x}}$ **10.** $\sqrt{\dfrac{2}{8y}}$ **11.** $\sqrt{\dfrac{9}{5z}}$ **12.** $\sqrt{\dfrac{25}{6w}}$

Simplify each expression and use the table of square roots to find an approximation correct to hundredths for the simplified expression.

EXAMPLE $\sqrt{245} = \sqrt{49} \cdot \sqrt{5} = 7\sqrt{5} \approx 7(2.236) \approx 15.65$

13. $\sqrt{252}$ 14. $\sqrt{224}$ 15. $\sqrt{300}$ 16. $\dfrac{2}{\sqrt{11}}$ 17. $\dfrac{13}{\sqrt{10}}$ 18. $\sqrt{\dfrac{13}{3}}$

Simplify. Assume that all variables denote positive real numbers and that the value of any expression under a radical sign is a positive number.

B 19. $\dfrac{2\sqrt{3a}}{\sqrt{a^3}}$ 20. $\dfrac{3\sqrt{5b}}{\sqrt{b^3}}$ 21. $2\sqrt{\dfrac{a^3}{ab}}$ 22. $\dfrac{4\sqrt{2r}}{\sqrt{r^3}}$ 23. $3\sqrt{\dfrac{2s^2}{3s}}$ 24. $\dfrac{r}{s}\sqrt{\dfrac{3s^2}{5r}}$

25. $2x\sqrt{\dfrac{y}{x}}$ 26. $\dfrac{3x}{y}\sqrt{\dfrac{7y^2}{2x^3}}$ 27. $\sqrt{x} \cdot \sqrt{\dfrac{1}{x^2y}}$ 28. $\dfrac{\sqrt{32x^3}}{\sqrt{\frac{1}{8}x^2}}$ 29. $\dfrac{7x\sqrt{5x}}{\sqrt{2x^5}}$ 30. $5\sqrt{\dfrac{3z^2}{7z}}$

C 31. $\dfrac{1}{\sqrt{m+n}}$ 32. $\dfrac{5}{\sqrt{2m-3n}}$ 33. $\dfrac{m}{\sqrt{(m-n)^3}}$ 34. $\sqrt{\dfrac{r^2+3r+2}{r+1}}$ 35. $\sqrt{\dfrac{r}{(r+s)^3}}$

36. Prove the quotient property of square roots on page 407.

11-8 Sums and Products of Radical Expressions

Radical expressions such as $3\sqrt{2}$ and $\frac{1}{2}\sqrt{2}$ are said to be *similar* because the same radical, in this case $\sqrt{2}$, appears in both.

Similar	Nonsimilar
$3\sqrt{5}$ and $5\sqrt{5}$	$2\sqrt{5}$ and $5\sqrt{2}$
$x\sqrt{y}$ and $z\sqrt{y}, \quad y \geq 0$	$3\sqrt{z}$ and $3\sqrt{2z}, \quad z \geq 0$

By using the distributive axiom, you can express the sum of two or more similar radicals as a single term.

EXAMPLES   1. $6\sqrt{3} + 9\sqrt{3} = (6+9)\sqrt{3} = 15\sqrt{3}$
2. $3\sqrt{x} - \sqrt{x} + 2\sqrt{x} = (3-1+2)\sqrt{x} = 4\sqrt{x}, \quad x \geq 0$

You should always write each radical in simplified form before deciding whether the radical expressions are similar.

EXAMPLES   3. $\sqrt{12} + \sqrt{27} - 2\sqrt{3} = 2\sqrt{3} + 3\sqrt{3} - 2\sqrt{3}$
$= (2+3-2)\sqrt{3} = 3\sqrt{3}$
4. $(6 + 2\sqrt{5}) + (5 - \sqrt{20}) = 6 + 2\sqrt{5} + 5 - 2\sqrt{5} = 11$

Example 4 illustrates the fact that the sum of irrational numbers may be a rational number.

The distributive axiom also enables you to simplify products of radical expressions.

EXAMPLES 5. $\sqrt{5}(3 + \sqrt{2}) = (\sqrt{5}) \cdot 3 + (\sqrt{5})\sqrt{2} = 3\sqrt{5} + \sqrt{10}$

6. $(3 + \sqrt{2})(1 - \sqrt{2}) = 3 - 3\sqrt{2} + 1 \cdot \sqrt{2} - (\sqrt{2})(\sqrt{2})$
$$= 3 - 3\sqrt{2} + \sqrt{2} - 2 = 1 - 2\sqrt{2}$$

7. $(\sqrt{5} + \sqrt{2})(\sqrt{5} - \sqrt{2}) = (\sqrt{5})(\sqrt{5}) - (\sqrt{5})(\sqrt{2})$
$$+ (\sqrt{2})(\sqrt{5}) - (\sqrt{2})(\sqrt{2}) = 5 - 2 = 3$$

8. $(\sqrt{5x} + \sqrt{3})(\sqrt{5x} - \sqrt{3}) = 5x^2 - 3$

Example 7 illustrates the fact that the product of irrational numbers may be a rational number.

In Chapter 8 you factored polynomials into factors that had integral coefficients. If the factor set is extended to include polynomials with irrational coefficients, additional polynomials may be factored, as illustrated in Example 8 above, where $5x^2 - 3 = (\sqrt{5x} + \sqrt{3})(\sqrt{5x} - \sqrt{3})$.

Two expressions of the form $a + \sqrt{b}$ and $a - \sqrt{b}$ may be called **conjugate expressions**. In particular, $\sqrt{r} + \sqrt{s}$ and $\sqrt{r} - \sqrt{s}$ are conjugate expressions. Notice that $(\sqrt{r} + \sqrt{s})(\sqrt{r} - \sqrt{s}) = r - s$. You may use this fact to rationalize some binomial denominators.

EXAMPLES 9. $\dfrac{\sqrt{3}}{5 - \sqrt{2}} = \dfrac{\sqrt{3}(5 + \sqrt{2})}{(5 - \sqrt{2})(5 + \sqrt{2})} = \dfrac{5\sqrt{3} + \sqrt{6}}{25 - 2} = \dfrac{5\sqrt{3} + \sqrt{6}}{23}$

10. $\dfrac{\sqrt{2}}{\sqrt{7} + \sqrt{2}} = \dfrac{\sqrt{2}(\sqrt{7} - \sqrt{2})}{(\sqrt{7} + \sqrt{2})(\sqrt{7} - \sqrt{2})} = \dfrac{\sqrt{14} - 2}{7 - 2} = \dfrac{\sqrt{14} - 2}{5}$

Oral Exercises

Simplify.

1. $5\sqrt{3} + 4\sqrt{3}$
2. $9\sqrt{5} - 6\sqrt{5}$
3. $\sqrt{2}(4 + \sqrt{3})$
4. $\sqrt{5}(\sqrt{2} - \sqrt{3})$
5. $(5\sqrt{2})(4\sqrt{2})$
6. $\sqrt{3} + \sqrt{12}$
7. $3\sqrt{5} - \sqrt{20}$
8. $3\sqrt{2} + 2\sqrt{18}$

9. Give an example that shows that the sum of two irrational numbers can be a rational number.

10. Find values for a and b to show that $\sqrt{a^2 + b^2} = a + b$ is not true for all positive numbers.

Written Exercises

Assume that each variable denotes a positive real number. Simplify the given expression.

A 1. $2\sqrt{3} + 3\sqrt{3} - \sqrt{3}$
2. $7\sqrt{2} - \sqrt{2} - 5\sqrt{2}$
3. $2\sqrt{2} + \sqrt{8} - 2\sqrt{18}$

4. $\sqrt{12} + 4\sqrt{3} - 2\sqrt{27}$ 5. $3\sqrt{2} + 2\sqrt{8} - 3\sqrt{32}$ 6. $3\sqrt{5} - \sqrt{20} + \sqrt{45}$

7. $\sqrt{8a} + \sqrt{9a} - 3\sqrt{a}$ 8. $\sqrt{xy^2} - y\sqrt{x} + \sqrt{16xy^2}$ 9. $(2\sqrt{8})(3\sqrt{3})$

10. $(4\sqrt{12})(-2\sqrt{5})$ 11. $\sqrt{2}(6 - \sqrt{18})$ 12. $\sqrt{3}(7 + \sqrt{12})$

13. $(\sqrt{2} + 3)(\sqrt{2} - 1)$ 14. $(\sqrt{6} - 3)(\sqrt{6} + 2)$ 15. $(3\sqrt{5} - 2)(\sqrt{5} - 3)$

16. $(2\sqrt{7} - 2)(3\sqrt{7} + 2)$ 17. $(\sqrt{7} - \sqrt{5})(\sqrt{7} + \sqrt{5})$ 18. $(\sqrt{10} - 4)(2\sqrt{10} + 7)$

B 19. $(2\sqrt{7} + 4)^2$ 20. $(3\sqrt{5} - 5)^2$ 21. $(2\sqrt{x} - 2\sqrt{y})^2$

22. $\dfrac{\sqrt{2}}{\sqrt{3} - 5}$ 23. $\dfrac{5}{\sqrt{5} + 2}$ 24. $\dfrac{\sqrt{3} + 1}{\sqrt{3} - 1}$

25. $\dfrac{\sqrt{2} + \sqrt{3}}{\sqrt{2} + 3}$ 26. $\dfrac{\sqrt{6} - 1}{1 - \sqrt{6}}$ 27. $\dfrac{\sqrt{x} - \sqrt{y}}{\sqrt{x} + \sqrt{y}}$

28. $12\sqrt{\frac{2}{3}} + \sqrt{54} - 2\sqrt{1\frac{1}{2}}$ 29. $\sqrt{\frac{12}{5}} - \frac{1}{5}\sqrt{60} + 6\sqrt{\frac{5}{12}}$ 30. $2\sqrt{12k} - \dfrac{2k^2}{3}\sqrt{\dfrac{108}{k}}$

31. $m^5\sqrt{\dfrac{n}{m^3}} + 3\sqrt{nm^7}$ 32. $\sqrt{\dfrac{r^2}{64} + \dfrac{r^2}{36}}$ 33. $\sqrt{\dfrac{3s^2}{5} - \dfrac{s^2}{20}}$

34. If $f(x) = 2x^2 - 5$, and $f(a) = 1$, find a.

35. Write an expression in simplified form for the area of a square whose perimeter is $(16\sqrt{3} - 8\sqrt{2})$ cm.

If $f(x) = x^2 - 2x - 3$, find:

C 36. $f(\sqrt{2})$ 37. $f(-\sqrt{3})$ 38. $f(\sqrt{3} + 1)$ 39. $f(\sqrt{2} - 3)$

40. Show that $(2 + \sqrt{2})$ and $(2 - \sqrt{2})$ are solutions of $x^2 - 4x + 2 = 0$.

41. Show that $(-3 + \sqrt{5})$ and $(-3 - \sqrt{5})$ are solutions of $x^2 + 6x + 4 = 0$.

Self-Test 3

VOCABULARY   product property of square roots (p. 406)
quotient property of square roots (p. 407)

Simplify.

1. $\sqrt{48}$ 2. $\sqrt{150}$ 3. $\sqrt{\frac{3}{7}}$ 4. $\sqrt{\frac{11}{10}}$ *Obj. 1, p. 405*

5. $3\sqrt{2} + 3\sqrt{18}$ 6. $5\sqrt{24} - 3\sqrt{54}$ *Obj. 2, p. 405*

7. $\sqrt{2}(7 + 3\sqrt{18})$ 8. $(\sqrt{5} - 3)(\sqrt{5} + 2)$

Check your answers with those at the back of the book.

*n*th Roots

If n is a positive integer and a is a real number, then any real number whose nth power equals a is called an **nth root** of a.

For example, 2 is a fifth root of 32 because $2^5 = 32$. In general, *if n is odd*, then there is only *one* real nth root of a, no matter whether a is negative, zero, or positive. In that case, the real nth root of a is denoted by

$$\sqrt[n]{a}.$$

There are two real fourth roots of 16, namely -2 and 2, since $2^4 = 16$ and $(-2)^4 = 16$. In general, *if n is even and a is positive*, then there are *two* real nth roots of a; one is positive and the other is negative. In that case, the *positive* nth root of a is denoted by

$$\sqrt[n]{a}$$

and is called the **principal *n*th root of *a*.** The *negative* nth root is denoted by

$$-\sqrt[n]{a}.$$

There are no real fourth roots of -16, for there are no real numbers that satisfy $x^4 = -16$. In general, *if n is even and a is negative*, then there are *no* real nth roots of a. In that case the symbols $\sqrt[n]{a}$ and $-\sqrt[n]{a}$ do not represent real numbers.

If $a = 0$, then 0 is the only real nth root of a, no matter whether n is odd or even: $0^n = 0$. Thus,

$$\sqrt[n]{0} = 0$$

for every positive integer n.

Keep in mind that whenever the symbol $\sqrt[n]{a}$ represents a real number, it is true that

$$(\sqrt[n]{a})^n = a.$$

The integer n is called the **root index.** In a radical such as $\sqrt{3}$ the root index is understood to be 2, but it is not written.

Properties of *n*th Roots

Whenever the radicals in the expressions represent real numbers and $c \neq 0$,

$$\sqrt[n]{a} \cdot \sqrt[n]{b} = \sqrt[n]{ab} \quad \text{and} \quad \frac{\sqrt[n]{a}}{\sqrt[n]{c}} = \sqrt[n]{\frac{a}{c}}.$$

Since $64 = (4)^3$, $-64 = (-4)^3$, and $\frac{8}{27} = (\frac{2}{3})^3$, the rational numbers 64, -64, and $\frac{8}{27}$ are called *perfect cubes*. Similarly, 81 is a perfect fourth power $(81 = 3^4)$, and 32 and -32 are perfect fifth powers. If a real number is not the nth power of some rational number, its nth root, if it is a real number, is an irrational number.

EXAMPLE Simplify: **a.** $\sqrt[4]{16}$;   **b.** $\sqrt[3]{16}$;   **c.** $\dfrac{3}{\sqrt[3]{4}}$

SOLUTION   **a.** $\sqrt[4]{16} = \sqrt[4]{(2)^4} = 2$

b. $\sqrt[3]{16} = \sqrt[3]{8} \cdot \sqrt[3]{2} = 2\sqrt[3]{2}$

c. $\dfrac{3}{\sqrt[3]{4}} = \dfrac{3\sqrt[3]{2}}{\sqrt[3]{4}\sqrt[3]{2}} = \dfrac{3\sqrt[3]{2}}{\sqrt[3]{8}} = \dfrac{3\sqrt[3]{2}}{2}$

Exercises

Simplify.

1. $\sqrt[3]{27}$ 2. $\sqrt[5]{32}$ 3. $\sqrt[7]{0}$ 4. $-\sqrt[8]{1}$

5. $\sqrt[5]{-1}$ 6. $\sqrt[3]{-125}$ 7. $\sqrt[3]{54}$ 8. $\sqrt[4]{48}$

9. $\dfrac{3}{\sqrt[3]{2}}$ 10. $\sqrt[4]{\frac{5}{27}}$ 11. $\sqrt[3]{\frac{5}{16}}$ 12. $\sqrt[4]{\frac{3}{32}}$

Chapter Summary

1. The nonnegative square root of a $(a \geq 0)$ is denoted by the symbol $\sqrt{a}$; the negative square root of a is denoted by the symbol $-\sqrt{a}$. If $b = \sqrt{a}$ or $b = -\sqrt{a}$, then $b^2 = a$.

2. For any real number, $\sqrt{a^2} = |a|$.

3. **a.** A number is a *rational number* if and only if it can be represented by a terminating or repeating decimal.
   b. *Irrational numbers* are those represented by nonrepeating, nonterminating decimals.
   c. The rational and irrational numbers together comprise the *real numbers*.

4. The Pythagorean Theorem is a statement of an important relationship among the sides of any right triangle: If the lengths of the legs are a and b, and the length of the hypotenuse is c, then $c^2 = a^2 + b^2$. A special form of the Pythagorean Theorem is the distance formula: The distance d between two points (x_1, y_1) and (x_2, y_2) is
$$d = \sqrt{(x_2 - x_1)^2 + (y_2 - y_1)^2}.$$

5. The following two properties are used to simplify radicals.
   a. **The Product Property of Square Roots:** For any nonnegative real numbers a and b, $\sqrt{a} \cdot \sqrt{b} = \sqrt{ab}$.
   b. **The Quotient Property of Square Roots:** For any nonnegative real number a and any positive real number b,

   $$\frac{\sqrt{a}}{\sqrt{b}} = \sqrt{\frac{a}{b}}.$$

Chapter Review

1. Which statement is true for all $x \in \Re$?   *11-1*
   a. $\sqrt{x^2} = x$
   b. $\sqrt{(x + 1)^2} = x + 1$
   c. $\sqrt{x^2} = |x|$
   d. $\sqrt{x}$ is a real number.

2. Simplify $\sqrt{(-5)^2}$.
   a. -5 b. 5   c. 25   d. 2.236

3. Which of these is a rational number?   *11-2*
   a. $\sqrt{3}$ b. $\sqrt{4}$ c. $\sqrt{5}$ d. π

4. Which of these is an irrational number?
   a. $\frac{2}{7}$ b. $3.\overline{48}$ c. $\sqrt{9}$ d. $3.484884888 \ldots$

5. $\sqrt{110}$ is between ___?___.   *11-3*
   a. 8 and 9   b. 9 and 10   c. 10 and 11   d. 11 and 12

6. Find $\sqrt{28.6}$ to the nearest tenth.
   a. 5.3   b. 5.4   c. 4.9   d. 5.2

7. If the hypotenuse of a right triangle is 8 and a leg is 5, then the other leg is: *11-4*
   a. 3   b. $\sqrt{89}$ c. $-\sqrt{39}$ d. $\sqrt{39}$

8. The distance between $(4, -2)$ and $(-3, 1)$ is   *11-5*
   a. $\sqrt{58}$ b. $\sqrt{10}$ c. $\sqrt{50}$ d. 10

Simplify. Assume that all variables denote positive real numbers.

9. $\sqrt{45}$ *11-6*
   a. $9\sqrt{5}$ b. $3\sqrt{5}$ c. $5\sqrt{3}$ d. $5\sqrt{9}$

10. $\sqrt{2a} \cdot \sqrt{12a}$
    a. $\sqrt{24a^2}$ b. $2a\sqrt{6a}$ c. $2a\sqrt{6}$ d. $2\sqrt{6a}$

11. $\dfrac{\sqrt{75}}{\sqrt{6x}}$ *11-7*
    a. $\dfrac{5\sqrt{3}}{\sqrt{6x}}$ b. $\dfrac{5}{\sqrt{2x}}$ c. $\dfrac{5\sqrt{x}}{2}$ d. $\dfrac{5\sqrt{2x}}{2x}$

12. $\sqrt{\dfrac{4}{3}}$

   a. $\dfrac{2\sqrt{3}}{3}$ **b.** $\dfrac{2}{3}$ **c.** $\dfrac{2}{9}$ **d.** $\dfrac{2}{\sqrt{3}}$

13. $\sqrt{50} - \sqrt{32}$ 11-8

   a. 1   **b.** $\sqrt{18}$ **c.** $\sqrt{2}$ **d.** 2

14. $(\sqrt{5} + \sqrt{6})(\sqrt{5} - \sqrt{6})$

   a. $-\sqrt{11}$ **b.** -1 **c.** -11 **d.** $-1 + 2\sqrt{30}$

Chapter Test

1. Simplify $\sqrt{\frac{49}{25}}$. 11-1

2. Use a radical sign to express $-\frac{1}{3}$.

3. Express $0.2\overline{3}$ in the form $\dfrac{a}{b}$, where a and b are integers. 11-2

4. Solve over $\mathfrak{R}$: $m^2 - 4 = 10$.

5. Find $12 + 2\sqrt{30}$ correct to hundredths. 11-3

6. Which is the greater number, $4 + \sqrt{5}$ or $8 - \sqrt{3}$?

7. A ladder 10 m long leans against a wall. If the ladder meets the wall at a point 9 m above the ground, how far from the base of the wall is the foot of the ladder? 11-4

8. Find the distance between the points $(-3, -6)$ and $(2, 6)$. 11-5

Simplify. Assume that all variables denote positive real numbers.

9. $\sqrt{75a^3}$ 10. $\sqrt{6c} \cdot \sqrt{12c^3}$ 11-6

11-7

11. $\dfrac{3}{\sqrt{12}}$ 12. $\sqrt{\dfrac{8}{7x}}$

13. $(\sqrt{5} - 2)^2$ 14. $\sqrt{2}(\sqrt{24} + \sqrt{54})$ 11-8

ON THE CALCULATOR

Using a calculator, find the following.

1. $\sqrt[3]{4913}$ 2. $\sqrt[3]{36,926,037}$ 3. $\sqrt[5]{248,832}$ 4. $\sqrt[5]{28,629,151}$

This vehicle is propelled by a linear induction motor and is capable of much higher speeds than vehicles propelled by conventional traction motors.

12 Quadratic Equations and Functions

Quadratic Equations

***OBJECTIVES** for Sections 12-1 and 12-2:*
1. Solve quadratic equations by completing the square.
2. Solve quadratic equations by using the quadratic formula.

12-1 Completing the Square

Any equation which can be written equivalently in the form (called the *standard form*)

$$ax^2 + bx + c = 0, \quad a, b, c \in \Re, a \neq 0$$

is called a **quadratic equation.** In Chapter 8, you learned to solve some equations of this form by factoring the left member. To discover a method you can use to solve *any* quadratic equation, you can begin by recalling this:

$$(x + r)^2 = x^2 + 2rx + r^2$$

Notice that the last term, r^2, is just the square of one-half the coefficient of x; that is, r^2 is the square of one-half of $2r$.

EXAMPLE 1 Find a value for r so that $x^2 - 4x + r^2$ is a perfect square, and write the expression equivalently as a perfect square.

SOLUTION The coefficient of x is -4, so that $r = \frac{1}{2}(-4) = -2$.

$$x^2 - 4x + (-2)^2 = x^2 - 4x + 4 = (x - 2)^2$$

You can use the process illustrated in Example 1, called completing the square, to solve any quadratic equation.

EXAMPLE 2 Solve $x^2 - 2x - 2 = 0$ by completing the square.

SOLUTION

1. Write the equation with the constant term in the right member. $x^2 - 2x \quad = 2$

2. Add to each member the square of one-half the coefficient of x.
$$x^2 - 2x + 1 = 2 + 1$$
$$x^2 - 2x + 1 = 3$$

3. Write the left member as a perfect square. $(x - 1)^2 = 3$

4. By definition, $x - 1$ is one of the square roots of 3. Write a disjunction stating this fact.
$$x - 1 = \sqrt{3} \text{ or } x - 1 = -\sqrt{3}$$
$$x = 1 + \sqrt{3} \text{ or } x = 1 - \sqrt{3}$$

5. Check the solutions.

$$(1 + \sqrt{3})^2 - 2(1 + \sqrt{3}) - 2 \overset{?}{=} 0$$
$$4 + 2\sqrt{3} - 2 - 2\sqrt{3} - 2 \overset{?}{=} 0$$
$$0 = 0 \checkmark$$

$$(1 - \sqrt{3})^2 - 2(1 - \sqrt{3}) - 2 \overset{?}{=} 0$$
$$4 - 2\sqrt{3} - 2 + 2\sqrt{3} - 2 \overset{?}{=} 0$$
$$0 = 0 \checkmark$$

The solution set is $\{1 + \sqrt{3}, 1 - \sqrt{3}\}$. **Answer.**

If the coefficient of x^2 in a quadratic equation is a number other than 1, you can use the multiplicative property of equality to find an equivalent equation in which the coefficient of x^2 is 1. Thus, before completing the square to solve the equation

$$3x^2 - 5x + 2 = 0,$$

you would first multiply each member by $\frac{1}{3}$ to obtain

$$x^2 - \tfrac{5}{3}x + \tfrac{2}{3} = 0,$$

which could then be solved by completing the square.

Oral Exercises

State the term necessary in each case to complete the square.

EXAMPLE $x^2 - 5x + \underline{\ ?\ }$

SOLUTION $\frac{1}{2}$ of -5 is $-\frac{5}{2}$; $\therefore$ the needed term is $\left(-\frac{5}{2}\right)^2$, or $\frac{25}{4}$.

1. $x^2 + 6x + \underline{\ ?\ }$ **2.** $y^2 - 10y + \underline{\ ?\ }$ **3.** $z^2 + z + \underline{\ ?\ }$ **4.** $x^2 + 7x + \underline{\ ?\ }$

5. $y^2 + \frac{1}{2}y + \underline{\ ?\ }$ **6.** $z^2 + \frac{3}{4}z + \underline{\ ?\ }$ **7.** $x^2 + bx + \underline{\ ?\ }$ **8.** $y^2 - \frac{b}{a}y + \underline{\ ?\ }$

Written Exercises

Solve by completing the square. Write any radicals involved in simplified form.

A   **1.** $p^2 + 2p - 7 = 0$ **2.** $q^2 - 6q + 8 = 0$ **3.** $m^2 + 4m = 14$

4. $n^2 + 10n + 16 = 0$ **5.** $r^2 + r - 6 = 0$ **6.** $s^2 + 7s = -2$

7. $t^2 + 6t + 4 = 0$ **8.** $u^2 - u - 3 = 0$ **9.** $x^2 = 20x - 19$

B   **10.** $w^2 - \frac{1}{3}w - \frac{1}{3} = 0$ **11.** $z^2 - \frac{2}{3}z - \frac{1}{3} = 0$ **12.** $k^2 + \frac{2}{3}k = 0$

13. $l^2 - \frac{3}{2}l = 0$ **14.** $3p^2 + 9p - 81 = 0$ **15.** $5q^2 + q - 3 = 0$

16. $\dfrac{3}{r+1} + \dfrac{1}{r-1} = 1$ **17.** $\dfrac{1}{s} + \dfrac{1}{s-2} = 2$ **18.** $\dfrac{1}{m} + \dfrac{1}{m-1} = 3$

Solve for x.

C   **19.** $x^2 + bx + c = 0$ **20.** $ax^2 + bx + c = 0$ **21.** $\frac{1}{4}x^2 - \frac{1}{2}kx = -\frac{9}{4}$

22. $\dfrac{x^2}{a} + x + c = 0$ **23.** $\dfrac{a}{x} + \dfrac{b}{x+1} = b$ **24.** $9x^2 - 6cx + c^2 - 4a = 0$

25. $\dfrac{x^2}{a} + \dfrac{x}{b} + 1 = 0$ **26.** $b^2x^2 + 4abx + 4a^2 = 9$ **27.** $x^2 + 2\sqrt{2}x + 2 + a = b$

DIVERSION

Complete the division problem shown at the right by replacing each question mark with a digit.

```
           ? ? 8 ? ?
? ? ?) ? ? ? ? ? ? ? ?
       ? ? ? ?
       ‾‾‾‾‾‾‾
         ? ? ? ?
           ? ? ?
           ‾‾‾‾‾
             ? ? ? ?
             ? ? ? ?
             ‾‾‾‾‾‾‾
```

12-2 The Quadratic Formula

By completing the square using the standard form $ax^2 + bx + c = 0$, you can derive a formula that expresses the roots of a quadratic equation in terms of its coefficients.

Compare the following solution processes.

$3x^2 + 5x + 1 = 0$

$ax^2 + bx + c = 0$

$x^2 + \frac{5}{3}x + \frac{1}{3} = 0$

$x^2 + \frac{b}{a}x + \frac{c}{a} = 0$

$x^2 + \frac{5}{3}x \quad\quad = -\frac{1}{3}$

$x^2 + \frac{b}{a}x \quad\quad = -\frac{c}{a}$

$x^2 + \frac{5}{3}x + \frac{25}{36} = -\frac{1}{3} + \frac{25}{36}$

$x^2 + \frac{b}{a}x + \frac{b^2}{4a^2} = -\frac{c}{a} + \frac{b^2}{4a^2}$

$\left(x + \frac{5}{6}\right)^2 = \frac{25 - 12}{36}$

$\left(x + \frac{b}{2a}\right)^2 = \frac{b^2 - 4ac}{4a^2}$

Since $25 - 12 > 0$, you have

$x + \frac{5}{6} = \frac{\sqrt{13}}{6}$

or

$x + \frac{5}{6} = -\frac{\sqrt{13}}{6}$

If $b^2 - 4ac \geq 0$, you have

$x + \frac{b}{2a} = \frac{\sqrt{b^2 - 4ac}}{2a}$

or

$x + \frac{b}{2a} = -\frac{\sqrt{b^2 - 4ac}}{2a}$

Thus

$x = -\frac{5}{6} + \frac{\sqrt{13}}{6}$

or

$x = -\frac{5}{6} - \frac{\sqrt{13}}{6}$

Thus

$x = -\frac{b}{2a} + \frac{\sqrt{b^2 - 4ac}}{2a}$

or

$x = -\frac{b}{2a} - \frac{\sqrt{b^2 - 4ac}}{2a}$

The check is left for you.

The solution set is

$$\left\{ \frac{-5 + \sqrt{13}}{6}, \frac{-5 - \sqrt{13}}{6} \right\}$$

The check is left for you.

The solution set is

$$\left\{ \frac{-b + \sqrt{b^2 - 4ac}}{2a}, \right.$$

$$\left. \frac{-b - \sqrt{b^2 - 4ac}}{2a} \right\}$$

Do you see from the foregoing work that the general quadratic equation,

$$ax^2 + bx + c = 0, \quad a, b, c \in \mathcal{R}, a \neq 0,$$

is equivalent to the disjunction

$$x = \frac{-b + \sqrt{b^2 - 4ac}}{2a} \quad \text{or} \quad x = \frac{-b - \sqrt{b^2 - 4ac}}{2a}$$

when $b^2 - 4ac \geq 0$? This open sentence, often written in the form

$$x = \frac{-b \pm \sqrt{b^2 - 4ac}}{2a},$$

is called the **quadratic formula.** The sign "$\pm$" here is read "plus or minus." If $b^2 - 4ac < 0$, then $\sqrt{b^2 - 4ac}$ is not a real number, and the quadratic equation has no real solutions.

EXAMPLE   Use the quadratic formula to solve

$$3y^2 - 8y = -2.$$

SOLUTION   After rewriting the given equation in standard form,

$$3y^2 - 8y + 2 = 0,$$

you compare it to the standard form

$$ay^2 + by + c = 0$$

to find that $a = 3$, $b = -8$, and $c = 2$.
Substitution in the quadratic formula gives:

$$y = \frac{-(-8) \pm \sqrt{(-8)^2 - 4(3)(2)}}{2(3)} = \frac{8 \pm \sqrt{64 - 24}}{6}$$

$$= \frac{8 \pm \sqrt{40}}{6} = \frac{8 \pm 2\sqrt{10}}{6} = \frac{2(4 \pm \sqrt{10})}{2 \cdot 3} = \frac{4 \pm \sqrt{10}}{3}$$

The check is left for you.

$\therefore$ the solution set is $\left\{ \dfrac{4 + \sqrt{10}}{3}, \dfrac{4 - \sqrt{10}}{3} \right\}$.   Answer.

Notice that the solution set in this example contains two members. Do you see the role of the value of $b^2 - 4ac$ in the quadratic formula?
When $b^2 - 4ac > 0$, $\sqrt{b^2 - 4ac}$ is a positive real number and the quadratic formula gives *two different* real roots for the equation.
When $b^2 - 4ac = 0$, $\sqrt{b^2 - 4ac} = 0$ and the quadratic formula gives *only one* root:

$$\frac{-b \pm 0}{2a} = \frac{-b}{2a}.$$

When $b^2 - 4ac < 0$, there is *no* real value for $\sqrt{b^2 - 4ac}$ and, hence, no real root.

Because the value of $b^2 - 4ac$ distinguishes the three possibilities, it is called the **discriminant** of the quadratic equation.

In general:

A quadratic equation with real coefficients can have

1. two different real roots $(b^2 - 4ac > 0)$,

2. one real root, which is sometimes called a *double root* or a *repeated root* $(b^2 - 4ac = 0)$, or

3. no real root $(b^2 - 4ac < 0)$.

Oral Exercises

Identify values for a, b, and c to be used in solving the given equation using the quadratic formula.

1. $3p^2 + 5p + 1 = 0$ **2.** $2q^2 - q - 3 = 0$ **3.** $7m - 8m^2 = -1$ **4.** $n^2 = 6n$

5. $2r^2 = 5$ **6.** $s^2 - 4s = -2$ **7.** $5t^2 = 8t + 2$ **8.** $6u^2 = 0$

9. $4x^2 + 25 = 20x$ **10.** $0 = 4y^2 - y$ **11.** $3w^2 = w$ **12.** $20z^2 = -17z$

Written Exercises

A   **1–12.** Use the quadratic formula to solve each equation in Oral Exercises 1–12. Express all radicals in simplified form.

In Exercises 13–21 use the discriminant to determine the number of real roots for the given equation.

EXAMPLE $5x^2 + 10x = -4$

SOLUTION   Writing the equation in standard form, you have

$$5x^2 + 10x + 4 = 0.$$

Hence $a = 5$, $b = 10$, and $c = 4$. Then

$$b^2 - 4ac = (10)^2 - 4(5)(4) = 100 - 80 = 20.$$

Since $20 > 0$, the equation has two real roots.

13. $p^2 - 4p - 3 = 0$ **14.** $q^2 - 6q + 2 = 0$ **15.** $2m^2 - 8m = -3$

16. $0 = 5 + 3n + 2n^2$ **17.** $4r^2 + 4r + 1 = 0$ **18.** $3s^2 + 3 = 6s$

19. $10t^2 - 17t + 3 = 0$ **20.** $2u^2 - 3u + 9 = 0$ **21.** $9x^2 = 7 - 3x$

422 | *Chapter 12*

In Exercises 22–36 use the quadratic formula to solve each equation.
Express all radicals in simplified form.

B **22.** $\dfrac{9}{p^2} - \dfrac{12}{p} + 4 = 0$ **23.** $\dfrac{6q + 10}{q} + \dfrac{3}{q^2} = 0$ **24.** $\dfrac{2}{m - 2} + \dfrac{1}{m + 2} = 3$

25. $\dfrac{3}{n - 1} - \dfrac{1}{n + 1} = 4$ **26.** $\dfrac{1}{r - 4} + \dfrac{1}{r + 4} = 4$ **27.** $\dfrac{1}{s + 2} = \dfrac{s + 5}{s + 6}$

28. $\dfrac{2t}{t - 1} + \dfrac{19}{1 - t^2} = \dfrac{t + 3}{t + 1}$ **29.** $\dfrac{3}{2 - u} + 2 = \dfrac{2}{1 - u}$ **30.** $\dfrac{z}{z - 3} + \dfrac{2z}{z + 3} = \dfrac{1}{z^2 - 9}$

C **31.** $x^2 + 1 = \sqrt{6}x$ **32.** $\frac{1}{2}x^2 + \frac{1}{4}x - \frac{1}{2} = 0$ **33.** $z^2 + \sqrt{3}z - 2 = 0$

34. $(w + 2) - (w - 5)^2 = 4$ **35.** $w^2 - \sqrt{5}w - 2 = 0$ **36.** $(k + 1)^2 = (2k + 1)^2$

37. Find the value of k so that the equation $kx^2 + 4x + 1 = 0$ has a single solution.

38. Find the value of k so that the equation $x^2 - kx + 9 = 0$ has a single solution.

Given the equation $ax^2 + bx + c = 0$ with $a \neq 0$ and $b^2 - 4ac > 0$, prove:

39. The sum of the roots of the equation is $-\dfrac{b}{a}$.

40. The product of the roots of the equation is $\dfrac{c}{a}$.

Problems

A **1.** The sum of a number and its reciprocal is $\frac{13}{6}$. What is the number?

2. The difference of a number and twice its reciprocal is $\frac{2}{3}$. What is the number?

3. A painting is 20 cm wider than it is high. Its area is 2400 cm². Find the width and height of the painting.

4. Gwen and Celeste live 13 km apart. If Gwen drives directly south, and Celeste drives directly west, they will meet. If Celeste will have driven 7 km more than Gwen when they meet, how far will each have driven?

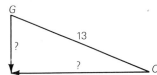

B **5.** Lloyd rowed a boat 4 km up a stream and then returned. If the round trip took him 1 h 20 min and if the current flowed at a rate of 4 km/h, how fast can Lloyd row in still water?

6. Pamela earned $180 in a certain number of days. If her daily wage had been $2 less, she would have taken one more day to earn the same amount. How many days did she work?

7. The average speed of a plane is 680 km/h. With a constant wind blowing, the plane flies 1440 km into the wind and returns. The round trip takes 4 h 15 min. What is the speed of the wind?

8. Alice takes 2 h less time than Rose to paint a room. Working together, they can paint the room in 2.4 h. How long would it take each of them alone to paint the room?

C   9. Bonnie Gurney bought a certain number of shares of stock for $900. As soon as the price of a share increased $5, she sold all but 2 shares for $800. How many shares did she buy?

10. The outside dimensions of a rectangular frame for a painting are 72 cm by 108 cm high. If the area of the painting excluding the frame is 6400 cm², and the frame has a uniform width, how wide is the frame?

Self-Test 1

VOCABULARY   quadratic equation (p. 417)   quadratic formula (p. 421)
                 completing the square (p. 418)   discriminant of a quadratic
                                                          equation (p. 422)

1. Solve $x^2 + 2x - 5 = 0$ over $\mathcal{R}$ by completing the square.   *Obj. 1, p. 417*
2. Solve $2x^2 - 3x + 1 = 0$ over $\mathcal{R}$ using the quadratic formula.   *Obj. 2, p. 417*

Check your answers with those at the back of the book.

---

programming in BASIC

The computer has a special built-in function for finding square roots. To see how it works, try this program:

```
10   PRINT SQR(16), SQR(5)
20   END
```

Exercise

Write a program to find the solution set of a quadratic equation by using the quadratic formula. Be sure to include a special print-out if $b^2 - 4ac < 0$.

Quadratic and Other Polynomial Functions

OBJECTIVES *for Sections 12-3 through 12-5:*
1. *Graph quadratic functions and estimate zeros from the graphs.*
2. *Graph simple polynomial functions and estimate zeros from the graphs.*
3. *Graph quadratic inequalities.*

12-3 Quadratic Functions

Any function of the type

$$f: x \rightarrow ax^2 + bx + c, \quad \text{where } a \neq 0,$$

is a **quadratic function.** If the domain of such a function is $\mathfrak{R}$, then its graph in a coordinate plane is a smooth curve called a **parabola.**

The simplest such function is

$$f: x \rightarrow x^2$$

whose graph is specified by $y = x^2$. Figure 1 shows the curve, together with the table of values used to construct it.

x	y
0	0
1	1
−1	1
2	4
−2	4
3	9
−3	9

Do you see from Figure 1 that the origin is the minimum point (or lowest point) on the graph of $y = x^2$? In general, a point on a curve is said to be a **minimum point** of the curve if its y-coordinate is less than or equal to the y-coordinate of every other point on the curve.

Recall from Chapter 6 that any member of the range of a function is called a *value* of the function. If point (j, k) is a minimum point of the graph of a function, then k is called the **minimum value** of the function. The definitions for a **maximum point** (or highest point) and the **maximum value** are similar.

It can be proved that each quadratic function with domain $\mathfrak{R}$ has one maximum value or one minimum value, but not both, and that the graph of any quadratic function has one maximum point or one minimum point, but not both. Such a point is called the **vertex** of the parabola.

If you imagine that the graph of $y = x^2$ in Figure 1 were folded along the y-axis, then the part of the curve in the first quadrant would coincide with the part in the second quadrant. The mathematical reason is that no matter what point (u, t) you choose on the curve, the point $(-u, t)$ will also be on the curve (for if $t = u^2$, then $t = (-u)^2$ also).

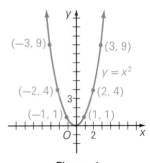

Figure 1

You say that the y-axis is the **axis of symmetry** of the curve, and you say that the curve is **symmetric with respect to the y-axis**. It is obvious that the graph of every quadratic function with domain $\mathscr{R}$ has an axis of symmetry.

The graphs of several other quadratic functions are shown in Figure 2.

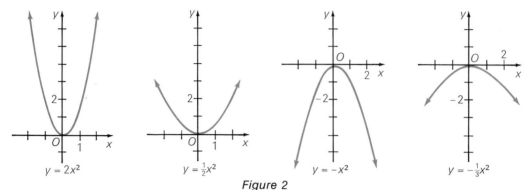

Figure 2

The equations specifying the functions pictured in Figure 2 are of the form

$$y = ax^2,$$

where a is 2, $\frac{1}{2}$, -1, and $-\frac{1}{3}$, respectively. Notice that

when $a > 0$, the parabola opens upward;
when $a < 0$, the parabola opens downward.

Notice also that the less $|a|$ is, the broader the parabola appears to be when the same scale is used. However, if the scale is changed suitably, all parabolas can be made to look alike, as suggested by Figure 3.

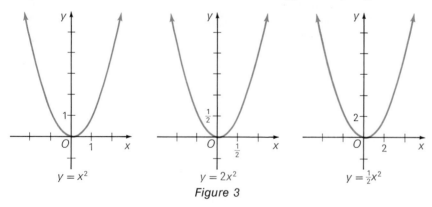

Figure 3

The remarks about a above also apply to the general quadratic function

$$f: x \rightarrow ax^2 + bx + c.$$

Any member of the domain of a function for which the value of the function is 0 is called a **zero of the function.** The zeros of the function $f: x \rightarrow x^2 - 3x - 4$ are 4 and -1, because for these values of x,

$$x^2 - 3x - 4 = (x - 4)(x + 1) = 0.$$

Do you see that the values of x at which the graph of a function intersects the x-axis are the zeros of the function?

EXAMPLE 1   Graph $f: x \rightarrow x^2 - 2x - 3$ and give the real zeros of the function, if any.

SOLUTION   Set $y = x^2 - 2x - 3$ and compare this equation with $y = ax^2 + bx + c$. Since $a = 1$ and $1 > 0$, the parabola opens upward. Make a table of values for x and y.

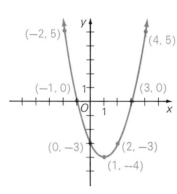

x	y
0	-3
1	-4
-1	0
2	-3
-2	5
3	0
4	5

The function has two zeros, -1 and 3.   **Answer.**

EXAMPLE 2   Graph $f: x \rightarrow 4x - x^2$ and give the real zeros of the function, if any.

SOLUTION   Set $y = 4x - x^2$ and compare this equation with $y = ax^2 + bx + c$. Since $a = -1$ and $-1 < 0$, the parabola opens downward. Make a table of values for x and y.

x	y
0	0
1	3
-1	-5
2	4
3	3
4	0
5	-5

The zeros of the function are 0 and 4.   **Answer.**

Do you see that you can find the zeros of the quadratic function specified by the equation

$$y = ax^2 + bx + c$$

by replacing y with 0 and solving the resulting quadratic equation? Also, if you know two values of x, say x_1 and x_2, which are paired with the same value of y, then

$$x = \frac{x_1 + x_2}{2}$$

is the equation of the axis of symmetry since the graphs of x_1 and x_2 are equidistant from this line.

Thus, in Example 1 above, $f(0) = -3$ and $f(2) = -3$, so that the axis of symmetry has the equation

$$x = \frac{0 + 2}{2}, \text{ or } x = 1.$$

The minimum point on the curve is then

$$(1, f(1)), \text{ or } (1, -4).$$

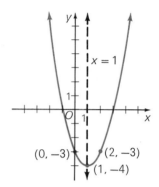

Figure 4

Oral Exercises

Tell whether the graph of the function opens upward or downward.

1. $f: x \rightarrow 3x^2$　　2. $g: x \rightarrow -\frac{3}{4}x^2$　　3. $h: x \rightarrow 2x + x^2$　　4. $f: x \rightarrow 4x - x^2$
5. $g: x \rightarrow (x + 1)^2$　6. $h: x \rightarrow -(1 - x)^2$　7. $f: x \rightarrow 6 + x - x^2$　8. $g: x \rightarrow 2x^2 - 5x + 7$

Estimate the zeros of the functions whose graphs are sketched in Exercises 9–11.

9.
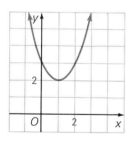

10.
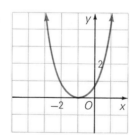

11.
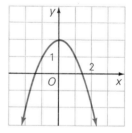

12. Which of these three quadratic functions has the narrowest graph and which has the broadest graph when plotted with the same scale?
   (a) $f: x \rightarrow -\frac{1}{3}x^2$　　　　(b) $g: x \rightarrow \frac{1}{2}x^2$　　　　(c) $h: x \rightarrow 3x^2$

Written Exercises

A　　1–8. Sketch the graph of each function specified in Oral Exercises 1–8.

B 9. Write a definition of a maximum point of a curve. (*Hint:* See the definition of a minimum point on page 425.)

10. Write a definition of a maximum value of a function.

11. Let A be $\{(x, y): x = y^2\}$. Plot the graph of A. Does the curve have a minimum point? a maximum point? Is A a function? Explain.

12. Follow the instructions in Exercise 11 for $A = \{(x, y): x = -y^2\}$.

13. Find, and plot the graph of, a linear function that has no minimum points and no maximum points.

14. Find, and plot the graph of, a linear function that has infinitely many minimum points and infinitely many maximum points. What is the minimum value of the function and what is the maximum value?

15. Find the coordinates of two points on the graph of the equation $y = x^2$ such that the x-coordinates differ by 1 and the y-coordinates differ by at least 1,000,000.

C 16. Find a value of k so that the line with equation $y = \dfrac{1}{1{,}000{,}000}\, x$ intersects the parabola with equation $y = kx^2$ at a point whose y-coordinate is 1.

17. A farmer had 200 m of wire fencing and wished to enclose a rectangular plot of land that would have the greatest possible area. Find the dimensions of the rectangular plot. [*Hint:* If you let x represent one of the dimensions, then the other dimension must be $100 - x$. (Why?) Then the area, $A(x)$, equals $x(100 - x)$. Now find the maximum value of the function specified by $A(x) = x(100 - x)$.]

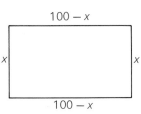

18. If the point $P(u, v)$ is on the graph of $y = ax^2 + bx + c,\ a \neq 0$, prove that the point $Q\left(-\dfrac{b}{a} - u,\ v\right)$ is also on the graph. Use that result to explain why the axis of symmetry passes through the point $\left(-\dfrac{b}{2a},\ 0\right)$.

12-4 Polynomial Functions

Any function of the form $x \to P(x)$, where $P(x)$ represents a polynomial, is called a **polynomial function.** For example, these are polynomial functions:

$$f: x \to 7 \qquad\qquad h: x \to 4 - 2x - 3x^2$$
$$g: x \to 6x - 5 \qquad f: x \to 2x^5 - 4x^2 + 3x - 1$$

Do you see that every constant function, linear function, and quadratic function is a polynomial function?

The graph of each polynomial function with domain $\Re$ is a smooth curve, which may require considerable effort to plot accurately. In more advanced courses you can learn many special plotting techniques. For the present, if you plot enough points, you will be able to make a rough sketch of the curve.

EXAMPLE 1 Sketch the graph of the function $f: x \rightarrow x^3 - x^2 - 6x$ and state the zeros.

SOLUTION Prepare a table of values for x and $f(x)$.

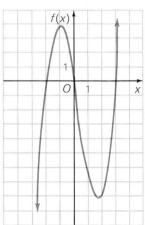

x	f(x)
0	0
1	−6
−1	4
2	−8
−2	0
3	0
−3	−18

The zeros are 0, -2, and 3.

EXAMPLE 2 Make a rough sketch of the graph of the function $f: x \rightarrow -x^3 + 5$ and estimate the zeros, if any.

SOLUTION

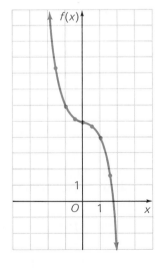

x	f(x)
0	5
$\frac{1}{2}$	$4\frac{7}{8}$
1	4
$1\frac{1}{2}$	$1\frac{5}{8}$
2	−3
3	−22

x	f(x)
$-\frac{1}{2}$	$5\frac{1}{8}$
−1	6
$-1\frac{1}{2}$	$8\frac{3}{8}$
−2	13
−3	32

There is one zero, and it lies between $1\frac{1}{2}$ and 2. Estimate it from the graph as $1\frac{3}{4}$.

Written Exercises

Sketch the graph of the given function and estimate the zeros, if any.

A   **1.** $f: x \to x^3$ **2.** $g: x \to -x^3$ **3.** $h: x \to x^3 - 1$ **4.** $f: x \to 1 - x^3$

     5. $g: x \to x^3 - 2x$ **6.** $h: x \to x^3 + 2x$ **7.** $f: x \to x^3 - x + 1$ **8.** $g: x \to x^3 + x + 1$

     9. $h: x \to x^4$ **10.** $f: x \to -x^4$ **11.** $g: x \to x^4 - x^2$ **12.** $h: x \to x^4 + x$

12-5 Quadratic Inequalities

Can you describe the graph of the inequality

$$y \geq x^2 - 1?$$

You can begin by looking at Figure 5, which pictures the graph of

$$y = x^2 - 1,$$

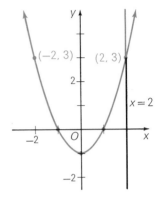

and considering the value of y when $x = 2$. For the point of intersection of the line whose equation is $x = 2$ and the graph, you have

$$y = x^2 - 1 = 2^2 - 1 = 3.$$

Of the remaining points on the line, those lying *above* the graph will have y-coordinates greater than 3. That is, for these points $y > x^2 - 1$. Therefore all points on this line which are on or above the parabola will be on the graph of $y \geq x^2 - 1$. Since a similar argument can be made for *any* value of x, it follows that the graph of $y \geq x^2 - 1$ is just the parabola and the region of the plane *above* the parabola. You can graph the solution set of any quadratic inequality by using similar reasoning.

Figure 5

EXAMPLE   Graph $\{(x, y): y < x^2 - 2x - 3\}$.

SOLUTION   1. First graph

$$y = x^2 - 2x - 3$$

         using a dashed line (because solutions of $y = x^2 - 2x - 3$ are not on the graph).

     2. Shade the region *below* the dashed curve.

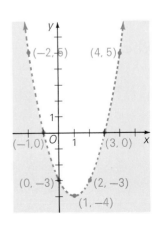

Oral Exercises

State (a) whether the graph of the given inequality will involve a solid or a dashed curve; and (b) whether the graph contains the region above or the region below the curve.

1. $y \leq x^2$ 2. $y > x^2$ 3. $y > x^2 + 4$ 4. $y \leq x^2 - 4$

5. $y > x^2 - 2x$ 6. $y \geq 2x - x^2$ 7. $y < x^2 - 4x$ 8. $y < x^2 - x - 6$

9. $y \geq x^2 - 4x + 4$ 10. $y > 4 - 3x - x^2$ 11. $y > 2 - x - x^2$ 12. $y \leq -4 + 5x - x^2$

Written Exercises

A   **1–12.** Graph each inequality in Oral Exercises 1–12.

Graph each set.

B   13. $\{(x, y): y > x^2\} \cap \{(x, y): y \leq 4\}$

14. $\{(x, y): y \geq x^2\} \cap \{(x, y): x < 1\}$

15. $\{(x, y): y < -x^2\} \cap \{(x, y): y > -2\}$

16. $\{(x, y): y < x^2 + 2\} \cap \{(x, y): y = 0\}$

17. $\{(x, y): y \leq -\frac{1}{2}x^2\} \cap \{(x, y): y \geq 2x - 3\}$

18. $\{(x, y): y \geq x\} \cap \{(x, y): y \leq x^2 + 2\}$

19. $\{(x, y): y < x - 1\} \cap \{(x, y): y \geq -x^2 + 4x - 4\}$

20. $\{(x, y): y \leq x\} \cap \{(x, y): y \geq x^2 - 3\}$

C   21. $\{(x, y): y \geq x^2 - 1\} \cap \{(x, y): y \leq x^2 + 2\}$

22. $\{(x, y): y \leq x^2 + 2\} \cap \{(x, y): y \leq 2 - x^2\}$

23. $\{(x, y): y > x^2 - 3x - 10\} \cap \{(x, y): y < x^2 + 3x - 10\}$

24. $[\{(x, y): y \geq x^2\} \cup \{(x, y): y \leq -x^2\}] \cap \{(x, y): |x| \leq 2\}$

Self-Test 2

VOCABULARY   parabola (p. 425)   zero of a function (p. 427)
                         vertex of a parabola (p. 425)

1. Graph $f: x \rightarrow x^2 + 2x - 3$ and estimate the zeros of the function, if any.   *Obj. 1, p. 425*

2. Sketch the graph of $f: x \rightarrow 3x - x^3$ and estimate the zeros of the function, if any.   *Obj. 2, p. 425*

3. Graph $y \geq x^2 + 1$.   *Obj. 3, p. 425*

Check your answers with those at the back of the book.

programming in BASIC

Up to now when we have used a FOR-NEXT loop, we have used one of the form

$$\text{FOR X} = 1 \text{ TO } 10$$

where X took on the successive values 1, 2, 3, . . . , 10. However, we can let the variable take on values in steps other than 1. For example, we can use a form such as

$$\text{FOR X} = 1 \text{ TO } 9 \text{ STEP } 2$$

where X will take on the values 1, 3, 5, 7, 9. Or we can use a form such as

$$\text{FOR X} = 1 \text{ TO } 5 \text{ STEP } 0.5$$

where X will take on the values 1, 1.5, 2, 2.5, . . . , 5.

Exercises

1. Write a program that will compute six ordered pairs, beginning with $x = 0$, for:
   a. $y = x^2$ (STEP 1)
   b. $y = 2x^2$ (STEP 0.5)
   c. $y = 0.5x^2$ (STEP 2)

2. In each part of Exercise 1, which ordered pairs have equal x- and y-coordinates? In general, for
$$y = ax^2$$
   which ordered pair besides (0, 0) has equal coordinates?

EXTRA FOR EXPERTS
Rational Exponents

When you simplify $2^3 \cdot 2^2$ by writing $2^3 \cdot 2^2 = 2^{3+2} = 2^5$, you apply the law of exponents for products of powers:

$$a^m \cdot a^n = a^{m+n}$$

Can you determine what value n must have if this law is to be true for

$$2^n \cdot 2^n = 2?$$

Since
$$2^n \cdot 2^n = 2^{n+n} = 2^{2n},$$

you have
$$2^{2n} = 2.$$

But if powers of the same base are equal, the exponents of the powers must be equal (provided the base is not $-1, 0,$ or 1). This means, in this case, that $2n = 1$, or $n = \frac{1}{2}$. If

$$2^{\frac{1}{2}} \cdot 2^{\frac{1}{2}} = 2,$$

what meaning can be given to the symbol $2^{\frac{1}{2}}$? Because you know that

$$\sqrt{2} \cdot \sqrt{2} = 2 \quad \text{and} \quad (-\sqrt{2})(-\sqrt{2}) = 2,$$

it makes sense to *define* the symbol $2^{\frac{1}{2}}$ to represent one of the square roots of 2. Selecting the positive square root, we say that

$$2^{\frac{1}{2}} = \sqrt{2}.$$

More generally, we make the definition:

If $a \in \Re$, $n \in N$, and $n \geq 2$, then:

$$a^{\frac{1}{n}} = \sqrt[n]{a}$$

The symbols $a^{\frac{1}{n}}$ and $\sqrt[n]{a}$ represent a real number provided a is a real number and n is an odd natural number, or else a is a nonnegative real number and n is an even natural number. If a is a negative real number and n is an even natural number, $\sqrt[n]{a}$ does not name a real number, and hence neither does $a^{\frac{1}{n}}$.

EXAMPLE 1 Simplify.

     a. $49^{\frac{1}{2}}$ **b.** $81^{\frac{1}{4}}$ **c.** $(-27)^{\frac{1}{3}}$ **d.** $(-81)^{\frac{1}{4}}$

SOLUTION   **a.** $49^{\frac{1}{2}} = \sqrt{49} = 7$ **b.** $81^{\frac{1}{4}} = \sqrt[4]{81} = 3$

     c. $(-27)^{\frac{1}{3}} = \sqrt[3]{-27} = -3$ **d.** $(-81)^{\frac{1}{4}}$ does not name a real number.

Now, consider the expression $(\sqrt{9})^3$. By the definition of a power with a natural-number exponent,

$$(\sqrt{9})^3 = \sqrt{9} \cdot \sqrt{9} \cdot \sqrt{9}$$
$$= 3 \cdot 3 \cdot 3 = 27.$$

Also, if the law of powers of powers, $(a^m)^n = a^{mn}$, is to be true for $(9^{\frac{3}{2}})^2$, you must have

$$(9^{\frac{3}{2}})^2 = 9^3 = 729 = 27^2, \quad \text{or} \quad 9^{\frac{3}{2}} = 27.$$

Thus, a reasonable definition of $9^{\frac{3}{2}}$ would be

$$9^{\frac{3}{2}} = (\sqrt{9})^3.$$

More generally, we have the definition:

> If $a^{\frac{1}{n}} \in \Re$, $m \in J$, and $n \in N$, then:
> $$a^{\frac{m}{n}} = (a^{\frac{1}{n}})^m$$

Of course, if $a^{\frac{1}{n}}$ does not name a real number, this definition gives no meaning to $a^{\frac{m}{n}}$. Thus, $(-9)^{\frac{3}{2}}$ is meaningless in this context.

EXAMPLE 2 Simplify.

       a. $4^{\frac{3}{2}}$ **b.** $(-27)^{-\frac{2}{3}}$ **c.** $(-9)^{\frac{5}{2}}$

SOLUTION **a.** $4^{\frac{3}{2}} = (4^{\frac{1}{2}})^3 = (2)^3 = 8$

        b. $(-27)^{-\frac{2}{3}} = [(-27)^{\frac{1}{3}}]^{-2} = (-3)^{-2} = \dfrac{1}{(-3)^2} = \dfrac{1}{9}$

        c. $(-9)^{\frac{5}{2}} = [(-9)^{\frac{1}{2}}]^5$ is not defined because $(-9)^{\frac{1}{2}}$ is not a real number.

It can be shown that under the definitions given here for $a^{\frac{1}{n}}$ and $a^{\frac{m}{n}}$, all the laws of powers listed on pages 283, 284, and 325 for positive integral exponents may be applied to powers with rational exponents, as can the definition of a negative exponent (page 339).

EXAMPLE 3 Simplify.

       a. $2^{\frac{1}{2}} \cdot 2^{\frac{1}{4}} \cdot 2^{\frac{5}{4}}$ **b.** $\dfrac{9^{\frac{3}{4}}}{9^{\frac{1}{4}}}$ **c.** $[3^{-\frac{2}{3}}]^6$

SOLUTION **a.** $2^{\frac{1}{2}} \cdot 2^{\frac{1}{4}} \cdot 2^{\frac{5}{4}} = 2^{\frac{1}{2}+\frac{1}{4}+\frac{5}{4}} = 2^{\frac{8}{4}} = 2^2 = 4$

        b. $\dfrac{9^{\frac{3}{4}}}{9^{\frac{1}{4}}} = 9^{\frac{3}{4}-\frac{1}{4}} = 9^{\frac{1}{2}} = 3$

        c. $[3^{-\frac{2}{3}}]^6 = 3^{(-\frac{2}{3})(6)} = 3^{-4} = \dfrac{1}{3^4} = \dfrac{1}{81}$

When the restriction $a > 0$ is placed on a, it can also be shown that
$$(a^{\frac{1}{n}})^m = (a^m)^{\frac{1}{n}}$$

for $m \in J$ and $n \in N$. For example, $(16^2)^{\frac{1}{4}}$ and $(16^{\frac{1}{4}})^2$ name the same number, for
$$(16^2)^{\frac{1}{4}} = (256)^{\frac{1}{4}} = 4$$

and
$$(16^{\frac{1}{4}})^2 = (2)^2 = 4.$$

Do you see that the form $(16^{\frac{1}{4}})^2$ is much easier to simplify?

Exercises

Simplify. If the expression does not represent a real number, so state. Assume that the domain of every variable is the set of positive real numbers.

1. $4^{\frac{1}{2}}$

2. $16^{\frac{1}{2}}$

3. $(-8)^{\frac{1}{3}}$

4. $81^{\frac{1}{4}}$

5. $(-4)^{\frac{1}{2}}$

6. $(9)^{\frac{3}{2}}$

7. $(16)^{\frac{3}{4}}$

8. $(8)^{\frac{2}{3}}$

9. $(36)^{-\frac{1}{2}}$

10. $(-8)^{-\frac{1}{3}}$

11. $3^{\frac{1}{2}} \cdot 3^{\frac{3}{2}}$

12. $5^{\frac{8}{3}} \div 5^{\frac{2}{3}}$

13. $(2y^2)^{\frac{1}{2}}(2y^2)^{\frac{3}{2}}$

14. $(27x)^{\frac{1}{3}} \div (27x)^{-\frac{2}{3}}$

15. $[(2x)^{\frac{2}{3}}]^3$

16. $[(81y)^{-\frac{1}{4}}]^2$

Chapter Summary

1. Any equation which can be written in the standard form $ax^2 + bx + c = 0$ (where a, b, and c are real numbers and $a \neq 0$) is called a *quadratic equation*.

2. Any quadratic equation with real roots can be solved by:
   a. *completing the square;*
   b. using the *quadratic formula*, producing the roots
   $$x = \frac{-b + \sqrt{b^2 - 4ac}}{2a} \text{ and } x = \frac{-b - \sqrt{b^2 - 4ac}}{2a}.$$

3. The *discriminant* $b^2 - 4ac$ can be used to determine the number of roots of a quadratic equation. If the discriminant is
   a. positive, there are two real roots;
   b. zero, there is one real root, called a double root;
   c. negative, there is no real root.

4. A function of the form $f: x \rightarrow ax^2 + bx + c$, $a \neq 0$, is a *quadratic function*. The graph of a quadratic function is a *parabola*. Every parabola has a *vertex* (maximum or minimum point) and an *axis of symmetry*.

5. The graph of every polynomial function is a smooth curve.

6. A parabola separates the coordinate plane into two regions, the set of points above the parabola and the set of points below the parabola. These regions are used in graphing quadratic inequalities.

Chapter Review

1. Find the term necessary to complete the square in $x^2 - 8x + \underline{\ ?\ }$.   *12-1*
   a. 8   b. 4   c. 16   d. -16

2. If $(n - 2)^2 = 5$ then $n = \underline{\ ?\ }$.
   a. 27   b. -3 or 7
   c. $2 + \sqrt{5}$ or $-2 + \sqrt{5}$ d. $2 + \sqrt{5}$ or $2 - \sqrt{5}$

3. Use the quadratic formula to solve $c^2 + 2c - 5 = 0$.   12-2

   a. $\{-1 + \sqrt{6}, -1 - \sqrt{6}\}$ **b.** $\{-1 + 2\sqrt{6}, -1 - 2\sqrt{6}\}$

   c. $\{1 + \sqrt{6}, 1 - \sqrt{6}\}$ **d.** no real roots

4. How many real roots are there for the equation $x^2 + 25 = 10x$?

   a. none   **b.** 1   **c.** 2   **d.** 3

5. For which of these functions does the graph open downward?   12-3

   a. $f: x \rightarrow (2 - x)^2$ **b.** $g: x \rightarrow x^2 - 3x$

   c. $h: x \rightarrow -(2x - x^2)$ **d.** $k: x \rightarrow 2x - x^2$

6. Which function is graphed at the right?

   a. $f: x \rightarrow 2 + x^2$ **b.** $g: x \rightarrow 2 - x^2$

   c. $h: x \rightarrow x^2 + 2$ **d.** $k: x \rightarrow -x^2 - 2$

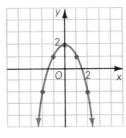

7. Which of these is *not* a zero of the function $f: x \rightarrow x^3 - x$?   12-4

   a. -2 **b.** -1 **c.** 0   **d.** 1

8. Which inequality is graphed at the right?   12-5

   a. $y \geq 3 - x^2$ **b.** $y \leq x^2 - 3$

   c. $y \leq 3 - x^2$ **d.** $y < 3 - x^2$

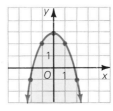

Chapter Test

1. Find a value of k so that $x^2 - 12x + k$ is a perfect square.   12-1
2. Solve $y^2 - 4y + 1 = 0$ using the quadratic formula.   12-2
3. Sketch the graph of $f: x \rightarrow x^2 + 2x - 3$.   12-3
4. Sketch the graph of $f: x \rightarrow 4x - x^3$.   12-4
5. Sketch the graph of $y > 4 - x^2$.   12-5

Cumulative Review (Chapters 9–12)

1. Simplify: $\dfrac{8x^3y^2 - 12x^2y}{4x^2}$

   a. $8x^3y^2 - 3y$ **b.** $\dfrac{4xy^2 - 6y}{2}$ **c.** $2xy^2 - 3y$ **d.** $2xy^2 - 12x^2y$

2. Find the quotient: $a^3 + 27 \div (a + 3)$

    a. $a^2 + 9$ b. $a^2 - 3a + 9$ c. $a^2 + 3a + 9$ d. $a^2 - 9$

3. Reduce to lowest terms: $\dfrac{a^2 + 2a}{a^2 + 5a + 6}$

    a. $\dfrac{a}{a + 3}$ b. $\dfrac{2a}{5a + 6}$ c. $\dfrac{2}{11}$ d. $\dfrac{a^2 + 2a}{a^2 + 5a + 6}$

4. Evaluate: $[(-2)^3(4^{-2})]^{-3}$

    a. 4 b. 8 c. $-\frac{1}{8}$ d. -8

5. Simplify: $\dfrac{5a}{2 - a} \div \dfrac{3a}{a^2 - 4}$

    a. $\dfrac{3}{-5a - 10}$ b. $\dfrac{-5a - 10}{3}$ c. $\dfrac{-5a^2 - 10a}{3a}$ d. $\dfrac{5a + 10}{3}$

6. Simplify: $\dfrac{a}{a - 4} + \dfrac{a + 1}{a + 4}$

    a. $\dfrac{2a + 1}{2a}$ b. $\dfrac{2a + 1}{a^2 - 16}$ c. $\dfrac{2a^2 + a - 4}{a^2 - 16}$ d. $\dfrac{a^2 + 5a + 1}{a^2 - 16}$

7. Find the slope of the line $\dfrac{x}{5} - \dfrac{y}{3} = 2$

    a. $\frac{1}{5}$ b. $\frac{3}{5}$ c. $\frac{5}{3}$ d. $-\frac{1}{3}$

8. $5000 is invested, part at 6% interest and part at 9%, and the total annual interest paid is $390. Which equation expresses these facts correctly?

    a. $0.06x + 0.09(5000 - x) = 390$ b. $6x + 9(5000 - x) = 390$

    c. $0.06x + 0.09x = 390$ d. $\dfrac{x}{0.06} + \dfrac{5000 - x}{0.09} = 390$

9. Solve: $1 + \dfrac{6}{x^2 - 9} = \dfrac{1}{x - 3}$

    a. $\{3, -2\}$ b. $\{4\}$ c. $\{2, -2\}$ d. $\{-2\}$

10. One valve can lower the level in a large chemical tank 1 m in 8 h, and another valve can lower the level 1 m in 6 h. The tank is 50 m deep. Which equation would you solve to find how long it takes the valves working together to empty the tank?

    a. $8x + 6x = 50$ b. $\dfrac{x}{8} + \dfrac{x}{6} = 1$ c. $\dfrac{x}{8} + \dfrac{x}{6} = 50$ d. $\dfrac{8}{x} + \dfrac{6}{x} = 1$

11. If y varies inversely as x, and $y = \frac{1}{2}$ when $x = 7$, find x when $y = 14$.

    a. $\frac{1}{4}$ b. 98 c. 1 d. 2

12. Which term best describes $\sqrt{-4}$?

    a. real and rational   b. integer

    c. real and irrational   d. not real

13. Which statement is true?

    a. $2 < \sqrt{3} < 4$ **b.** $2.24 < \sqrt{5} < 2.25$

    c. $\sqrt{3} + \sqrt{6} > \sqrt{9}$ **d.** $\sqrt{(-4)^2} = -4$

14. Which set of numbers represents the lengths of the sides of a right triangle?

    a. $\{2, 3, 4\}$ **b.** $\{\sqrt{2}, \sqrt{3}, \sqrt{5}\}$

    c. $\{\sqrt{5}, \sqrt{6}, 11\}$ **d.** $\{5, 5, \sqrt{10}\}$

15. Simplify: $\sqrt{72a^4b^3}$

    a. $3a^2\sqrt{8b^3}$ **b.** $6ab\sqrt{2a^2b}$ **c.** $6a^2\sqrt{2b^3}$ **d.** $6a^2b\sqrt{2b}$

16. Simplify: $\dfrac{\sqrt{6}}{4 + \sqrt{3}}$

    a. $\dfrac{4\sqrt{6} - 3\sqrt{2}}{13}$ **b.** $\dfrac{\sqrt{2}}{4}$ **c.** $\dfrac{4\sqrt{6} - \sqrt{18}}{13}$ **d.** $\dfrac{4\sqrt{6} - 3\sqrt{2}}{7}$

17. For which values of p and q is the expression $x^2 + px + q$ *not* a perfect square?

    a. $p = 8$, $q = 16$ **b.** $p = -5$, $q = 6\frac{1}{4}$

    c. $p = 4$, $q = 8$ **d.** $p = -2$, $q = 1$

18. How many real roots are there for the equation $2x^2 - 5x = -3$?

    a. none   **b.** 1   **c.** 2

19. Solve: $2x^2 - 4 = 4x$

    a. $\{-1 + \sqrt{3}, -1 - \sqrt{3}\}$ **b.** $\{1 + \sqrt{3}, 1 - \sqrt{3}\}$

    c. $\{0, 2\}$ **d.** $\{2 + 2\sqrt{3}, 2 - 2\sqrt{3}\}$

20. Which statement about the function $f\colon x \to 2 - (x + 1)^2$ is *not* true?

    a. the graph has a maximum point   **b.** $f(-3) = f(1)$

    c. the vertex is $(-1, 2)$ **d.** the graph opens upward

21. The graph of which inequality is shown here?

    a. $y \le x^2 - 9$

    b. $y \ge x^2 - 9$

    c. $y \le 9 - x^2$

    d. $y \ge 9 - x^2$

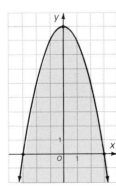

This photograph of the San Francisco harbor is a product of aerial surveys carried out by the NASA Earth Resources Aircraft Program. From it, areas and heights of buildings and objects can be determined with great accuracy.

13 Trigonometry and Vectors

Trigonometry

OBJECTIVES *for Sections 13-1 through 13-4:*
1. *Picture an angle as a rotation.*
2. *Determine the sine, cosine, and tangent of an angle in standard position.*
3. *Use trigonometric tables to find the sine, cosine, and tangent of an angle.*
4. *Solve right triangles.*

13-1 Angles

When an airport beacon revolves (Figure 1), you can think of its beam of light as generating an *angle.* For example, since the measure of a right angle is ninety degrees (90°), when the beacon turns through one complete revolution, the beam generates an angle of measure 360°. In two revolutions, it generates an angle of measure 720°, and so on.

In geometry an angle is defined as the union of two rays which are not in the same line but which have the same endpoint. To develop the extended notion of an angle described above, you can call one ray of an angle the *initial*

Figure 1

side (at the starting position of the generating ray) and the other ray the *terminal side* (at the final position after rotation), as shown in part (a) of Figure 2. A **(directed) angle** is then defined as the union of two ordered rays with a common endpoint, called the **vertex** of the angle, *together with a rotation* from the initial side to the terminal side. **Coterminal angles,** or angles with the same initial side and the same terminal side, are shown in part (b) of Figure 2.

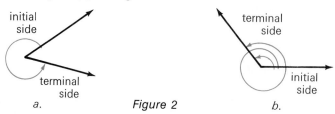

a.   Figure 2   b.

The number of degrees through which a ray rotates in turning from the initial side to the terminal side is the **measure** of the angle. Ordinarily, the measure is considered to be *positive* if the rotation is *counterclockwise,* and *negative* if the rotation is *clockwise.*

In Figure 3 you can use a protractor to verify that $m° \angle AOB$ (read "the degree measure of angle A, O, B") is approximately $145°$, and that $m° \angle COD$ is approximately $-30°$.

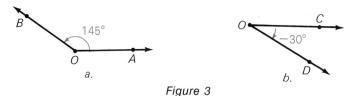

a.   b.

Figure 3

In a coordinate plane, an angle that has its vertex at the origin and has the positive x-axis as its initial side is said to be in **standard position.** In Figure 4, $\angle A$ and $\angle B$ are in standard position.

Recall from Section 6-1 how you number the four quadrants into which the coordinate axes separate the rest of the plane. The terminal side of an angle in standard position determines the quadrant in which the angle is said to *be,* or to *lie.* For example, $\angle A$ in Figure 4 lies in the fourth quadrant (or Quadrant IV). Do you agree that an angle in standard position having a measure of $200°$ lies in Quadrant III?

An angle in standard position whose terminal side coincides with an axis, such as $\angle B$ in Figure 4, is called a **quadrantal angle.** Angles in standard position having measures of $0°, 90°, 180°, 270°, 360°, -90°, -180°,$ and so on, are quadrantal angles. A quadrantal angle does *not* lie in any quadrant.

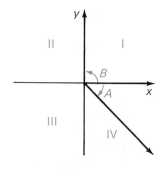

Figure 4

Oral Exercises

In Exercises 1–6 state $m°\angle A$.

1.

2.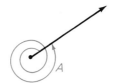

3.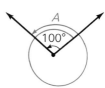

4.

5.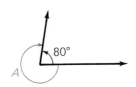

6.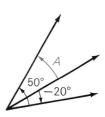

In Exercises 7–16 state $m°\angle A$ if $\angle A$ is coterminal with the angle whose measure is given, with $0° \le m°\angle A < 360°$.

7. $370°$ **8.** $-15°$ **9.** $-180°$ **10.** $540°$ **11.** $450°$

12. $-450°$ **13.** $-160°$ **14.** $-250°$ **15.** $-90°$ **16.** $720°$

17–26. State the quadrant, if any, in which an angle in standard position with the given measure lies in Exercises 7–16 above.

In Exercises 27–29 state the measure of the quadrantal angle A that satisfies the given conditions.

27. $100° < m°\angle A < 200°$ **28.** $-60° < m°\angle A < 60°$ **29.** $-120° < m°\angle A < -80°$

Written Exercises

In Exercises 1–6 determine $m°\angle A$.

A

1.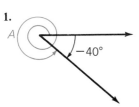

2.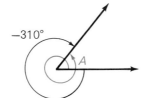

3.

4.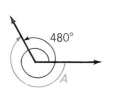

5.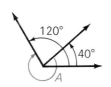

6.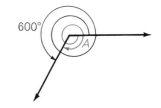

In Exercises 7–14 assume that $\angle A$ is in standard position and that $0° < m° \angle A < 360°$.

7. If $m° \angle A < 180°$ and $\angle A$ is not a quadrantal angle, then $\angle A$ is in Quadrant ? or ?.

8. If $270° < m° \angle A < 360°$, then $\angle A$ is in Quadrant ?.

9. If $\angle A$ is in Quadrant II, then $? < m° \angle A < ?$.

10. If $\angle A$ is in Quadrant III, then $? < m° \angle A < ?$.

11. If $\angle A$ is a quadrantal angle and $0° < m° \angle A < 180°$, then $m° \angle A = ?$.

12. If $\angle A$ is a quadrantal angle and $180° < m° \angle A < 270°$, then $m° \angle A = ?$.

13. If $\angle A$ is coterminal with an angle measuring $-15°$, then $m° \angle A = ?$.

14. If $\angle A$ is coterminal with an angle measuring $400°$, then $m° \angle A = ?$.

B   15. If $\angle A$ is coterminal with an angle measuring $-630°$, and if $500° < m° \angle A < 900°$, then $m° \angle A = ?$.

16. If $\angle A$ is coterminal with an angle measuring $840°$, and if $-300° < m° \angle A < -100°$, then $m° \angle A = ?$.

In Exercises 17–22 assume that all angles are in standard position.

17. If $90° < |m° \angle A| < 180°$, then $\angle A$ is in Quadrant ? or ?.

18. If $270° < |m° \angle B| < 360°$, then $\angle B$ is in Quadrant ? or ?.

19. If $|m° \angle C| < 90°$ and $\angle C$ is not in Quadrant I or IV, then $m° \angle C = ?$.

20. If $|90° - m° \angle A| < 90°$ and $\angle A$ is not in Quadrant I or II, then $m° \angle A = ?$.

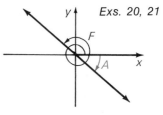

Exs. 20, 21

21. If the terminal sides of $\angle A$ and $\angle F$ are collinear, $m° \angle F = 490°$, and $|m° \angle A| < 90°$, then $m° \angle A = ?$.

22. If the terminal sides of $\angle B$ and $\angle G$ are collinear, $m° \angle B = -300°$, and $90° < |m° \angle G| < 180°$, then $m° \angle G = ?$.

23. A wagon wheel makes 20 revolutions per minute. Through how many degrees does a spoke of the wheel turn in 5 s?

24. The beam of a radarscope sweeps through an angle of $72°$ each second. How many revolutions does it make in 5 min?

25. The beam of a radarscope rotates counterclockwise at a constant rate. If the beam makes one rotation in 5 s, through how many degrees does it turn in 3.5 s?

26. A revolving aircraft beacon turns through $15°$ in 0.5 s. How long does it take the beacon to complete 4 revolutions?

13-2 Trigonometric Functions

A circle with unit radius (radius 1) and center at the origin such as is shown in Figures 5 and 6, below, is called the **unit circle.** It follows from the distance formula derived in Section 11-5 that a point (x, y) is on the unit circle if and only if $\sqrt{x^2 + y^2} = 1$. Hence an equation of the unit circle is $x^2 + y^2 = 1$.

You can see from Figure 5 that the terminal side of each angle in standard position intersects the unit circle in exactly one point. For example, the terminal side of $\angle AOB$ intersects the circle in point B, and the terminal side of $\angle AOC$ intersects it in point C.

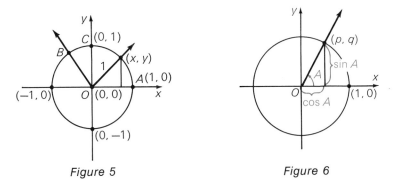

Figure 5   Figure 6

Special names are given to the ordinate and abscissa of the intersection point for each angle, and also to the quotient of the ordinate and the abscissa. Namely, if (p, q) is the point of intersection of the unit circle and the terminal side of an angle A in standard position (Figure 6), then

$$\sin A = q \text{ (read "the sine of } A \text{ is (equal to) } q\text{"),}$$
$$\cos A = p \text{ (read "the cosine of } A \text{ is (equal to) } p\text{"),}$$

and if $p \neq 0$,

$$\tan A = \frac{q}{p} \text{ (read "the tangent of } A \text{ is (equal to) } q \text{ divided by } p\text{").}$$

These definitions tell you what is meant by the sine, cosine, and tangent of an angle in *standard position.* But any angle can be put into standard position by choosing the coordinate system so that the origin is the vertex of the angle and the positive x-axis is the initial side of the angle. Therefore, you can speak of the sine, cosine, and tangent of an angle whether or not it is given in standard position.

Although the symbol $\angle$ is omitted in the notation "sin A," the letter A denotes an angle, not the measure of an angle. It is customary, however, to use notation such as "tan 45°" to mean "the tangent of an angle of measure 45°." You should realize that the terminal side of an angle is determined when its initial side and its measure are known.

The definitions of the numbers sin A, cos A, and tan A suggest three new functions:

$$\text{sine:} \quad \angle A \rightarrow \sin A$$
$$\text{cosine:} \quad \angle A \rightarrow \cos A$$
$$\text{tangent:} \quad \angle A \rightarrow \tan A$$

Since the numbers sin A and cos A are defined for every angle A, the sine and cosine functions have the same domain: the set of all angles. Further, since sin A and cos A are coordinates of points on the unit circle, these functions also have the same range: the set of all real numbers between -1 and 1, inclusive.

Notice that, for an angle A in standard position, tan A is not defined if the terminal side is on the y-axis, since division by zero is not defined. Therefore the domain of the tangent function is the set of all angles whose degree measure is not an odd-numbered multiple of 90. The range of tangent is the set of all real numbers.

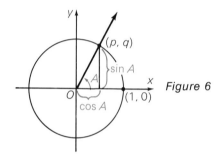

Figure 6

Sine, cosine, and tangent are examples of *trigonometric functions*. The word *trigonometric* comes from two Greek words: *trigonon* (triangle) and *metron* (measure).

Here are two very useful trigonometric formulas:

1. For each point (p, q) on the unit circle,

$$q^2 + p^2 = 1.$$

But (Figure 6) $q = \sin A$ and $p = \cos A$, and therefore

$$(\sin A)^2 + (\cos A)^2 = 1.$$

In place of the symbols $(\sin A)^2$ and $(\cos A)^2$, you ordinarily write $\sin^2 A$ and $\cos^2 A$; thus:

$$\sin^2 A + \cos^2 A = 1.$$

2. If $\cos A \neq 0$, then

$$\tan A = \frac{\sin A}{\cos A},$$

because $\tan A = \dfrac{q}{p}$, and $q = \sin A$ and $p = \cos A$.

Oral Exercises

In the figure below, state the sine, cosine, and tangent (if it exists) of the angle whose terminal side contains point:

1. A
2. B
3. C
4. D
5. E
6. F
7. G
8. H

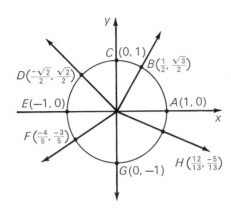

Use the figure above to state the value (if it exists) of each of the following.

9. $\sin 0°$
10. $\cos 0°$
11. $\sin 90°$
12. $\cos 90°$
13. $\tan 0°$
14. $\tan 90°$
15. $\cos 180°$
16. $\sin 270°$
17. $\cos 360°$
18. $\sin 180°$
19. $\cos 450°$
20. $\tan 180°$

Explain why there is no angle A for which:

21. $\sin A = 2$
22. $\cos A = 3$
23. $\sin A = 1$ and $\cos A = -1$
24. $0 < \tan A < \sin A$
25. $\cos A = 0$ and $\sin A = 0$
26. $\sin A < 0$, $\cos A > 0$, and $\tan A > 0$

Written Exercises

Give the sine, cosine, and tangent of an angle A in standard position whose terminal side contains the given point on the unit circle.

A
1. $\left(\frac{3}{5}, \frac{4}{5}\right)$
2. $\left(\frac{4}{5}, \frac{3}{5}\right)$
3. $\left(-\frac{4}{5}, \frac{3}{5}\right)$

4. $\left(\frac{3}{5}, -\frac{4}{5}\right)$
5. $\left(\frac{5}{13}, \frac{12}{13}\right)$
6. $\left(-\frac{5}{13}, \frac{12}{13}\right)$

7. $\left(\frac{\sqrt{2}}{2}, -\frac{\sqrt{2}}{2}\right)$
8. $\left(-\frac{\sqrt{2}}{2}, -\frac{\sqrt{2}}{2}\right)$
9. $\left(\frac{\sqrt{2}}{2}, \frac{\sqrt{2}}{2}\right)$

10. $\left(-\frac{\sqrt{3}}{2}, -\frac{1}{2}\right)$
11. $\left(\frac{\sqrt{3}}{2}, \frac{1}{2}\right)$
12. $\left(\frac{1}{2}, -\frac{\sqrt{3}}{2}\right)$

In Exercises 13–20 the quadrant of angle A is given, and also the value of $\sin A$, $\cos A$, or $\tan A$. Find the remaining two values, using $\sin^2 A + \cos^2 A = 1$ and $\tan A = \dfrac{\sin A}{\cos A}$ as needed.

EXAMPLE $\sin A = -\dfrac{\sqrt{2}}{2}$; Quadrant III

SOLUTION Substitute $-\dfrac{\sqrt{2}}{2}$ for $\sin A$ in

$$\sin^2 A + \cos^2 A = 1.$$

$$\left(-\frac{\sqrt{2}}{2}\right)^2 + \cos^2 A = 1$$

$$\cos^2 A = 1 - \frac{2}{4} = \frac{1}{2}$$

$$\cos A = \frac{\sqrt{2}}{2} \text{ or } \cos A = -\frac{\sqrt{2}}{2}.$$

Since A is in Quadrant III, $\cos A$ is negative.

Therefore, $\cos A = -\dfrac{\sqrt{2}}{2}.$

Then $\tan A = \dfrac{\sin A}{\cos A} = \dfrac{-\dfrac{\sqrt{2}}{2}}{-\dfrac{\sqrt{2}}{2}} = 1.$

Therefore, $\cos A = -\dfrac{\sqrt{2}}{2}$ and $\tan A = 1$.   Answer.

13. $\cos A = \dfrac{\sqrt{3}}{2}$; Quadrant I

14. $\sin A = \dfrac{3}{5}$; Quadrant I

15. $\sin A = -\dfrac{\sqrt{2}}{2}$; Quadrant IV

16. $\cos A = -\dfrac{4}{5}$; Quadrant II

17. $\tan A = 1$; Quadrant III

18. $\tan A = -\sqrt{3}$; Quadrant IV

19. $\cos A = -\dfrac{12}{13}$; Quadrant II

20. $\sin A = -\dfrac{5}{13}$; Quadrant III

Find $\sin A$, $\cos A$, and $\tan A$ if:

B   **21.** $\sin A = \cos A$ and $\sin A < 0$

22. $\sin A = \tan A$ and $\cos A > 0$

In each of Exercises 23–26 k is a positive real number. Find the sine, cosine, and tangent of the angle A in standard position whose terminal side contains the given point.

23. $(3k, 4k)$ **24.** $(5k, 12k)$ **25.** $(k, k\sqrt{3})$ **26.** $(k\sqrt{2}, k\sqrt{2})$

Find the value of k if the terminal side of the acute angle A in standard position contains the point whose coordinates are given.

27. $(k, 4)$, if $\tan A = 3$

28. $(3, k)$, if $\tan A = \frac{1}{2}$

29. $(6, 3k)$, if $\tan A = -1$

30. $(6, -2k)$, if $\tan A = -\frac{5}{12}$

13-3 Trigonometric Tables

You can use the table on page 504 to find approximations for sin A, cos A, and tan A for angles *in the first quadrant* when you know the measure of angle A in degrees. A part of that table is shown below.

$m° \angle A$	$\sin A$	$\cos A$	$\tan A$
16	0.2756	0.9613	0.2867
17	0.2924	0.9563	0.3057
18	0.3090	0.9511	0.3249

For example, to find an approximation for tan 17° from this table, you use the entry where the *row* containing **17** intersects the *column* headed **tan A**. That entry is shown here in color and is 0.3057. The word "approximation" is used in describing the entries in this table because, except for sin 30°, tan 45°, cos 60°, sin 90°, and cos 90°, the entries represent rational-number approximations to irrational numbers.

If you are given a four-digit numeral for a value of sin A, cos A, or tan A, you can reverse the procedure described above to find an approximation for $m° \angle A$. For example, if you are given cos $A = 0.9511$, you locate 0.9511 in the column headed **cos A** and read $m° \angle A$ from the left-hand column in the same (horizontal) row. You find $m° \angle A = 18°$.

If a given value for sin A, cos A, or tan A is not an entry in the table, you can use the nearest table entry. Thus, if you are given sin $A = 0.2806$, you find that 0.2806 is not an entry in the **sin A** column, but that 0.2756 and 0.2924 are. Because 0.2806 is closer to 0.2756 than it is to 0.2924, you use 0.2756. Therefore, if sin $A = 0.2806$, you can say that $m° \angle A \approx 16°$, where the approximation is to the nearest degree.

Oral Exercises

Use the table on page 504. In Exercises 1–8 state the approximation given in the table for the given item. If the value is exact, state this.

1. cos 40°

2. cos 60°

3. sin 55°

4. sin 90°

5. tan 25°

6. tan 45°

7. sin 30°

8. tan 60°

In Exercises 9–17 state the measure of angle A, $1° \leq m° \angle A \leq 90°$, to the nearest degree.

9. $\sin A = 0.1736$

10. $\cos A = 0.5\bar{0}$

11. $\sin A = 0.5\bar{0}$

12. $\tan A = 0.6009$

13. $\cos A = 0.1219$

14. $\tan A = 2.2500$

15. $\cos A = 0.6500$

16. $\sin A = 0.9826$

17. $\tan A = 25.6127$

In Exercises 18–25 let A denote an angle for which $1° \leq m° \angle A \leq 90°$.

18. For what value of $m° \angle A$ is $\sin A = \cos A$?

19. $\sin 20° = \cos A$

20. $\cos 60° = \sin A$

21. $\sin 65° = \cos A$

22. $\cos 80° = \sin A$

23. The greater the degree measure of $\angle A$ is, the ? (greater/lesser) the value of $\sin A$ is.

24. The greater the degree measure of $\angle A$ is, the ? (greater/lesser) the value of $\cos A$ is.

25. The greater the degree measure of $\angle A$ is, the ? (greater/lesser) the value of $\tan A$ is.

Written Exercises

In Exercises 1–12 find the approximation in the table on page 504 for the given item.

A

1. $\cos 53°$

2. $\sin 28°$

3. $\tan 9°$

4. $\sin 39°$

5. $\cos 11°$

6. $\tan 89°$

7. $\tan 41°$

8. $\cos 27°$

9. $\sin 35°$

10. $\cos 25°$

11. $\sin 75°$

12. $\tan 1°$

In Exercises 13–21 find the measure of angle A, $1° \leq m° \angle A \leq 90°$, to the nearest degree.

13. $\cos A = 0.7193$

14. $\sin A = 0.8910$

15. $\tan A = 0.5317$

16. $\tan A = 12$

17. $\cos A = 0.9500$

18. $\sin A = 0.9500$

19. $\tan A = 42.9431$

20. $\sin A = 0.2250$

21. $\cos A = 0.1250$

13-4 Solving Triangles

Suppose that you know the coordinates (x, y), $x \neq 0$, of some point B (other than the origin) on the terminal side of an angle A in standard position. Can you find $\sin A$, $\cos A$, and $\tan A$?

Figure 7 pictures an angle A in standard position. It also pictures points $(p, q), p \neq 0$, and $B(x, y), x \neq 0$, on the terminal side of A; (p, q) is on the unit circle, and $B(x, y)$ is not on the unit circle.

You learned in Section 6-9 that the slope of a line is constant. The slope of line OB is given by $\dfrac{q}{p}$ and also by $\dfrac{y}{x}$. There-fore,

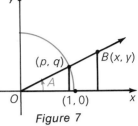

Figure 7

$$\frac{q}{p} = \frac{y}{x}.$$

But by definition of tangent, $\tan A = \dfrac{q}{p}$, and so

$$\tan A = \frac{y}{x}.$$

Since $\dfrac{q}{p} = \dfrac{y}{x}$, you have $q = \dfrac{y}{x}p$, and since $\sin A = q$ and $\cos A = p$, you have

$$\sin A = \frac{y}{x}\cos A.$$

Replacing $\sin A$ with $\dfrac{y}{x}\cos A$ in $\sin^2 A + \cos^2 A = 1$, you have:

$$\left(\frac{y}{x}\right)^2 \cos^2 A + \cos^2 A = 1$$

$$\left(\frac{y^2}{x^2} + 1\right)\cos^2 A = 1$$

$$\cos^2 A = \frac{1}{\dfrac{x^2 + y^2}{x^2}} = \frac{x^2}{x^2 + y^2}$$

Therefore:

$$\cos A = \frac{x}{\sqrt{x^2 + y^2}} \quad \text{or} \quad \cos A = \frac{-x}{\sqrt{x^2 + y^2}}$$

Since $\sqrt{x^2 + y^2} > 0$ and, for $\angle A$ in any quadrant, $\cos A$ and x are both positive or both negative, only the first of these equations applies.

$$\therefore \cos A = \frac{x}{\sqrt{x^2 + y^2}}$$

Since $\sin A = \dfrac{y}{x}\cos A$, you also have $\sin A = \dfrac{y}{x}\left(\dfrac{x}{\sqrt{x^2 + y^2}}\right)$, that is:

$$\sin A = \frac{y}{\sqrt{x^2 + y^2}}$$

For the case $x = 0$, you know that the terminal side of the angle is on the y-axis, and the above formulas give $\sin A = 1$ and $\cos A = 0$ if $y > 0$, and $\sin A = -1$ and $\cos A = 0$ if $y < 0$. These are also the values resulting from the definitions of sine and cosine on page 445.

Thus the formulas for $\sin A$, $\cos A$, and $\tan A$ shown on page 451 hold in every case, except that $\tan A$ is not defined if $x = 0$.

Notice that $\sqrt{x^2 + y^2}$ is the distance from the origin to point $B(x, y)$, because $\overline{OB}$ is the hypotenuse of a right triangle having remaining sides of lengths $|x|$ and $|y|$. (See Figure 8.)

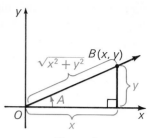

Figure 8

EXAMPLE 1 Find $\sin A$, $\cos A$, and $\tan A$ for an angle A in standard position whose terminal side contains the point $(3, -2)$.

SOLUTION Since $x = 3$ and $y = -2$, you have:

$$\sin A = \frac{y}{\sqrt{x^2 + y^2}} = \frac{-2}{\sqrt{3^2 + (-2)^2}}$$

$$= \frac{-2}{\sqrt{9 + 4}} = \frac{-2}{\sqrt{13}}.$$

$$\cos A = \frac{x}{\sqrt{x^2 + y^2}} = \frac{3}{\sqrt{13}}.$$

$$\tan A = \frac{y}{x} = \frac{-2}{3}.$$

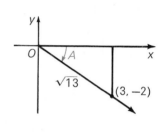

Any right triangle ACB in which $\angle C$ is the right angle can be placed so that $\angle A$ is in standard position, $\overline{AC}$ lies along the x-axis, and the point B is in the first quadrant. By the formulas we have just found, you have:

$$\tan A = \frac{a}{b} = \frac{\text{length of side opposite } \angle A}{\text{length of side adjacent to } \angle A}$$

$$\sin A = \frac{a}{\sqrt{a^2 + b^2}} = \frac{a}{c} = \frac{\text{length of side opposite } \angle A}{\text{length of hypotenuse}}$$

$$\cos A = \frac{b}{\sqrt{a^2 + b^2}} = \frac{b}{c} = \frac{\text{length of side adjacent to } \angle A}{\text{length of hypotenuse}}$$

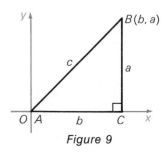

Figure 9

Finding measures (or approximations to measures) of various parts (angles and sides) of a triangle when measures of other parts are given is called *solving the triangle*. It is customary in labeling triangles to use capital letters for vertices and the corresponding lowercase letters for the lengths of the sides opposite the vertices. (See Figure 10.) The capital letters are also used to denote the angles of the triangle. When A, B, and C and a, b, and c are used in this way, the right angle is ordinarily labeled C, and the length of the hypotenuse is represented by c.

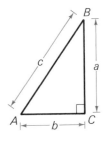

Figure 10

EXAMPLE 2   Solve the right triangle ABC given that $m° \angle A = 34°$ and $b = 5$. Give angle measures to the nearest degree and lengths to the nearest tenth of a unit.

SOLUTION   Make a sketch and label it. To find $m° \angle B$, notice that

$$m° \angle A + m° \angle B = 90°.$$

Therefore:

$$34° + m° \angle B = 90°$$
$$m° \angle B = 90° - 34° = 56°$$

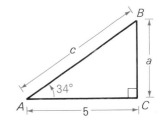

To find a, use the fact that $\tan A = \dfrac{a}{b}$, or $a = b \tan A$. Then

$$a = 5 \tan 34°,$$

and from the table on page 504 you find that $\tan 34° \approx 0.6745$; so

$$a \approx 5 \times 0.6745 \approx 3.4.$$

To find c, you can use

$$\frac{b}{c} = \sin B \quad \text{or} \quad c = \frac{b}{\sin B} = \frac{5}{\sin 56°}.$$

From the table, $\sin 56° \approx 0.8290$; so

$$c \approx \frac{5}{0.8290} \approx 6.0.$$

As a check in terms of the Pythagorean Theorem, you might verify that the values a, b, and c satisfy $a^2 + b^2 = c^2$. (*Note:* Because a, b, and c are rounded off, they are only approximations. Thus, $a^2 + b^2$ will only approximate c^2.)

Therefore, $m° \angle B = 56°$, $a \approx 3.4$, $c \approx 6.0$.   Answer.

Trigonometry and Vectors | **453**

Oral Exercises

In Exercises 1–10 state the value of the given item.

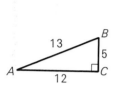

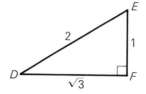

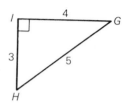

1. $\sin A$ **2.** $\cos B$ **3.** $\tan E$ **4.** $\cos D$ **5.** $\sin G$

6. $\cos H$ **7.** $\tan G$ **8.** $\sin D$ **9.** $\tan B$ **10.** $\sin H$

11. In triangle *MNP* below, which would be a more convenient way to find an approximation for *x*, using tan 35° or using tan 55°? Why?

12. In triangle *XYZ* below, which would be a more convenient way to find an approximation for *x*, using tan 20° or using tan 70°? Why?

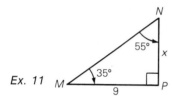

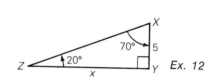

Ex. 11

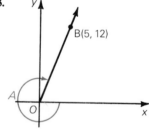

Ex. 12

Written Exercises

In Exercises 1–6 <u>find</u>:
a. the length of $\overline{OB}$;   b. $\sin A$;   c. $\cos A$;   d. $\tan A$.

A   **1.**

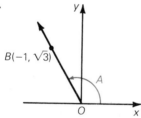

2.

3.

4.

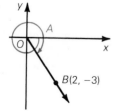

5.

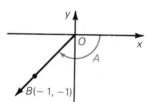

6.

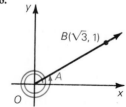

In Exercises 7–12 find sin *A*, cos *A*, and tan *A* for an angle *A* in standard position whose terminal side contains the given point.

7. $(3, 3)$ **8.** $(3, 1)$ **9.** $(-5, 3)$ **10.** $(5, -12)$ **11.** $(-\sqrt{2}, \sqrt{2})$ **12.** $(0, -2)$

In Exercises 13–20 find an approximation for the length *x* or the measure of ∠*A*. Use the table on page 504 as necessary. Give angle measures to the nearest degree and lengths to the nearest tenth of a unit.

13.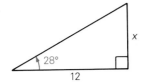

14.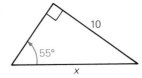

15.

16.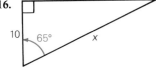

17.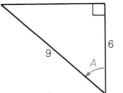

18.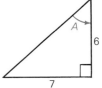

19.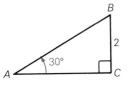

20.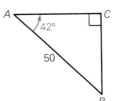

In Exercises 21–26 solve the right triangle. Use the table on page 504 as necessary. Give angle measures to the nearest degree and lengths to the nearest tenth of a unit.

B **21.**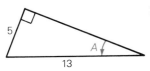

22.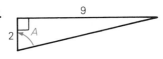

23.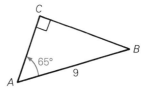

24.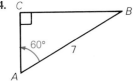

25.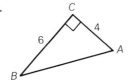

26.

In Exercises 27–31 copy the figure, and by drawing an additional line segment introduce some right triangles to help you find a value for x to the nearest tenth of a unit. (*Hint for Exercise 27:* Copy the figure and draw a segment from D perpendicular to $\overline{AB}$. Determine z, then y, and note that $x = y + z$.)

C **27.**

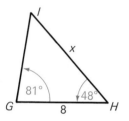

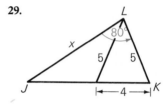

28.

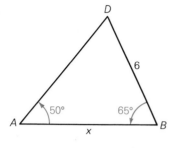

29.

30.

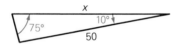

31.

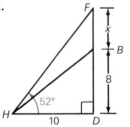

Problems

Solve each problem using the table on page 504 as necessary. Give angle measures to the nearest degree and lengths to the nearest tenth of a unit.

A **1.** After takeoff, an airplane maintained a flight angle of 8° with the ground as shown at the right. Find its elevation x after it had covered a ground distance of 500 m.

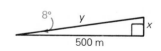

2. For the airplane in Problem 1, find the distance y it traveled in the air along the flight path while covering the ground distance of 500 m.

3. If an airplane attains an elevation of 110 m after take-off while covering a ground distance of 500 m, what is the degree measure of its flight angle A with the ground?

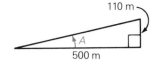

4. Diving at a constant angle A, as shown at the right, a submarine descends 92 m while traveling 300 m. Find the degree measure of A.

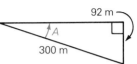

5. A tree is 40 m tall. When the angle of elevation of the sun is 64°, how long is its shadow?

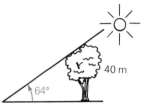

6. A submarine maintains a diving angle of 22°, as shown at the right. How far has it traveled when it is directly under a point 400 m along the surface from the point where it submerged?

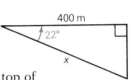

B 7. At a point 60 m from a building, the angle of elevation of the top of the building is 32° and the angle of the top of a television antenna at the edge of the building is 34°, as shown below. What is the height x of the antenna?

8. At one point along a straight road the direction toward Mount Krasha makes an angle of 33° with the direction of the road. At another point 16 km farther along the road, the angle is 35°. Find the perpendicular distance x of Mount Krasha from the road.

9. Two surveying transits are located 200 m apart. Both transits are sighted on the same rock. For each transit, the angle between the line of sight to the rock and the line of sight to the other transit is 45°. What is the distance x from the rock to the line connecting the two transits?

Ex. 7

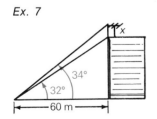

Ex. 8

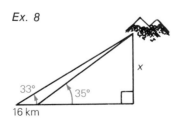

Ex. 9

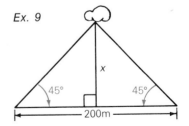

10. If the transits described in Problem 9 are 100 m apart, and if each angle described in Problem 9 measures 60°, how far is the rock from the line connecting the transits?

Trigonometry and Vectors | **457**

C 11. A radar set at a point A sights a UFO at point B and tracks it to a point C along a straight and level path $\overline{BC}$. The distance from A to B is 12 km and the distance from A to C is 14 km. If $m°\angle BAC$ is $5°$, what is the distance from B to C? How fast was the UFO traveling (in km/h) if it took 10 s to move from B to C? (*Hint:* Use the dashed segment in the diagram.)

12. On a coordinate plane, line t passes through the origin and forms a $45°$ angle with the positive x-axis. Line m passes through the point $(3, 0)$ and forms a $70°$ angle with the positive x-axis. Find the coordinates of the point of intersection of lines t and m.

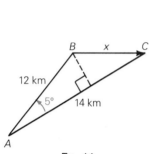

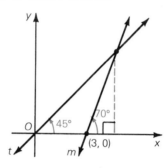

Ex. 11   *Ex. 12*

Self-Test 1

VOCABULARY   standard position of an angle (p. 442)
sine of an angle (p. 445)
cosine of an angle (p. 445)

coterminal angles (p. 442)
tangent of an angle (p. 445)
solving a triangle (p. 453)

1. Find the measure of $\angle A$, if $\angle A$ is coterminal with the given   *Obj. 1, p. 441*
angle and $0° \le m°\angle A < 360°$:
   a. $500°$ **b.** $-135°$ **c.** $1000°$

2. If $\angle A$ is in standard position, and the terminal side of $\angle A$ con-   *Obj. 2, p. 441*
tains the point $\left(\frac{2\sqrt{2}}{3}, -\frac{1}{3}\right)$, find:

   a. $\sin A$ **b.** $\cos A$ **c.** $\tan A$

3. Use the table on page 504 to find: *Obj. 3, p. 441*

   a. $\sin 32°$ **b.** $\cos 49°$ **c.** $\tan 81°$

4. Solve the right triangle ABC given that $m°\angle A = 60°$ and $b = 5$.   *Obj. 4, p. 441*
Give lengths to the nearest tenth of a unit.

Check your answers with those at the back of the book.

Vectors

OBJECTIVES *for Sections 13-5 through 13-7:*
1. Sketch a vector in standard position in the coordinate plane.
2. Use the distance formula to find the norm of a vector.
3. State the x- and y-components of a vector.
4. Find the norm of the resultant of two given vectors.

13-5 Vectors in the Plane

The red arrow, or *directed line segment,* in Figure 11 represents a *vector quantity,* namely, the *displacement* of a boat that has traveled 9 km northeast from port. Such an arrow indicating both a magnitude (a measure of distance in this case) and a direction is called a *geometric vector* or simply a **vector.** Force and velocity are other examples of vector quantities.

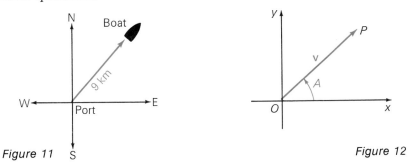

Figure 11

Figure 12

Figure 12 shows a vector **v** in *standard position* in the coordinate plane, that is, with *initial point* at the origin. The *terminal point P* is at the tip of the arrowhead.

The magnitude, or **norm,** of a vector **v** is represented by the symbol $\|\mathbf{v}\|$. The direction of a vector in standard position is the measure of the angle of rotation from the positive x-axis to the vector. The direction of vector **v** in Figure 12 is $m°\angle A$.

EXAMPLE 1 For the vector **v**, find $\|\mathbf{v}\|$ to the nearest tenth and its direction to the nearest degree.

SOLUTION From the Pythagorean Theorem, you have

$$\|\mathbf{v}\|^2 = 3^2 + 2^2 = 13,$$
$$\|\mathbf{v}\| = \sqrt{13} \approx 3.6.$$

Then to find the direction of **v**, you observe that $\tan A = \frac{2}{3} \approx 0.6666$.

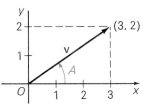

(Solution continued on next page.)

From the table on page 504 and the fact that $\angle A$ is in the first quadrant, you find that $m° \angle A \approx 34°$.

Therefore $\|\mathbf{v}\| \approx 3.6$ and the direction of $\mathbf{v} \approx 34°$.   Answer.

You can find the norm and direction of a vector located anywhere in the plane if you know the coordinates of its initial and terminal points.

EXAMPLE 2   For the vector $\mathbf{u}$, find the norm $\|\mathbf{u}\|$ to the nearest tenth and its direction to the nearest degree.

SOLUTION   Using the distance formula, you have:

$$\|\mathbf{u}\| = \sqrt{(x_2 - x_1)^2 + (y_2 - y_1)^2}$$
$$= \sqrt{(3 - 2)^2 + (4 - 1)^2}$$
$$= \sqrt{1^2 + 3^2} = \sqrt{1 + 9} = \sqrt{10}$$
$$\approx 3.2$$

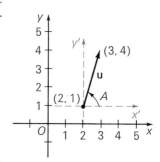

Further, to find the direction you have:

$$\tan A = \frac{y_2 - y_1}{x_2 - x_1} = \frac{3}{1} = 3$$

so that, by the table on page 504,

$$m° \angle A \approx 72°.$$

Therefore, $\|\mathbf{u}\| \approx 3.2$ and the direction of $\mathbf{u} \approx 72°$.   Answer.

For any vector $\mathbf{v}$, the lengths of the horizontal and vertical displacements from the initial to the terminal point are called the *x*- and *y-components* of $\mathbf{v}$. In Figure 13, the *x*- and *y*-components of $\mathbf{v}$ are 1 and 3; for $\mathbf{u}$, the *x*-component is $4 - 2$, or 2, and the *y*-component is $4 - 1$, or 3.

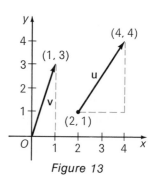

Figure 13

Oral Exercises

A navigator gives the *bearing*, or *heading*, of a vector by measuring *clockwise* the angle from north to the vector. Notice that this is different from our definition of *direction* of a vector.

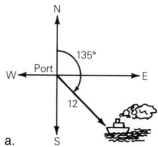

Diagram a.

Thus the bearing and magnitude of the ship's displacement from port in diagram (a) are 135° and 12 respectively. Give the bearing in diagram (b) of the vectors:

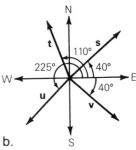

1. **v**   2. **t**   3. **s**   4. **u**

Diagram b.

5. What is the bearing, from its initial point, of a ship that has traveled 4 km due east and then 4 km due north?

6. If the ship in Exercise 5 next travels 8 km due west, what is the final bearing from its initial point?

In Exercises 7–11 state the norm if the terminal point of a vector in standard position has coordinates:

7. $(2, 2)$ 8. $(0, 4)$ 9. $(2, 3)$ 10. $(-1, -4)$ 11. (r, s)

In the figure at the right, state the x- and y-components of the vectors:

12. **m**   13. **n**

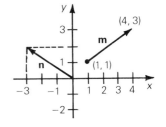

Written Exercises

Each of Exercises 1–4 describes a vector in standard position in the coordinate plane. Draw a sketch of each one.

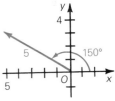

EXAMPLE $\|v\| = 5$; direction 150°

SOLUTION   Sketch the axes and scale them. Then sketch the vector and write labels as shown in red.

A   1. $\|v\| = 5$; direction 90°   2. $\|v\| = 3$; direction 135°
    3. $\|v\| = 4\frac{1}{2}$; direction 240°   4. $\|v\| = 10$; direction 330°

From the information in the given figure, find:

5. the norm of **s**   6. the norm of **t**
7. the x-component of **s**   8. the x-component of **t**

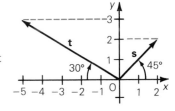

In Exercises 9–12 find the norm and direction of the vector **u**.

9.

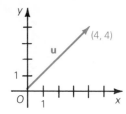

10.

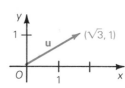

11.

12.

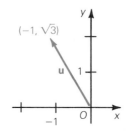

In Exercises 13–16 find ‖**u**‖ to the nearest tenth and the direction of **u** to the nearest degree.

B   13. **u** is in standard position, terminal point is $(-4, 5)$.

14. **u** is in standard position, terminal point is $(5, 3)$.

15. **u** has initial point $(2, 1)$ and terminal point $(3, 5)$.

16. **u** has initial point $(1, 3)$ and terminal point $(6, 1)$.

13-6 The Sum of Two Vectors

Figure 14 pictures two successive displacements in the plane, s and t. From the figure you can see that the x- and y-components of **s** are 3 and 2, and those of **t** are 1 and 4. The total resulting displacement from O is represented by the vector **v**, with initial point $O(0, 0)$ and terminal point $P(4, 6)$.

The x-component of **v** is the sum of the x-components of **s** and **t**, namely, $3 + 1$, or 4. The y-component of **v** is the sum of the y-components of **s** and **t**, namely, $2 + 4$, or 6. Therefore, **v** is called the **sum**, or **resultant**, of the vectors **s** and **t**.

Vectors that have the same magnitude and direction are called **equivalent vectors**. Figure 15 shows five equivalent vectors, each having a norm of 2 units and directed 60° counterclockwise from the positive x-direction.

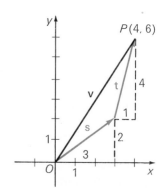

Figure 14

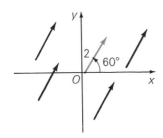

Figure 15

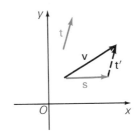

Figure 16

Figure 16 shows two vectors **s** and **t** in the plane. In order to find the sum of **s** and **t,** you would first draw a vector **t′** equivalent to **t** and positioned as shown, with its tail at the head of vector **s.** Then **v** represents the sum **s** + **t.**

EXAMPLE   Draw a vector diagram showing the sum **v** of **s** and **t,** and find the x- and y-components and the norm of **v.**

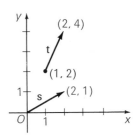

SOLUTION   The red vector **v** represents **s** + **t.** The x-component of **s** is 2, and that of **t** is 2 − 1, or 1. Hence the x-component of **v** is 2 + 1, or 3.

The y-component of **s** is 1, and that of **t** is 4 − 2, or 2. Hence the y-component of **v** is 1 + 2, or 3.

Then

$$\|\mathbf{v}\| = \sqrt{3^2 + 3^2} = \sqrt{18} = 3\sqrt{2}.$$

Thus, the x-component of **v** is 3, the y-component is 3, and the norm is $3\sqrt{2}$.   Answer.

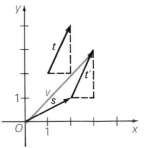

Oral Exercises

State which one of the vectors **p, q, s** is the resultant of the other two.

1.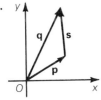

2.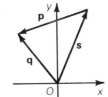

3.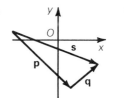

4.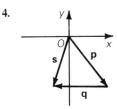

In Exercises 5–8 the vector **v** is the sum of **s** and **t**. In each case, state (a) the x- and y-components of **t**; and (b) ‖**v**‖.

5.

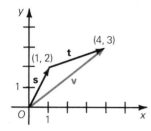

6.

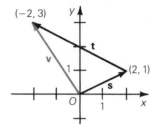

7.

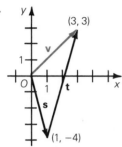

8.

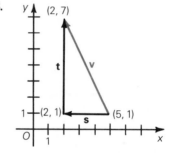

9. Tell which vectors in the figure appear to be equivalent, and give a reason for your answer.

10. In the figure for Exercise 9, if you were to add **p** and **q**, what would be the magnitude of the vector for the sum?

11. What are the coordinates of the terminal point of the vector in standard position that is equivalent to the vector from (2, 3) to (5, 1)?

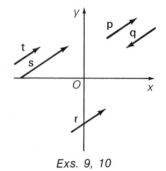

Exs. 9, 10

Written Exercises

In Exercises 1–4 find (a) the x- and y-components of **t**; and (b) the norm of the resultant vector **v**.

A   **1.**

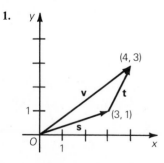

2.

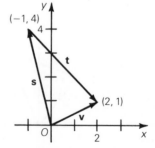

3.

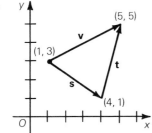

(5, 5)

v

(1, 3)

t

s

(4, 1)

4.

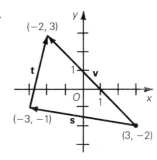

(−2, 3)

t

v

(−3, −1) **s**

(3, −2)

5–8. Find the direction of vector **v** in each of Exercises 1–4, to the nearest degree.

In Exercises 9 and 10, the vectors **u** and **v** are equivalent. Find the coordinates of the terminal point *P* of the vector **u**.

9.

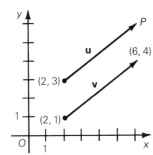

P

u

(6, 4)

(2, 3)

v

(2, 1)

10.

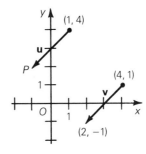

(1, 4)

u

P

(4, 1)

v

(2, −1)

In Exercises 11 and 12:

a. Make a copy of the given vector diagram and on it show the sum **u** = **s** + **t** and the sum **v** = **t** + **s**.

b. What can you say about **u** and **v**?

11.

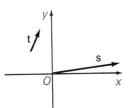

t

s

12.

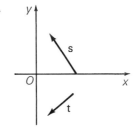

s

t

In Exercises 13–15, determine *p* and *q* so that the given points will be the initial and terminal points, respectively, of a vector equivalent to the vector from (5, 2) to (1, 4).

B **13.** $(-3, 1), (p, q)$ **14.** $(p, q), (-3, 1)$ **15.** $(-1, q), (p, 7)$

In Exercises 16–18 the norms and directions of vectors **u** and **v** are given. Sketch **u, v,** and their resultant **u + v.** Then find the norm of **u + v,** and its direction to the nearest degree.

C 16. **u:** 6; 0°
    v: 7; 90°

17. **u:** 8; 60°
    v: 10; 300°

18. **u:** 20; 30°
    v: 10; 60°

ON THE CALCULATOR

The norms and directions of **s** and **t** are given. Sketch **s, t,** and their resultant, **s + t.** Use a calculator to find the norm of **s + t** to the nearest tenth and its direction to the nearest degree.

1. **s:** 5.6; 30°
   t: 9.7; 45°

2. **s:** 4.7; 65°
   t: 9.1; 36°

3. **s:** 10.5; 24°
   t: 6.7; 82°

DIVERSION

Detective Watson was given 9 coins to test. He knew that one of them was counterfeit and that it weighed a little less than the others, all of which weighed the same. He used a balance scale and in just two weighings found the counterfeit coin. How did he do it?

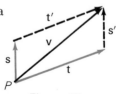

13-7 Applications of Vectors

Figure 17 pictures the resultant force **v** when two different forces, **s** and **t,** are operating on an object at a point *P.* Physicists refer to the fact that the resultant force can be represented in this way as the *parallelogram law of forces,* because the resultant force **v** can be represented geometrically as the diagonal of the parallelogram formed by **s** and **t** and the equivalent vectors **s′** and **t′.**

You can see from the figure that addition of vectors is a commutative operation since

$$\mathbf{v} = \mathbf{s} + \mathbf{t'} = \mathbf{t} + \mathbf{s'},$$

or

$$\mathbf{v} = \mathbf{s} + \mathbf{t} = \mathbf{t} + \mathbf{s}.$$

Figure 17

EXAMPLE 1   Find **(a)** the magnitude and **(b)** the direction of the resultant force when an object is acted on at the point *A* by two forces at right angles to each other, each of magnitude 20 kg.

SOLUTION   First draw a *vector diagram*, such as the one shown.

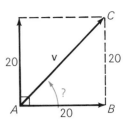

a. Since the two forces acting at A are at right angles to each other, you can use the Pythagorean Theorem to determine $\|\mathbf{v}\|$.

$$\|\mathbf{v}\| = \sqrt{20^2 + 20^2}$$
$$= \sqrt{800} = 20\sqrt{2}. \quad \text{Answer.}$$

b. Since the two forces are equal and at right angles, triangle ABC in the diagram is an isosceles right triangle.

Therefore, $m°\angle CAB = 45°$.   Answer.

EXAMPLE 2   A man standing on a dock pulls with a force of 30 kg on a rope attached to a boat. If the rope makes an angle of 20° with the water, what are the horizontal component x and the vertical component y of the force, to the nearest tenth of a kilogram?

SOLUTION   From the diagram, you can see that

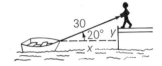

$$\frac{x}{30} = \cos 20° \qquad \frac{y}{30} = \sin 20°,$$

so that

$$x = 30 \cos 20° \approx 30(0.9397) \approx 28.19,$$

and

$$y = 30 \sin 20° \approx 30(0.3420) \approx 10.26.$$

Therefore, $x \approx 28.2$ kg and $y \approx 10.3$ kg.   Answer.

Problems

Give magnitudes to the nearest tenth and angle measures to the nearest degree.

A   1. A boat moves due east at a speed of 10 km/h in a current flowing due south at a rate of 3 km/h. Describe the speed and bearing of the boat over the surface of the earth.

2. An airplane flies at an air speed of 400 km/h on a bearing of 60° through a wind blowing due north at 40 km/h. Find the speed and bearing of the plane with respect to the ground.

3. Beth walks across fields from her home, first 3 km due north and then 1 km due east. To return home by the shortest path, how far and on what heading must she walk?

4. A weather balloon is released vertically at a speed of 5 m/s in a wind blowing horizontally at a speed of 3 m/s. What angle does the path of the balloon form with the level ground?

5. From a tractor equipped with a winch, a cable is attached to a tree stump. If the cable makes a 30° angle with the ground, what is the magnitude of the force applied vertically to the stump when a force of magnitude 450 kg is applied along the cable?

6. Two people push an object across a smooth surface. One pushes with a force of 36 kg due south and the other with a force of 45 kg due west. Describe the force acting on the object.

B 7. Nathan has a mass of 120 kg and sits in the center of a hammock. Find the pull on each supporting rope if each is 45° from the horizontal.

8. A ship sails 90 km due north from a harbor, and then turns 30° toward the east and sails 30 km. How far from the harbor is the ship at that time, and what is the bearing from the harbor to the ship?

C 9. A river flows from north to south. To cross from the east bank directly to the west bank a boat captain finds that she must keep on a course of 280°. If the trip takes 15 min when the boat travels 19.8 km/h, how wide is the river? What is the speed of the current?

Self-Test 2

VOCABULARY   vector (p. 459)   equivalent vectors (p. 462)
                  norm of a vector (p. 459)

1. Sketch the vector **v** in standard position with $\|\mathbf{v}\| = 6$ and direction 135°.   *Obj. 1, p. 459*

2. Find the norm and direction of a vector **v** whose initial and terminal points have the coordinates (2, 4) and (4, 7), respectively.   *Obj. 2, p. 459*

3. Find the *x*- and *y*-components of the vector **v** in Exercise 2.   *Obj. 3, p. 459*

4. Find the norm of the resultant of **s** and **t** in the diagram.   *Obj. 4, p. 459*

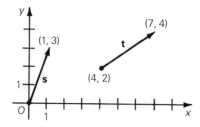

Check your answers with those at the back of the book.

Chapter Summary

1. A *directed angle* is the union of two ordered rays with the same endpoint, together with a rotation from one ray, called the initial side, to the other ray, called the terminal side.

2. The number of degrees through which a ray rotates from the initial side to the terminal side is the *measure* of an angle.

3. An angle with its vertex at the origin and with the positive x-axis as its initial side is in *standard position*.

4. If the terminal side of $\angle A$ in standard position intersects the unit circle at (p, q), then the *cosine* of $\angle A$ (cos A) is p, the *sine* of $\angle A$ (sin A) is q, and the *tangent* of $\angle A$ (tan A) is $\dfrac{q}{p}$. The sine, cosine and tangent of an angle in standard position can be determined from any point on the terminal side.

5. Given certain parts (sides and angles) of a triangle, *solving a triangle* means to find its remaining parts.

6. A *vector* is a directed line segment, having both magnitude and direction. The *magnitude*, or *norm*, of a vector, $\|\mathbf{v}\|$, is its length. The *direction* of a vector in standard position is the measure of the angle of rotation from the positive x-axis to the vector.

7. *Equivalent vectors* have the same magnitude and direction.

8. The *sum*, or *resultant*, of two vectors is found by adding the x- and y-components of the vectors.

Chapter Review

1. Which angle is *not* coterminal with an angle of $110°$ in standard position?

   a. $-250°$ **b.** $470°$ **c.** $900°$ **d.** $-610°$ *13-1*

2. If sin $A < 0$ and cos $A > 0$ and angle A is in standard position, then the terminal side of angle A is in Quadrant ? .   *13-2*

   a. I   **b.** II   **c.** III   **d.** IV

Use the table on page 504 to complete Exercises 3 and 4.

3. sin $65° =$?   *13-3*

   a. 0.9603   **b.** 0.8192   **c.** 0.9063   **d.** 0.8290

4. cos ? $= 0.2250$

   a. $13°$ **b.** $77°$ **c.** $76°$ **d.** $58°$

5. The terminal side of $\angle A$ is in Quadrant II and $\cos A = -\frac{3}{5}$. Then $\sin A = \underline{\ ?\ }$.

a. $-\frac{4}{5}$ b. $-\frac{3}{4}$ c. $\frac{4}{5}$ d. $\frac{4}{3}$

6. The terminal side of $\angle A$ in standard position contains the point $(-5, 12)$. Then $\cos A = \underline{\ ?\ }$.   *13-4*

a. $-\frac{5}{13}$ b. $\frac{12}{13}$ c. $-\frac{12}{5}$ d. $-\frac{13}{5}$

7. In right triangle ABC, $m°\angle A = 40°$ and $c = 12$. Find b to the nearest tenth.

a. 7.7   b. 9.2   c. 10.1   d. 15.7

8. If a vector has initial point $(1, 2)$ and terminal point $(5, 4)$, find its direction to the nearest degree.   *13-5*

a. 39°   b. 30°   c. 63°   d. 27°

9. If vector $\mathbf{v}$ is in standard position, with terminal point $(-3, 4)$, then $\|\mathbf{v}\| = \underline{\ ?\ }$.

a. 25   b. 5   c. $\sqrt{7}$ d. -5

10. If vector $\mathbf{s}$ has terminal point $(3, 4)$, vector $\mathbf{t}$ has terminal point $(-3, 4)$, and $\mathbf{v} = \mathbf{s} + \mathbf{t}$, then $\|\mathbf{v}\| = \underline{\ ?\ }$.   *13-6*

a. 8   b. 6   c. 10   d. 0

Chapter Test

1. If A is a quadrantal angle and $-300° \le m°\angle A \le -200°$, find $m°\angle A$.   *13-1*

2. If $\angle A$ is in standard position, $90° \le m°\angle A \le 180°$, and $\angle A$ is coterminal with an angle of $-210°$, find $m°\angle A$.

3. Evaluate $\cos 630°$.   *13-2*

4. Angle A is in standard position and its terminal side contains the point $(-\frac{\sqrt{3}}{2}, \frac{1}{2})$. Find $\tan A$.

5. If $1° \le m°\angle A \le 90°$ and $\cos A = 0.2588$, find $m°\angle A$ to the nearest degree.   *13-3*

6. In right triangle ABC, $c = 22$ and $b = 16$. Find $m°\angle A$ to the nearest degree.   *13-4*

7. At a point 20 m from a building, the angle of elevation of the top of the building is 66°. How tall is the building?

8. If a vector in standard position has direction 135° and y-component 4, find its norm.   *13-5*

9. If vectors $\mathbf{s}$ and $\mathbf{v}$ are in standard position with terminal points $(-3, 4)$ and $(1, 7)$, respectively, and $\mathbf{v} = \mathbf{s} + \mathbf{t}$, find $\|\mathbf{t}\|$.   *13-6*

10. A sailor walks west at 5 km/h on a ship that is sailing north at 30 km/h. What is the sailor's speed with respect to Earth's surface, to the nearest tenth?

Richard Dedekind
1831–1916

Dedekind began his career in chemistry and physics, but early turned toward mathematics, seeking the order and logical structure he found lacking in physics. He had the good fortune to be associated with the giants of mathematics of his time. His name is linked with Gauss, Dirichlet, Riemann, and Weierstrass, all leaders in a new era of mathematical research.

Dedekind is best known for his "Dedekind cuts," a method for giving an exact description of the irrational numbers. Such a task had been attempted unsuccessfully by mathematicians since the discovery of the irrational numbers by the Pythagoreans.

Dedekind belonged to those mathematicians with great musical talent. An accomplished pianist and cellist, he also tried his hand at composing.

Careers

in Broadcasting

Competition for jobs in the television and radio announcing fields is keen. Those in the industry tend to specialize in specific areas of announcing, such as sports, news, or weather. Broadcasters may become involved in the actual research and writing of the material used. A news reporter, for example, may collect and analyze information about newsworthy events, organize the material, verify the information through observation and research, determine emphasis, and write the story according to prescribed editorial style and format. A college liberal arts education provides an excellent background for this type of job. Courses in English, public speaking, dramatics, foreign languages, and electronics are all helpful.

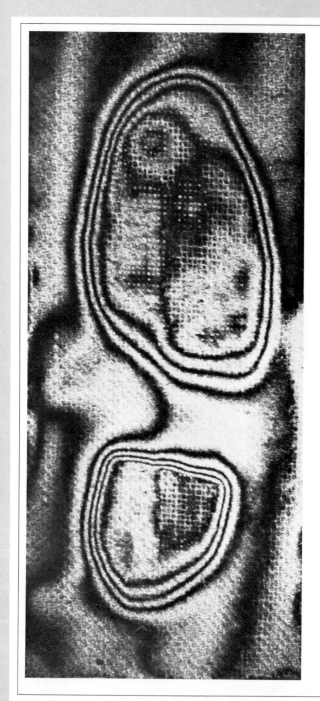

Invisible to the naked eye, this footprint, made on a carpet 24 hours earlier, was revealed by a photographic technique called holography. This technique uses laser light and is so accurate that it can even reveal a footprint made on a bare wooden floor.

14

Statistics and Probability

Graphical Representation of Data

OBJECTIVES *for Sections 14-1 through 14-3:*
1. *Draw and interpret a dot frequency diagram and determine relative frequencies for a given set of data.*
2. *Draw and interpret a histogram and a frequency polygon for a given set of data.*
3. *Draw and interpret a cumulative frequency polygon for a given set of data.*

14-1 Dot Frequency Diagrams and Relative Frequency

Statistics is the science of organizing and analyzing a set of numerical facts, or *data,* so that probable conclusions can be drawn from the data.

Suppose that 25 thirteen-year-old girls are measured and that their heights to the nearest centimeter are given in the following array:

158	160	155	159	160
162	161	158	163	157
156	165	160	157	165
161	157	163	160	160
158	161	159	161	158

In trying to analyze such a distribution of heights, it may be helpful to display the data as a **dot frequency diagram** (Figure 1).

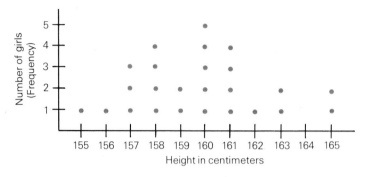

Figure 1

By observing the dots you can quickly see the **frequency,** or number of occurrences, of each measurement. You can also see, for example, that 21 of the 25 students have heights between 157 and 163 cm, inclusive.

You can obtain the **relative frequency** of each measurement if you divide its particular frequency by the total number of measurements. Converting these ratios to percents makes the data easier for most people to interpret.

Height	Frequency	Relative frequency	Percent
155	1	$\frac{1}{25}$	4
156	1	$\frac{1}{25}$	4
157	3	$\frac{3}{25}$	12
158	4	$\frac{4}{25}$	16
159	2	$\frac{2}{25}$	8
160	5	$\frac{5}{25}$	20
161	4	$\frac{4}{25}$	16
162	1	$\frac{1}{25}$	4
163	2	$\frac{2}{25}$	8
165	2	$\frac{2}{25}$	8
Total: 25		$\frac{25}{25} = 1$	100

From this table can you see that the heights of 72%, or about three fourths of the class, are between 157 and 161 cm, inclusive?

Oral Exercises

In Exercises 1–3 state the frequency, the relative frequency, and the percent of the given letter from the word "data."

1. d   **2.** a   **3.** t

In Exercises 4–8 state the frequency, the relative frequency, and the percent of the given letter from the word "bookkeeper."

4. b   **5.** e   **6.** p   **7.** k   **8.** o

9. Do you think that the set of waistline measurements of all the members of your algebra class would show the same pattern of distribution as the set of heights; that is, with most of the readings clustered near the middle reading?

10. Can you think of a word that begins with "s," in which "s" has a frequency of 2 and a relative frequency of $\frac{2}{5}$, or 40%?

Written Exercises

Make a dot frequency diagram for the given data. (Label the axes "Number" and "Frequency.")

A **1.**
2	1	4	4	6
2	4	7	5	3
9	4	3	4	2

2.
27	28	25	28	26
23	25	28	27	29
27	28	31	26	27

Make a table showing the frequencies, relative frequencies, and percents for the data in the given exercise.

3. Exercise 1   **4.** Exercise 2

5. Make a dot frequency diagram for the letters in the word "mammal."

6. Make a dot frequency diagram for the letters in the word "Mississippi."

Make a table for the data in the given exercise, showing frequencies, relative frequencies, and percents.

7. Exercise 5   **8.** Exercise 6

B **9.** To express $\frac{1}{17}$ as a decimal, we write $\frac{1}{17} = 0.\overline{0588235294117647}$. (You might recall that the fraction $\frac{1}{n}$ can have at most $n - 1$ digits in the repeating block of digits in its decimal equivalent.)

   a. Make a table showing the frequency, relative frequency, and percent of each of the ten digits.

   b. Make a dot frequency diagram. (Label the horizontal axis "Digit" and the vertical axis "Frequency.")

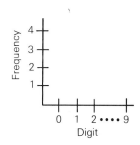

10. The birth dates of the first 20 United States presidents are shown.
   a. Make a table showing the frequency, relative frequency, and percent of each month.
   b. Make a dot frequency diagram for the data. (Label the horizontal axis "Month" and the vertical axis "Frequency.")

No.	Name	Date of Birth
1	George Washington	Feb. 22, 1732
2	John Adams	Oct. 30, 1735
3	Thomas Jefferson	Apr. 13, 1743
4	James Madison	Mar. 16, 1751
5	James Monroe	Apr. 28, 1753
6	John Quincy Adams	July 11, 1767
7	Andrew Jackson	Mar. 15, 1767
8	Martin Van Buren	Dec. 5, 1782
9	William Henry Harrison	Feb. 9, 1773
10	John Tyler	Mar. 29, 1790
11	James K. Polk	Nov. 2, 1795
12	Zachary Taylor	Nov. 24, 1784
13	Millard Fillmore	Jan. 17, 1800
14	Franklin Pierce	Nov. 23, 1804
15	James Buchanan	Apr. 23, 1791
16	Abraham Lincoln	Feb. 12, 1809
17	Andrew Johnson	Dec. 29, 1808
18	Ulysses S. Grant	Apr. 27, 1822
19	Rutherford B. Hayes	Oct. 4, 1822
20	James A. Garfield	Nov. 19, 1831

DIVERSION

The plywood square at the right measures 20 cm by 20 cm. How would you cut it into two equal pieces which could be fitted together to form a 16 cm by 25 cm rectangle?

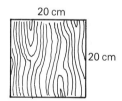

14-2 Histograms and Frequency Polygons

For a large set of data with a wide range of values, a dot frequency diagram is not a practical device for visualizing frequencies. Instead you can make a table showing a frequency distribution in which the data are grouped in equal intervals, and the frequency is shown for each interval. From this you can make a type of *bar graph* called a **histogram** to help visualize the distribution.

The table at the top of the next page shows the distribution of the masses in kilograms of 100 high-school sophomore boys.

Interval	Frequency	Relative frequency	Percent
45–50	2	$\frac{2}{100}$	2
50–55	7	$\frac{7}{100}$	7
55–60	12	$\frac{12}{100}$	12
60–65	28	$\frac{28}{100}$	28
65–70	23	$\frac{23}{100}$	23
70–75	16	$\frac{16}{100}$	16
75–80	8	$\frac{8}{100}$	8
80–85	3	$\frac{3}{100}$	3
85–90	1	$\frac{1}{100}$	1
Total: 100		$\frac{100}{100}$	100

The masses here are grouped into nine intervals, each of length 5. A large collection of data is usually compressed into anywhere from 10 to 20 such *class intervals,* depending on the number and range of the measurements.

A boundary value ordinarily is included in the interval on its *left.* For example, in Figure 2 a mass of 50 kg would be included in the 45–50 interval, *not* in the 50–55 interval.

The histogram of the given distribution (Figure 2) indicates that the greatest clustering of masses occurs between 60 and 70 kg (51%).

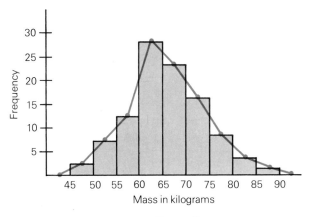

Figure 2

The red, broken-line graph joining the midpoints of the intervals is called a **frequency polygon.**

Notice that the frequency polygon extends a half-interval beyond the histogram at each end, starting and ending on the horizontal axis. For a reason why this is done, see Oral Exercise 9.

A frequency distribution can be displayed equally well by a frequency polygon or a histogram.

Oral Exercises

The histogram at the right shows a frequency distribution of the waistline measurements in an algebra class.

1. How many had waistlines measuring between 60 and 65 cm?

2. How many between 70 and 75 cm?

3. How many between 75 and 80 cm?

4. How many between 65 and 80 cm?

5. How many students were measured?

6. What was the relative frequency of measurements between 80 and 85 cm?

7. What percent of the measurements were between 65 and 70 cm?

8. What percent of the measurements were between 65 and 75 cm?

9. For the figure shown at the right, explain why the area between the horizontal axis and the frequency polygon is equal to the sum of the areas of the rectangles in the histogram. Refer to triangles *ABC* and *CDE* in giving your answer.

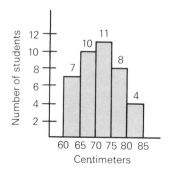

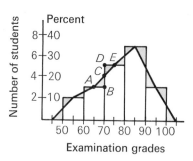

Written Exercises

A 1. **a.** Make a frequency distribution table showing the data at the right. Group the data in the intervals (5–35), (35–65), (65–95), and show relative frequencies and percents.
   b. Make a histogram from the table in part a.
   c. Draw a frequency polygon for the distribution in part a, using the histogram in part b.

30	36	36	48
6	72	48	42
48	60	24	66
90	18	36	48
24	54	42	78

2. Make a histogram and a frequency polygon for the frequency distribution of Scholastic Aptitude Test–Mathematics scores shown at the right.

Class interval	Frequency
450–500	12
500–550	7
550–600	8
600–650	4
650–700	4

B **3.** For the set of measurements shown at the right:

   a. Make a table, grouping measurements in the intervals 4.5–34.5, 34.5–64.5, 64.5–94.5, and showing frequencies, relative frequencies, and percents.

   b. Draw a histogram for the data shown in the table of part a.

   c. Draw a frequency polygon for the data shown in the table of part a.

29	72	17	53	84
55	14	93	65	29
25	82	16	84	92
39	48	72	71	35

4. Repeat Exercise 3 using these measurements:

52	68	44	16	64
86	67	86	46	67
86	63	46	48	67
25	25	88	91	42
23	25	86	67	32

5. The lengths in kilometers of some of the major rivers of the world are listed below.

River	Length (km)	River	Length (km)
Albany	976	Loire	1014
Amazon	6400	Mekong	4160
Amur	4320	Mississippi	3757
Brahmaputra	2880	Nile	6632
Colorado	2320	Orinoco	2560
Columbia	1989	Ottawa	1264
Congo	4349	Po	648
Dnieper	2272	Rio Grande	3016
Elbe	1158	St. Lawrence	1280
Euphrates	3576	Thames	344
Garonne	571	Yangtze	5440
Indus	2880	Zambezi	2720
Jordan	320		

   a. Make a table of frequencies, relative frequencies, and percents, grouping the lengths in the intervals 0–1000, 1000–2000, 2000–3000, and so on.

   b. Draw a histogram from the data in your table.

   c. Use your histogram to draw a frequency polygon for the data.

14-3 Cumulative Frequency

In analyzing a set of numerical data, it is often helpful to tabulate **cumulative frequencies** and **cumulative percents,** that is, the number and percent of measurements that are *less than or equal to* a given value.

The table below shows the frequency distribution for the set of 25 heights given in Section 14-1, along with the *cumulative frequency* and the *cumulative percent.* Notice that any measurement that falls on an interval boundary is included in the interval to its left; for example, 157 is included in the 155–157 interval.

Interval	Frequency	Percent	Cumulative frequency	Cumulative percent
153–155	1	4	1	4
155–157	4	16	5	20
157–159	6	24	11	44
159–161	9	36	20	80
161–163	3	12	23	92
163–165	2	8	25	100

The table tells us, for example, that 80% of the students have heights equal to or less than 161 cm. In the **cumulative frequency polygon** (Figure 3) displaying the facts in the table, the ordinates of the red dots are cumulative frequencies, and the abscissas are the right-hand endpoints of the corresponding intervals. (In using cumulative percents, we sometimes find that the final value is not exactly 100%, because of rounding off in our computations.)

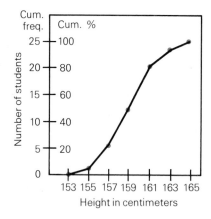

Figure 3

Oral Exercises

Exercises 1–9 refer to the table of heights at the right.

1. What number does the entry *a* represent?

2. Explain how the third entry in the cumulative frequency column, 9, was determined, and then interpret what it means.

3. What number does *b* represent?

4. What number does *c* represent?

5. What number does *d* represent?

6. How many people were 1.55 m tall?

7. How many were at most 1.57 m tall?

8. How many people were measured in all?

9. How many people were more than 1.54 m but less than 1.57 m tall? State two ways of obtaining the answer to this question.

Height (m)	Frequency	Cumulative frequency
1.50	1	1
1.51	3	*a*
1.52	5	9
1.53	6	15
1.54	*b*	20
1.55	7	27
1.56	8	*c*
1.57	5	40
1.58	*d*	44
1.59	3	47
1.60	1	48

10. Roll two dice 50 times and record each score. Make a table showing the frequency, percent, cumulative frequency, and cumulative percent. Which score occurred the most often?

Written Exercises

A

1. Make a table showing the frequency, percent, cumulative frequency, and cumulative percent for the following set of data: 2, 2, 4, 4, 5, 6, 6, 6, 1, 3.

2. Copy the table below and replace each ? with the appropriate entry.

Cost in dollars	Frequency	Percent	Cumulative frequency	Cumulative percent
20	2	8	?	?
21	4	?	?	?
22	5	?	?	?
23	8	?	?	?
24	5	?	?	?
25	1	?	?	?
Total: ?		?		

The cumulative frequency polygon shown at the right concerns tuition fees per student per semester at 20 private colleges.

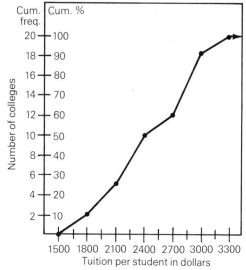

3. How many colleges had a tuition fee no greater than $1800 per semester?

4. How many colleges had tuition fees between $1500 and $3000 per semester, inclusive?

5. What percent of the colleges charged more than $2700 per semester?

6. What percent of the colleges charged at most $2400 per semester?

7. What percent of the colleges charged more than $2400 but less than $3000 per semester?

8. Draw the cumulative frequency polygon for the data in Exercise 2.

For Exercises 9 and 10, copy the table, replacing each ? with the appropriate entry, and draw the cumulative frequency polygon.

B 9.

Interval	Frequency	Percent	Cumulative frequency	Cumulative percent
115–130	3	?	?	?
130–145	1	?	?	?
145–160	6	?	?	?
160–175	5	?	?	?
175–190	8	?	?	?
190–205	2	?	?	?
	Total: ?	?		

10.

Interval	Frequency	Percent	Cumulative frequency	Cumulative percent
0–100	7	?	?	?
100–200	11	?	?	?
200–300	13	?	?	?
300–400	15	?	?	?
400–500	4	?	?	?
	Total: ?	?		

11. Following is a table showing the number of tax returns filed in 1976 with adjusted gross incomes of $30,000 or less. The number of returns in each interval has been recorded to the nearest million. Copy and complete the table and draw the cumulative frequency polygon. (Express all percents to the nearest tenth of a percent.)

Adjusted gross income	Frequency	Percent	Cumulative frequency	Cumulative percent
0–$5000	24	29.6	?	?
$5000–$10,000	20	?	?	?
$10,000–$15,000	15	?	?	?
$15,000–$20,000	11	?	?	?
$20,000–$25,000	7	?	?	?
$25,000–$30,000	4	?	?	?

⊐IVERSION

In a bottom bureau drawer there are 16 gray socks and 12 blue socks. If it is too dark to distinguish the colors, how many socks would you have to take to be sure you get a matching pair?

Self-Test 1

VOCABULARY   relative frequency of a measurement (p. 474)
cumulative frequency (p. 480)
cumulative percent (p. 480)

1. Make a dot frequency diagram for the given data and find the relative frequency of the measurement 5.   *Obj. 1, p. 473*

$$
\begin{array}{cccc}
1 & 2 & 4 & 2 \\
4 & 5 & 1 & 4 \\
5 & 6 & 3 & 3
\end{array}
$$

2. In the histogram shown at the right, _?_ students received grades between 60 and 90, and _?_ students in all took the test.   *Obj. 2, p. 473*

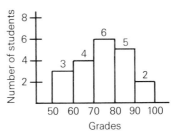

3. Make a table showing the frequency, percent, cumulative frequency, and cumulative percent for the following set of data: *Obj. 3, p. 473*

$$2, 3, 4, 3, 1, 2, 5, 4$$

Then draw a cumulative frequency polygon for these data.

Check your answers with those at the back of the book.

Arithmetical Description of Data

OBJECTIVES *for Sections 14-4 and 14-5:*
1. Find the median, mode, and arithmetic mean of a frequency distribution.
2. Find the range, variance, and standard deviation of a frequency distribution.

14-4 Statistical Averages

You have seen how a graph can give a quick visual summary of the distribution of values in a large collection of data. A compact description of how the data are "centered" can be given by obtaining certain "averages" from the data: the *median*, the *mode*, and the *arithmetic mean*.

To obtain the median from a set of data, such as the 25 heights in Section 14-1, the data must first be arranged in order, as at the right.

The **median** in an ordered set of n values is the *middle* entry if n is odd. If n is even, then there are two middle measurements, and the median is half their sum. In this example, the median is 160.

The **mode** is the value that occurs with the *greatest frequency*. In this example, the mode happens to be the same as the median, 160. There may be more than one mode in a list of data. For example, in 1, 3, 3, 4, 4, 6, there are *two* modes: 3 and 4.

The **arithmetic mean**, often called the *average*, or simply the *mean*, is the sum of the n values divided by n. Here the mean is $\frac{3994}{25} = 159.76$.

In some cases the median may give a more accurate picture of a distribution than the mean would because one or two extreme values can greatly affect the mean. For example, the annual income of the owner of a small business might be $25,000, while the earnings of his four employees might be $5000, $5000, $5000, and $7000. The owner would perhaps want to point to the *mean income* of the entire group, which is

	155
	156
	157
	157
	157
	158
	158
	158
	158
	159
	159
	160
	160 ← **Median**
Mode	160
	160
	160
	161
	161
	161
	161
	162
	163
	163
	165
	165
Sum:	**3994**

$$\frac{\$25,000 + \$5000 + \$5000 + \$5000 + \$7000}{5} = \frac{\$47,000}{5}, \text{ or } \$9400.$$

The employees, on the other hand, would probably feel that the *median income,*

$$\$5000,$$

is a more meaningful figure as far as they are concerned.

In most cases, however, the arithmetic mean is the most reliable measure, and also the most useful one for computational work.

The mode is of very limited value.

EXAMPLE   Find the mean of the 25 heights in the table in Section 14-1 by using their frequency count.

SOLUTION $155(1) + 156(1) + 157(3) + 158(4) + 159(2)$
$+ 160(5) + 161(4) + 162(1) + 163(2) + 165(2)$

$= 155 + 156 + 471 + 632 + 318 + 800 + 644 + 162 + 326 + 330$

$= 3994$

$$\text{Mean} = \tfrac{3994}{25} = 159.76. \quad \text{Answer.}$$

Oral Exercises

For the list of data 2, 3, 3, 3, 4, 4, 8, 9, 9, state:

1. the mode(s)   **2.** the median   **3.** the mean

4. For the data pictured in the dot frequency diagram at the left below, tell why the mode, median, and mean are all 20.

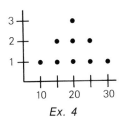

Ex. 4

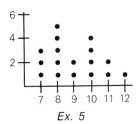

Ex. 5

5. For the data in the dot frequency diagram at the right above, state the mode, the median, and the mean.

6. Does the median have to be a member of the set of data? Does the mode?

7. Does every set of data have a mode? Explain.

8. Suppose you received the following grades on four tests: 86, 39, 92, 91. Which average, the mean or the median, would appear to be the "fairer" one to use as your test average? Why?

Written Exercises

In Exercises 1–4, find for the given list of data:
a. the mode(s)   b. the median   c. the mean

A   **1.** 3, 4, 5, 8, 8, 9, 12   **2.** 23, 27, 30, 30, 34, 34, 34, 35

3. 15, 29, 15, 30, 25, 36, 29, 25, 15, 29   **4.** 10, 17, 9, 15, 17, 21, 17, 20, 15, 17

5. Find the mean from this table of scores on an algebra test given to two classes. If necessary, round off to the nearest tenth.

Score	Frequency
80	10
70	15
60	12
50	8
40	6
30	2

B   **6.** Find the median in Exercise 5.

7. If $r_1, r_2, r_3, \ldots, r_n$ is a list of n measurements, write a formula for their mean, M.

8. If the list in Exercise 7 is ordered from least to greatest, and $n = 25$, what symbol would you use to denote the median measurement?

9. If the sum of n measurements is 2000 and their mean is 5, how many measurements are there?

10. If the mean of 5, x, 19, 14, 9, 11 is 11, find the value of x.

11. Find the value of x if the mean of 4 items is 6, of 5 other items is x, and of all 9 items is 11.

12. Dwight paid $3.25 for 100 items and $1.25 for 50 more items. What was the mean amount he paid per item?

C   **13.** When the owner of a shoe store is putting in an order for more shoes, which descriptive measure of the foot sizes of her regular customers would be most useful to her—the mean, median, or mode? Explain.

14. Show that if x and y are integers, and one is odd and one is even, then their mean is not an integer.

15. Show that if $x < y$, and if the mean of x and y is z, then $x < z < y$.

16. Show that if y^2 is the mean of x^2 and z^2, then $\dfrac{1}{z + x}$ is the mean of

$\dfrac{1}{y + z}$ and $\dfrac{1}{x + y}$ when $x^2 \neq z^2$.

14-5 Measures of Variation

The **range** of a collection of data is the difference between the greatest and least values in the list. It is the simplest *measure of the variation* in the data. The range tells us nothing, however, about how the data are scattered or clustered together, in particular, around the mean. For example, consider these two lists of data, A and B.

$$A: 2, 5, 6, 7, 10$$
$$B: 2, 2, 6, 10, 10$$

They have the same range, 8, and the same mean, 6. But in list A the numbers are clustered much more closely about the mean than in list B.

You can compare the degree of scattering in two data samples by finding the *variance* of each one. The **variance** is the average of the squares of the deviations (differences) of *all* the measurements from their mean.

EXAMPLE 1   Find the variance for each of lists A and B above, and interpret the results.

SOLUTION   For list A, the deviations from the mean are: $2 - 6$, or -4; $5 - 6$, or -1; $6 - 6$, or 0; $7 - 6$, or 1; $10 - 6$, or 4. Then the variance is:

$$\frac{(-4)^2 + (-1)^2 + 0^2 + 1^2 + 4^2}{5} = \frac{34}{5}$$
$$= 6.8$$

For list B, the deviations are: $2 - 6$, or -4; $2 - 6$, or -4; $6 - 6$, or 0; $10 - 6$, or 4; $10 - 6$, or 4.
The variance is:

$$\frac{(-4)^2 + (-4)^2 + 0^2 + (4)^2 + (4)^2}{5} = \frac{64}{5}$$
$$= 12.8$$

Since the variance in A is 6.8 whereas in B it is 12.8, the data in B are considerably more scattered from their mean than are the data in A.
Answer.

If the observations are measured in units, such as meters, then the variance is given in terms of square units. When you take the *nonnegative square root* of the variance, you then have a measure of variation that is in the same units as those of the given data. This measure is called the **standard deviation**, usually denoted by s.

EXAMPLE 2   Find the standard deviation s for lists A and B.

SOLUTION   For A, the variance is 6.8, so $s = \sqrt{6.8} \approx 2.61$.   Answer.

For B, the variance is 12.8, so $s = \sqrt{12.8} \approx 3.58$.   Answer.

Oral Exercises

For Exercises 1–3, use the list of data 3, 4, 6, 7, and find:

1. the range   **2.** the mean   **3.** the deviation of each value from the mean

4. The mean of each of the following lists of data, **a** and **b**, is 8. Which
one do you think has the greater variance?
     a. 5, 7, 8, 9, 11   **b.** 4, 5, 5, 6, 7, 9, 10, 11, 15

5. Check your answer to Exercise 4 by finding the variance of each list.

Written Exercises

For the list 4, 7, 8, 9, find to the nearest tenth:

1. the variance, s^2 **2.** the standard deviation, s

For the list 10, 11, 13, 14, find to the nearest tenth:

3. s^2 **4.** s

If $r_1, r_2, r_3, \ldots, r_n$ denotes a list of n observations and m denotes their
mean, write an expression for each of the following:

B   **5.** the deviation of r_1 from the mean

      6. the average of the set of all the deviations from the mean

      7. the variance, s^2

      8. the standard deviation, s

Self-Test 2

VOCABULARY   median (p. 484)   variance (p. 487)
                                    range (p. 486)   standard deviation (p. 487)

1. Find the median, mode, and mean of the following data: 7, 2, 2,   *Obj. 1, p. 484*
6, 4, 2, 7, 2, 4.

2. Find the range, variance, and standard deviation of the data in   *Obj. 2, p. 484*
Exercise 1.

Check your answers with those at the back of the book.

Probability

OBJECTIVES for Sections 14-6 and 14-7:
1. **Find the probability of a specified event in an experiment with equally likely outcomes.**
2. **Calculate an experimental probability.**

14-6 The Probability of an Event

The main use for statistical theory is to help people make decisions on the basis of incomplete information. For example, businessmen and politicians often use marketing surveys and opinion polls to assess the reaction of the *population* (the universe of elements) by studying a *sample* (a subset) of it. In choosing a sample of data and then trying to draw conclusions about the population, the statistician must be guided by the theory of *probability*.

Simple games of chance offer ideal situations for understanding the notion of probability. Suppose that you perform the experiment of tossing a die and observing the number of dots on the top face. The six possible outcomes are these:

Figure 4

If the die is not loaded you can assume that each outcome is equally likely. That is, each outcome has an equal chance of occurring, namely, 1 out of 6. Then the *measure* of chance, or **probability**, that a particular outcome will occur is $\frac{1}{6}$.

If an experiment has n possible, equally likely outcomes, then the probability that any one of them will occur is $\dfrac{1}{n}$.

An **event** is a *specified subset* of the set of all possible outcomes in an experiment. When you toss a die, the probability P of the event "the top face shows *either* 3 *or* 4 dots" is $\frac{2}{6}$. That is, there are 2 out of 6 chances that either one or the other of these outcomes will occur. If any *one* of the outcomes in an event occurs, you say that the event **occurs.**

If an experiment has n possible, equally likely outcomes and an event E consists of e of these outcomes, then the probability P that E will occur is $\frac{e}{n}$. That is:

$$P(E) = \frac{\text{number of outcomes in } E}{\text{number of possible outcomes}} = \frac{e}{n}$$

EXAMPLE 1   There are 4 queens in a standard bridge deck of 52 cards. What is the probability of the event that a card drawn at random is a queen?

SOLUTION   Since there are 4 queens and 52 cards in the deck, $e = 4$ and $n = 52$. Then

$$P(E) = \frac{e}{n} = \frac{4}{52} = \frac{1}{13}. \quad \text{Answer.}$$

EXAMPLE 2   A jar contains 2 red, 4 blue, and 5 white marbles. If a marble is drawn from the jar at random, what is the probability of the event "the marble is not blue"?

SOLUTION   Of the 11 marbles in the jar, 7 are not blue. Hence

$$P(E) = \frac{7}{11}. \quad \text{Answer}$$

All the experiments described in this chapter are assumed to be *random* ones, that is, experiments conducted in such a way that the outcomes are strictly a matter of chance.

Oral Exercises

1. When a die is tossed, what is the probability that the side with just two dots will be on top?

2. When you toss a coin, what is the probability that it will land with the head showing?

3. When you draw a marble, while blindfolded, from a jar containing 4 marbles—1 red, 1 blue, 1 white, 1 green—what is the probability of drawing a blue marble?

4. In Exercise 3, what is the probability of the event "either a red or a green marble is drawn"?

5. When a coin is tossed, what is the probability of the event "it will land either heads or tails"?

Assume that the arrow on the spinner will not stop on a dividing line.

6. What are the possible outcomes of a spin? Are they all equally likely?

7. What is the probability that the arrow will stop on the region labeled 1? 2? 4?

8. What is the probability that the arrow will not stop on the region labeled 3?

9. Suppose that in the state lottery 200,000 tickets are sold one week, and that you buy a single ticket.
   a. What is the probability that your ticket wins the grand prize, if only one winner is drawn at random?
   b. If your friend also has a single ticket, how does your chance of winning compare to your friend's?

Written Exercises

Exercises 1–7 refer to the spinner at the right. Assume that the arrow will neither stop on a dividing line nor favor a particular numbered region.

A   1. List the possible outcomes and give the probability that any particular one will occur when the arrow is spun.

What is the probability that after a spin the arrow will stop on:

2. region 3?

3. region 2 or 6?

4. an even-numbered region?

5. a region with a number less than 3?

6. a region other than region 1?

7. a region with a number greater than 2 but less than 3?

A jar contains 2 white, 3 red, 1 green, and 4 blue marbles. If a marble is drawn at random from the jar, what is the probability that the marble is:

8. white?

9. red?

10. green?

11. white or blue?

12. not blue?

13. yellow?

In a standard deck of 52 cards, there are four of each kind of card: ace, king, queen, jack, ten, nine, . . . , two. Half the cards are black and half are red, with two black and two red aces, two black and two red kings, and so on. If a card is drawn at random from the deck, what is the probability that it will be:

14. a black card?

15. a king?

16. a four?

17. a red jack?

18. a black five?

19. a ten?

20. not a queen?

21. not a red queen?

B 22. List the set of possible outcomes when a penny and a nickel are tossed, using the symbols H and T for heads and tails. (For example, for the outcome when the penny lands heads and the nickel tails, you would write HT; but for the outcome when the penny lands tails and the nickel heads, you would write TH.)

23. In Exercise 22, do the events "they both land heads" and "they do not both land heads" have the same probability? What is the probability of each of these events?

What is the probability that after a spin the arrow will stop on:

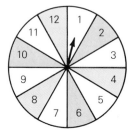

24. a region corresponding to a multiple of 2?

25. a region corresponding to a prime number?

26. an unshaded region?

27. a region corresponding to a factor of 12?

14-7 Experimental Probability

Suppose that in a particular year, a school with 1000 students has 40 who are left-handed. Then there are 40 chances in 1000, or 1 in 25, that a student chosen at random will be left-handed. If there are 25 seats in each classroom, would you therefore conclude that every classroom should have exactly one seat with an arm for left-handed writers? Your intuition and experience tell you, "Of course not! For example there might be three left-handed students in one room and none in another."

What is the probability that exactly 40 of the school's 1000 students the next year will be left-handed? Again, experience based on repeated observations would indicate a relatively small probability of this event. Nevertheless, in the absence of further information regarding the percent of left-handed people in the population, you would have to give $\frac{1}{25}$ as your best estimate of the probability that a randomly chosen student in that school next year will be left-handed.

A certain baseball player may have a *batting average* (that is, a ratio of safe hits to times at bat) of 0.400 for a particular season. Then the *experimental probability* of his making a safe hit the next time he is at bat is $\frac{4}{10}$, or $\frac{2}{5}$. This does not mean, however, that he will surely make 2 hits any day that he bats 5 times.

In many real-life situations such as these, you have only experimental probabilities from which to predict future occurrences of events.

If an experiment is conducted n times, and an event E occurs e of these times, then the experimental probability P that E will occur in another trial is $\frac{e}{n}$. That is:

$$P(E) = \frac{\text{number of occurrences of } E}{\text{number of trials}} = \frac{e}{n}$$

EXAMPLE A thumbtack was tossed 100 times. It landed "point up" 70 times and "point down" 30 times.

Point up Point down

a. What is the experimental probability that on the next toss it will land point up?

b. If it were tossed 50 additional times, about how many times would you expect it to land point down?

c. If on 50 additional tosses it actually landed point down 20 times, what would be the experimental probability, based on all 150 tosses, that on the next toss it would land point down?

SOLUTION **a.** For the event E that the thumbtack lands point up, you have $e = 70$ and $n = 100$. Therefore,

$$P(E) = \tfrac{70}{100} = 0.7. \quad \textsf{Answer.}$$

b. Since the experimental probability that the thumbtack will land point down is $\tfrac{30}{100}$, or 0.3, you would expect it to land point down about 3 times out of every 10. Therefore, you would expect it to land point down about

$$5 \times 3, \text{ or } 15, \text{ times out of 50 additional tosses.} \quad \textsf{Answer.}$$

c. If the thumbtack lands point down 30 times out of the first 100 throws, and 20 times out of the next 50 throws, then in all you have $e = 30 + 20 = 50$ and $n = 100 + 50 = 150$. Therefore the experimental probability that the thumbtack will land point down is now

$$\frac{e}{n} = \frac{50}{150} = \frac{1}{3} \approx 0.33. \quad \textsf{Answer.}$$

Oral Exercises

1. If a baseball player gets 2 hits out of 3 times at bat in one game, what is the experimental probability that he will get a hit the next time at bat?

2. If Sharon sinks 8 free throws out of 10, what is the experimental probability that she will make a basket on the next free throw?

3. The dietician in a company cafeteria made an informal survey one day, finding that 130 employees bought lunch in the cafeteria, 50 brought their lunches with them and ate in the cafeteria, and 20 did not come to the cafeteria at all. What is the experimental probability that an employee will buy lunch in the cafeteria? What is the experimental probability that an employee will not bring a lunch to the cafeteria?

Written Exercises

A 1. If a football quarterback is completing passes at the rate of 81%, what is the experimental probability that he will complete his next pass?

2. In Exercise 1, what is the experimental probability that he will not complete the next pass?

3. If the experimental probability is $\frac{5}{8}$ that a softball player will get a hit the next time she goes to bat, does this mean that she is sure to get 5 hits in the next 8 times she bats?

Suppose that the results of 36 successive random drawings of one marble from a jar containing an unknown number of colored marbles are as follows: 4 blue, 14 red, 12 yellow, 6 green. What is the experimental probability that the next marble drawn will be:

4. red?   5. blue?   6. green?   7. yellow?

8. black?   9. yellow or green?   10. not yellow?

B 11. If it is found that there are 60 yellow marbles in the jar used in Exercises 4–10, how many green ones would you expect to find? How many blue marbles would you expect to find?

12. If two carriers of the disease called sickle-cell anemia marry, the experimental probability of any child of theirs having the disease is $\frac{1}{4}$. If they have 4 children, is it true that:
a. one of them is certain to have the disease?
b. none of them might have the disease?
c. all of them might have it?

C 13. United States mortality tables show that about 95% of people born now can expect to be alive at age 40, but only about 75% of those born now will still be alive at age 65. What is the experimental probability that a person born now who reaches age 40 will still be alive at 65? (Assume that life expectancies do not change in the next 65 years.)

494

Self-Test 3

VOCABULARY   event (p. 489)   experimental probability of an event (p. 493)

When you toss a die, what is the probability that:

1. the top face will show 4 dots?   *Obj. 1, p. 489*
2. the top face will not show 4 dots?
3. the top face will show an odd number of dots?
4. the top face will show either an even number or an odd number of dots?

5. If it had rained on Thanksgiving Day in New York City for 35   *Obj. 2, p. 489* out of the past 65 years, what would be the probability that it would rain there next Thanksgiving Day?

Suppose the results, when you drew 30 slips of paper at random from a hat containing 100 slips, were as follows: 6 purple, 9 red, 8 yellow, 7 white. What is the probability that the next slip you draw will be:

6. purple?   7. red or yellow?   8. yellow or white?

9. How many red slips would you expect to find in the hat?
10. How many purple slips would you expect to find in the hat?

Check your answers with those at the back of the book.

Chapter Summary

1. Statistics is the science of organizing and analyzing a set of facts, or data, so that probable conclusions can be drawn from the data.
2. The number of occurrences of a particular measurement is called the frequency of the measurement. The relative frequency is the ratio of the frequency of a particular measurement to the total number of measurements.
3. Histograms and frequency polygons are used to help visualize frequency distributions for large sets of data.
4. The number and percent of measurements that are less than or equal to a given value are called cumulative frequency and cumulative percent.

5. The median, the mode, and the arithmetic mean are three descriptions of the "center" of a set of data.

6. The range, the variance, and the standard deviation are three measures of the variation in a set of data.

7. If an experiment has n possible, equally likely outcomes, then the probability that any one of them will occur is $\frac{1}{n}$.

8. An event is a specified subset of the set of all possible outcomes in an experiment.

Chapter Review

1. What is the relative frequency of the letter "p" in the word "purple"?

    a. $\frac{2}{5}$ *b.* 2   *c.* $\frac{1}{3}$ *d.* $\frac{1}{6}$

14-1

2. What is the relative frequency of data that are in the interval 10–15 in the following set of data: 23, 5, 12, 14, 21, 27, 6, 13, 13, 15?

    a. $\frac{30}{100}$ *b.* $\frac{40}{100}$ *c.* $\frac{50}{100}$ *d.* $\frac{60}{100}$

14-2

3. What is the cumulative frequency of data in the interval 35–40 for the following set of figures: 32, 53, 47, 38, 43, 37, 36, 29, 31, 28, 46, 50?

    a. 7   *b.* 8   *c.* 10   *d.* 11

14-3

4. Find the mean for the following list of data: 12, 25, 4, 18, 32, 6, 22, 17.

    a. 20   *b.* 16   *c.* 23   *d.* 17

14-4

5. Find the standard deviation for the following list of data: 9, 10, 10, 15, 16, 18.

    a. 3.53   *b.* 3.46   *c.* 5.53   *d.* 3.64

14-5

6. A jar contains 3 white, 2 red, and 5 blue marbles. If a marble is drawn at random from the jar what is the probability that it is red or blue?

    a. $\frac{3}{10}$ *b.* $\frac{1}{2}$ *c.* $\frac{1}{5}$ *d.* $\frac{7}{10}$

14-6

7. An archer has hit the bull's eye 8 out of 72 tries. Approximately what is the experimental probability that the next arrow she shoots will hit the bull's eye?

    a. 0.11   *b.* 0.99   *c.* 0.01   *d.* 0.09

14-7

Chapter Test

1. Make a table showing the frequencies, relative frequencies, and percents for the following data: 16, 16, 19, 11, 12, 17, 16, 13, 18, 12. ·   *14-1*

2. Make a histogram and a frequency polygon for the following data, grouping them in the intervals (20–25), (25–30), (30–35): 33, 26, 30, 22, 21, 33, 23, 34, 27, 21, 32, 35, 25, 26, 34.   *14-2*

3. Make a table showing the frequency, percent, cumulative frequency, and cumulative percent for the data given in Exercise 2.   *14-3*

4. Find the median, the mode, and the mean for the following data: 12, 25, 16, 19, 9, 23, 27, 25, 25, 26, 18, 19.   *14-4*

5. Find the standard deviation for the following data: 6, 8, 9, 11, 12.   *14-5*

6. What is the probability of drawing at random a red king from a standard deck of 52 cards?   *14-6*

7. Philip predicted correctly 18 out of 200 times which way a tossed coin would fall: that is, heads up or tails up. What is the experimental probability that his next prediction will be correct?   *14-7*

Careers

in Farming

Today many farmers raise both crops and livestock. Responsibilities of the farmer include determining the kind and amount of crops to be grown and livestock to be bred, the selection and purchase of seed, fertilizer, farm machinery, livestock and feed, and the sale of crop and livestock products. The success of today's large-scale farms depends on the management skills and the technical knowledge of the farmer. Thus many young commercial farmers today have a college education which includes courses in economics, accounting, and biology, as well as soil science and agronomy.

Comprehensive Review (Chapters 1–14)

1. Simplify $6[(12 - 4) + (8 \div 2)]$
   - **a.** 48
   - **b.** 52
   - **c.** 72
   - **d.** 64

2. Find the solution set of $x - 4 = 7$ if the domain of x is {the positive real numbers}.
   - **a.** 11
   - **b.** 3
   - **c.** -3
   - **d.** -11

3. Simplify $2(3x - 4) + 5x^2$.
   - **a.** $5x^2 + 6x - 4$
   - **b.** $11x^2 - 8$
   - **c.** $5x^2 + 6x - 8$
   - **d.** $x^2 - 8$

4. Simplify $|-6 + 4| - 3$.
   - **a.** -1
   - **b.** 1
   - **c.** 7
   - **d.** -5

5. Simplify $(-6)(-8) + (-25)(3)$.
   - **a.** -120
   - **b.** 120
   - **c.** 27
   - **d.** -27

6. Simplify $-\frac{1}{6}(72m^2 - 24m)$.
   - **a.** $-12m^2 - 4m$
   - **b.** $-12m^2 + 4m$
   - **c.** $12m^2 - 4m$
   - **d.** $12m^2 + 4m$

7. Solve $12x = -\frac{72}{3}$.
   - **a.** 2
   - **b.** -2
   - **c.** 6
   - **d.** -6

8. One number is 2 more than 6 times another and their sum is 23. What are the numbers?
   - **a.** 1 and 8
   - **b.** 2 and 14
   - **c.** 3 and 20
   - **d.** 4 and 26

9. Solve $6x - 8 > 16$.
   - **a.** $x > -4$
   - **b.** $x > 4$
   - **c.** $x < -4$
   - **d.** $x < 4$

10. A piggy bank holds some dimes and nickels worth $12.75. There are 13 times as many nickels as dimes. How many of each kind of coin are in the piggy bank?
    - **a.** 17 dimes, 221 nickels
    - **b.** 221 dimes, 17 nickels
    - **c.** 117 dimes, 9 nickels
    - **d.** 9 dimes, 117 nickels

11. The point $P(-3, 6)$ is in Quadrant $\underline{\ ?\ }$.
    - **a.** I
    - **b.** II
    - **c.** III
    - **d.** IV

12. The equation $5x + 2y = -3$ is equivalent to:
    - **a.** $y = \frac{5}{2}x - \frac{3}{2}$
    - **b.** $y = -\frac{5}{2}x + \frac{3}{2}$
    - **c.** $y = \frac{5}{2}x + \frac{3}{2}$
    - **d.** $y = -\frac{5}{2}x - \frac{3}{2}$

13. A line with slope 5 which passes through the point $(1, 7)$ also passes through the point $\underline{\ ?\ }$.
    - **a.** $(2, 10)$
    - **b.** $(1, 5)$
    - **c.** $(5, 3)$
    - **d.** $(3, 17)$

14. Solve the system $\quad x - 2y = 7$
$$4x - 2y = 4$$

    a. $(-3, -5)$ **b.** $(3, 5)$ **c.** $(-1, -4)$ **d.** $(1, 4)$

15. Solve the system $2x + 3y = 3$
$$3x - 4y = 13$$

    a. $(3, -1)$ **b.** $(3, 1)$ **c.** $(6, -3)$ **d.** $(2, -2)$

16. Simplify $3mn^2(2m^2n^2 - 4m^2n + 6n^2)$.

    a. $6m^3n^3 - 12m^2n^3 + 18mn^4$ **b.** $6m^3n^4 - 12m^3n^2 + 18m^3n^2$
    c. $6m^3n^4 - 12m^2n^3 + 18mn^4$ **d.** $6m^3n^4 - 12m^3n^3 + 18mn^4$

17. Factor $3x^2 - 6x - 45$ completely.

    a. $3(x^2 - 2x - 15)$ **b.** $3(x + 3)(x - 5)$
    c. $3(x + 5)(x - 3)$ **d.** $3(x - 3)(x - 5)$

18. Divide $12a^2 + 7a - 10$ by $3a - 2$.

    a. $4a - 5$ **b.** $5a - 4$ **c.** $4a + 3$ **d.** $4a + 5$

19. Simplify $\dfrac{9x^2 - 4y^2}{3x^2 - 8xy + 4y^2} \div \dfrac{6x^2 - 2xy - 4y^2}{3x^2 - 5xy + 2y^2}$.

    a. $\dfrac{3x - 2y}{2x - 4y}$ **b.** $\dfrac{3x + 2y}{2x - 4y}$ **c.** $\dfrac{3x - 2y}{x - 2y}$ **d.** $\dfrac{x - y}{x - 2y}$

20. Find the solution set of $\dfrac{5}{x} + \dfrac{1}{3} = \dfrac{2}{x}$.

    a. $\{-3\}$ **b.** $\{1\}$ **c.** $\{6\}$ **d.** $\{-9\}$

21. If d varies inversely as t and $d = 7$ when $t = 4$, find d when $t = 2$.

    a. $\{12\}$ **b.** $\{14\}$ **c.** $\{4\}$ **d.** $\{8\}$

22. The legs of a right triangle are 6 and 2 units long. How long is its hypotenuse to the nearest tenth?

    a. 12.7   **b.** 4.4   **c.** 6.3   **d.** 5.6

23. Simplify $(\sqrt{3} - \sqrt{6})(2\sqrt{3} + \sqrt{2})$.

    a. $6 - 6\sqrt{2} + \sqrt{6} - 2\sqrt{3}$ **b.** $6 - 8\sqrt{3} + \sqrt{6}$
    c. $6 - 4\sqrt{3} + \sqrt{6} - 6\sqrt{2}$ **d.** $6 + 4\sqrt{3} + \sqrt{6}$

24. Use the quadratic formula to solve $x^2 + 4x - 48 = 0$.

    a. $\{-2 + \sqrt{208}, -2 - \sqrt{208}\}$ **b.** $\{4 + \sqrt{208}, 4 - \sqrt{208}\}$
    c. $\{-4 + \sqrt{13}, -4 - \sqrt{13}\}$ **d.** $\{-2 + 2\sqrt{13}, -2 - 2\sqrt{13}\}$

25. Which description best fits the graph of $y > 3x^2 + 4$?

    a. A solid curve and the region under it.
    b. A dashed curve and the region under it.
    c. A solid curve and the region above it.
    d. A dashed curve and the region above it.

26. What is the measure of the angle A that is coterminal with $-290°$, with $0° \leq m° \angle A < 360°$?

    a. $70°$ **b.** $110°$ **c.** $150°$ **d.** $200°$

27. Find $\tan A$ if angle A is in Quadrant II and $\sin A = \frac{7}{25}$.

    a. $\frac{24}{25}$ **b.** $\frac{7}{24}$ **c.** $-\frac{7}{24}$ **d.** $-\frac{24}{25}$

28. Find the measure of angle A, $1° \leq m° \angle A \leq 90°$, to the nearest degree if $\cos A = 0.7336$.

    a. $48°$ **b.** $36°$ **c.** $42°$ **d.** $43°$

29. What is the relative frequency of 5 in the following data: 4, 3, 5, 6, 5, 2, 4, 5, 2, 2

    a. $\frac{4}{10}$ **b.** $\frac{3}{10}$ **c.** $\frac{2}{10}$ **d.** $\frac{1}{10}$

30. A jar contains 3 green, 2 blue, and 7 red marbles. If a marble is drawn at random from the jar what is the probability of the event "the marble is blue or green"?

    a. $\frac{1}{4}$ **b.** $\frac{5}{12}$ **c.** $\frac{1}{6}$ **d.** $\frac{7}{12}$

Tables

Table 1 Formulas

Circle	$A = \pi r^2, C = 2\pi r$	Cube	$V = s^3$
Parallelogram	$A = bh$	Rectangular Box	$V = lwh$
Right Triangle	$A = \frac{1}{2}bh, c^2 = a^2 + b^2$	Cylinder	$V = \pi r^2 h$
Square	$A = s^2$	Pyramid	$V = \frac{1}{3}Bh$
Trapezoid	$A = \frac{1}{2}h(b + b')$	Cone	$V = \frac{1}{3}\pi r^2 h$
Triangle	$A = \frac{1}{2}bh$	Sphere	$V = \frac{4}{3}\pi r^3$
Sphere	$A = 4\pi r^2$		

Table 2 Metric Units of Measure

Base Units	**Time**	**Temperature**
Length: meter (m)	second (s), minute (min)	degree Celsius (°C)
Mass: kilogram (kg)*	day (d), month (mo), year (yr)	degree Kelvin (°K)

Capacity	**Force**	**Pressure**
liter (L)	Newton (N)	Pascal (Pa)
$1\,L = 1000\,cm^3$		

Prefixes

Factor	Prefix	Symbol	Factor	Prefix	Symbol
10^{18}	exa	E	10^{-1}	deci	d
10^{15}	peta	P	10^{-2}	centi	c
10^{12}	tera	T	10^{-3}	milli	m
10^{9}	giga	G	10^{-6}	micro	μ
10^{6}	mega	M	10^{-9}	nano	n
10^{3}	kilo	k	10^{-12}	pico	p
10^{2}	hecto	h	10^{-15}	femto	f
10	deka	da	10^{-18}	atto	a

A prefix multiplies a unit by the factor given in the table.

Examples   gigameter: $1\,Gm = 10^9\,m = 1{,}000{,}000{,}000\,m$

milligram: $1\,mg = 10^{-3}\,g = 0.001\,g^*$

Compound units may be formed by division or multiplication.

Examples   kilometers per hour: km/h   square centimeters: cm^2 cubic meters: m^3

---

*Although the kilogram is defined as the base unit, the gram (g) is used with the prefixes to name other units of mass.

Table 3 Squares of Integers from 1 to 100

Number	Square	Number	Square	Number	Square	Number	Square
1	1	26	676	51	2601	76	5776
2	4	27	729	52	2704	77	5929
3	9	28	784	53	2809	78	6084
4	16	29	841	54	2916	79	6241
5	25	30	900	55	3025	80	6400
6	36	31	961	56	3136	81	6561
7	49	32	1024	57	3249	82	6724
8	64	33	1089	58	3364	83	6889
9	81	34	1156	59	3481	84	7056
10	100	35	1225	60	3600	85	7225
11	121	36	1296	61	3721	86	7396
12	144	37	1369	62	3844	87	7569
13	169	38	1444	63	3969	88	7744
14	196	39	1521	64	4096	89	7921
15	225	40	1600	65	4225	90	8100
16	256	41	1681	66	4356	91	8281
17	289	42	1764	67	4489	92	8464
18	324	43	1849	68	4624	93	8649
19	361	44	1936	69	4761	94	8836
20	400	45	2025	70	4900	95	9025
21	441	46	2116	71	5041	96	9216
22	484	47	2209	72	5184	97	9409
23	529	48	2304	73	5329	98	9604
24	576	49	2401	74	5476	99	9801
25	625	50	2500	75	5625	100	10,000

Table 4 Square Roots of Integers from 1 to 100

Exact square roots are shown in red. For the others, rational approximations are given correct to three decimal places.

Number	Positive Square Root	Number	Positive Square Root	Number	Positive Square Root	Number	Positive Square Root
N	$\sqrt{N}$	N	$\sqrt{N}$	N	$\sqrt{N}$	N	$\sqrt{N}$
1	1	26	5.099	51	7.141	76	8.718
2	1.414	27	5.196	52	7.211	77	8.775
3	1.732	28	5.292	53	7.280	78	8.832
4	2	29	5.385	54	7.348	79	8.888
5	2.236	30	5.477	55	7.416	80	8.944
6	2.449	31	5.568	56	7.483	81	9
7	2.646	32	5.657	57	7.550	82	9.055
8	2.828	33	5.745	58	7.616	83	9.110
9	3	34	5.831	59	7.681	84	9.165
10	3.162	35	5.916	60	7.746	85	9.220
11	3.317	36	6	61	7.810	86	9.274
12	3.464	37	6.083	62	7.874	87	9.327
13	3.606	38	6.164	63	7.937	88	9.381
14	3.742	39	6.245	64	8	89	9.434
15	3.873	40	6.325	65	8.062	90	9.487
16	4	41	6.403	66	8.124	91	9.539
17	4.123	42	6.481	67	8.185	92	9.592
18	4.243	43	6.557	68	8.246	93	9.644
19	4.359	44	6.633	69	8.307	94	9.695
20	4.472	45	6.708	70	8.367	95	9.747
21	4.583	46	6.782	71	8.426	96	9.798
22	4.690	47	6.856	72	8.485	97	9.849
23	4.796	48	6.928	73	8.544	98	9.899
24	4.899	49	7	74	8.602	99	9.950
25	5	50	7.071	75	8.660	100	10

Table 5 Values of Sine, Cosine, and Tangent for Angles A such that 1° ≤ m° ∠ A ≤ 90°

m° ∠ A	sin A	cos A	tan A	m° ∠ A	sin A	cos A	tan A
1	0.0175	0.9998	0.0175	46	0.7193	0.6947	1.0355
2	0.0349	0.9994	0.0349	47	0.7314	0.6820	1.0724
3	0.0523	0.9986	0.0524	48	0.7431	0.6691	1.1106
4	0.0698	0.9976	0.0699	49	0.7547	0.6561	1.1504
5	0.0872	0.9962	0.0875	50	0.7660	0.6428	1.1918
6	0.1045	0.9945	0.1051	51	0.7771	0.6293	1.2349
7	0.1219	0.9925	0.1228	52	0.7880	0.6157	1.2799
8	0.1392	0.9903	0.1405	53	0.7986	0.6018	1.3270
9	0.1564	0.9877	0.1584	54	0.8090	0.5878	1.3764
10	0.1736	0.9848	0.1763	55	0.8192	0.5736	1.4281
11	0.1908	0.9816	0.1944	56	0.8290	0.5592	1.4826
12	0.2079	0.9781	0.2126	57	0.8387	0.5446	1.5399
13	0.2250	0.9744	0.2309	58	0.8480	0.5299	1.6003
14	0.2419	0.9703	0.2493	59	0.8572	0.5150	1.6643
15	0.2588	0.9659	0.2679	60	0.8660	0.50	1.7321
16	0.2756	0.9613	0.2867	61	0.8746	0.4848	1.8040
17	0.2924	0.9563	0.3057	62	0.8829	0.4695	1.8807
18	0.3090	0.9511	0.3249	63	0.8910	0.4540	1.9626
19	0.3256	0.9455	0.3443	64	0.8988	0.4384	2.0503
20	0.3420	0.9397	0.3640	65	0.9063	0.4226	2.1445
21	0.3584	0.9336	0.3839	66	0.9135	0.4067	2.2460
22	0.3746	0.9272	0.4040	67	0.9205	0.3907	2.3559
23	0.3907	0.9205	0.4245	68	0.9272	0.3746	2.4751
24	0.4067	0.9135	0.4452	69	0.9336	0.3584	2.6051
25	0.4226	0.9063	0.4663	70	0.9397	0.3420	2.7475
26	0.4384	0.8988	0.4877	71	0.9455	0.3256	2.9042
27	0.4540	0.8910	0.5095	72	0.9511	0.3090	3.0777
28	0.4695	0.8829	0.5317	73	0.9563	0.2924	3.2709
29	0.4848	0.8746	0.5543	74	0.9613	0.2756	3.4874
30	0.50	0.8660	0.5774	75	0.9659	0.2588	3.7321
31	0.5150	0.8572	0.6009	76	0.9703	0.2419	4.0108
32	0.5299	0.8480	0.6249	77	0.9744	0.2250	4.3315
33	0.5446	0.8387	0.6494	78	0.9781	0.2079	4.7046
34	0.5592	0.8290	0.6745	79	0.9816	0.1908	5.1446
35	0.5736	0.8192	0.7002	80	0.9848	0.1736	5.6713
36	0.5878	0.8090	0.7265	81	0.9877	0.1564	6.3138
37	0.6018	0.7986	0.7536	82	0.9903	0.1392	7.1154
38	0.6157	0.7880	0.7813	83	0.9925	0.1219	8.1443
39	0.6293	0.7771	0.8098	84	0.9945	0.1045	9.5144
40	0.6428	0.7660	0.8391	85	0.9962	0.0872	11.4301
41	0.6561	0.7547	0.8693	86	0.9976	0.0698	14.3007
42	0.6691	0.7431	0.9004	87	0.9986	0.0523	19.0811
43	0.6820	0.7314	0.9325	88	0.9994	0.0349	28.6363
44	0.6947	0.7193	0.9657	89	0.9998	0.0175	57.2900
45	0.7071	0.7071	1	90	1	0	Undefined

Extra Practice

For use after Chapter 1.

Find the value of each expression.

1. $189 + 556$ **2.** 47×31 **3.** $4896 \div 8$ *1-1*

4. $29 - x$ when $x = 12$ **5.** $\dfrac{7 + y}{2}$ when $y = 9$

6. $7(10 + 68)$ **7.** $[4(9) - 2(3)] \div 5$ **8.** $\dfrac{(12 \div 4)8}{1 + 5}$ *1-2*

9. $3(9 + t) - 4$ when $t = 8$ **10.** $8v + 42$ when $v = 6$

11. $6(7^2 - 1^3)$ **12.** $6^2 \div (2^3 + 1)$ **13.** $(7 + 4^2)^2$ *1-3*

14. $x^2 + 2x + 1$ when $x = 3$ **15.** $4(5w^2 - 7w)$ when $w = 2$

Draw a horizontal number line and on it show the graphs of the given numbers.

16. 3 **17.** $-1\frac{1}{2}$ **18.** 0 **19.** 2.5 *1-4*

20. State the coordinate of the point on the number line at which you would arrive if you were to start at the origin, move 6 units in the negative direction, and then move 4 units in the positive direction.

Find the decimal equivalent for each number.

21. $\frac{13}{25}$ **22.** $\frac{1}{18}$ **23.** $-\frac{31}{36}$ **24.** $-\frac{53}{80}$ *1-5*

25. Find the rational number that is halfway between -1.9 and -1.6.

Replace $\underline{?}$ by $\in$ or $\subset$ to make a true statement.

26. $1 \underline{?} \{1, 2\}$ **27.** $\{0\} \underline{?} \{0\}$ *1-6*

28. $\varnothing \underline{?} \{1, 3, 5\}$ **29.** $\{5\} \underline{?} \{1, 3, 5\}$

30. Draw a Venn diagram to illustrate $x \in A$, $x \notin B$, and $B \subset A$.

Draw the graph of each set on a number line. Is the set finite or infinite?

31. $\{-2\frac{1}{2}, -1, 0, 3\}$ **32.** $\{-5, -3\frac{1}{2}, -1\frac{1}{2}, 2\}$ *1-7*

33. {the negative real numbers greater than -4}

34. {the integers less than or equal to 2}

35. {the real numbers greater than 1 and less than -1}

Find the solution set of the given open sentence if $x \in \{0, 2, 4\}$.

36. $2x + 1 = 9$ **37.** $15 - 3x > 7$ **38.** $x^2 \leq 4x$ *1-8*

Find and graph the solution set of the given open sentence if the domain of y is {the positive real numbers}.

39. $3 \geq y$ **40.** $y^2 - 1 = 0$ **41.** $3 < 5y - 7$

Translate the word sentence into a mathematical sentence and find the solution set over the set of real numbers.

42. The difference of a and six is eight. *1-9*

43. The product of seven and b is seventy.

44. The sum of one and the quotient of c and four is nine.

Translate the open mathematical sentence into a word sentence.

45. $r + 8 = 11$ **46.** $\dfrac{x}{5} = 1$ **47.** $20 - w = 10$

For use after Chapter 2.

Show that each statement is true by finding a value of the variable for which the statement is true.

1. For some integer n, $2n - 1 > 8$. *2-1*

2. There is a whole number x such that $x^2 = 9$.

3. For at least one real number z, $3z \leq 2z$.

Show that each statement is false by finding a value of the variable for which the statement is false.

4. Any natural number y satisfies the equation $y^2 - 5y \neq 0$.

5. For all integers n, $3n + 1 \geq 4$.

6. For every real number a, $a^3 > a^2$.

Simplify each expression.

7. $14 + 8 + 6 + 22$ **8.** $23 + 19 + 37$ *2-2*

9. $\frac{1}{2} \cdot 27 \cdot 8 \cdot \frac{2}{3}$ **10.** $\frac{1}{8} \cdot \frac{1}{5} \cdot 16 \cdot 35$

11. $t + 9 + 3t + 2t^2$ **12.** $8b + 2d + 2d + 7b$ *2-3*

13. $2\frac{1}{2} \times 9 + \frac{1}{2} \times 9$ **14.** $1\frac{5}{6} \times 8 + 2\frac{1}{6} \times 8$

15. $5 + 3(1 + k) + 6k$ **16.** $4(a + b) + (a + 2b)$

Simplify each expression.

17. $(7 + {}^-8) + {}^-5$ **18.** ${}^-3 + ({}^-11 + {}^-5)$ **19.** $({}^-6 + 4) + 10$ *2-4*

20. ${}^-9 + ({}^-2 + 15)$ **21.** $({}^-1 + {}^-8) + 13$ **22.** ${}^-18 + (5 + {}^-5)$

Solve each equation over the set ℛ of real numbers.

23. $x + 10 = 2$ **24.** $^-3 + y = {^-1}$ **25.** $z + {^-4} = 3$

Simplify each expression.

26. $2 + [-(-9)]$ **27.** $-(-18 + 13)$ **28.** $-(-1) + (-7)$ 2-5
29. $-[-6 + (-6)]$ **30.** $-[5 + (-5)] + (-8)$
31. $4 + [-(10 + (-1))]$ **32.** $-(-4 + 6) + [-(-3)]$

Find the value of each expression.

33. $-2|-2|$ **34.** $|4| + |12 + (-12)|$ **35.** $|-3 + 8| + (-11)$ 2-6
36. $|-7| + |10 + (-15)|$
37. $-|-6 + (-5)|$ **38.** $-3|-9| + 3|-9|$

Solve each equation over the set ℛ of real numbers.

39. $|z| + (-8) = 0$ **40.** $-1 + |y| = 3$ **41.** $-|x| = 2$

For use after Chapter 3.

Simplify each expression.

1. $-3 + 18$ **2.** $29 + (-43)$ **3.** $31 + (-15)$ **4.** $-67 + 44$ 3-1
5. $19 + (-52) + (-21) + 37$ **6.** $-40 + (-118) + 74 + (-13)$
7. $-(-32 + 55) + [86 + (-29)]$ **8.** $-[-93 + 61 + 45 + (-12)]$
9. $2\frac{5}{8} + 1\frac{1}{8} + (-5\frac{3}{8}) + (-1\frac{7}{8}) + 4\frac{1}{8}$
10. $1.9 + (-1.4) + 3.8 + (-2.57) + (-0.42)$

11. $5r + (-2)s + (-9)r + 3s$ 3-2
12. $7b^2 + (-8)b + (-4)b + (-10)b^2 + 3b$
13. $(-1)m^2 + 4m + (-9) + m^2 + (-3)m + 7$
14. $3x^2 + (-4)xy + 10y^2 + x^2 + xy + (-11)y^2$
15. $(-6)k^5 + k + (-2)k^3 + (-14)k + 12k^3 + 6k^5$
16. $(-15)j^2 + 9jt + (-13)t^2 + 8t^2 + (-10)j^2 + 2jt$

Simplify each expression.

17. $[29 + (-53)](-8)$ **18.** $(-10)(13) + (-17)(-25)$ 3-3
19. $(-1)^7 + 4(-37)$ **20.** $(-6)(0) + (-14)(-60)$
21. $-2[5x + (1 - 9)y]$ **22.** $7[-8a + (-b)]$
23. $-1[-s + (-3)s^4]$ **24.** $-5z + 3m + [-(-z + 8m)]$
25. $-p^2 + 8p^4 + p^2 + (-10)p^4 + 3p^2$
26. $e^2 + (-12)ef + (-2)f^2 + (-e^2) + 7ef + (-f^2)$

Simplify each expression.

27. $6(-8c + 2d) + (-4)[-3d + (-c)]$

28. $-10(3g^2 + 8g) + (-9)[-g + 5g^2]$

29. $\frac{1}{7}(-35)$ **30.** $-\frac{1}{9}(-54)$ **31.** $-\frac{1}{2}(-50)(-\frac{1}{5})$ 3-4

32. $\frac{1}{-6}\left(-\frac{1}{2}\right)(36)$ **33.** $\frac{1}{15}(-15xy)$ **34.** $\left(-\frac{1}{3}\right)\left(\frac{1}{-7}\right)(-210)$

35. $\frac{1}{y}(-3xyz),\ y \neq 0$ **36.** $12pt\left(\frac{1}{p}\right),\ p \neq 0$

37. $\left(-\frac{1}{a^2}\right)(24a^2)\left(-\frac{1}{4}\right),\ a \neq 0$ **38.** $\frac{1}{5}(10x + 25)$

39. $-\frac{1}{9}[-90 + (-45t)]$ **40.** $(-12 + 21b)\frac{1}{3}$

41. $\frac{1}{4}[36s^2 + (-28t^2)]$ **42.** $[60g^2 + 6gh + (-18h^2)](-\frac{1}{6})$

For use after Chapter 4.

Solve each equation.

1. $a + 15 = 23$ **2.** $z + 12 = 4$ **3.** $17 = t + 17$ 4-1

4. $-10 = b + (-5)$ **5.** $-15 + q = 9$ **6.** $-1 = -1 + d$

Simplify each expression.

7. $-12 - 20$ **8.** $30 - (-20)$ **9.** $-[-9 - (-8)]$ 4-2

10. $(a + 5) - (a - 1)$ **11.** $(\frac{1}{4} - \frac{5}{6})24 - 8$

Solve each equation.

12. $18 - z = 20$ **13.** $y + 3 = -6$ **14.** $2h - (h - 3) = 0$

Solve each equation.

15. $-\frac{1}{6}x = 4$ **16.** $3 = \frac{1}{5}b$ **17.** $-8k = -72$ 4-3

18. $29r = -29$ **19.** $-3 = -\frac{1}{19}c$ **20.** $\frac{1}{7}s = -\frac{20}{7}$

Simplify each expression.

21. $10 \div (-\frac{1}{5})$ **22.** $-99 \div (-11)$ **23.** $-\frac{2}{9} \div \frac{1}{3}$ 4-4

24. $-42r^2 \div (-21)$ **25.** $-\frac{7}{8}a \div \frac{1}{40}$ **26.** $\frac{7}{6}x \div (-\frac{1}{3})$

Solve each equation.

27. $-108 = 4r$ **28.** $-17b = 323$ **29.** $f \div 12 = 30$

30. $3y - 8 = 7$ **31.** $-14 = 5z + 16$ **32.** $-4 + 2t - 3t = 9$ 4-5

33. $6n + 1 - 10n = -23$ **34.** $c - \frac{5}{6}c + 4 + \frac{1}{6}c = 4$

35. $3(z - 3) + 4z + 1 = -22$

Represent the English sentence(s) by an equation.

4-6

36. The perimeter of a square s units on a side is 96 units.

37. Tony saves \$3.20, or $\frac{1}{5}$ of the price P, when he buys a sweater on sale.

38. Nine less than 4 times a number x is -1.

39. The atomic number of iron is 2 less than the atomic number, n, of nickel. The sum of these atomic numbers is 54.

40–43. Solve each equation written in Exercises 36–39.

4-7

Solve each equation.

4-8

44. $5x - 48 = -3x$ **45.** $12 + 7r = 5r$

46. $9t - 5 = -t + 10t$ **47.** $1 + 2z = 16 - z$

48. $4b + 13 = -2b - 41$ **49.** $5(2 - b) = -3(b + 2)$

Solve for x, y, or z.

4-9

50. $a = 2x - b$ **51.** $A = \pi r y$ **52.** $bt + az = ab$

53. $s^2 = 2px$ **54.** $A = 2\pi r^2 + 2\pi r y$ **55.** $p = \dfrac{a}{a + z}$

For use after Chapter 5.

Supply the reasons for the proof of the following: If x, y, and z are real numbers and $x + y < z$, then $x < z + (-y)$.

5-1

1. x, y, and z are real numbers and $x + y < z$.

2. $(x + y) + (-y) < z + (-y)$ **3.** $x + [y + (-y)] < z + (-y)$

4. $x + 0 < z + (-y)$ **5.** $x < z + (-y)$

Solve each inequality and graph the solution set.

5-2

6. $a + 3 < 1$ **7.** $5x - 2 \geq 18$ **8.** $6r - 5 > -r + 2$

9. $\dfrac{-2z - 7}{5} \leq 1$ **10.** $-3 + \dfrac{t}{5} > -2$ **11.** $4m - 8 < -m - 8 + 3m$

Specify the intersection and the union of the given sets.

5-3

12. $\{-3, -1, 0, 1\}$, $\{0, 1, 2\}$ **13.** $\{1, 4\}$, $\{1, 5, 10\}$

14. $\{0\}$, {the integers}   **15.** $\{0\}$, {the odd integers}

16. $\{-1, 0, 1\}$, {the whole numbers less than 3}

17. {the real numbers greater than -2}, {the real numbers less than 3}

Solve each open sentence.

5-4

18. $-1 < x + 2 \leq 4$ **19.** $-1 \leq 3a + 2 \leq 11$ **20.** $-10 < 4 - 7z < 25$

Solve each open sentence.

21. $w - 5 > 1$ or $w - 5 < -1$
22. $-2b + 3 \geq 13$ or $-2b + 3 \leq -13$
23. $-7 - 4a \geq 9$ or $-7 - 4a < -9$

24. $|x + 2| = 3$ 25. $|y - 3| > 0$ 26. $|4 - z| \leq 2$ 5-5
27. $|2w - 5| = 5$ 28. $|5 - 3d| < 11$ 29. $4|t| - 9 > -5$

30. Find three consecutive odd integers whose sum is -3.   5-6

31. Find the greatest two consecutive integers such that three times the lesser is at most 11 more than the greater.

32. Find the least two consecutive even integers whose sum is greater than 40.

33. The degree measures of the angles of a triangle are consecutive even integers. What are the measures?   5-7

34. Find the measure of an angle for which three times the measure of the complement of the angle is 50° more than the measure of its supplement.

35. A disabled freight train is traveling toward a station 140 km away at a speed of 32 km/h. Another train, traveling 80 km/h, leaves the station to pick up the freight. When will the two trains meet?   5-8

36. A car traveling 87 km/h and a bus leave a toll booth at the same time. Twenty minutes later, the bus is 3 km farther from the toll booth than the car. What is the average speed of the bus?

37. At the Keppler Planetarium, adult tickets cost $2.50 each and children's tickets cost $1.50 each. One day 408 tickets were sold in all, and $733 was collected. How many of each type of ticket were sold?   5-9

38. One molecule of pentane contains 5 atoms of carbon and 12 atoms of hydrogen. How many atoms each of carbon and hydrogen are contained in pentane composed of 1513 atoms in all?

For use after Chapter 6.

Plot each ordered pair of numbers.

1. $(0, 3)$ 2. $(1, -2)$ 3. $(-3, -1)$ 4. $(-5, 0)$ 6-1

State the domain and range of each relation. Is the relation a function?

5. $\{(2, 4), (3, 6), (-1, -2)\}$ 6. $\{(1, 2), (0, 2), (-7, 2)\}$ 6-2

7. $\{(0, 0), (-1, 1), (-1, -1)\}$ 8. $\{(-2, 7), (-1, 4), (0, 1)\}$

Find the values of *a* and *b* for which the equation is true.

9. $(a + 1, -2b) = (-2, 6)$ **10.** $(3a - 5, -b + 4) = (1, 4)$

The domain of each function is $D = \{-2, 0, 2, 4\}$. Compute the table of ordered pairs of the function and graph the function.

11. $f: x \rightarrow -x + 1$ **12.** $g: x \rightarrow 2x - 5$ **13.** $h: x \rightarrow (x - 2)^2$ 6-3

Graph the solution set of each equation. The domain of *x* is $\{-2, -1, 0, 1, 2\}$.

14. $y = 2x - 3$ **15.** $y = 3x + 4$ **16.** $y = -x^2 + 1$ 6-4

17. $5x - 2y = 4$ **18.** $y = (x + 1)^2$ **19.** $y = 4 - x^2$

Graph each equation in a coordinate plane.

20. $x = -2$ **21.** $y = 0$ **22.** $y = x + 2$ 6-5

23. $y = \frac{3}{2}x$ **24.** $3x - 2y = 12$ **25.** $3x + 5y = 15$

y varies directly as *x*. *y* is 10 when *x* is 14. Find *y* when:

26. $x = 35$ **27.** $x = -21$ **28.** $x = 91$ 6-6

Find the value of the variable for which each proportion is true.

29. $\dfrac{x - 3}{4x} = \dfrac{1}{8}$ **30.** $\dfrac{5z - 10}{2z + 7} = \dfrac{2}{3}$ **31.** $\dfrac{-5}{m + 8} = \dfrac{2}{-m + 1}$

Graph the line whose slope and *y*-intercept are given. Write an equation for the line, in the form $Ax + By = C$, where *A*, *B*, and *C* are integers.

32. $m = 1, b = -2$ **33.** $m = -2, b = 3$ **34.** $m = 0, b = -2$ 6-7

Find an equation of the line having the given slope (if any) and passing through the given point.

35. $5; (3, -4)$ **36.** $-2; (0, 6)$ **37.** $-\frac{3}{7}; (-6, 1)$ 6-8

38. $\frac{4}{5}; (-10, -3)$ **39.** $0; (0, -1)$ **40.** No slope; $(8, 5)$

Find an equation of the line passing through the given points.

41. $(-4, 7), (2, 1)$ **42.** $(-6, 0), (3, -3)$ 6-9

43. $(-1, -3), (0, 1)$ **44.** $(4, -2), (8, 1)$

45. $(0, 0), (-5, 6)$ **46.** $(-6, -1), (0, -1)$

Graph each inequality.

47. $y \leq 2x + 3$ **48.** $5x - 2y < 10$ **49.** $x + 3y > 6$ 6-10

Solve each system graphically.

1. $-x + y = 3$
   $x + y = 5$

2. $y = -2x + 1$
   $2x + y = 3$

3. $y = -x$
   $3x + y = -6$

7-1

Use addition or subtraction to solve each system.

4. $2x + 7y = -24$
   $2x + 5y = -20$

5. $x - 3y = 18$
   $8x + 3y = -18$

6. $5x - 2y = 12$
   $3x - 2y = 4$

7-2

Use linear combinations to solve each system.

7. $6x + 11y = -12$
   $-4x + 7y = 8$

8. $x + 4y = -1$
   $-2x - 8y = 2$

9. $4x + 9y = 22$
   $3x + 5y = 6$

7-3

10. $5x - 3y = 4$
    $15x - 9y = 8$

11. $2x + 3y = 17$
    $3x - 7y = 14$

12. $7x - 5y = 1$
    $5x + 2y = -16$

Use substitution to solve each system.

13. $2x + 5y = -13$
    $x - 3y = 21$

14. $4x + 3y = 43$
    $y = 2x + 1$

15. $12x + y = -20$
    $11x - 4y = 21$

7-4

Solve each problem.

16. One line has slope 5 and y-intercept 2. Another line has slope -3 and y-intercept -6. At what point do the lines intersect?

7-5

17. The density in kilograms per cubic meter of nitrogen is 0.23 less than that of phosphine. If 5 times the density of phosphine is 0.1 less than 6 times the density of nitrogen, find the density of phosphine.

18. Tom rowed 6 km in 1.5 h against the current. He then rowed 10 km with the current in just 1.25 h. How fast can Tom row in still water?

19. In still air a plane can fly 725 km/h. The plane flew for 84 min with a tail wind. The return trip, with a head wind, took 90 min. Find the total distance traveled.

7-6

20. The ones digit of a two-digit numeral is 1 more than twice the tens digit. The sum of the digits is 10. Find the number.

7-7

21. The sum of the digits of a two-digit numeral is 7. The number with its digits reversed is 27 more than the original number. Find the original number.

Graph each system of inequalities.

22. $x \geq 3$
    $y < -2$

23. $y > -x$
    $x + y \leq 4$

24. $5x - 2y < 10$
    $4x + y > -5$

7-8

Shade the region that represents the solution set of the system: $x \geq 0$, $y \geq 0$, $y \leq 3x + 6$, $3x + 2y \geq 8$, $x + 2y \geq 4$.

7-9

25. Name the corner points.

26. Evaluate the expression $4x - y$ at each corner point.

27. Evaluate the expression $x + 2y$ at each corner point.

Find the solution set of each system.

7-10

28. $\begin{aligned} x + y + z &= 1 \\ x + 2y - z &= -1 \\ 2x - y - 3z &= 8 \end{aligned}$

29. $\begin{aligned} -x - y + z &= 7 \\ x + 2y + 3z &= 7 \\ 4x - 3y - 2z &= -6 \end{aligned}$

For use after Chapter 8.

Simplify.

8-1

1. $(3x^4 - 2x^2 + x) + (x^2 + 5x - x^3)$

2. $(a^2 + 4ab + 2b^2) - (a^2 - b^2)$

3. $(2s^2t - 3st - st^2) + (2s^2t + st + st^2)$

4. $(-z^4 + 6z^2 - 1) - (5z^3 - z^2 + 7z)$

8-2

5. $(-3x^2)(-9x^3)$

6. $(2r^4w)(-rw^3)$

7. $2c(5c^2)(-3c^2)$

8. $(8d^3)(3d)^2$

9. $(-j^2k)^2(jk)^3(-k)$

10. $(5y^2)^3(-3y^5)$

8-3

11. $f^5(2f^2 - f - 5)$

12. $g^2h(h^3 + 4h + 2)$

13. $-4mn(3m^4 - m^2n^2 + 7n^2)$

14. $3pv^4(-5p^3v^2 + 2p^2v^3 + p^4v)$

15. $(q + 4)(q - 3)$

16. $(r - 8)(r - 6)$

8-4

17. $(2x - 1)(3x + 4)$

18. $(7c + 2d)(3c + 5d)$

19. $(3t + 2y)^2$

20. $(5 + 2z^3)(2z^3 - 5)$

21. $(ab - c)^2$

22. $(2a + 1)^3$

Find the greatest common factor of the given monomials.

8-5

23. 462, 825

24. 648, 900

25. $36c^2d^3, -27c^3d$

26. $64xy^5z^4, -72x^2y^5z^3$

Factor completely.

8-6

27. $6a^2 + 2a - 4$

28. $-5x^3y + 7x^2y + 3x^2y^2$

29. $12a^3b^2 + 21a^2b^2c - 3abc^2$

30. $2m^5n^2 - 5m^4n^4 + 6m^3n^5$

31. $y(y + 8) - b(y + 8)$

32. $(5k + 1) + k^3(5k + 1)$

8-7

33. $36c^2 - 25d^2$

34. $x^2 + 4xy + 4y^2$

35. $50z^3 - 20z^2 + 2z$

36. $9h^3 - 3h^2 + 21h^2 - 7h$

37. $35x^3 + 7x + 40x^2 + 8$

38. $16a^3 - ab^2 + 16a^2b - b^3$

Factor completely.

8-8

39. $t^2 + 8t + 12$ **40.** $m^2 + 3m - 28$

41. $p^3 - 11p^2 + 18p$ **42.** $-4z^2 + 4z + 48$

43. $-s^4 + 4s^3 - 3s^2$ **44.** $y^3 + 8y^2 - 20y$

8-9

45. $3x^2 + 11x + 6$ **46.** $5a^2 - 38a - 63$

47. $12n^2 - 7n - 10$ **48.** $4k^2 + 16k + 7$

49. $15y^2 + y - 2$ **50.** $36b^2 - 48b + 7$

Solve each equation.

8-10

51. $5x^2 = 35x$ **52.** $r^2 - 6r + 5 = 0$ **53.** $q^2 - 2q - 24 = 0$

54. $25z^2 - 9 = 0$ **55.** $24b^2 + 2b = 15$ **56.** $4m^3 + 13m^2 + 3m = 0$

8-11

57. The base of a triangle is 5 cm longer than the height. The area is 12 cm². Find the dimensions of the triangle.

58. Find all pairs of consecutive integers such that twice their product is 12.

59. Find two positive numbers whose difference is 5 and whose product is 84.

For use after Chapter 9.

Simplify each expression.

9-1

1. $\dfrac{24a^5b^2}{4ab^3}$ **2.** $\dfrac{-18t^2z}{6t^2z^2}$ **3.** $\dfrac{-7r^3s^4}{-35r^4s^7}$ **4.** $\dfrac{12gh^2}{15hr}$

5. $\dfrac{30x^8y^{11}}{-18x^2y^3}$ **6.** $\dfrac{72w^3p^8}{40w^9p^6}$ **7.** $\dfrac{2(a+2b)}{22(a+2b)^3}$ **8.** $\dfrac{n^5(n+8)^4}{-n^2(n+8)^2}$

9-2

9. $\dfrac{6x^2y^2 + 8xy - 14y}{2y}$ **10.** $\dfrac{27m^4 - 30m^3 - 12m^2}{-3m}$

11. $\dfrac{-28d^2 + 7d + 21}{7d^2}$ **12.** $\dfrac{8r^5s^5 + 5r^3s^7 + 9rs^9}{r^2s^3}$

13. $\dfrac{4ab - 16a^2b^2 - 32b^4}{8ab^2}$ **14.** $\dfrac{45c^3j^3 + 65c^2j^2 - 35c^6j^4}{5c^2j^2}$

9-3

15. $\dfrac{8y^2 + 6y - 5}{4y + 5}$ **16.** $\dfrac{t^2 - 64}{t + 8}$

17. $\dfrac{9y^2 - 36y + 32}{3y - 8}$ **18.** $\dfrac{2p^2 + 9p + 3}{2p - 7}$

19. $(n^3 + 1) \div (n + 1)$ **20.** $(z^3 - 3z^2 - z + 3) \div (z - 3)$

21. $\dfrac{3x^2 + 17x + 20}{x + 4}$ **22.** $\dfrac{15r^2 - 8r - 7}{5r - 1}$

23. $\dfrac{4a + 12}{a^2 + 3a}$

24. $\dfrac{3x}{2x^2 - 5x}$

25. $\dfrac{z^2 + 2z}{z^2 - 4}$

9-4

26. $\dfrac{6a - 1}{12a^2 + 16a - 3}$

27. $\dfrac{y^2 - 25}{2y^2 + 15y + 25}$

28. $\dfrac{m^2 + 5m - 24}{m^2 + m - 12}$

Express each result using only positive exponents.

29. $2^{-3} \cdot 3^{-2}$

30. $(5ax^2)^{-2}$

31. $(r^{-3}s^0t^2)^{-3}$

9-5

32. $\dfrac{7z^{-1}}{42z^{-4}}$

33. $\dfrac{3^{-4}j^{-1}k^2r^{-5}}{9^{-1}jk^{-3}r^2}$

34. $\dfrac{2^4b^{-3}c^3d^4}{-8b^{-2}cd^0}$

Simplify.

35. $\dfrac{-15a}{4} \cdot \dfrac{2}{a^2}$

36. $\dfrac{2y - 14}{-5y} \div \dfrac{3y - 21}{5y^3}$

9-6

37. $\dfrac{z^2 + 3z}{-z^2 + 2z} \cdot \dfrac{7z - 14}{7z + 21}$

38. $(2r^2 + 3rs) \div \dfrac{2rs + 3s^2}{rs^2 - s^3}$

39. $\dfrac{16c^2 - d^2}{6c + 3d} \cdot \dfrac{2cd + d^2}{4cd - d^2}$

40. $\dfrac{x^2 + x}{-x^2 - 2x - 1} \div \dfrac{x^2 - 3x}{2x^2 - 2}$

41. $\dfrac{4t - 1}{3t} + \dfrac{-t + 7}{3t}$

42. $\dfrac{m + 1}{m - 5} - \dfrac{m - 6}{m - 5}$

9-7

43. $\dfrac{-a + 5b}{3a + 3b} + \dfrac{2a - 4b}{3a + 3b}$

44. $\dfrac{y + 3z}{y - 4z} - \dfrac{5y - z}{y - 4z}$

45. $\dfrac{-7}{8n - 4} + \dfrac{-4n + 9}{8n - 4}$

46. $\dfrac{3x + 1}{2w + 10x} - \dfrac{x + 5}{2w + 10x}$

47. $\dfrac{5a}{b^2} + \dfrac{3b}{a^2}$

48. $\dfrac{j - k}{2j} - \dfrac{j + 4k}{3k}$

9-8

49. $\dfrac{p + 1}{4p} + \dfrac{2 - p}{4p^2}$

50. $\dfrac{9}{9r + 3s} - \dfrac{2}{12r + 4s}$

51. $\dfrac{1}{g - 4h} + \dfrac{1}{20h - 5g}$

52. $\dfrac{3}{m - n} - \dfrac{1}{m + n}$

For use after Chapter 10.

Find the solution set of each open sentence.

1. $\dfrac{3z}{4} + \dfrac{z}{7} = 1$

2. $\dfrac{a}{6} + \dfrac{5a}{2} > \dfrac{1}{3}$

10-1

3. $\dfrac{x - 7}{8} < \dfrac{x + 3}{6}$

4. $\frac{3}{5}p - 2 = \frac{1}{2}p$

5. $\dfrac{6r + 5}{9} - \dfrac{r - 3}{6} = \dfrac{r}{2}$

6. $\dfrac{2m}{3} + \dfrac{7}{10} \ge \dfrac{11}{15} + \dfrac{5m}{6}$

Solve.

7. 18% of 25 = a **8.** 2.7% of 65 = b **9.** 138% of c = 151.8   *10-2*

10. 24% of d = 108   **11.** e% of 128 = 48   **12.** f% of 75 = 3

Solve each equation over $\mathcal{R}$.

13. $\dfrac{7}{x} - \dfrac{5}{2x} = 3$ **14.** $\dfrac{1}{q} + \dfrac{4}{5q} = \dfrac{3}{10}$ *10-3*

15. $\dfrac{2}{z} = \dfrac{9}{z+1} + \dfrac{11}{2z}$ **16.** $\dfrac{4}{2x+11} = \dfrac{2}{x-8}$

17. $\dfrac{2}{z+11} = \dfrac{7}{-9z+1}$ **18.** $\dfrac{3x}{x+4} - \dfrac{2}{x-3} = 3$

19. The sum of two numbers is 48 and their quotient is $\frac{5}{7}$. Find the   *10-4*
numbers.

20. The denominator of a fraction is 7 less than three times the numerator, and the fraction is equal to $\frac{2}{5}$. Find the fraction.

21. Twice the sum of a number and its reciprocal is equal to four times the number. Find the number.

22. It takes Janet 4 h to take inventory of the stock in her store. Her   *10-5*
assistant needs 6 h to do the job. How long would it take to do the job if they work together?

23. A bank machine requires 45 s to process each transaction. A newer model can process each transaction in 30 s. How long will it take for the machines to process 40 transactions, working together?

24. Jogging at a constant rate, Miguel needs 6 min less than twice as long   *10-6*
to jog 7.5 km as 4 km. Find Miguel's rate in kilometers per hour.

25. The O'Keefe family drove 126 km at a certain speed and then took a scenic route of 28 km at two thirds their original speed. If the entire trip was 2 h long, what was the original speed?

If a varies inversely as b and $a = 15$ when $b = 12$, find b when:

26. $a = 18$ **27.** $a = 4$ **28.** $a = 22.5$ **29.** $a = 5\frac{5}{8}$ *10-7*

30. If x varies directly as y and inversely as z, and $x = 8$ when $y = 12$ and   *10-8*
$z = 5$, find x when $y = 9$ and $z = 2$.

31. If r varies jointly as s and t^3, and $r = 54$ when $s = 7$ and $t = 3$, find r when $s = 21$ and $t = 2$.

For use after Chapter 11.

Simplify each expression.

1. $\sqrt{49}$ **2.** $-\sqrt{(-5)^2}$ **3.** $\sqrt{\frac{1}{100}}$ **4.** $-\sqrt{13^2}$ *11-1*

Express the number using a radical symbol.

5. 14   **6.** $-\frac{3}{7}$ **7.** -11 **8.** $\frac{9}{4}$

Express in the form $\dfrac{a}{b}$, where a and b are integers.

9. $0.\overline{62}$ **10.** 5.025 **11.** $0.0\overline{08}$ **12.** $0.\overline{345}$ *11-2*

Solve over $\mathfrak{R}$.

13. $x^2 + 1 = 50$ **14.** $y^2 - 3 = 19$ **15.** $4z^2 + 1 = 53$

Use the method of successive approximations to find each square root correct to three digits.

16. $\sqrt{74}$ **17.** $\sqrt{9.5}$ **18.** $\sqrt{304}$ **19.** $\sqrt{771}$ *11-3*

Use the table of square roots to evaluate each expression correct to hundredths.

20. $\sqrt{12} + 3\sqrt{7}$ • **21.** $8 + 2\sqrt{15}$ **22.** $\dfrac{\sqrt{79} - 2}{5}$ **23.** $\dfrac{-4\sqrt{3}}{9}$

State whether or not each set is a Pythagorean triple.

24. $\{9, 12, 15\}$ **25.** $\{8, 15, 18\}$ **26.** $\{6, 9, 11\}$ **27.** $\{20, 21, 29\}$ *11-4*

Find the distance between the two points having the given coordinates. Leave each answer in radical form.

28. $(-2, -5), (1, 0)$ **29.** $(8, 3), (4, -2)$ *11-5*
30. $(0, -7), (1, -3)$ **31.** $(-6, -9), (-1, 4)$
32. $(3, 1), (5, 7)$ **33.** $(2, -8), (-5, 3)$

Simplify. Assume that all variables denote positive real numbers.

34. $\sqrt{54}$ **35.** $\sqrt{8n^2}$ **36.** $\sqrt{81z^6}$ *11-6*
37. $\sqrt{5} \cdot \sqrt{35}$ **38.** $\sqrt{2x} \cdot \sqrt{32x}$ **39.** $\sqrt{20t}$
40. $\sqrt{3q^3} \cdot \sqrt{6q^2}$ **41.** $\sqrt{80c^2d^3}$ **42.** $\sqrt{72fg^4}$

43. $\dfrac{8}{\sqrt{a}}$ **44.** $\sqrt{\dfrac{5k}{7}}$ **45.** $\dfrac{\sqrt{39}}{\sqrt{12}}$ *11-7*

46. $\dfrac{\sqrt{30}}{\sqrt{75}}$ **47.** $\sqrt{\dfrac{40}{3y}}$ **48.** $\sqrt{\dfrac{7s}{44}}$

49. $5\sqrt{48} - 9\sqrt{3} + \sqrt{75}$ **50.** $-2\sqrt{2} + \sqrt{72} - 5\sqrt{18}$ *11-8*
51. $\sqrt{9r^3} + 8r\sqrt{r} - \sqrt{r^3}$ **52.** $2\sqrt{5}(7 + 3\sqrt{30})$
53. $(6\sqrt{2} + 1)(3\sqrt{2} - 4)$ **54.** $(4\sqrt{3} - \sqrt{7})(4\sqrt{3} + \sqrt{7})$

For use after Chapter 12.

Solve by completing the square. Express all radicals in simplified form.

1. $x^2 + 2x - 35 = 0$ 2. $z^2 + 7z + 12 = 0$ 12-1

3. $a^2 + 6a + 7 = 0$ 4. $b^2 - b - 4 = 0$

5. $d^2 - 8d + 3 = 0$ 6. $m^2 + 5m + 1 = 0$

7. $y^2 - 12y - 7 = 0$ 8. $k^2 + 9k - 2 = 0$

Use the quadratic formula to solve each equation. Express all radicals in simplified form.

9. $2r^2 + r - 5 = 0$ 10. $3c^2 + 20c - 32 = 0$ 12-2

11. $5g^2 - 3g = 0$ 12. $x^2 - 7x + 4 = 0$

13. $25j^2 + 10j + 1 = 0$ 14. $4y^2 + 9y + 2 = 0$

15. $4q^2 - 3q - 3 = 0$ 16. $5w^2 + 3w - 1 = 0$

Sketch the graph of each function and estimate the zeros, if any.

17. $f: x \rightarrow -x^2$ 18. $g: x \rightarrow \frac{1}{3}x^2$ 12-3

19. $h: x \rightarrow x^2 + 3x$ 20. $F: x \rightarrow (x - 3)^2$

21. $G: x \rightarrow 12 + 4x - x^2$ 22. $H: x \rightarrow 2x^2 - 5x - 3$

23. $f: x \rightarrow x^3 + 2$ 24. $g: x \rightarrow x^2 - x^3$ 12-4

25. $h: x \rightarrow x^3 - 9x$ 26. $F: x \rightarrow \frac{1}{2}x^4$

27. $G: x \rightarrow x^3 - 2x^2 - x + 2$ 28. $H: x \rightarrow x^3 - 7x - 6$

Graph each inequality.

29. $y > x^2 - 4$ 30. $y \leq -5x - x^2$ 12-5

31. $y < x^2 + 4x + 4$ 32. $y \geq 4 - 3x - x^2$

33. $y \leq 2x^2 + 11x + 15$ 34. $y > 3x^2 - 4x - 4$

For use after Chapter 13.

Picture each angle with the given measure as a rotation.

1. $90°$ 2. $-120°$ 3. $390°$ 4. $-560°$ 13-1

Give the sine, cosine, and tangent of an angle A in standard position whose terminal side contains the given point on the unit circle.

5. $(\frac{8}{17}, -\frac{15}{17})$ 6. $(-\frac{21}{29}, -\frac{20}{29})$ 7. $\left(-\frac{\sqrt{2}}{2}, \frac{\sqrt{2}}{2}\right)$ 13-2

The quadrant of $\angle A$ is given, and also the value of $\sin A$, $\cos A$, or $\tan A$. Find the remaining two values.

8. $\cos A = -\frac{3}{5}$; II   9. $\sin A = -\frac{1}{2}$; IV   10. $\tan A = \sqrt{3}$; III

11. $\sin A = \frac{4}{5}$, I   12. $\cos A = -\frac{1}{2}$, III   13. $\tan A = -1$; II

Find the approximation in the table on page 504 for each value.

14. $\sin 83°$ 15. $\cos 12°$ 16. $\tan 65°$ 17. $\cos 48°$ *13-3*

Find the measure of $\angle A$, $1° \le m°\angle A \le 90°$, to the nearest degree.

18. $\sin A = 0.5909$ 19. $\cos A = 0.3611$ 20. $\tan A = 0.4837$

21. $\sin A = 0.1405$ 22. $\cos A = 0.2950$ 23. $\tan A = 5.6378$

Solve right triangle ABC using the given information and the table on page 504. Give angle measures to the nearest degree and lengths to the nearest tenth of a unit.

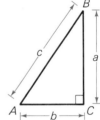

24. $m°\angle B = 25°$; $c = 4$ *13-4*

25. $m°\angle A = 50°$; $b = 3$

26. $m°\angle A = 65°$; $c = 10$

27. $m°\angle B = 45°$; $a = 7$

28. $a = 2$; $c = 8$

29. $a = 3$; $b = 5$

Find the norm and direction of **v** in standard position for the given terminal point of **v**.

30. $(-3, 0)$ 31. $(-4, -4)$ 32. $(12, -5)$ 33. $(-3, 5)$ *13-5*

The initial and terminal points of vectors **s** and **t** are given. Find the x- and y-components of **t** and the norm of the resultant of **s** and **t**.

34. **s:** $(1, 3)$, $(7, 4)$ 35. **s:** $(0, 0)$, $(5, 2)$ *13-6*
   t: $(0, 1)$, $(0, 8)$ **t:** $(0, 0)$, $(7, 3)$

36. **s:** $(0, 0)$, $(3, 8)$ 37. **s:** $(2, 0)$, $(-1, -5)$
   t: $(0, 0)$, $(-5, 2)$ **t:** $(-1, -1)$, $(-3, 0)$

Give answers to the nearest tenth of a unit for magnitudes and to the nearest degree for angle measures.

38. Greg hiked 8 km east and 4 km northeast. Where was he then in relation to his starting point?   *13-7*

39. Find the magnitude and the direction of the resultant force when an object is acted on by two forces at right angles to each other, one of 12 kg and the other of 35 kg.

Make a dot frequency diagram for the letters in each word. Find the frequency, relative frequency, and percent for the data.

1. initiate   **2.** dissension   **3.** successful   *14-1*

Use the data in the frequency distribution at the right. The data give the maximum speeds in kilometers per hour of some electric cars.

Speed	Frequency
70–90	2
90–110	5
110–130	3
130–150	1

4. Draw a histogram for the data.

5. Draw a frequency polygon for the data.

14-2

6–8. Find the cumulative frequency and cumulative percent for the letters in the words given in Exercises 1–3. (Put the letters in alphabetical order.)

14-3

Find the mode(s), the median, and the mean for the data given.

9. 3, 4, 6, 6, 7, 9, 12, 12   **10.** 2, 5, 6, 8, 9, 12, 15, 16, 17   *14-4*

11.

Number	Frequency
2	5
3	2
4	1
5	4

12.

Number	Frequency
1	2
3	3
5	1
7	3
9	1

Find the range, the variance, and the standard deviation for the data given.

13. 2, 6, 7, 9   **14.** 1, 2, 5, 9, 13   **15.** 4, 5, 5, 6   *14-5*

A jar contains 8 pennies, 5 nickels, 10 dimes, and 7 quarters. If a coin is drawn at random from the jar, what is the probability that the coin is:

16. a penny?   **17.** a nickel?   **18.** not a dime?   *14-6*

19. a nickel or a quarter?   **20.** a nickel or a dime?

An examination of 300 sample bicycles showed that 8 were defective, 285 were acceptable, and 7 exceeded standards. What is the experimental probability that a bicycle:

21. is defective?   **22.** is not defective?   **23.** exceeds standards?   *14-7*

Challenge Exercises

For use with Chapter 1.

1. Use each of the digits 0, 1, 2, . . . , 8, 9 exactly once, together with *1-2* addition symbols and fraction bars as needed, to write an expression for:

   a. 100   **b.** 200

2. Use exactly four 4's and any of the symbols $+$, $-$, $\times$, $\div$, $\sqrt{}$, and ! *1-3* to write an expression for each of the integers 0, 1, 2, . . . , 19, 20. Use grouping symbols as needed. (*Recall:* For any positive number n, $\sqrt{n}$ (read "square root of n") is the positive number whose square is n. For any positive integer n, $n!$ (read "n factorial") is the integer $n \times (n-1) \times (n-2) \times \cdots \times 1$. For example, $\sqrt{9} = 3$ and $5! = 5 \times 4 \times 3 \times 2 \times 1 = 120$.)

3. The decimal equivalent of a rational number is purely periodic if it *1-5* does not terminate and the first digit after the decimal point is part of the first repeating block. For example, $0.\overline{54}$ is purely periodic, but $0.8\overline{3}$ is not purely periodic. Devise a method for predicting whether or not the decimal equivalent of a rational number is purely periodic. (*Hint:* What is true of the denominator of such a rational number?)

For use with Chapter 5.

1. At a certain school 40 students belong to the science club, 36 belong to *5-3* the drama club, and 42 belong to the math club. Seventeen students are in both the science and drama clubs, 7 are in both the drama and math clubs, and 14 are in both the science and math clubs. Five students belong to all three clubs. How many students belong to at least one club? (*Hint:* Use a Venn diagram.)

For use with Chapter 6.

1. A function f assigns to the ordered pair (x, y) the greater of the two *6-3* numbers x and y if $x \neq y$, and x if $x = y$. Write a variable expression in x and y to complete the following statement of a rule for f:

   $$f: (x, y) \rightarrow \underline{?}$$

2. A certain airline allows each passenger k kilograms of baggage free *6-6* but charges c cents per kilogram for baggage in excess of the limit. On a particular flight, two passengers had a combined total of 52 kg of baggage. One of these passengers had to pay $1.00 for excess baggage; the other had to pay $1.40. A third passenger had 52 kg of baggage and had to pay an excess-baggage charge of $6.40.

   a. How many kilograms of baggage is each passenger allowed free?
   b. What is the excess-baggage charge?

3. Thermometers A, B, and C have different scales. When thermometer A registers a rise from 14° to 38°, thermometer B registers a rise from 10° to 26°. When thermometer B registers a rise from 15° to 27°, thermometer C registers a rise from 55° to 82°. If thermometer A registers a drop of 30°, what will be the drop on thermometer C? (*Hint:* What will be the drop on thermometer B?)

For use with Chapter 7.

1. Frieda bought some pears for 24¢ apiece and an equal number of apples for 18¢ apiece. When she arrived home, her roommate said, "Had you spent equal amounts for each of the two types of fruit, you would have bought two more pieces of fruit for the same amount of money." How much money did Frieda spend?   7-5

2. Leslie was rowing upstream and passed a branch floating downstream as his boat went under a bridge. Five minutes later Leslie started rowing back downstream and caught up with the branch 1 km downstream from the bridge.   7-6

   a. How many minutes after turning around did it take Leslie to catch up to the branch?
   b. What is the rate of the current?

3. Following his dance class each Saturday, Eddie was picked up by his sister who always left home in just enough time to arrive promptly. One Saturday the class was over one hour early, and Eddie started walking home at 2:00 o'clock. When he saw his sister driving by, he waved, she stopped the car, he got in, and they arrived home one-half hour earlier than usual. For how long did Eddie walk?   7-6

For use with Chapter 8.

1. Show that the sum of the squares of three consecutive integers cannot end in 1, 3, 6, or 8.   8-7

2. In how many ways can 288 be represented as a product of three different positive integers?   8-7

3. If n is an odd number, prove that 8 is a factor of $n^2 - 1$.   8-7

4. As a scout troop started on a hike, a scout advanced from the rear of the troop to the front to deliver a set of keys to the troop leader and returned immediately to the rear of the troop. The troop is 400 m long, and it advanced 400 m during the time it took the scout to complete the errand. How far had the troop advanced when the scout reached the front?   8-11

For use with Chapter 9.

1. The figure below is called a multiplication magic square when the products of the entries in each row, column, and diagonal are equal to a constant, say k. For example, $pqr = qtw = vtr = k$.

9-1

p	q	r
s	t	u
v	w	x

For what values of k between 0 and 100 can the figure be a multiplication magic square with positive integers as its entries?

For use with Chapter 10.

1. A store sold a certain suit at a gross profit of 10%; that is, the store sold the suit for 10% more than it had paid for the suit. Had the suit cost the store 10% less, the store would have been able to sell the suit for $2.50 less yet at a gross profit of 20%. For what price did the store sell the suit?

10-2

2. Viola competed in a boat race that consisted of going 50 km upstream and then returning to the starting point. She ran her boat, which is capable of going 25 km/h in still water, at its top speed throughout the race. If her average speed for the whole race was 24 km/h, what is the rate of the current?

10-3

For use with Chapter 11.

1. How many different Pythagorean triples have 60 as one of the two smaller integers?

11-4

For use with Chapter 12.

1. Find all the positive integers x and y such that
$$\frac{1}{x} + \frac{1}{y} = \frac{1}{6}.$$

12-1

For use with Chapter 14.

1. If four coins are tossed, what is the probability of tossing exactly two heads?

14-6

Answers to Self-Tests

Chapter 1, Self-Test 1, page 15
1. 82   2. 203   3. 19   4. 65
5.–8.

-3 -2.5 0.4 $1\frac{1}{4}$

9. 10   10. -5

Chapter 1, Self-Test 2, page 27
1. rational   2. rational   3. irrational
4. rational   5. {E, N, P, T, U}
6. months that begin with letter "J"   7. $\in$
8. $\subset$
9.

10.

Chapter 1, Self-Test 3, page 33
1.

2. $2(n + 3) = 12$ 3. $x - 6 = 3\left(\dfrac{x}{9}\right)$

Chapter 2, Self-Test 1, page 52
1. $n = 5$ 2. Answers may vary. Examples:
$y = {}^-1, y = {}^-2, y = {}^-3$ 3. a   4. $7m^2 + 12$
5. $12x + 22$ 6. (a) Substitution principle
(b) Distributive axiom   (c) Closure axiom
(d) Transitive property

Chapter 2, Self-Test 2, page 64
1.

2. -24 3. -7
4. $4\frac{1}{2}$ blocks north and 2 blocks east
5. -6 6. -1
7. $y = 1$ or $y = -1$

8. $-1 \leq z \leq 1$

Chapter 3, Self-Test 1, page 78
1. 30   2. 0   3. Substitution principle
4. Associative axiom of addition
5. Additive inverse axiom
6. Additive identity axiom
7. Transitive property

Chapter 3, Self-Test 2, page 91
1. -5 2. -176 3. Substitution
4. Associative axiom of multiplication
5. Multiplicative inverse axiom
6. Multiplicative identity axiom
7. Transitive property

Chapter 4, Self-Test 1, page 106
1. {4}   2. {4}   3. -326 4. $2p + 5$

Chapter 4, Self-Test 2, page 118
1. {312}   2. {$\frac{1}{3}$}   3. -14 4. 72
5. $\dfrac{rs^2}{2}$ 6. {42}   7. {-7}   8. {6}

Chapter 4, Self-Test 3, page 135
1. Let x be Celeste's age now. Then
$(x + 1) + (x - 3) = 24$ 2. {3}   3. -28
4. $l = 15$ cm, $w = 9$ cm   5. 25 min
6. quarters   7. 14 m

Chapter 5, Self-Test 1, page 158
1. b. Additive axiom of inequality
e. Multiplicative axiom of inequality
2. $x \leq 2$

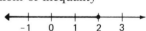

3. $A \cup B = \{-3, -2, -1, 0, 1, 2, 3, 4\}$,
$A \cap B = \{0, 1, 2, 3\}$
4. $0 \leq y < 6$

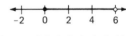

5. $z > 2$ or $z < -2$

6. $k > 7$ or $k < 1$

Chapter 5, Self-Test 2, page 178
1. 7, 9, and 11   2. 88° and 50°
3. 2:00 P.M.   4. 16 notebooks at 35¢ and 29
notebooks at 50¢

Chapter 6, Self-Test 1, page 199

1.

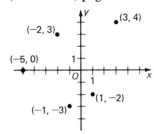

2. $A(2, 2)$; $B(-2, -3)$; $C(-3, 2)$; $D(1, -4)$
3. Domain = $\{0, 1, -1, -2\}$;
range = $\{-1, 0, -2, -3\}$

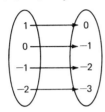

4. $(-2, 2), (0, -1), (3, 2), (3, 4)$ 5. a

6.

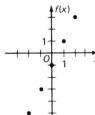

7.

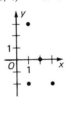

Chapter 6, Self-Test 2, page 217

1.

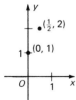

2.

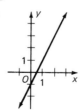

3. 4   4. 4.5 cm

Chapter 6, Self-Test 3, page 229

1. Slope = $-\frac{3}{2}$; y-intercept = $(0, -2)$

2. $3x + y = 2$

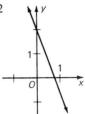

3. a. $x + y = -2$
   b. $x = -2$
4. $m = 1$;
   $x - y = -2$

5.

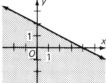

Chapter 7, Self-Test 1, page 250

1. inconsistent
2. $x = 5$,
   $y = 13$
   See figure
   at right.
3. $x = \frac{3}{5}$,
   $y = \frac{8}{5}$
4. $x = -2$,
   $y = 6$

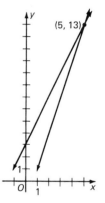

Chapter 7, Self-Test 2, page 260

1. 8   2. Rate in still water, 7 km/h; rate of current, 5 km/h   3. 36

Chapter 7, Self-Test 3, page 271

1.

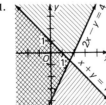

2. 435
3. Coos Oil, 3;
   Sullivan Oil, 4
4. $x = 1$,
   $y = 2$,
   $z = -1$

Answers to Self-Tests | **525**

Chapter 8, Self-Test 1, page 292
1. $7a^2 - 3a$ 2. $5x^3 + 15x^2 - 5x + 12$
3. $x^3 - 6x^2 - 2x$ 4. $7x^3 - 9x^2 + 19$
5. $3x^5$ 6. $4x^7$ 7. $r^7 \cdot s^5$ 8. $3k^3 - 15k$
9. $m^2 - 3m - 10$ 10. $12n^2 + 7n - 12$
11. $4p^2 + 12p + 9$ 12. $36r^2 - 25$
13. $9s^4 - 1$

Chapter 8, Self-Test 2, page 305
1. $2^5 \cdot 7$ 2. $12a^2b^2$ 3. $3x(2 - x)$
4. $4(y + 2)(y - 2)$ 5. $(y - 5)^2$
6. $(2w - 1)(3w + 5)$ 7. $5(v + 4)(v - 4)$
8. $(m - 9)(m + 7)$ 9. $(p + 6)(9p - 2)$
10. $-(4n - 3)(3n - 5)$

Chapter 8, Self-Test 3, page 313
1. $\{0, \frac{5}{2}\}$ 2. $\{0, -4\}$ 3. $\{7, -4\}$
4. $\{\frac{2}{3}, \frac{9}{4}\}$ 5. 7 and 8

Chapter 9, Self-Test 1, page 334
1. $-7a$ 2. $\dfrac{4d^2}{c}$ 3. $5e^2 - 6e + 3$

4. $\dfrac{4}{g} + \dfrac{3}{f} - \dfrac{7}{2fg}$ 5. $2x - 1$
6. $y^2 - 2y + 2$

Chapter 9, Self-Test 2, page 341
1. $\dfrac{-2a}{3z}$ 2. $\dfrac{2}{3}$ 3. $2x + 1$ 4. $\dfrac{1}{p^2}$

5. $\dfrac{x^5z^2}{y^3}$

Chapter 9, Self-Test 3, page 352
1. $\dfrac{2h^2}{7}$ 2. $\dfrac{-(a + 2)}{2}$ 3. $\dfrac{3x}{x^2 + 2x + 2}$

4. $\dfrac{15p^2 + 2}{5p^3}$

Chapter 10, Self-Test 1, page 365
1. $\{12\}$ 2. $\{-4\}$ 3. $z \le 10$ 4. $w < 8$
5. 43.7   6. 43.75%   7. 37.5%   8. 1 L

Chapter 10, Self-Test 2, page 376
1. $\{2\}$ 2. $\{14\}$ 3. $\{-8\}$ 4. $\{-3, -\frac{5}{16}\}$
5. 72 and 45   6. $1\frac{7}{8}$ h
7. 32 km/h and 64 km/h

Chapter 10, Self-Test 3, page 384
1. 15   2. 12   3. 8.208 kg/m³

Chapter 11, Self-Test 1, page 398
1. 6   2. -11 3. $\frac{1}{5}$ 4. $\frac{8}{9}$ 5. $\sqrt{144}$
6. $-\sqrt{25}$ 7. $\sqrt{\frac{1}{9}}$ 8. $-\sqrt{\frac{25}{36}}$ 9. $\frac{83}{100}$
10. $\frac{31}{25}$ 11. $\frac{17}{99}$ 12. $\frac{679}{99}$ 13. 3.873
14. 0.794   15. -2.87 16. 7.11

Chapter 11, Self-Test 2, page 405
1. 10.39   2. 9.487 m   3. $\sqrt{58}$

Chapter 11, Self-Test 3, page 411
1. $4\sqrt{3}$ 2. $5\sqrt{6}$ 3. $\dfrac{\sqrt{21}}{7}$ 4. $\dfrac{\sqrt{110}}{10}$

5. $12\sqrt{2}$ 6. $\sqrt{6}$ 7. $7\sqrt{2} + 18$
8. $5 - \sqrt{5} - 6$

Chapter 12, Self-Test 1, page 424
1. $\{-1 + \sqrt{6}, -1 - \sqrt{6}\}$ 2. $\{\frac{1}{2}, 1\}$

Chapter 12, Self-Test 2, page 432
1. Zeros at 1 and -3

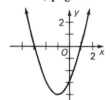

2.

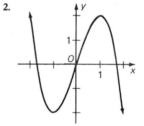

Zeros at $-\sqrt{3}, 0, \sqrt{3}$

3.

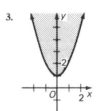

Chapter 13, Self-Test 1, page 458

1. a. $140°$ b. $225°$ c. $80°$ 2. a. $-\dfrac{1}{3}$

b. $\dfrac{2\sqrt{2}}{3}$ c. $-\dfrac{\sqrt{2}}{4}$ 3. a. 0.5299 b. 0.6561

c. 6.3138 4. $m°\angle B = 30°$, $a = 8.7$, $c = 10$

Chapter 13, Self-Test 2, page 468

1. 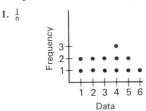   2. $3.61, 56°$

3. 2 and 3

4. 6.4

Chapter 14, Self-Test 1, page 483

1. $\frac{1}{6}$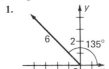

2. 15; 20

3. Data	Freq.	%	Cum. Freq.	Cum. %
1	1	12.5	1	12.5
2	2	25	3	37.5
3	2	25	5	62.5
4	2	25	7	87.5
5	1	12.5	8	100

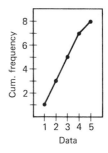

Chapter 14, Self-Test 2, page 488

1. Median, 4; mode, 2; mean, 4   2. Range, 5; variance, 4.2; standard deviation, 2.05

Chapter 14, Self-Test 3, page 495

1. $\frac{1}{6}$ 2. $\frac{5}{6}$ 3. $\frac{1}{2}$ 4. 1   5. $\frac{7}{13}$ 6. $\frac{1}{5}$

7. $\frac{17}{30}$ 8. $\frac{1}{2}$ 9. 30   10. 20

Glossary

abscissa (p. 188). Coordinate of the point where a vertical line from P meets the x-axis. Also called *x-coordinate*.

absolute value (p. 61). The positive number of any pair of opposite real numbers is the absolute value of each of the numbers. The absolute value of 0 is 0.

addends (p. 42). In a sum, the numbers that are added.

additive axiom of inequality (p. 144). For all real numbers a, b, and c: 1. If $a < b$, then $a + c < b + c$ and $a - c < b - c$. 2. If $a > b$, then $a + c > b + c$ and $a - c > b - c$.

additive inverse (p. 58). The additive inverse of the real number a is the unique real number $-a$ such that $a + (-a) = 0$ and $-a + a = 0$. Also called *negative of a, opposite of a*.

additive property of equality (p. 69). If a, b, and c are real numbers and $a = b$, then $a + c = b + c$ and $c + a = c + b$.

angle (p. 441). The union of two ordered rays with a common endpoint, together with a rotation from the first ray (the initial side) to the second (the terminal side).

arithmetic mean (p. 484). The arithmetic mean of n values is the sum of the values divided by n. Also called *average* and *mean*.

associative axioms (p. 42). For all real numbers a, b, and c: (Addition) $(a + b) + c = a + (b + c)$; (Multiplication) $(ab)c = a(bc)$.

assumption (p. 42). *See* axiom.

average (p. 484). *See* arithmetic mean.

axiom (p. 42). A statement accepted as true without proof; an assumption. Also called *postulate*.

axiom of additive inverses (p. 58). For every real number a there is a unique real number $-a$ such that $a + (-a) = 0$ and $(-a) + a = 0$.

axiom of comparison (p. 143). For all real numbers a and b, one and only one of the following statements is true: $a < b$, $a = b$, $b < a$.

axiom of multiplicative inverses (p. 84). For every real number a except zero, there is a unique real number $\frac{1}{a}$ such that $a \cdot \frac{1}{a} = 1$ and $\frac{1}{a} \cdot a = 1$.

axis (p. 188). In a coordinate plane, the two number lines which intersect at right angles at the origin of each are the axes.

base (in percent problems) (p. 361). *See under* percentage.

base (in a power) (p. 7). One of the equal factors.

basic property of quotients (p. 324). For all real numbers r and s, and all nonzero real numbers t and u, $\frac{rs}{tu} = \frac{r}{t} \cdot \frac{s}{u}$.

bearing of a vector (p. 460). The clockwise angle from north to the direction of the vector.

binary operation (p. 42). An operation which assigns to any two real numbers a unique real number.

binomial (p. 281). A polynomial of two terms.

boundary of a half-plane (p. 227). A line which separates the plane into two half-planes.

closed half-plane (p. 228). The union of an open half-plane and its boundary.

closure axioms (p. 42). For all real numbers a and b: (Addition) the sum $a + b$ is a unique real number; (Multiplication) the product ab is a unique real number.

coincide (p. 238). Two lines coincide if they have all their points in common.

combined variation (p. 381). When a variable z varies directly as one variable x and inversely as another variable y, the variation is called combined variation.

commutative axioms (p. 42). For all real numbers a and b: (Addition) $a + b = b + a$; (Multiplication) $ab = ba$

complementary angles (p. 164). Two angles whose degree measures have the sum 90. Each angle is the *complement* of the other.

completing the square (p. 418). Adding a term to an expression in the form "$x^2 + bx$" to produce a trinomial square.

conclusion (p. 48). A statement that follows from another statement assumed to be true.

conjugate expressions (p. 410). Two expressions of the form $a + \sqrt{b}$ and $a - \sqrt{b}$, b not a perfect square.

conjunction (p. 153). A sentence formed by joining two sentences by the word *and*.

consecutive integers (p. 159). Numbers obtained by counting by ones from any given integer.

consecutive multiples of n (p. 160). Numbers obtained by multiplying consecutive integers by n.

consistent equations (p. 239). Simultaneous equations that have a common root or common roots.

constant of variation (pp. 211, 377). In a direct variation expressed by $y = kx$ or in an inverse variation expressed by $y = \dfrac{k}{x}$ ($k \neq 0$), k is the constant of variation. Also called *constant of proportionality*.

constant term (p. 301). A numerical term with no variable factor.

converse (p. 306). A statement formed by interchanging the hypothesis and conclusion of a given statement.

convex polygonal region (p. 263). A region of the plane which is the intersection of a finite number of closed half-planes.

convex region (p. 263). A region which contains the line segment between any two points of the region.

coordinate (p. 12). The number paired with a point on the number line.

coordinate axes (p. 188). The axes of a coordinate system set up in a plane.

coordinate plane (p. 188). A plane in which a coordinate system has been set up.

coordinates of a point (p. 188). The abscissa and the ordinate of the given point, written as an ordered pair of numbers.

cosine of an angle (pp. 445, 452). For an angle in standard position, the abscissa of the point at which the terminal side intersects the unit circle. For an angle of a right triangle, the ratio of the length of the side adjacent to the angle to the length of the hypotenuse.

coterminal angles (p. 442). Angles with the same initial sides and terminal sides.

cumulative frequency (p. 480). The number of measurements less than or equal to a given value is the cumulative frequency of that value.

data (p. 473). A set of numerical facts.

degree of a monomial (p. 280). The number of times that a variable occurs as a factor in a monomial is the degree of the monomial in that variable. The sum of the degrees in each of the variables is the degree of the monomial. A nonzero constant has degree 0. 0 has no degree.

degree of a polynomial (p. 281). For a polynomial in simple form, the greatest of the degrees of its terms.

denominator (p. 336). In the fraction $\dfrac{r}{s}$, s is the denominator.

difference (p. 99). For any two real numbers a and b, the difference $a - b$ is the number whose sum with b is a.

difference of squares (p. 299). A quadratic binomial of the form $a^2 - b^2$, the product of two factors, $a + b$ and $a - b$.

direct proof (p. 48). Logical reasoning from given assumptions and known facts to conclusions.

direct variation (p. 211). *See* linear direct variation.

directed numbers (p. 12). Positive and negative numbers.

direction of a vector (p. 459). The measure of the angle of rotation from the positive x-axis to the vector.

discriminant (p. 422). The expression "$b^2 - 4ac$" is called the discriminant of the quadratic equation $ax^2 + bx + c = 0$.

disjoint sets (p. 150). Sets that have no members in common.

disjunction (p. 154). A sentence formed by joining two sentences by the word *or*.

distance formula (p. 404). For points $P_1(x_1, y_1)$ and $P_2(x_2, y_2)$: $d(P_1, P_2) = \sqrt{(x_2 - x_1)^2 + (y_2 - y_1)^2}$.

distributive axiom of multiplication with respect to addition (p. 47). For all real numbers a, b, and c, $a(b + c) = ab + ac$ and $(b + c)a = ba + ca$.

domain of a function (p. 192). *See under* function.

domain of a relation (p. 192). The set of first coordinates in the ordered pairs that form the relation.

domain of a variable (p. 2). The set whose members may be used as replacements for the variable. Also called *replacement set*.

dot frequency diagram (p. 474). A graph picturing frequency of measurements by using a dot for each measurement.

element of a set (p. 19). Any object in the set.

empty set (p. 20). The set with no members.

equal ordered pairs (p. 192). Two ordered pairs in which the first coordinates are equal and the second coordinates are equal.

equal sets (p. 19). Sets having the same members.

equation (p. 29). Any statement of equality.

equivalent equations (inequalities) (pp. 98, 148). Equations (inequalities) having the same solution set over a given set.

equivalent expressions (p. 48). Expressions which represent the same number for all values of the variables that they contain.

equivalent systems (p. 242). Systems of equations that have the same solution set.

equivalent vectors (p. 462). Vectors that have the same magnitude and direction.

event (p. 489). A specified subset of the set of all possible outcomes in an experiment.

experimental probability of an event (p. 492). If an experiment is conducted n times, and an event E occurs e of these times, then the experimental probability that E will occur in another trial is $\dfrac{e}{n}$.

exponent (p. 7). In a power, the number of times the base occurs as a factor.

extremes of a proportion (p. 212). In the proportion $\dfrac{y_1}{x_1} = \dfrac{y_2}{x_2}$, y_1 and x_2 are the extremes.

factor (p. 42). When two or more numbers are multiplied, each of the numbers is called a factor of the product.

factoring (p. 293). To factor a number over a given set of numbers means to express the number as the product of two or more members of the given set.

finite set (p. 23). If there is a whole number which is the number of members in a set, then the set is finite.

first coordinate (p. 188). The first number of an ordered pair.

formula (p. 132). An equation stating numerical relationships between physical or other measurements.

fractional equation (p. 365). An equation which has a variable in the denominator of one or more terms.

frequency (p. 474). The number of occurrences of a given measurement in a set of data is the frequency of that measurement.

frequency polygon (p. 477). A broken-line graph representing the frequency of measurements in given intervals.

function (p. 192). A function consists of two sets, D, the domain, and R, the range, together with a rule which assigns to each element of D exactly one element of R.

graph of a number (p. 12). The point on the number line that is paired with the number.

graph of an open sentence (p. 29). The graph of the solution set of the sentence.

graph of an ordered pair (p. 188). The point in the coordinate plane paired with an ordered pair of real numbers.

graph of a relation (p. 195). The graphs in a coordinate plane of all the ordered pairs that form the relation.

graph of a set of numbers (p. 22). The set of points corresponding to the numbers.

greatest common factor of integers (p. 294). The greatest integral factor of each of two or more given integers.

greatest common factor of monomials (p. 294). The monomial with the greatest integral numerical coefficient and the greatest degree which is a factor of each of the given monomials.

grouping symbol (p. 4). A pair of parentheses, brackets, braces, or any other symbol that includes an expression for a particular number.

half-plane (p. 227). Part of a plane, bounded by a line in the plane.

heading of a vector (p. 460). *See* bearing of a vector.

histogram (p. 476). A bar graph representing the frequency of measurements in given intervals.

hyperbola (p. 377). The graph of an inverse variation.

hypotenuse (p. 399). The longest side of a right triangle; the side opposite the right angle.

hypothesis (p. 48). A given assumption.

identity axioms (p. 42). There are real numbers 0 and 1 $(0 \neq 1)$, such that for any real number a,
$$a + 0 = a \quad \text{and} \quad 0 + a = a$$
$$a \cdot 1 = a \quad \text{and} \quad 1 \cdot a = a.$$

identity (p. 128). An equation which is a true statement for every numerical replacement of the variable(s).

identity element for addition (p. 42). Zero. When 0 is added to any given number, the sum is identical with the given number.

identity element for multiplication (p. 42). One. When any given real number and 1 are multiplied, the product is the given real number.

inconsistent equations (p. 239). Simultaneous equations that have no common root.

inequality (p. 29). A mathematical sentence containing an inequality symbol.

inequality symbols (p. 28). $>$, $<$, $\geq$, $\leq$ and $\neq$.

infinite set (p. 23). A set which is not finite.

initial side (of an angle) (p. 441). The ray at which the rotation starts.

integer (p. 23). A member of the set
$$\{ \ldots, -3, -2, -1, 0, 1, 2, 3, \ldots \}$$

intersection (p. 150). The set consisting of the elements belonging to both of two given sets.

inverse operations (p. 115). Operations that "undo" each other; for example, addition and subtraction.

inverse variation (p. 377). A function specified by an equation of the form $y = \dfrac{k}{x}$, k a nonzero constant.

irrational number (p. 17). A real number named by a nonrepeating, nonterminating decimal numeral.

irreducible polynomial (p. 297). A polynomial which cannot be factored into two or more nonconstant polynomials of lower degree belonging to a designated set.

joint variation (p. 380). When a variable z varies directly as the product of variables x and y, the variation is said to be joint.

laws of exponents (pp. 283–284, 325). For all real numbers a and b and all positive integers m and n: (Multiplication): 1. $a^m \cdot a^n = a^{m+n}$

2. $(a^m)^n = a^{mn}$

3. $(ab)^m = a^m b^m$

(Division): 1. If $m = n$ than $\dfrac{a^m}{a^n} = 1$

2. If $m > n$ then $\dfrac{a^m}{a^n} = a^{m-n}$

3. If $m < n$ then $\dfrac{a^m}{a^n} = \dfrac{1}{a^{n-m}}$

least common denominator (L.C.D.) (p. 349). The smallest positive common multiple of the denominators of two or more fractions. For $\dfrac{7}{10}$ and $\dfrac{4}{15}$ the L.C.D. is 30.

like monomials (p. 280). *See* similar monomials.

linear combination (p. 245). An equation obtained by multiplying one equation of a system by a nonzero constant and another equation of the system by another nonzero constant, and adding the resulting equations is called a linear combination of the two equations.

linear direct variation (p. 211). A function in which the ratio between a number y of the range and the corresponding number x of the domain is the same for all pairs of the function other than (0, 0). The equation $y = kx$ $(k \neq 0)$ is associated with a linear direct variation. Also called *direct variation*.

linear equation in two variables (p. 206). Any equation which can be written equivalently in the form $ax + by = c$ where a, b, and c are real numbers.

linear programming (p. 262). A problem solving technique using systems of linear inequalities to maximize or minimize some quantity.

linear term (p. 301). A term of degree one.

lowest terms (p. 336). A fraction is in lowest terms if the greatest common factor of its numerator and denominator is 1.

mathematical sentence (p. 29). A sentence that involves the symbols $>$, $<$, $\geq$, $\leq$, $=$ or $\neq$.

maximum point of a curve (p. 425). A point on the curve whose y-coordinate is greater than or equal to the y-coordinate of every other point on the curve.

mean (p. 484). *See* arithmetic mean.

means of a proportion (p. 212). In the proportion $\dfrac{y_1}{x_1} = \dfrac{y_2}{x_2}$, x_1 and y_2 are the means.

measure of an angle (p. 442). The number of degrees through which the ray rotates from the initial to the terminal side of the angle.

median (p. 484). In an ordered set of n values, the median is the middle entry if n is odd, and is half the sum of the two middle entries if n is even.

member of a set (p. 19). Any object in the set.

members of an equation (inequality) (p. 29). The expressions joined by the symbol of equality (inequality).

minimum point of a curve (p. 425). A point on the curve whose y-coordinate is less than or equal to the y-coordinate of every other point on the curve.

mode (p. 484). The value with greatest frequency.

monomial (p. 279). A term which is either a numeral or a variable, or a product of a numeral and one or more variables, raised to positive powers.

multiplicative axiom of inequality (p. 145). For all real numbers a, b, and c:

1. If $a < b$ and $c > 0$, then $ac < bc$ and $\dfrac{a}{c} < \dfrac{b}{c}$.

If $a > b$ and $c > 0$, then $ac > bc$ and $\dfrac{a}{c} > \dfrac{b}{c}$.

2. If $a < b$ and $c < 0$, then $ac > bc$ and $\dfrac{a}{c} > \dfrac{b}{c}$.

If $a > b$ and $c < 0$, then $ac < bc$ and $\dfrac{a}{c} < \dfrac{b}{c}$.

multiplicative inverse (p. 84). For a nonzero real number a, the real number $\dfrac{1}{a}$, for which $a \cdot \dfrac{1}{a} = 1$ and $\dfrac{1}{a} \cdot a = 1$. Also called *reciprocal*.

multiplicative property of equality (p. 70). If a, b, and c are real numbers and $a = b$, then $ac = bc$ and $ca = cb$.

multiplicative property of -1 (p. 79). For all real numbers a, $a(-1) = -a$ and $(-1)a = -a$.

multiplicative property of zero (p. 79). For each real number a, $a \cdot 0 = 0$ and $0 \cdot a = 0$.

natural numbers (p. 24). The numbers in the set $\{1, 2, 3, \ldots\}$.

negative of a number (p. 58). *See* additive inverse.

negative numbers (p. 12). The numbers paired with points on the negative side of the number line.

norm of a vector (p. 459). The magnitude of the vector.

number line (p. 12). A line on which a coordinate system has been established.

numeral (p. 1). A name, or symbol, for a number. Also called *numerical expression*.

numerator (p. 336). In the fraction $\dfrac{r}{s}$, r is the numerator.

numerical coefficient (p. 280). In the monomial ax^n, with a a nonzero real number, the numerical coefficient is a.

open half-plane (p. 227). A half-plane without its boundary.

open sentence (p. 29). An equation or an inequality which contains a nonconstant variable.

opposite of a number (p. 58). *See* additive inverse.

ordered pair (p. 188). A pair of elements in which the order is specified.

ordinate (p. 188). The ordinate of a point P is the coordinate of the point where a horizontal line from P meets the y-axis. Also called *y-coordinate*.

origin (pp. 11, 188). The starting point, labeled "O" on a number line; the zero point of both of two number lines that intersect at right angles.

parabola (p. 425). The graph of a quadratic function with domain $\Re$.

parallel lines (p. 238). Lines that lie in the same plane, but have no point in common.

percent (p. 361). Hundredths; divided by 100. The symbol for "percent" is "%".

percentage (p. 361). A number equal to the product of a rate (percent) and another number, called the base.

perfect square (p. 393). The square of a rational number.

polynomial (p. 280). A string of monomials connected by plus signs.

polynomial function (p. 429). A function of the form $x \rightarrow P(x)$, where $P(x)$ is a polynomial whose only nonconstant variable is x.

positive numbers (p. 11). The numbers paired with points on the positive side of the number line.

postulate (p. 42). *See* axiom.

power (p. 7). The number named by an expression in the form a^n, where n denotes the number of times a is used as a factor.

prime number (p. 293). An integer greater than one which has no positive integral factor other than itself and one.

principal square root (p. 390). The positive square root.

probability of an event (p. 489). If an experiment has n possible, equally likely outcomes and an event E consists of e of these outcomes, then the probability P that E will occur is $\dfrac{e}{n}$.

product property of square roots (p. 406). For any nonnegative real numbers a and b: $\sqrt{a} \cdot \sqrt{b} = \sqrt{ab}$.

program (p. 10). A list of instructions given to a computer.

property of the opposite of a sum (p. 74). The opposite of a sum of real numbers is the sum of the opposites of the numbers; that is, for all real numbers a and b, $-(a + b) = (-a) + (-b)$.

property of opposites in products (p. 80). For all real numbers a and b, $(-a)b = -ab$, $a(-b) = -ab$, $(-a)(-b) = ab$.

property of the reciprocal of a product (p. 86). The reciprocal of a product of real numbers, each different from zero, is the product of the reciprocals of the numbers; that is, for all real numbers a and b such that $a \neq 0$ and $b \neq 0$, $\dfrac{1}{ab} = \dfrac{1}{a} \cdot \dfrac{1}{b}$.

proportion (p. 212). An equation of the form $\dfrac{y_1}{x_1} = \dfrac{y_2}{x_2}$.

Pythagorean Theorem (p. 399). In any right triangle, the square of the length of the hypotenuse equals the sum of the squares of the lengths of the other two sides.

Pythagorean triple (p. 401). Any set $\{a, b, c\}$ of positive integers satisfying $c^2 = a^2 + b^2$. Also called a set of *Pythagorean numbers*.

quadrant (p. 189). One of the four regions into which the plane is separated by the two axes.

quadrantal angle (p. 442). An angle in standard position whose terminal side coincides with an axis.

quadratic equation (p. 417). Any equation which can be written equivalently in the form $ax^2 + bx + c = 0$, $a, b, c \in \Re$, $a \neq 0$.

quadratic formula (p. 421). The equation $$x = \frac{-b \pm \sqrt{b^2 - 4ac}}{2a}.$$

quadratic function (p. 425). Any function of the type $f: x \rightarrow ax^2 + bx + c$, where $a \neq 0$.

quadratic polynomial (p. 301). A polynomial of degree two that contains a single variable.

quadratic term (p. 301). A term of degree two in the variable.

quantifier (p. 40). An expression involving the idea of "how many" or of "quantity". For example, *some*, *all*, and *every* are quantifiers.

quotient (p. 109). The quotient $a \div b$ of a real number a by a nonzero real number b is the number whose product with b is a.

quotient property of square roots (p. 407). For any nonnegative real number a and positive real number c, $\sqrt{\dfrac{a}{c}} = \dfrac{\sqrt{a}}{\sqrt{c}}$.

radical (p. 405). An expression of the form $\sqrt{a}$.

radicand (p. 390). Any expression occurring within a radical sign.

range of a function (p. 192). *See under* function.

range of a relation (p. 192). The set of second coordinates in the ordered pairs that form the relation.

range of a set of data (p. 486). The difference between the greatest and least values.

rational expression (p. 332). Any expression that is the quotient of two polynomials.

rational number (p. 15). Any real number that is the quotient of two integers (the second integer not zero).

rationalizing the denominator (p. 407). The process of changing the form of a fraction with an irrational denominator to an equal fraction with a rational denominator.

real number (p. 12). Any number paired with a point on the number line.

reciprocal (p. 84). *See* multiplicative inverse.

reducible polynomial (p. 297). A polynomial that can be expressed as the product of two or more nonconstant polynomials of lower degree.

reducing to lowest terms (p. 336). Dividing the numerator and denominator of a fraction by their greatest common factor.

reflexive axiom of equality (p. 43). $a = a$.

relation (p. 192). Any set of ordered pairs of elements.

relationship between addition and subtraction (p. 101). For all real numbers a and b, $a - b = a + (-b)$.

relationship between multiplication and division (p. 110). For all real numbers a and all nonzero real numbers b, $a \div b = a \times \dfrac{1}{b}$.

relative frequency (p. 474). The relative frequency of a measurement is the frequency of that measurement divided by the total number of measurements.

repeating decimal (p. 16). A nonterminating decimal in which the same digit or block of digits repeats unendingly. Also called *periodic decimal*.

replacement set (p. 2). *See* domain of a variable.

resultant (p. 462). The vector representing the sum of two or more vectors.

right angle (p. 164). An angle with degree measure 90.

root of an equation (p. 29). A solution of the equation.

root of a number (p. 412). For any positive integer n, a number x is an nth root of the number a if it satisfies $x^n = a$.

root index (p. 412). A numeral signifying the root to be taken; in $\sqrt[n]{a}$, it is n.

roster (p. 19). A list of the members of a set.

satisfy (p. 201). Each member of the solution set of an open sentence satisfies that sentence.

second coordinate (p. 188). The second number of an ordered pair.

set (p. 19). A collection of objects so well described that it is always possible to tell whether or not an object belongs to the set.

similar monomials (p. 280). Two monomials are similar, or like, if they differ only in their numerical coefficients (except for the order of the factors).

simple form of a polynomial (p. 281). A polynomial is in simple form if no two of its terms are similar.

sine of an angle (pp. 445, 452). For an angle in standard position, the ordinate of the point at which the terminal side intersects the unit circle. For an angle of a right triangle, the ratio of the length of the side opposite the angle to the length of the hypotenuse.

slope of a line (p. 218). The steepness of a nonvertical line as defined by the quotient:
$$\frac{\text{difference of ordinates}}{\text{difference of abscissas}}.$$
A horizontal line has slope 0; a vertical line has no slope.

slope-intercept form (p. 219). An equation in the form $y = mx + b$ is in the slope-intercept form. The value of the slope is given by m and that of the y-intercept by b.

solution (p. 29). A member of the solution set of an open sentence.

solution of an open sentence in two variables (p. 201). Any ordered pair of numbers that makes an open sentence a true statement.

solution set of an open sentence (p. 201). The set that consists of the members of the domain of the variable for which the sentence is true is called the solution set of the sentence over that domain.

solution set of a system in two variables (p. 238). The set of all ordered pairs that satisfy all equations of a system of simultaneous equations.

solve (p. 29). To determine the solution set of an open sentence over a given domain.

solving a triangle (p. 453). Finding the measures of various parts of a triangle when the measures of other parts are given.

square root (p. 389). A number b is a square root of a number a if $b^2 = a$. The positive square root of a is denoted by $\sqrt{a}$; the negative square root by $-\sqrt{a}$.

standard deviation (p. 487). The nonnegative square root of the variance.

standard position of an angle (p. 442). An angle is in standard position if its vertex is at the origin and its initial side is the positive x-axis of a coordinate system.

standard position of a vector (p. 459). A vector is in standard position if its initial point is the origin of a coordinate system.

subset (p. 20). If every member of a set A is also a member of set B, then A is a subset of B.

substitution principle (p. 2). Changing the numeral by which a number is named in an expression does not change the value of the expression.

supplementary angles (p. 164). Two angles whose degree measures have the sum 180. Each angle is the *supplement* of the other.

symmetric axiom of equality (p. 43). If $a = b$, then $b = a$.

system of linear equations (p. 238). A set of linear equations in the same variables.

tangent of an angle (pp. 445, 452). For an angle in standard position whose terminal side intersects the unit circle in the point (p, q), with $p \neq 0$, the ratio $\frac{q}{p}$. For an angle in a right triangle, the length of the side opposite the angle divided by the length of the side adjacent to the angle.

terminal side of an angle (p. 442). The ray at which the rotation stops.

terminating decimal (p. 16). A decimal in which the division process stops because a remainder of 0 is reached.

terms of a polynomial (p. 280). The monomials in the expression for the polynomial.

theorem (p. 48). An assertion that is proved.

transformation Each of the following always produces an equation equivalent to the original equation:

by addition (p. 98). Adding the same real number to each member.

by division (p. 112). Dividing each member by the same nonzero real number.

by multiplication (p. 107). Multiplying each member by the same nonzero real number.

by substitution (p. 98). Substituting for either member an expression equivalent to that member.

by subtraction (p. 102). Subtracting the same real number from each member.

transitive axiom of equality (p. 43). If $a = b$ and $b = c$, then $a = c$.

transitive axiom of inequality (p. 144). For all real numbers a, b, and c:
1. If $a < b$ and $b < c$, then $a < c$.
2. If $a > b$ and $b > c$, then $a > c$.

trinomial (p. 281). A polynomial of three terms.

trinomial square (p. 299). A trinomial obtained by squaring a binomial. The pattern of the terms is $a^2 + 2ab + b^2$ or $a^2 - 2ab + b^2$.

union (p. 150). The set of all the elements belonging to at least one of two given sets.

unit circle (p. 404). The circle with radius 1 and center at the origin.

value of a numerical expression (p. 1). The number named by the expression.

value of a variable (p. 2). Any member of the replacement set of the variable.

values of a function (p. 195). Members of the range of the function.

variable (p. 2). A symbol which may represent any of the members of a specified set.

variable expression (p. 2). An expression which contains a nonconstant variable.

variance (p. 487). The average of the squares of the deviations of all the measurements from their arithmetic mean.

vector (p. 459). A quantity that has both magnitude and direction.

Venn diagram (p. 20). A diagram that pictures the relationships among some sets.

vertex of an angle (p. 442). The common endpoint of the rays in the angle.

vertex of a parabola (p. 425). The maximum or minimum point of the parabola.

x-axis (p. 188). The horizontal axis, ordinarily.

x-coordinate (p. 188). *See* abscissa.

y-axis (p. 188). The vertical axis, ordinarily.

y-coordinate (p. 188). *See* ordinate.

y-intercept (p. 219). The ordinate of the point where a graph intersects the y-axis.

zero of a function (p. 427). A member of the domain of the function for which the value of the function is 0.

zero-product property of real numbers (p. 307). For all real numbers a and b, $ab = 0$ if and only if $a = 0$ or $b = 0$.

Index

CREDITS

Answers to Selected Exercises

Chapter 1 Numbers, Sets, and Variables

Pages 3–4, Written Exercises

1. 578   **3.** 789   **5.** 237   **7.** 2795   **9.** 428
11. 2268   **13.** 12   **15.** 13   **17.** 5   **19.** 0
21. 13   **23.** 1   **25.** 7   **27.** 8   **29.** $1\frac{1}{12}$
31. 1.92   **33.** $2\frac{1}{6}$ **35.** $\frac{2}{3}$ **37.** 1.4   **39.** 8
41. $\frac{1}{3} + \frac{1}{5} = \frac{5}{15} + \frac{3}{15} = \frac{8}{15}; \frac{1}{5} + \frac{1}{3} = \frac{3}{15} + \frac{5}{15} = \frac{8}{15}$
43. Choose values so that $x = y$.

Page 6, Written Exercises

1. 300   **3.** 34   **5.** 200   **7.** 2   **9.** 4   **11.** 116
13. 25   **15.** 7   **17.** 10   **19.** 57   **21.** 215
23. 15   **25.** 8; 18; 28   **27.** 51   **29.** 0   **31.** 63
33. 425   **35.** 55   **37.** 50   **39.** $14\frac{4}{5}$
41. 2; 6; 10; 14; 18   **43.** $\frac{17}{18}$

Pages 8–9, Written Exercises

1. 89   **3.** 12,100   **5.** 10   **7.** 14   **9.** 5   **11.** 19
13. 45   **15.** 4   **17.** 96   **19.** 190   **21.** 2   **23.** $\frac{1}{2}$
25. 6   **27.** 172   **29.** 8   **31.** 0   **33.** 1   **35.** 0
37. $3\frac{21}{25}$ **39.** $5\frac{1}{9}$ **41.** 1.125

Pages 10–11, Programming in BASIC

1. 3*5+8   **3.** (6+7)*10   **5.** 4*6+3↑2
7. 8↑2/4↑2   **9.** (16+2)/9   **11.** (4+8*5)/(4*5+2)
13. (4*3↑2−3*2↑2)/(6*4−2*9)
15. (4*5↑2*3−6↑3)/(3↑5−37*6)   **17.** 5↑5   **19.** 5↑7
21. 5↑9   **23.** 244,140,625   **25.** 6,103,515,625
27. 152,587,890,625

Pages 13–14, Written Exercises

1, 3, 5, 7, 9.

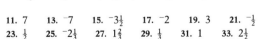

11. 7   **13.** −7   **15.** −$3\frac{1}{2}$ **17.** −2   **19.** 3   **21.** −$\frac{1}{2}$
23. $\frac{1}{2}$ **25.** −$2\frac{1}{3}$ **27.** $1\frac{2}{3}$ **29.** $\frac{1}{3}$ **31.** 1   **33.** $2\frac{1}{2}$

Pages 18–19, Written Exercises

1. 0.125   **3.** 0.$\overline{6}$ **5.** 0.8$\overline{3}$
7. 0.41$\overline{6}$ **9.** −1.1$\overline{6}$ **11.** 1.4$\overline{6}$ **13.** 0.$\overline{230769}$
15. 2.3125   **17.** 0.91, 0.92, 0.93 (many answers
possible)   **19.** −0.00211, −0.00212, −0.00213 (many

answers possible)   **21.** 4.5   **23.** −0.5   **25.** $\frac{3}{8}$
27. −$\frac{3}{4}$ **29.** 4.085   **31.** −$2\frac{4}{15}$ **33.** some
35. all

Page 19, On the Calculator

1. 0.$\overline{285714}$, 0.$\overline{428571}$, 0.$\overline{571428}$, 0.$\overline{714285}$,
0.$\overline{857142}$ **3.** 0.09, 0.18, 0.27, 0.36, 0.45, 0.$\overline{54}$, 0.$\overline{63}$,
0.72, 0.81, 0.90

Page 21, Written Exercises

1. ∈   **3.** ⊂   **5.** ∈   **7.** ∈   **9.** ∈
11. true   **13.** false
15.

17.

19.

    or

21. $A = B$ **23.** Yes

Page 22, Diversion

1. ∅; {p}, {q}, {r}; {p, q}, {p, r}, {q, r};
{p, q, r}; 1 + 3 + 3 + 1 = 8 = 2³
3. ∅; {p}, {q}, {r}, {s}, {t}; {p, q}, {p, r}, {p, s},
{p, t}, {q, r}, {q, s}, {q, t}, {r, s}, {r, t}, {s, t};
{p, q, r}, {p, q, s}, {p, q, t}, {p, r, s}, {p, r, t},
{p, s, t}, {q, r, s}, {q, r, t}, {q, s, t}, {r, s, t};
{p, q, r, s}, {p, q, r, t}, {p, q, s, t}, {p, r, s, t},
{q, r, s, t}; {p, q, r, s, t};
1 + 5 + 10 + 10 + 5 + 1 = 32 = 2⁵

Pages 25–26 Written Exercises

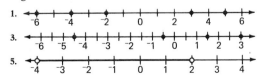

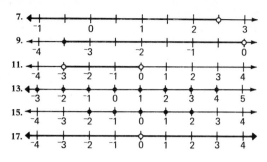

7. [number line with open circle at 2½ region, from -1 to 3]

9. [number line from -4 to 0]

11. [number line from -4 to 4, open circle at -2]

13. [number line from -3 to 5, closed circles]

15. [number line from -4 to 4]

17. [number line from -4 to 4, open circle at 0]

19. ∅ **21.** {⁻1, 9} **23.** {2, 7} **25.** {⁻1, 9}
27. ∅ **29.** {1, 3, 4, 5, 6, 8, 10, 11, 12, . . .} **31.** all
33. no **35.** some **37.** all **39.** all **41.** some

Page 27, Diversion

1. She could form a one-to-one correspondence between the dimes and the quarters. **3.** Every line that passes through P and either segment AB or segment CD passes through both segments. Thus every point Q on segment CD has a corresponding point on segment AB: the point in which PQ intersects AB. Since there is a one-to-one correspondence between the points, neither segment has more points than the other.

Pages 30–31, Written Exercises

1. {0} **3.** {1, 2} **5.** {2, 6} **7.** {4, 6}

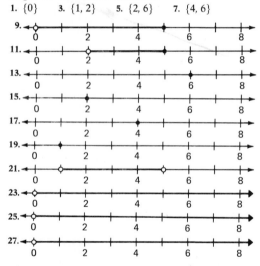

9. [number line 0 to 8, open circle at 0, closed at 4]

11. [number line 0 to 8, open circle at 2, closed at 4]

13. [number line 0 to 8]

15. [number line 0 to 8, closed at 2]

17. [number line 0 to 8, closed at 4]

19. [number line 0 to 8, closed at 1]

21. [number line 0 to 8, open circles at 1 and 5]

23. [number line 0 to 8, open circle at 0]

25. [number line 0 to 8, open circle at 0]

27. [number line 0 to 8, open circle at 0]

29. $4 + 4 = 8$ **31.** None **33.** $5 + 2 > 3(2)$
35. $(0)(1) + 2 = 0 + (1)(2)$; $(0)(1) + 3 = 0 + (1)(3)$
37. $4 + 2(3) = 10$ **39.** $4 < 2(2) + 1$
41. $2(0) + 2 \neq 0 + 3(2)$; $2(0) + 4 \neq 0 + 3(4)$;
$2(1) + 0 \neq 1 + 3(0)$; $2(1) + 2 \neq 1 + 3(2)$;
$2(1) + 4 \neq 1 + 3(4)$; $2(2) + 0 \neq 2 + 3(0)$;
$2(2) + 2 \neq 2 + 3(2)$; $2(2) + 4 \neq 2 + 3(4)$
43. Many answers are possible. Example:
$2x - 1 = 0$ **45.** Many answers are possible.
Example: $x + 3 < x$ **47.** Many answers are
possible. Example: $2(x - 1) = 2x - 2$

Pages 32–33, Written Exercises

1. $5 + x = 11$; {6} **3.** $2g = 10$; {5} **5.** The difference of x and four is five. **7.** The sum of w and eight is seventeen. **9.** The difference of fourteen and y is eight. **11.** $2x + 5 = 25$; {10}
13. $16 + 3f = 25$; {3} **15.** $\frac{m}{7} - 5 = 1$; {42}
17. $31 = 3 + 7t$; {4} **19.** $x + 3x = 2(12)$; {6}
21. $9 + 3x = 24$; {5}

Page 34, Programming in BASIC

1. a. {0, 2, 4, 6, 8} **b.** {1, 3, 5, 7, 9}
c. {2, 4, 6, 8, 10} **3. a.** {0, 2, 8, 18, 32}
b. {1, 3, 9, 19, 33} **c.** {1, 4, 11, 22, 37}
5. a. {0, 1, 8, 27, 64} **b.** {1, 2, 9, 28, 65}
c. {1, 4, 13, 34, 73}
7. a. 10 INPUT X **b.** 10 INPUT X
      20 PRINT 2*X   20 PRINT 2*X↑2
      30 END   30 END
  c. 10 INPUT X
      20 PRINT 2*X↑3
      30 END
9. a. 10 INPUT X **b.** 10 INPUT X
      20 PRINT 64*X↑3   20 PRINT 64*X↑3+2
      30 END   30 END
  c. 10 INPUT X
      20 PRINT 64*X↑3+16*X↑2+4*X↑2
      30 END

Pages 35–36, Chapter Review

1. a **3.** d **5.** a **7.** b **9.** a **11.** c **13.** c
15. a

Chapter 2   Basic Properties of Real Numbers

Page 41, Written Exercises

1. $k = {}^-2$ **3.** $t = 0$ or $t = 1$ **5.** $p = 0$

7. $n = 1$ (other answers possible)   **9.** $y = 2x$

11. $y = 2x + 6$ **13.** $\dfrac{x}{y} = y$

Pages 44–45, Written Exercises

1. 6750   **3.** $11\frac{1}{2}$ **5.** 1080   **7.** $20\frac{3}{8}$

9. 27   **11.** $\emptyset$ **13.** $\emptyset$ **15.** $\Re$ **17.** $\Re$

19. $\{0\}$ **21.** $\{0\}$ **23.** $\emptyset$ **25. a.** $^-1$ **b.** false

c. neither   **d.** yes, 0   **27. a.** 11   **b.** true

c. neither   **d.** no   **29. a.** 13   **b.** true

c. commutative   **d.** no   **31. a.** 25   **b.** true

c. commutative   **d.** no   **33. a.** 108   **b.** true

c. neither   **d.** no   **35. a.** 18   **b.** false   **c.** neither

d. no

Page 45, Diversion

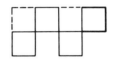

Page 47, Programming in BASIC

1. Answers will vary.

```
3.  10  INPUT A
    20  INPUT B
    30  INPUT C
    40  PRINT (A+B)*C
    50  END
5.  10  INPUT L
    20  INPUT W
    30  INPUT H
    40  PRINT L*W*H
    50  END
7.  10  INPUT R
    20  PRINT 2*3.14159*R
    30  END
```

Pages 50–51, Written Exercises

1. $23x^2$ **3.** $32a^2 + 16$ **5.** $28c + 21$

7. $3l^2 + 4m + 5l$ **9.** $4p^3 + p^2 + 3p$

11. $3x^2 + 11x + 4$ **13.** $7y^3 + 2y^2 + 6y$

15. $8s^4 + 3s^3 + s^2 + 10s + 10$

17. $7k^2 + 7k + 10h + 4h^2$ **19.** $9x + 5y$

21. $17a^2 + 16a + 15$ **23.** $34z^2 + 37z$

25. $15k + 14$ **27.** $6h + 12i + 10j$

29. $26y^2 + 32y + 24$ **31.** $11d^2 + 20d + 15$

33. $6r + 8s$ **35.** $3t^2 + 8$ **37.** $13z^2 + 6$

39. $13 + k + k^5$ **41.** $24y + 39$ **43.** $145m^2 + 334$

45. $37k + 44l + 70$ **47.** $7x + 14y + 14z + 28$

Page 52, Diversion

KT mows lawns, KC washes windows, and KG rakes leaves.

Pages 55–56, Written Exercises

1. $^-1$ **3.** $^-14$ **5.** $^-9.2$ **7.** $^-8$ **9.** $2\frac{3}{4}$ **11.** $^-1$

13. 4   **15.** $\{^-7\}$ **17.** $\{^-2\}$ **19.** $\{0\}$ **21.** $\{3\}$

23. $\{^-13\}$ **25.** $\{^-2\frac{1}{4}\}$ **27.** $\{4\}$ **29.** $\{^-3\}$

31. $\{2\}$ **33.** $\{^-1\}$ **35.** $\{^-16\}$ **37.** $\{\frac{5}{8}\}$

39. $\{1.5\}$ **41.** $\{^-81.8\}$ **43.** $\{9\}$ **45.** $\{0\}$

47. $\{1\frac{1}{2}\}$

Pages 56–57, Problems

1. a. $36 + {}^-13 + 10 + {}^-6 + 5$ **b.** 32   **c.** At the 32nd floor   **3. a.** $48 + {}^-53$ **b.** $^-5$ **c.** 5 km south of Eagle Point   **5. a.** $^-12.15 + {}^-13.05 + {}^-13.40 + {}^-12.85 + 12.25 + 12.85 + 13.15 + 12.75$ **b.** $^-0.45$

c. A 45¢ loss   **7. a.** $550(^-8) + 45(9.5) + 325(9) + 180(7.75)$ **b.** 347.50   **c.** \$347.50

9. a. $19.25 + 37.50 + {}^-23.20 + 74.40 + {}^-54.70$

b. 53.25   **c.** An increase of \$53.25   **11. a.** $35 + 2(2.5) + 3(^-1.85)$ **b.** 34.45   **c.** \$34.45

13. a. $4.184(25)(147)$ **b.** 15,376.2   **c.** 15,376.2 newton-meters

Pages 59–60, Written Exercises

1. 23   **3.** -4 **5.** 2   **7.** $-\frac{2}{3}$ **9.** 0.1

11. a. 3   **b.** 3   **13. a.** 8   **b.** 10   **15. a.** 6

b. -54 **17.** $\{1\}$ **19.** $\emptyset$ **21.** $\{1\}$ **23.** $\{2\}$

25. $\{1, 2\}$ **27.** $\{-2, -1, 0, 1\}$ **29.** $\{-1\}$

31. $\{-2, -1, 0, 1, 2\}$ **33.** $\{-1\}$ **35.** $\{-1, 0\}$

37. $\{-2, -1, 0, 1, 2\}$ **39.** $\emptyset$ **41.** $\{-1, 0, 1\}$

43. $\{1\}$ **45.** $3\frac{1}{4}$ **47.** $3\frac{3}{4}$ **49.** $-\frac{1}{4}$ **51.** $1\frac{1}{4}$

53. $-3\frac{3}{4}$

Pages 63–64, Written Exercises

1. false   **3.** true   **5.** false   **7.** true   **9.** 24

11. 56   **13.** 1   **15.** -9 **17.** -19 **19.** $\emptyset$

21. $\{2, -2\}$ **23.** $\{10, -10\}$ **25.** $\{3, -3\}$

27. $\{2, -2\}$ **29.** $\{6, -6\}$

31.
33. $\emptyset$
35.
37.
39.
41.
43.

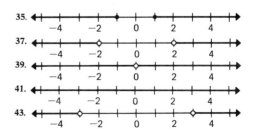

45.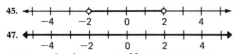

47. (number line from −4 to 4)

49. No real values   **51.** Many answers are possible. Example: $|2 + (-1)| = |2| + |-1|$
53. $\leq$

1. b   **3.** a   **5.** c   **7.** a   **9.** c

Page 67, Programming in BASIC
1. a. −1, 0, 1 **b.** −2, 2

Chapter 3 Using Number Properties

Pages 72–73, Written Exercises
1. 12   **3.** 36   **5.** 122   **7.** 44
9. (1) Given; (2) Axiom of additive inverses;
(3) Commutative axiom of addition; (4) Associative axiom of addition; (5) Axiom of additive inverses;
(6) Identity axiom of addition; (7) Transitive axiom of equality
11. (1) Given; (2) Axiom of additive inverses;
(3) Associative axiom of addition; (4) Axiom of additive inverses; (5) Identity axiom of addition;
(6) Transitive axiom of equality
13. (1) Given; (2) Substitution principle;
(3) Commutative axiom of addition
15. (1) Given; (2) Symmetric axiom of equality;
(3) Transitive axiom of equality
17. 1. a and b are real numbers.   Given
  2. $-a$ and $-b$ are real numbers.   Ax. add. inv.
  3. $a + b + [(-a) + (-b)]$ Assoc. ax.
    $= a + [b + (-a)] + (-b)$ of add.
  4. $= a + [(-a) + b] + (-b)$ Comm. ax. of add.
  5. $= [a + (-a)] + [b + (-b)]$ Assoc. ax. of add.
  6. $= 0 + 0$ Ax. of add. inv.
  7. $= 0$ Identity ax. of add.
  8. $a + b + [(-a) + (-b)] = 0$ Trans. ax. of equality

Pages 76–77, Written Exercises
1. 6   **3.** −238   **5.** 3   **7.** −65   **9.** 46   **11.** 12
13. −33   **15.** 2.7   **17.** $-8\frac{3}{4}$ **19.** $a^2 + 7b$
21. $2n^4 + (-9)n$ **23.** $(-8)y^2 + (-5)xy + 2x^2$
25. (1) Given; (2) Axiom of additive inverses;
(3) Property of the opposite of a sum; (4) Additive property of equality; (5) Associative axiom of addition; (6) Axiom of additive inverses;
(7) Identity axiom of addition; (8) Transitive axiom of equality

27. (1) Given; (2) Axiom of additive inverses;
(3) Property of the opposite of a sum;
(4) Transitive axiom of equality

Note: In order to save space, the proofs that follow are given in a shortened arrangement. Students' proofs should be arranged as illustrated in the text and in the answer for Exercise 17, text page 73. The shortened arrangement will be used throughout these answers.

29. $[(-a) + (a + 1)] + (-1) = [(a + 1) + (-a)] + (-1)$, Comm. ax. of add.; $(a + 1) + [(-a) + (-1)]$, Assoc. ax. of add.; $(a + 1) + [-(a + 1)]$, Prop. of opp. of a sum; 0, Ax. of add. inv.;
$\therefore$ $[(-a) + (a + 1)] + (-1) = 0$, Trans. ax. of eq.
31. $-[(-a) + (-b)] + (-a) = [-(-a) + [-(-b)]] + (-a)$, Prop. of opp. of a sum;
$[a + b] + (-a)$, since $-(-a) = a$; $(-a) + [a + b]$, Comm. ax. of add.; $[(-a) + a] + b$, Assoc. ax. of add.; $0 + b$, Ax. of add. inv;
b, Identity ax. of add.;
$\therefore -[(-a) + (-b)] + (-a) = b$, Trans. ax. of eq.
33. $-(a + b) + [(a + b) + (-c)] = [-(a + b) + (a + b)] + (-c)$, Assoc. ax. of add.; $0 + (-c)$, Ax. of add. inv.; $-c$, Iden. ax. of add.;
$\therefore$ $-(a + b) + [(a + b) + (-c)] = -c$, Trans. ax. of add.
35. $(-a) + [a + (-b)] + b = [(-a) + a] + [(-b) + b]$, Assoc. ax. of add.; $0 + 0$, Ax. of add. inv.; 0, Iden. ax. of add.;
$\therefore$ $(-a) + [a + (-b)] + b = 0$, Trans. ax. of eq.

Page 78, Problems
1. $144.56   **3.** $180.54   **5.** At least 7

Pages 81–83, Written Exercises
1. −39   **3.** 60   **5.** −750   **7.** 0   **9.** −1166
11. −16,669   **13.** $-15a + (-6b)$
15. $-8p + (-32q)$ **17.** $15y + 20y^2$ **19.** $4l + 8h$
21. $3d + (-e)$ **23.** $-ef$ **25.** $-r + 3s$
27. $4n^2 + n^3$ **29.** $-17x + (-2)$ **31.** $-13y + 13t$

33. $25a + (-29b)$ **35.** $-14y^2 + (-31w)$
37. $-18u^2 + 42v + (-15)$ **39.** $-40bc + 80$
41. $31x + 20y$ **43.** $-12r + (-4s)$ **45.** $9a + 3b$
47. -9 **49.** 0   **51.** 8   **53.** -1 **55.** 144
57. (1) Given; (2) Multiplicative property of -1; (3) Multiplicative property of -1; (4) Multiplicative property of equality or substitution principle; (5) Associative axiom of multiplication; (6) Multiplicative property of -1 and associative axiom of multiplication; (7) Identity axiom of addition; (8) Transitive axiom of equality
59. $(-a)b = [(-1)a]b$, Mult. prop. of -1; $(-1)[ab]$, Assoc. ax. of mult.; $-ab$, Mult. prop. of -1; $\therefore (-a)b = -ab$, Trans. prop. of equality

Page 84, On the Calculator
1. -1919.61 **3.** 31.9   **5.** 163.42805
7. 5.252864

Pages 88–89, Written Exercises
1. 63   **3.** 9   **5.** -10 **7.** 15   **9.** $-xy$ **11.** 1
13. $-7x^2$ **15.** $2h$ **17.** -1 **19.** $3p + 6$
21. $3k + (-5h)$ **23.** $12y + (-4xu)$
25. $-r^2 + s^2$ **27.** $10x + (-6y)$

29. $-3p^2 + 3p + (-7)$ **31.** $-3w + (-2)$
33. $-10a^2 + 2b^2$ **35.** $3k + (-4h)$
37. $1 + 6x + 12y$ **39.** $-2x^2 + 10y^2 + (-z^2)$
41. x and y are real numbers, Given; $y \neq 0$,

Given; $\dfrac{1}{y}$ is a real number, Ax. of mult. inv.;

$\left(-\dfrac{1}{y}\right)(-yx) = \left[(-1)\left(\dfrac{1}{y}\right)\right][(-1)(yx)]$, Mult. prop.

of -1; $(-1)\left[\dfrac{1}{y}(-1)\right](yx)$, Assoc. prop. of mult.;

$(-1)\left[(-1)\dfrac{1}{y}\right](yx)$, Comm. prop. of mult.;

$[-1(-1)]\left[\dfrac{1}{y}(yx)\right]$, Assoc. prop. of mult.; $1 \cdot \dfrac{1}{y} \cdot yx$,

Mult. prop. of -1; $\left(\dfrac{1}{y} \cdot y\right)x$, Iden. prop. and

Assoc. ax. of mult.; $1 \cdot x$, Ax. of mult. inv.;

x, Iden. prop. of mult.; $\therefore \left(-\dfrac{1}{y}\right)(-yx) = x$, Trans.

ax. of eq.

Pages 93–94, Chapter Review
1. c   **3.** c   **5.** c   **7.** a

Chapter 4   Solving Equations and Problems

Page 99, Written Exercises
1. $\{57\}$ **3.** $\{65\}$ **5.** $\{0\}$ **7.** $\{44\}$ **9.** $\{-23\}$
11. $\{320\}$ **13.** $\{4\frac{3}{4}\}$ **15.** $\{8.43\}$ **17.** $\{0\}$
19. $\{-16\}$ **21.** $\{2\}$ **23.** $\{9\}$ **25.** $\{-15\}$
27. $\{12\}$ **29.** $\{5\}$ **31.** $\{-8\}$ **33.** $\{0\}$
35. $\{4, -4\}$ **37.** $\emptyset$ **39.** $\{6, -6\}$

Pages 103–105, Written Exercises
1. 446   **3.** -3174 **5.** -1.01 **7.** $\frac{7}{8}$ **9.** -20
11. -585 **13.** $1\frac{4}{7}$ **15.** 169   **17.** -3 **19.** 12
21. -43 **23.** $4x^2 - 5x - 3$ **25.** $-4m^3 - 2m$
27. $\{-4\}$ **29.** $\{25\}$ **31.** $\{17\}$ **33.** $\{9\}$
35. $\{0\}$ **37.** $\{2\}$ **39.** $-8 - 3$; -11
41. $x - (x - 7)$; 7   **43.** $(3x + 1) - (3x - 1)$; 2
45. $(1 - k) - (6 - 2k)$; $-5 + k$
47. $4(1 - e) - (5e + 1)$; $3 - 9e$
49. (1) Given; (2) Multiplicative property of -1; (3) Distributive property of multiplication with respect to subtraction; (4) Multiplicative property of -1; (5) Relationship between addition and subtraction; (6) $-(-a) = a$, p. 59

55. Not closed; $0 - 1 \notin \{0, 1\}$ **57.** Closed;
$\dfrac{a}{b} - \dfrac{c}{d} = \dfrac{ad - bc}{bd}$; $ad - bc$ and bd are integers, so

$\dfrac{ad - bc}{bd}$ is a rational number.   **59.** Closed;

$a - b = a + (-b)$, an integer   **61.** Not closed;
$3 - 1 \notin \{\text{the odd integers}\}$

Pages 105–106, Problems
1. 168 m   **3.** $-395.47°C$ **5.** 84 blocks   **7.** 100°
9. The first diver   **11.** 90 m

Page 106, Diversion
Figures A, B, D, and E

Page 109, Written Exercises
1. $\{36\}$ **3.** $\{-96\}$ **5.** $\{13\}$ **7.** $\{-8\}$
9. $\{-80\}$ **11.** $\{29.9\}$ **13.** $\{1\}$ **15.** $\{0\}$
17. $\{9\}$ **19.** $\{-5\}$ **21.** $\{-2\}$ **23.** $\{30\}$
25. $\{-39\}$ **27.** $\{-10\}$ **29.** $\{0.8\}$ **31.** $\{-10\}$
33. $\emptyset$ **35.** $\{18, -18\}$ **37.** $\{320, -320\}$
39. $\{140, -140\}$

Pages 113–115, Written Exercises
1. -5 3. 16   5. -54 7. $3\frac{2}{3}$ 9. -0.16
11. -56 13. $-4ab$ 15. $10d$ 17. 4   19. -4
21. $\{-234\}$ 23. $\{6\}$ 25. -6 27. $-1\frac{3}{8}$
29. $-\frac{1}{4}$ 31. -51 33. $\frac{1}{13}$
35. (1) Given; (2) Relationship between
multiplication and division; (3) Distributive axiom
of multiplication with respect to addition;
(4) Relationship between multiplication and
division; (5) Transitive axiom of equality

37. Closed; $1 \div 1 = 1$ 39. Closed; $a \div b = a\left(\frac{1}{b}\right)$;
$\frac{1}{b} > 0$, so $a\left(\frac{1}{b}\right)$ is a positive real number.
41. Not closed; $-1 \div (-1) = 1$ 43. Not closed;
$2 \div \frac{1}{2} = 4$ 45. Closed; $a \div b = a\left(\frac{1}{b}\right)$, a real

number

Pages 117–118, Written Exercises
1. $\{8\}$ 3. $\{6\}$ 5. $\{32\}$ 7. $\{-5\}$
9. $\{-9\}$ 11. $\{8\}$ 13. $\{-4\}$ 15. $\{-2\}$
17. $\{4\}$ 19. $\{12\}$ 21. $\{-12\}$ 23. $\{-17.5\}$
25. $\{3\}$ 27. $\{15\}$ 29. $\{-1\}$ 31. $\{-4\}$
33. $\{-11\}$ 35. $\{-1\}$ 37. $\{-19\}$ 39. $\{5\}$
41. $\{8\}$ 43. $\{-3\}$ 45. $\{0\}$ 47. $\{-3\}$
49. $\{-3\}$ 51. $\{-10\}$ 53. $\{0\}$ 55. $\{-9\}$
57. $\{2, -2\}$ 59. $\emptyset$

Pages 120–121, Written Exercises
Answers may vary.   1. $n + 2n = -15$
3. $\frac{1}{2}n - 9 = 10$ 5. $4s = 64$ 7. $x + 3x = 90$
9. $(w - 6) + (w - 6) + w + w = 88$
11. $k + 3k + k + 3k = 64$ 13. $15w = 210$

Pages 121–122, Problems
Answers may vary.   1. **a.** Let n = lesser number.
Then $n + 3$ = greater number.
b. $n + (n + 3) = 42$ 3. **a.** Let t = number of
oxygen atoms. Then $2t$ = number of hydrogen
atoms and $t + 1$ = number of carbon atoms.
b. $t + 2t + (t + 1) = 45$ 5. **a.** Let x = number
of minutes to return home. Then $x + 10$ =
number of minutes to drive to school.
b. $x + (x + 10) = 50$ 7. **a.** Let y = number of
words that the slower secretary can type. Then the
other secretary can type $y + 68$ words.
b. $y + (y + 68) = 256$ 9. **a.** Let g = number of
girls. Then $\frac{2}{3}g$ = number of boys.
b. $g + \frac{2}{3}g = 500$ 11. **a.** Let d = number of
dimes. Then $d + 6$ = number of quarters.
b. $10d + 25(d + 6) = 3000$ (or $0.1d +$
$0.25(d + 6) = 30$)   13. **a.** Let n = number

of dollars invested at 8%. Then $n + 250$ = number
invested at 9%.   **b.** $0.08n + 0.09(n + 250) = 75$ (or
$8n + 9(n + 250) = 7500$)   15. **a.** Let m = Maria's
earnings in dollars. Then $2m - 35$ = sister's
earnings in dollars.   **b.** $m + (2m - 35) = 850$

Pages 125–127, Problems
1. 38 m   3. 27 m   5. $10,000   7. $84
9. 19 m   11. 18 m by 14 m   13. -14 15. 146
voted for the loser and 249 voted for the
winner.   17. 22   19. 60 and 85
21. 10 min   23. Yes; $6(10) + 12(8) + 10(15) = 306$
25. 11 m by 7 m   27. 9 nickels, 21 quarters
29. 17 m   31. 16 years old

Page 129, Written Exercises
1. $\{8\}$ 3. $\{8\}$ 5. $\{-7\}$ 7. $\{4\}$ 9. $\{12\}$
11. $\{-48\}$ 13. $\{-9\}$ 15. $\{9\}$ 17. $\{-3\}$
19. $\{2\frac{2}{5}\}$ 21. $\{5\}$ 23. $\{4\}$ 25. $\Re$; identity
27. $\{-8\}$ 29. $\{-9\}$ 31. $\{2\}$ 33. $\{11\}$
35. $\{0\}$ 37. $\emptyset$ 39. $\{11\}$ 41. $\{-2\}$ 43. $\emptyset$
45. $\{-15\}$ 47. $\{-1\frac{1}{2}\}$ 49. $\{-17\frac{1}{2}\}$
51. $\{-1\frac{3}{7}\}$ 53. $\{2\}$

Pages 130–132, Problems
1. -5 3. 17   5. 9 years old   7. 5 years
9. Circus: 8000; hockey game: 18,000; basketball
game: 14,000   11. Every real number satisfies
the problem.   13. 11   15. Quarters   17. 70°,
70°, 40°   19. 24 dimes   21. -3 23. 7 m
25. Washington Monument: 169.5 m; Gateway
Arch: 195 m   27. Potatoes: 8¢; onions: 12¢; green
beans: 24¢   29. 1706

Page 134, Written Exercises
1. $b = p - a - c$ 3. $b = \dfrac{2A}{h}, h \neq 0$
5. $r = \dfrac{A}{2\pi h}, h \neq 0$ 7. $m = \dfrac{E}{c^2}, c \neq 0$
9. $t = \dfrac{A - P}{Pr}, Pr \neq 0$ 11. $y = \dfrac{6 - 2x}{3}$
13. $t = \dfrac{d}{a} + \dfrac{1}{2} = \dfrac{2d + a}{2a}, a \neq 0$
15. $T = \dfrac{E + amt}{am}, am \neq 0$ 17. 5 m   19. 3.14
21. 9000 kg   23. 1,065,312,000°K   25. 81 N

Page 135, On the Calculator
1. About 57,000 kg

Pages 137–139, Chapter Review
1. c   3. a   5. c   7. a   9. c   11. c   13. a
15. c   17. a

Pages 140–141 Cumulative Review, Chapters 1–4
1. c   3. d   5. b   7. d   9. d   11. c
13. c   15. b   17. d   19. b   21. c

Chapter 5 Solving Inequalities and Problems

Pages 146–147, Written Exercises

1. (1) Given; (2) Additive axiom of inequality; (3) Axiom of additive inverses; (4) Identity axiom of addition; (5) Substitution principle; (6) Definition of *greater than*

3. (1) Given; (2) Additive axiom of inequality; (3) Relationship between addition and subtraction; (4) Axiom of additive inverses; (5) Substitution principle

5. (1) Given; (2) and (3) Multiplicative axiom of inequality; (4) Transitive axiom of inequality

7. (1) Given; (2) and (3) Additive axiom of inequality; (4) Transitive axiom of inequality

9. Many answers are possible. Example: $x = 7$, $y = 5$, $t = 6$, $w = 1$.

Page 149, Written Exercises

1. $x \geq -7$

3. $m < -8$

5. $r > 8$

7. $p < -3$

9. $h < -2$

11. $A \geq -10$

13. $C > 0$

15. $x < 26$

17. $k \leq 2$

19. $s \geq 3$

21. $m \geq 9$ 23. $p > 9$ 25. $x \leq 7$
27. $h < -79$ 29. $k \leq 2$ 31. $A \leq 0$
33. $x < 1.5$ 35. $m < \frac{10}{3}$ 37. $y < b$
39. $w \geq \dfrac{7d}{9}$ 41. $w \geq 4s$

Pages 151–153, Written Exercises

1. $\{-1, 6\}$; $\{-1, 0, 5, 6, 8\}$ 3. $\{-2, 7\}$; $\{-2, 0, 2, 7\}$ 5. $\emptyset$; $\{3, 4, 5, 6, 7, 8, 9, 10\}$

7. $\{1, 3\}$; $\{0, 2, \text{the odd whole numbers}\}$ 9. $\emptyset$; $\Re$

11. a.
b.
c.
d.

13. a.
b.
c.
d.

15. a.
b.
c.
d.

17. a.
b.
c.
d.

19. a.
b.
c.
d.

21. a.
b.
c.
d.

23. a.
b.
c.
d.

25. a.
b.
c.
d. $\emptyset$

27. a.
b.

27. c.

d. ∅

29. a.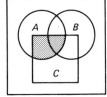

b.

c.

d.

31. a.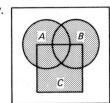

b.

c.

d.

33.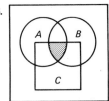

35.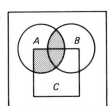

37.

39.

41. $B = \{0, 3, 4, 6\}$ **43.** $B = \{0, 1, 2, 3, 4, 5\}$

Pages 155–156, Written Exercises

1. $-4 \le x < 2$

3. $-3 \le z \le 1$

5. $1 < t < 6$

7. $p > 4$ or $p < -8$

9. $s \ge 1$ or $s \le 0$

11. $-4 \le n \le 7$

13. $a > 3$ or $a \le -2$

---

15. $\Re$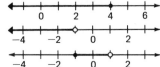

17. $m \le 4$

19. $p < -1$

21. $-1 \le a < 1$

23. $c < 7$

25. $x < -3$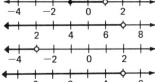

27. $w < 5$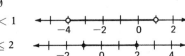

29. ∅   **31.** ∅

33. $-4 < w < 1$

35. $-1 \le k \le 2$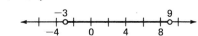

37. ∅

Pages 157–158, Written Exercises

1. $\{-2, 10\}$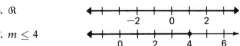

3. $\{t: t \ge 1 \text{ or } t \le -1\}$

5. $\{r: -3 < r < 9\}$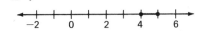

7. $\{t: t \ge 11 \text{ or } t \le 5\}$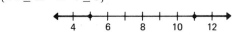

9. $\{-\frac{1}{2}, 1\}$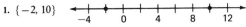

11. $\{h: 4 \le h \le 5\}$

13. $\{f: f \le -\frac{4}{3} \text{ or } f \ge \frac{10}{3}\}$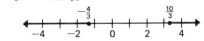

15. ∅   **17.** $\{-2, 6\}$ **19.** $\{z: -3 < z < 5\}$
21. $\{-3, 4\}$ **23.** $\{z: z \le -2 \text{ or } z \ge 2\}$ **25.** ∅
27. $\Re$ **29.** $\Re$ **31.** $\{m: m \le -6 \text{ or } m \ge 0\}$
33. $\{-3, 1\}$ **35.** ∅   **37.** $\{t: t \le \frac{1}{4} \text{ or } t \ge 2\}$
39. $\{k: k \le \frac{1}{4} \text{ or } k \ge 2\}$
41. $\{a: 2 < a < 6 \text{ or } -6 < a < -2\}$
43. $\{c: -1 < c < 1\}$
45. $\{e: 5 \le e < 6 \text{ or } -4 < e \le -3\}$ **47.** ∅

Pages 162–163, Problems
1. 187, 188   **3.** −11, −12, −13   **5.** −50,
−51, −52   **7.** 122, 123, 124   **9.** 28 m by 26 m
11. 15, 17   **13.** 8, 10   **15.** 46, 47, 48, 49
17. 4, 6   **19.** 10, 11, 12   **21.** −3, −2, −1
23. 8, 12, 16, 20   **25.** $\{-4, -8, -12\}$;
$\{-8, -12, -16\}$; $\{-12, -16, -20\}$;
$\{-16, -20, -24\}$ **27.** 48, 56, 64, 72

Page 166, Written Exercises
1. 105°   **3.** 60°   **5.** 30°   **7.** 138°
9. $\left(180 - \dfrac{3n}{2}\right)^{\circ}$ **11.** $(190 - 3k)^{\circ}$ **13.** $x = 20$
15. $x = 35$ **17.** $n = 7$ **19.** $k = 14$

Pages 167–168, Problems
1. 76°   **3.** 26°, 64°   **5.** 8°   **7.** 23.25°, 69.75°,
87°   **9.** 60°   **11.** 96°, 42°, 42°
13. Since consecutive angles are supplementary
and opposite angles are equal, the sum of the
measures is $2 \cdot 180°$, or 360°. (Or: Draw a diagonal.
Since the sum of the measures of the angles of
each triangle is 180°, the total sum is $2 \cdot 180°$, or
360°.)
15. 60°   **17.** 17.5°   **19.** Any measure between
72° and 75° inclusive

Pages 171–174, Problems
1. 3.5 h   **3.** 650 km/h   **5.** 96 km/h; 480 km
7. 10 min   **9.** 400 m   **11.** 96 km/h
13. 100 km/h   **15.** Mark: 21.25 km/h; Tom:
16.25 km/h   **17.** $59\frac{3}{7}$ km   **19.** 1124 km
21. 32 km   **23.** 96 km   **25.** 20 s; 60 s
27. $7\frac{1}{27}$ min   **29.** $4\frac{1}{12}$ km/h, $4\frac{11}{12}$ km/h
31. 6:30 P.M.

Pages 176–178, Problems
1. Inconsistent   **3.** 18 first-class passengers
5. 11 nickels, 21 quarters   **7.** 45°, 45°, 90°
9. Part A: 7 questions; Part B: 8 questions

11. 38 kg   **13.** 6 dimes; 12 quarters;
3 nickels   **15.** \$3   **17.** 4 months at
\$950/month; 8 months at \$1130/month
19. Inconsistent   **21.** Insufficient
information   **23.** 1408 km   **25.** 7 bulbs
27. 48 cm by 12 cm

Page 178, Diversion
$\frac{1}{2}$ and $1\frac{1}{2}$

Page 181, Extra for Experts
1. T   **3.** T   **5.** T   **7.** T   **9.** T
11. Since $p \longrightarrow q$ is false, p is true and q is false.
Then q' is true. Both p and q' are true, so $p \wedge q'$
is true.

13. No;

p	q	$p \vee q$	$(p \vee q) \to p$
T	T	T	T
T	F	T	T
F	T	T	F
F	F	F	T

15. Yes;

p	q	$p \vee q$	$q \vee p$	$(p \vee q) \leftrightarrow (q \vee p)$
T	T	T	T	T
T	F	T	T	T
F	T	T	T	T
F	F	F	F	T

17. Yes; see table below.

Pages 183–184, Chapter Review
1. b   **3.** d   **5.** c   **7.** c   **9.** c   **11.** a

p	q	r	$r \wedge (p \vee q)$	$(r \wedge p) \vee (r \wedge q)$	$r \wedge (p \vee q) \leftrightarrow (r \wedge p) \vee (r \wedge q)$
T	T	T	T	T	T
T	T	F	F	F	T
T	F	T	T	T	T
T	F	F	F	F	T
F	T	T	T	T	T
F	T	F	F	F	T
F	F	T	F	F	T
F	F	F	F	F	T

Chapter 6 Functions, Relations, and Graphs

Pages 190–191, Written Exercises

1–11.

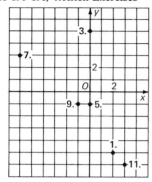

13–21. Graphs may vary.

13.

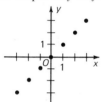

15.

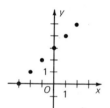

17.

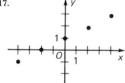

19.

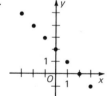

21.

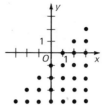

23.

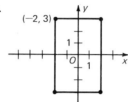

25.

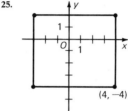

27.

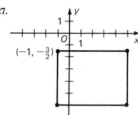

29. $(5, -2)$, $(9, 4)$, $(-1, 0)$ **31.** $(1, -2)$, $(-2, 1)$

Pages 193–194, Written Exercises

1. $\{(1, 6), (2, 12), (3, 18)\}$; yes   **3.** $\{(-2, 8), (0, 8),$
$(2, 8)\}$; yes   **5.** $\{(-2, 2), (0, -7), (2, 11)\}$; yes
7. $\{(-3, -3), (3, -3), (5, -3), (5, 6)\}$; no
9. $m = 4$; $n = -8$ **11.** $m = 4$; $n = -3$
13. $m = -\frac{1}{2}$; $n = 2$ or $n = -2$ **15.** $m = 2$ or
$m = -2$; $n = 0$
17. $D = \{0, 1, 2\}$;
   $R = \{1, 2, 3\}$;
   no

19. $D = \{1, 2\}$;
$R = \{0, 1, 2\}$;
no

21. $D = \{1, 2, 3, 4\}$;
$R = \{-1, -3, -6, -10\}$;
yes

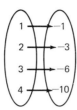

23. $D = \{3, 6, 9, 12\}$;
$R = \{4\frac{1}{2}, 9, 13\frac{1}{2}, 18\}$;
yes

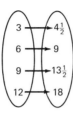

25. $\{((1, 3), 4), ((2, 4), 6), ((4, 6), 10), ((6, 8), 14),$
$((15, 0), 15)\}$; yes; every ordered pair (a, b) in the
domain is assigned to the number $a + b$ in the
range.

Page 197, Diversion

Pages 198–199, Written Exercises
1. $f = \{(x, f(x)): x \in D \text{ and } f(x) = 2x\}$;
$f = \{(x, 2x): x \in D\}$

x	-2	-1	0	1	2
$f(x)$	-4	-2	0	2	4

3. $h = \{(x, h(x)): x \in D \text{ and } h(x) = 2x + 1\}$;
$h = \{(x, 2x + 1): x \in D\}$

x	-2	-1	0	1	2
$h(x)$	-3	-1	1	3	5

5. $s = \{(x, s(x)): x \in D \text{ and } s(x) = 2 - x^2\}$;
$s = \{(x, 2 - x^2): x \in D\}$

x	-2	-1	0	1	2
$s(x)$	-2	1	2	1	-2

7. $w = \{(x, w(x)): x \in D \text{ and } w(x) = (x + 1)^2\}$;
$w = \{(x, (x + 1)^2): x \in D\}$

x	-2	-1	0	1	2
$w(x)$	1	0	1	4	9

9. $u = \{(x, u(x)): x \in D \text{ and } u(x) = x^3\}$;
$u = \{(x, x^3): x \in D\}$

x	-2	-1	0	1	2
$u(x)$	-8	-1	0	1	8

11.

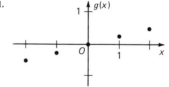

13.

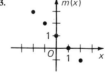

15.

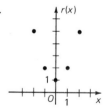

17.

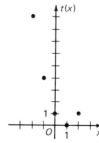

19. No;

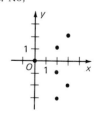

21. Yes;

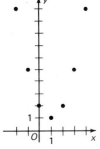

23. Yes;

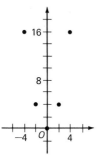

25. No;

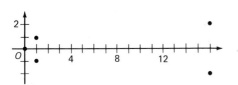

27. Yes;

29. Yes;

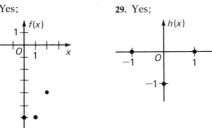

31. Yes;

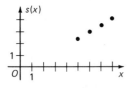

33.

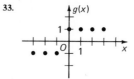

35.

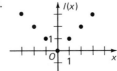

37.

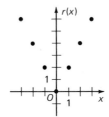

39.

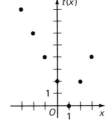

Pages 203–204, Written Exercises

1.

3.

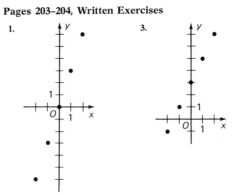

5.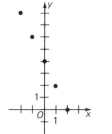

7.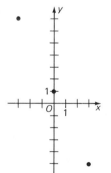

9.

11.

13.

15.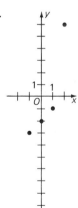

17.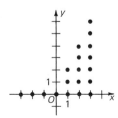

19.

21.

23.

25.

27.

29.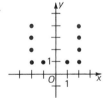

31.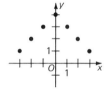

33.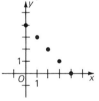

35.

37.

39.

41.

Page 205, Programming in BASIC

7. $\{(-10, 0), (-8, 6), (-8, -6), (-6, 8), (-6, -8),$
$(0, 10), (0, -10), (6, 8), (6, -8), (8, 6), (8, -6),$
$(10, 0)\}$ 9. $\{(-5, 0), (-4, 6), (-4, -6),$
$(-3, 8), (-3, -8), (0, 10), (0, -10), (3, 8), (3, -8),$
$(4, 6), (4, -6), (5, 0)\}$

23. Let a = no. of grams of
white flour and b = no.
of grams of wheat flour;
$\{(a, b): b = 4a$ and
$a \in D\}$

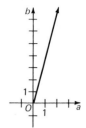

Pages 209–210, Written Exercises

1.

3.

25. Let f = no. of francs
and m = no. of marks;
$\{(m, f): f = 2.3m$ and
$m \in D\}$

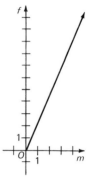

5.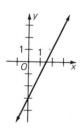

7.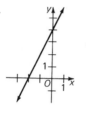

9.

11.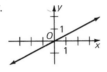

27.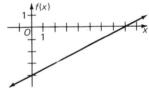

29.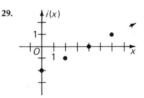

13. **a.** $(5, 0)$ **b.** $(0, -3)$ 15. **a.** $(\frac{12}{5}, 0)$
b. $(0, 12)$ 17. **a.** $(0, 0)$ **b.** $(0, 0)$
19. **a.** $(-\frac{2}{3}, 0)$ **b.** $(0, -2)$

21. Let n = no. of months;
$\{(n, t): t = 22 + 4.5n$
and $n \in D\}$

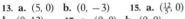

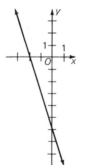

31.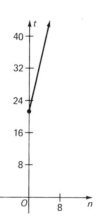

33.

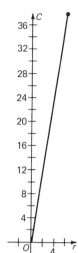

35.

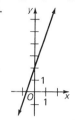

No

37.

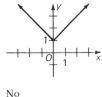

No

39.

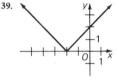

No

41.

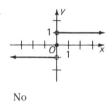

No

Pages 213–215, Written Exercises

1. **a.** 4 **b.** -8 **c.** 100
3. -9 5. -9 7. 36 9. 7 11. 40
13. $\frac{14}{9}$ 15. 4 17. -36 19. 6 21. $\frac{1}{2}$
23. $g = \frac{1}{4}h^2$ 25. $y_1 = 2c(0) = 0$;
$y_2 = 2c(1) = 2c$; $y_3 = 2c(-1) = -2c$ 27. Let
d = distance in kilometers and t = time in hours.
a. $d = 90t$ **b.** $\dfrac{d_1}{t_1} = \dfrac{d_2}{t_2}$ 29. Let d = distance in
centimeters on the map and D = actual distance in
kilometers. **a.** $d = 0.5D$ **b.** $\dfrac{d_1}{D_1} = \dfrac{d_2}{D_2}$ 31. Let
R = resistance and l = length. **a.** $R = 0.005l$
b. $\dfrac{R_1}{l_1} = \dfrac{R_2}{l_2}$

Pages 215–217, Problems

1. 240 votes 3. 48 g 5. $27\frac{7}{9}$ m 7. 25 g
9. 1680 g 11. $3\frac{3}{35}$ km 13. About $2845.41
15. $2\frac{1}{4}$ times as much 17. 6.3 cm $\times$ 8.25 cm
19. 81 times as much 21. 324π m^2

Page 221, Written Exercises

1. 8 3. -3 5. -1
7. **a.**

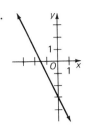

9. **a.**

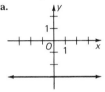

b. $3x - y = -2$ **b.** $2x + y = -3$
11. **a.**

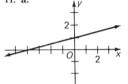

13. **a.**

b. $x - 4y = -4$ **b.** $y = -3$

Pages 222–223, Written Exercises

1. $x - y = -2$ 3. $2x + y = -6$
5. $3x - 4y = -15$ 7. $y = 2$ 9. $4x + 6y = 3$
11. -2 13. 7 15. -1 17. -2 19. 2
21. $2x - y = -2$ 23. $3x - 4y = 26$
25. $x = -1$ 27. $y = -2x + 8$
29. $x + 2y = -4$ 31. $r = 2, s = 3$
33. $y = mx + b$; $y_1 = mx_1 + b$; $b = y_1 - mx_1$
35. $y = mx + b$; $y_1 = mx_1 + b$; $b = y - mx$;
$b = y_1 - mx_1$; $y - mx = y_1 - mx_1$;
$y - y_1 = m(x - x_1)$; $\dfrac{y - y_1}{x - x_1} = m$. Similarly, using
equations $y_1 = mx_1 + b$ and $y_2 = mx_2 + b$, you
find that $\dfrac{y_1 - y_2}{x_1 - x_2} = m$. Thus $\dfrac{y_1 - y_2}{x_1 - x_2} = \dfrac{y - y_1}{x - x_1}$.

Pages 225–226, Written Exercises

1. 1 3. $-\frac{1}{2}$ 5. 2 7. -3 9. $-\frac{4}{3}$
11. $x = 2$ 13. $x - 2y = 0$
15. $y = 3$ 17. $3x - 5y = 0$
19. $f = \{(x, y): y = 7x - 2\}$
21. $f = \{(x, y): y = x + 1\}$
23. $f = \{(x, y): y = -\frac{1}{2}x - \frac{3}{2}\}$
25. $f = \{(x, y): y = \frac{1}{2}x + \frac{13}{2}\}$ 27. $f = \{(x, y): y = 0\}$
29. $\frac{1}{5}, \frac{3}{2}, -\frac{2}{3}$ 31. 3, $-\frac{1}{3}$, 1
33. $y = mx + b$; $y_1 = mx_1 + b$; $y_2 = mx_2 + b$;
$y_2 - y_1 = mx_2 - mx_1 = m(x_2 - x_1)$; $m = \dfrac{y_2 - y_1}{x_2 - x_1}$

Answers to Selected Exercises | **557**

35. $(x_1, f(x_1))$ and $(x_2, f(x_2))$ are different points on the graph of f. Then, by Exercise 33, the graph of f has slope $m = \dfrac{f(x_2) - f(x_1)}{x_2 - x_1}$.

Page 229, Written Exercises

1.

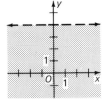

3.

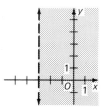

5.

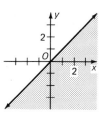

7.

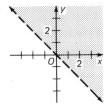

9.

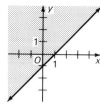

11.

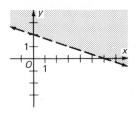

13.

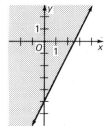

15.

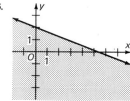

17.

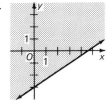

Page 231, Extra for Experts

1. a. $x' = x + 1;\ y' = y - 3$ **b.** $(-6, 0)$
3. a. $x' = x + 13;\ y' = y - 3$ **b.** $(6, 0)$
5. a. $x' = x + 7;\ y' = y + 2$ **b.** $(0, 5)$
7. a. $x' = x + 9;\ y' = y - 5$ **b.** $(2, -2)$
9. a. $x' = x;\ y' = y + k$ **b.** $x' = x + h;\ y' = y$

Pages 233–235, Chapter Review

1. c   **3.** b   **5.** a   **7.** d   **9.** b   **11.** b
13. a   **15.** a   **17.** a   **19.** b

Chapter 7   Systems of Open Sentences

Pages 240–241, Written Exercises

1. $\{(2, 2)\}$

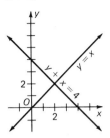

3. $\{(5, 2)\}$

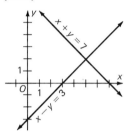

5. $\{(2, -1)\}$

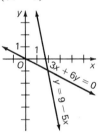

7. $\emptyset$

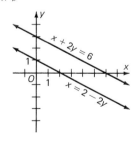

9. $\{(4, 0)\}$

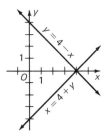

11. $\{(3, 3)\}$

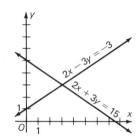

13. parallel lines; empty set   **15.** parallel lines; empty set   **17.** same line; an infinite set
19. $(-2, -2)$ **21.** $(2, -2)$ **23.** 25
25. 25   **27.** 8

Page 244, Written Exercises

1. $\{(5, 2)\}$ **3.** $\{(-2, -6)\}$ **5.** $\{(13, 21)\}$
7. $\{(6, -\frac{1}{2})\}$ **9.** $\{(2, -3)\}$ **11.** $\{(-\frac{11}{3}, -5)\}$
13. $\{(17, 5)\}$ **15.** $\{(-4, \frac{24}{5})\}$ **17.** $\{(\frac{15}{8}, \frac{9}{8})\}$
19. $\{(\frac{22}{7}, \frac{1}{7})\}$ **21.** $\emptyset$
23. (1) $ax + by = c$; (2) $ax - by = d$; adding
equations (1) and (2), $2ax = c + d$ and $x = \dfrac{c + d}{2a}$;
subtracting (2) from (1), $2by = c - d$ and $y = \dfrac{c - d}{2b}$;
the solution of the system is $\left(\dfrac{c + d}{2a}, \dfrac{c - d}{2b}\right)$.

Pages 246–247, Written Exercises

1. $\{(3, -1)\}$ **3.** $\{(-7, -8)\}$ **5.** $\{(-9, 3)\}$
7. $\{(-4, -4)\}$ **9.** $\{(2, 2)\}$ **11.** $\{(1, 1)\}$
13. $\{(5, 5)\}$ **15.** $\{(-3, 2)\}$ **17.** $\{(\frac{1}{3}, \frac{1}{9})\}$
19. $\{(\frac{1}{5}, \frac{1}{3})\}$ **21.** $\{(-9d, -12c)\}$

23. The graph of $ax + by = c$ crosses the axes at $\left(\dfrac{c}{a}, 0\right)$ and $\left(0, \dfrac{c}{b}\right)$. The graph of $by - ax = c$ crosses the axes at $\left(-\dfrac{c}{a}, 0\right)$ and $\left(0, \dfrac{c}{b}\right)$. Since the y-axis is perpendicular to the base and bisects the base, the triangle with vertices $\left(\dfrac{c}{a}, 0\right)$, $\left(0, \dfrac{c}{b}\right)$, and $\left(-\dfrac{c}{a}, 0\right)$ is isosceles.   **25. a.** $\{(\frac{1}{3}, \frac{1}{4})\}$
b. $e(9 \cdot \frac{1}{3} - 8 \cdot \frac{1}{4} - 1) + f(6 \cdot \frac{1}{3} + 12 \cdot \frac{1}{4} - 5) =$
$e(3 - 2 - 1) + f(2 + 3 - 5) = 0$

Page 249, Written Exercises
1. $\{(1, 2)\}$ **3.** $\{(-3, 1)\}$ **5.** $\{(-2, -3)\}$
7. $\{(-\frac{3}{4}, \frac{13}{12})\}$ **9.** $\{(1, 0)\}$ **11.** $\{(3, 1)\}$
13. $\{(\frac{9}{4}, \frac{1}{2})\}$ **15.** $\{(-8, 4)\}$ **17.** $\{(31, 54)\}$

Pages 253–254, Problems
1. Bill's lot: 15; Fran's lot: 12   **3.** 69 m by 23 m
5. 19 \$5 bills; 17 \$10 bills   **7.** \$2500 at 5%;
\$7500 at 6%   **9.** Kurt: 14; sister: 12   **11.** 3 large
boxes; 2 medium boxes   **13.** 64°, 116°
15. 12 smaller cars and 7 larger cars   **17.** 24 cm
by 10 cm   **19.** 6 blocks   **21.** $\frac{19}{36}$ and $-\frac{5}{36}$

Pages 256–257, Problems
1. Wind speed: 70 km/h; air speed: 490 km/h
3. $35\frac{5}{7}$ km/h   **5.** 7 km/h   **7.** $2\frac{11}{12}$ km/h
9. 0.48 km/h   **11.** $s = 3c$ **13.** 560 km/h

Pages 258–259, Problems
1. 65   **3.** 45   **5.** 86   **7.** 236
9. $(10t + u) - (10u + t) = 9t - 9u = 9(t - u)$
11. 49   **13.** 0.18   **15.** 642

9.
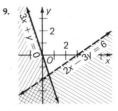

11.
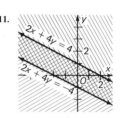

13.
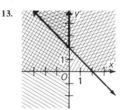

15.
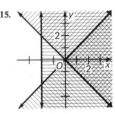

17.
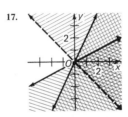

19.
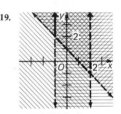

21.
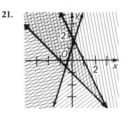

23.
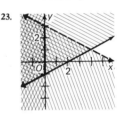

25.
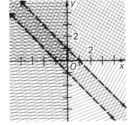

Pages 261–262, Written Exercises

1.
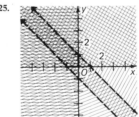

3.
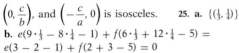

5.
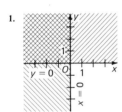

7.
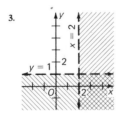

Pages 264–265, Written Exercises
1. Not convex   **3.** Convex   **5.** (0, 8), 8; (1, 4), 7;
(4, 2), 14; (8, 0), 24   **7.** It has a greater value at
some points in the region than at the corner
points.   **9.** It has a lesser value at some points
in the region than at the corner points.
11. a. 27 **b.** 22 **c.** 20 **d.** 24

13. a.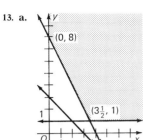

b. 10

15. a.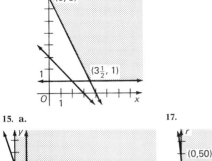

b. 11

17.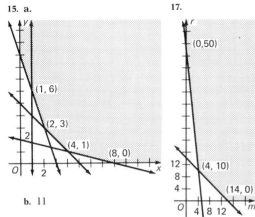

19. Answers may vary. Examples:
no monkeys and 50 rabbits,
1 monkey and 40 rabbits,
2 monkeys and 30 rabbits,
3 monkeys and 20 rabbits,
4 monkeys and 10 rabbits

21. a. Expenses would be lowered by \$1.65 if the post office used the older machine for 4 h and the newer one for 3 h. **b.** Yes; expenses would be lowered by 45¢.

Pages 269–271, Written Exercises
1. b and c   **3.** $(1, 0, 0)$, $(0, -1, 0)$, $(0, 0, 1)$. Other answers are possible.   **5.** $(-6, 0, 0)$, $(0, 4, 0)$, $(0, 0, 12)$. Other answers are possible.
7. $\{(3, 2, 1)\}$ **9.** 1, 2, and 4   **11.** 5 mm, 19 mm, 20 mm   **13.** $\{(-\frac{2}{3}, 1, \frac{1}{3})\}$ **15.** $\{(4, 2, 3)\}$
17. $\{(-5, 3, -2)\}$ **19.** $a = 3$, $b = -1$, $c = 2$
21. $a = 1$, $b = 2$, $c = -5$ **23.** $a = 2$, $b = 1$, $c = -1$

Page 274, Extra for Experts
1. $\{(1, 2)\}$ **3.** $\{(4, 2)\}$ **5.** $\{(4, -1)\}$

Pages 275–276, Chapter Review
1. d   **3.** c   **5.** d   **7.** b   **9.** b

Chapter 8   Polynomials and Their Factors

Pages 282–283, Written Exercises
1. $xy - 7 + 4xy = 5xy - 7$
3. $-4a^2 + 5a - 4a + 2a^2 = a - 2a^2$
5. $7k^4 - 2k^2 + 3k^4 - 5k = 10k^4 - 2k^2 - 5k$
7. $6 + 4i^2j^2 - 3ij - 8 = 4i^2j^2 - 3ij - 2$
9. $3ef^2 - 5ef + 7e^2f - 4e^2f + ef = 3ef^2 - 4ef + 3e^2f$
11. $9x^2yz - 2xy^2z + 3xyz - 7x^2yz + xy^2z - 1 = 2x^2yz - xy^2z + 3xyz - 1$
13. $-3uv^2w + uvw^2 + 2u^2vw - 4uvw^2 + 3u^2vw + 9uv^2w = 6uv^2w - 3uvw^2 + 5u^2vw$
15. $2^4x^3y - 3^2x^3y - 2^2x^3y + 2^3x^3y = 11x^3y$
17. 2, 3   **19.** 2, t **21.** e, 4   **23.** t
25. $p^3 + 7p^2 - p - 14$ **27.** $4x^2y + 3xy^2 - 3xy$
29. $-3g^2$ **31.** $6c^3 + c^2 - 4c$
33. $3g^4 - g^3 - 2g^2 + g - 5$
35. $z^6 - z^5 - z + 1$ **37.** $(3a^2 - 5ab + 7b^2)$
39. $3x^2 + 4xy - 6y^2$

Page 283, Diversion: A

Page 285, Written Exercises
1. $-6x^3$ **3.** $-15c^3$ **5.** x^3y^3 **7.** $-14p^8$
9. $4n^5$ **11.** s^6 **13.** $-8k^4$ **15.** $-r^9s^7$
17. $-90m^6n^5$ **19.** $8x^3$ **21.** $7x^4y^4$ **23.** $a^7 + 2a^3$
25. $-6x^3y - 15x^2y^2$ **27.** $4r^3s^2t^2$ **29.** $73a^6b^{10}c$
31. $-6x^4y^2 + 2x^3y^4$ **33.** $-12x^3y^2z^2 - 2x^3y^4$
35. k^{3n} **37.** x^{3n} **39.** $10^4 = 10,000$

Page 288, Written Exercises
1. $k^5 + 2k^3 - k^2$ **3.** $-y^5 + 4y^4 - 2y^3$
5. $r^3s + r^2s^2 - rs^3$ **7.** $-10p^5 + 20p^4 + 10p^3 - 5p^2$
9. $5x^3y^5 + 2x^4y^4 - x^3y^6$ **11.** $y^2 + 3y + 2$
13. $z^2 - z - 6$ **15.** $s^2 - 9s + 20$
17. $6m^2 - 7m - 3$ **19.** $4h^2 + 12h + 9$
21. $15r^2 - 34r + 15$ **23.** $x^3 - 27$ **25.** $x^3 - y^3$
27. $\{y: y = -3\}$ **29.** $\{p: p = \frac{7}{2}\}$ **31.** $\{r: r < 2\}$
33. $\{s: s \le -7\}$ **35.** $-j^4 - j^3 - j^2 - 11j + 24$
37. $2b^3$ **39.** $e^{3n+1} + fe^{n+1}$ **41.** $r^{4n} - 36$
43. Answers may vary.

Answers to Selected Exercises | **561**

Page 289, Problems

1. 8 cm, 10 cm   3. 27 and 28   5. Width, 26 cm; length, 32 cm; height, 11 cm   7. 8 and 10
9. 9 cm   11. 113.04 m²

Pages 291–292, Written Exercises

1. 396   3. 896   5. 3596   7. 2491   9. 8091
11. $10a^2 - a - 2$ 13. $35c^2 + 47c + 8$
15. $35 + 31e - 18e^2$ 17. $12g^2 + 7gh - 45h^2$
19. $t^2x^2 + 2txc + c^2$ 21. $169k^2 - 64$
23. $9m^4 - 64$ 25. $9 - 16p^4$
27. $8s^3 - 38s^2 + 24s$ 29. $9u^3 - 39u^2 + 40u - 12$
31. $-32w^4 - 80w^3 + 46w^2 - 6w$
33. $125y^3 + 150y^2 + 60y + 8$ 35. $\{x: x = -2\}$
37. $\{t: t = \frac{3}{4}\}$ 39. $\{x: x = \frac{1}{27}\}$
41. $2x^4 + 7x^3 + 11x^2 - x - 10$ 43. $p^5 + 1$
45. $2x^{2n} - 5x^n - 3$ 47. $r^{4n} - r^{2n}s^n - 2s^{2n}$
49. $e^{2n} + e^{n+m} - e^{n+2m} - e^{3m}$

Pages 295–296, Written Exercises

1. $3^2 \cdot 2^2$ 3. 29   5. $2^5 \cdot 3 \cdot 5$ 7. $2^3 \cdot 5 \cdot 7 \cdot 13$
9. $\{1, 2, 3, 4, 6, 8, 12, 24\}$ 11. $\{1, 2, 29, 58\}$
13. $\{1, 2, 3, 5, 6, 9, 10, 15, 18, 30, 45, 90\}$
15. $\{1, 11, 121\}$ 17. 2   19. 25   21. $4x^2y$
23. $38r^3s^2$ 25. $16d^2e^3f^2$ 27. $2x$ 29. $25a^6b^6$
31. None   33. $-4m^3np^2$ 35. $12x$ 37. $9e^2f^3$
39. $3x^{2n}$

Page 297, Programming in BASIC

1. $\{1, 5, 17, 85\}$, $\{1, 2, 43, 86\}$, $\{1, 3, 29, 87\}$,
$\{1, 2, 4, 8, 11, 22, 44, 88\}$, $\{1, 89\}$
3. $\{1, 2, 4, 6, 9, 12, 27, 36, 54, 108\}$,
$\{1, 2, 3, 4, 5, 6, 10, 12, 15, 20, 30, 60\}$, $\{1, 2, 3, 4, 6, 12\}$,
12

Pages 298–299, Written Exercises

1. $2(a - 4)$ 3. Irreducible   5. $x(x - 1 + y)$
7. $3(5r^2 - 6r + 15)$ 9. $4xy(3x + 4y - 1)$
11. $3ab^2(5ab - 6a^4 + 8b^2)$
13. $3p(p^3 - 4p^2 + 2p - 7)$
15. $6mn(5m - 4n + 6m^2)$ 17. $(a + 5)(a - 3)$
19. $(2c - 1)(3c + 1)$ 21. $(4x + 5)(3x - 7)$
23. $(x + 5)(x + 3)$ 25. $(4z + 17)(z + 5)$
27. $(10w + 11)(2w + 3)$ 29. $(e^2 + f)(e - f)$
31. $(5x - 1)(y^3 + z)$ 33. $a^n - 2$ 35. $2c - 1$
37. $1 + 4e$ 39. $n = 1$ 41. $n = 1$

Pages 300–301, Written Exercises

1. $(4a - 1)(4a + 1)$ 3. $(5 - 2c)(5 + 2c)$
5. $3(5e - 7f)(5e + 7f)$ 7. $(i - 9)^2$ 9. $2(7k - 1)^2$
11. $(2n + 3)^2$ 13. $(x + y)(6 + y)$
15. $(w + 3z)(w - 1)$ 17. $(3t - 2)(2t^2 - 1)$
19. $(x + 1)(x - 1)(y^3 + 2)$
21. $(5a + 1 - b)(5a + 1 + b)$
23. $(c + 2d - 5)(c + 4d + 5)$ 25. $(r - 5)^2$
27. $(a - b - c)(a + b + c)$
29. $(r + s)(d - e - f)$
31. $(4 - 7h + k)(4 + 7h - k)$

33. $(a - b + c)(a + b - c)$ 35. $(y^n - 1)(y^n + 1)$
37. $(x^n - 4)^2$ 39. $(x^n - w^n)(x^n + w^n)(y + m)$

Page 303, Written Exercises

1. $(x + 2)(x + 3)$ 3. $(x + 6)(x + 4)$
5. $(x + 7)(x - 5)$ 7. $(x - 7)(x - 3)$
9. $(x - 7)(x + 3)$ 11. $(3 + b)(2 - b)$
13. $2d(d + 10)(d - 2)$ 15. $f^2(f + 5)(f + 6)$
17. $h(h - 8)(h + 2)$ 19. $(a + 3b)(a - 2b)$
21. Irreducible   23. $p(x - 9p)(x + 5p)$
25. $r(y + 9r)(y + 5r)$ 27. $18x(x - y)(x - y)$
29. Irreducible   31. $(ab + 2)(ab - 1)$
33. $4(xy + 2)(xy + 3)$ 35. $-e^2f(8 + g)(2 + g)$

Pages 304–305, Written Exercises

1. $(3a + 1)(a + 1)$ 3. $(2c - 1)(c - 1)$
5. $(4e - 1)(e + 1)$ 7. $(3k - 5)(k + 1)$
9. $(8r - 5)(r + 1)$ 11. $(8y - 5)(2y - 5)$
13. $(4x + 1)(4x - 5)$ 15. $(16z + 5)(z - 1)$
17. $(2u + 3)(2u + 5)$ 19. $(3a + 2b)(5a + 2b)$
21. $(e + f)(6e - 9f)$ 23. $2rs(r - 3s)(2r + s)$
25. $-3u(3u - 2)(3u + 1)$ 27. $-3l(2l - 3)(3l + 2)$
29. $2(2m + 1)(2m - 1)(m^2 + 2)$
31. $e(8y - 3)(3y + 2)$ 33. $cn(2x - 3)(3x + 2)$
35. $(x^2 - x - 3)(x^2 + x - 3)$
37. $(c^2 + cd + d^2)(c^2 - cd + d^2)$
39. $x^2 + x + c = (x + r)(x - s)$; the absolute values of r and s must differ by one and c must be negative. For example, $c = (-3)(4)$, $c = (-4)(5)$, and so on.

Pages 309–310, Written Exercises

1. $\{-6, 3\}$ 3. $\{-8, -11\}$ 5. $\{0, 2, -2\}$
7. $\{-6, \frac{3}{2}\}$ 9. $\{6\}$ 11. $\{0, 4\}$ 13. $\{\frac{7}{3}, -\frac{7}{3}\}$
15. $\{-\frac{1}{4}, 1\}$ 17. $\{0, -\frac{1}{2}, 3\}$ 19. $\{4, -3\}$
21. $\{-4, -5\}$ 23. $\{\frac{4}{3}, -2\}$ 25. $\{-\frac{3}{2}, \frac{5}{2}\}$
27. $\{-1, \frac{10}{3}\}$ 29. $\{-2, -5\}$ 31. $\{-1, 4\}$
33. $x^2 + 2x - 8 = 0$ 35. $x^2 + 4x + 4 = 0$
37. $x^2 + 13x + 36 = 0$ 39. Multiplicative property of equality   41. Multiplicative property of zero

Pages 311–313, Problems

1. 6 and 7   3. -5 and -6 5. 2 m
7. 3 and 14   9. Yes; 16 cm   11. 23 cm, 7 cm
13. 23   15. Failure to transform the equation so that the right member is 0 before factoring left member   17. 6 cm, 8 cm
19. 2 m by 6 m or 3 m by 4 m   21. 16
23. 20 cm by 8 cm
25. Let $2n + 1$ be an odd integer; then
$(2n + 1)^2 = 4n^2 + 4n + 1 = 4n(n + 1) + 1$. If
$n = 2k$, then $4 \cdot 2k(2k + 1) + 1 = 8k(2k + 1) + 1$,
which is 1 more than a multiple of 8. If $n + 1$ is
even, then $n + 1 = 2k$ and $n = 2k - 1$; then
$4n(n + 1) + 1 = 4(2k - 1)(2k) + 1 =$
$(8k - 4)2k + 1 = 16k^2 - 8k + 1 = 8k(2k - 1) + 1$,
which is one more than a multiple of 8.

Page 315, Extra for Experts

1.

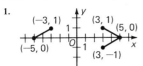

3.

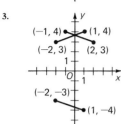

5.

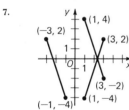

7.

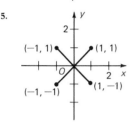

9. It must be parallel to the x-axis; parallel to the y-axis

Pages 316–318, Chapter Review

1. d   **3.** c   **5.** b   **7.** a   **9.** c   **11.** d
13. b   **15.** a   **17.** d   **19.** b

Page 318, On the Calculator

n	$(n - 1)! + 1$	Factor?
2	2	Yes
3	3	Yes
4	7	No
5	25	Yes
6	121	No
7	721	Yes
8	5041	No
9	40,321	No
10	362,881	No
11	3,628,801	Yes

n is a factor of $(n - 1)! + 1$ when n is a prime number.

Pages 319–321, Cumulative Review Chapters 5–8

1. c   **3.** d   **5.** a   **7.** b   **9.** a   **11.** c   **13.** d
15. b   **17.** d   **19.** c   **21.** b   **23.** b

Chapter 9   Polynomials and Rational Expressions

Pages 327–328, Written Exercises

1. $3a^3b^2$ **3.** $-3f$ **5.** $\dfrac{y}{3x^2}$ **7.** $\dfrac{3w}{z^2}$

9. $\dfrac{7}{x}$ **11.** $\dfrac{-c^3}{4}$, or $-\dfrac{c^3}{4}$ **13.** $\dfrac{6}{x}$ **15.** $\dfrac{12x}{y^3}$

17. $\dfrac{4}{(x - y)}$ **19.** $\dfrac{3x + 2}{k}$ **21.** $\{13\}$ **23.** $\{7\}$

25. $\{11\}$ **27.** $\{20\}$ **29.** $\dfrac{2b}{3a}$ **31.** $\dfrac{1}{9}$

33. $-\dfrac{7x^3}{2}$ **35.** $-\dfrac{y^4}{2}$ **37.** $\dfrac{4n^6}{m^3}$

39.
$$\left(\dfrac{a}{b}\right)^2 = \left(a \cdot \dfrac{1}{b}\right)^2 \qquad \text{Rel. between mult. and div.}$$
$$= a^2 \cdot \dfrac{1}{b^2} \qquad \text{Third law of exponents for mult.}$$
$$= \dfrac{a^2}{b^2} \qquad \text{Rel. bet. mult. and div.}$$

41.
$$\dfrac{a}{-b} = \dfrac{a}{(-1)b} \qquad \text{Mult. prop. of } -1$$
$$= \dfrac{1}{-1} \cdot \dfrac{a}{b} \qquad \text{Rel. bet. mult. and div.}$$
$$= -\dfrac{a}{b} \qquad \text{Mult. prop. of } -1$$

Pages 330–331, Written Exercises

1. $5a - 3$ **3.** $6c + 2$ **5.** $e^2 + 2e - \dfrac{f}{2}$

7. $-3y^2 + 2y - 4$ **9.** $z + \dfrac{3}{w}$ **11.** $\dfrac{4y^2}{3x} - 1 + \dfrac{6}{x}$

13. $-\dfrac{3}{q} - \dfrac{5}{p} + 2$ **15.** $\dfrac{21}{5d} - \dfrac{4}{c} + \dfrac{24e}{5cd}$

17. $x^n - 1 + \dfrac{1}{x}$ **19.** $3a^2 + a - 2$

21. $4c^3 + 5c^2 - 9$ **23.** $\dfrac{x}{y} = 20$

25. $\{\Re \text{ except } 0\}$
27. No, because 0 cannot be a solution of
$\dfrac{6m^2 - 10m}{2m} < 1.$

Pages 333–334, Written Exercises

1. $a - 5$ **3.** $c + 3$ **5.** $5e - 7$ **7.** $g - 2$

9. $6j - 7 + \dfrac{3}{2j + 3}$ **11.** $x^2 - 2x + 4$

13. $3z + w$ **15.** $p - 3 + \dfrac{18}{p + 3}$ **17.** $2b^2 - 6$

19. $d^2 - 2d - 3 + \dfrac{5d + 10}{d^2 + 3}$ **21.** $2f^2 + f - 2$

23. $(2y + 1), (y + 3)$ **25.** $p = -20$ **27.** $p = -4$
29. $p = 10$ **31.** $10x + 10y$

Pages 337–338, Written Exercises

1. $\dfrac{4b}{9c}$ **3.** $\dfrac{-1}{2}$, or $-\dfrac{1}{2}$ **5.** $\dfrac{8}{13(c - 1)}$

7. $\dfrac{(3e + 2f)}{4(2e + f)}$ **9.** $5 + \dfrac{8}{f}$ **11.** $\dfrac{4}{s + 4}$ **13.** $\dfrac{1}{p + 1}$

15. $\dfrac{p}{p + 1}$ **17.** $\dfrac{x + y}{x + 3y}$ **19.** $\dfrac{2w - 3}{3}$

21. $-(u + 7)$ **23.** $\dfrac{5(b - 2)}{-4(2b - 1)}$ **25.** $\dfrac{f + 2}{5(f - 2)}$

27. Given fraction is in lowest terms; k is not a *factor* of either term.
29. z is not a factor of either term.

31.
$\dfrac{ab - ac}{b - c} < 0$	Given
$b \neq c$	Given
$\dfrac{a(b - c)}{b - c} < 0$	Distributive prop.
$a < 0$	Theorem, p. 335

Page 338, Diversion

Answers may vary. Examples: $\dfrac{16}{64} = \dfrac{1\!\!/6}{6\!\!/4} = \dfrac{1}{4}$;

$\dfrac{2666}{6665} = \dfrac{26\!\!/6\!\!/6}{6\!\!/6\!\!/65} = \dfrac{2}{5}$

Pages 340–341, Written Exercises

1. 6 **3.** $-\dfrac{1}{64}$ **5.** $\dfrac{3}{a^2}$ **7.** $\dfrac{8b}{a^3}$ **9.** $\dfrac{d}{c^2}$

11. $\dfrac{1}{x^6y^3}$ **13.** 1 **15.** $\dfrac{1}{m + n}$ **17.** r^4 **19.** $\dfrac{f^2}{d^2e^2}$

21. 1 **23.** $\dfrac{a^2b^2}{256}$ **25.** $3 + y$ **27.** $\dfrac{w}{3} + \dfrac{3}{w^4}$

29. $c^{-3}d^2e^{-3}$ **31.** $\dfrac{u^{-3}v^5w^{-1}}{6}$ **33.** $-\dfrac{11x^3z^{-3}}{12}$

35. $\dfrac{6}{5}$ **37.** $\dfrac{1}{17}$ **39.** $\dfrac{1}{4}$

Pages 344–346, Written Exercises

1. $\dfrac{20}{21}$ **3.** $-\dfrac{33}{17}$ **5.** a **7.** $-\dfrac{4c}{5}$ **9.** $\dfrac{4e}{f}$

11. $\dfrac{w}{3}$ **13.** $\dfrac{3}{4}$ **15.** $\dfrac{a}{(a + 4)(2a - 1)}$

17. $\dfrac{c(2c - 1)}{6(c + 1)}$ **19.** $\dfrac{4}{9x}$ **21.** $\dfrac{4z}{z + 1}$ **23.** $\dfrac{5(x + 2)}{x^2}$

25. $x - 3$ **27.** $\dfrac{x^2 + 1}{(x - 1)(x + 1)(x - 5)}$ **29.** $\dfrac{x + 7}{x - 3}$

31. $\dfrac{k + 2}{k + 1}$ **33.** $\dfrac{w + 6}{w - 2}$ **35.** 1 **37.** $\dfrac{3n - 1}{n - 2}$

39. $\dfrac{1}{k - 1}$ **41.** $\dfrac{c - 1}{c - 3}$ **43.** $\dfrac{(e - 1)(e - 3)(e + 2)}{e(e - 7)(e + 1)}$

45. Given that $a, b, c \neq 0$, we have $\dfrac{a}{1} \div \dfrac{b}{c} = \dfrac{a}{1} \cdot \dfrac{c}{b} = \dfrac{ac}{b}$ (second theorem on page 343 and rule for naming products). The reciprocal of $\dfrac{ac}{b}$ is $\dfrac{b}{ac}$. Therefore the reciprocal of $a \div \dfrac{b}{c}$ is $\dfrac{b}{ac}$.

Page 348, Written Exercises

1. a. $\dfrac{3x + 6}{5}$ **b.** $\dfrac{x + 8}{5}$ **3. a.** $\dfrac{x + 2}{x^2 + 1}$ **b.** $\dfrac{3x - 2}{x^2 + 1}$

5. $\dfrac{1}{x - 3}$ **7.** $\dfrac{1}{y + 3}$ **9.** $\dfrac{3m + 1}{m^2 - m + 2}$ **11.** 0

13. $\dfrac{2r^2}{r^2 + 4}$

15.
$\dfrac{1}{x - y} + \dfrac{1}{y - x}$	
$= \dfrac{1}{x - y} + \dfrac{-1}{x - y}$	Mult. prop. of -1
$= \dfrac{1 + (-1)}{x - y}$	Rule for naming sums of real nos.
$= \dfrac{0}{x - y}$	Ax. of add. inv.
$= 0$	Property of 0

17. $\dfrac{8x^2 + 4x}{x^2 - 9}$

Pages 351–352, Written Exercises

1. $\dfrac{17}{6}$ 3. $\dfrac{19}{5}$ 5. $\dfrac{3b + 4a}{ab}$ 7. $\dfrac{2cd - 7}{c^2 d^2}$

9. $\dfrac{2s^2 + rs + 2r^2}{2rs}$ 11. $\dfrac{w + 2}{2(w + 3)}$ 13. $\dfrac{14}{5(x - 2)}$

15. $\dfrac{2c^2 - c + 5}{(c - 5)(c + 5)}$ 17. 0 19. $\dfrac{4s}{r - 4}$

21. a. $\dfrac{24x - 5}{3x(3x - 1)}$ b. $\dfrac{5 - 6x}{3x(3x - 1)}$ 23. a. $\dfrac{5x + 2}{6x(x - 1)}$

b. $\dfrac{2 - x}{6x(x - 1)}$ 25. $\dfrac{-2(4x + 3y)(x - y)}{(3x + y)(2x - y)}$

27. $\dfrac{2(k - 2)}{(k - 3)(k - 1)^2}$ 29. $\dfrac{2(r^2 + 2r + 2)}{(r - 4)(r + 4)(r - 1)}$

31. $\dfrac{1}{m^2 - 25}$ 33. $\dfrac{-4h^3 - 10h^2 + 27h - 9}{(2h - 3)(h + 4)}$

35. $\dfrac{d^2 + 2cd - c - d}{cd}$ 37. $\dfrac{p^3 - 8p - 9}{(p + 3)(p - 3)}$

39. $\dfrac{l - 25}{(3l - 4)(2l + 5)(l - 2)}$

41. $\dfrac{2(r^2 - 4rs - 3s^2)}{(r + 3s)(r - s)(r + s)}$ 43. $\dfrac{w}{w - 2}$ 45. $\dfrac{x + y}{x - y}$

47. $\dfrac{q^2(p + q)}{p(p^2 - pq + 2q^2)}$

Page 353, Programming in BASIC

```
1. 10   INPUT M,N
   20   FOR F=1 TO N
   30   LET L=M*F
   40   LET Q=L/N
   50   IF Q=INT(Q) THEN 70
   60   NEXT F
   70   PRINT "THE L.C.M OF"; M; "AND" ;N;
        "IS" ;Q "."
   80   END
```
2. That the G.C.F. of M and N is 1.

Pages 353–354, Chapter Review

1. b 3. a 5. d 7. d

Chapter 10 Rational Expressions in Open Sentences

Pages 359–360, Written Exercises

1. $\{4\}$ 3. $\{c: c \geq \frac{28}{3}\}$ 5. $\{\frac{2}{51}\}$ 7. $\{f: f > \frac{126}{7}\}$

9. $\{12\}$ 11. $\{c: c < \frac{16}{3}\}$ 13. $\{e: e \leq -\frac{6}{5}\}$

15. $\{-7\}$ 17. $\{2\}$ 19. $\{k: k < 12\}$ 21. $\{-2\}$

23. $\{w: w \leq 4\}$ 25. $\{600\}$ 27. $\{9\}$

29. $\{-\frac{85}{26}\}$ 31. $\frac{3}{5}$ 33. $-\frac{1}{6}$ 35. $-\frac{2}{5}$

37. $\{(6, 0)\}$ 39. $\{(0, -2)\}$ 41. $\{(3, -2)\}$

43. $\{(\frac{1}{6}, \frac{1}{2})\}$ 45. $\{(-5, 3)\}$

47. $\dfrac{a}{b} = 0$, given; $b \neq 0$, $b\left(\dfrac{a}{b}\right) = b(0)$, mult. prop. of equality; $a = 0$, mult. prop. of 0 and rel. of mult. to div.

Page 363, Written Exercises

1. 150.5 3. 67.4% 5. 311.1 7. 199.7

9. 116.9% 11. 34.5 13. 0.9 15. 0.2%

17. 4800

Pages 363–365, Problems

1. 312.5 t 3. $5500 5. 24 mL 7. 16 g of 35% alloy, 4 g of 65% alloy 9. 56% 11. $50.67

13. 80 L 15. $5800 17. $127.28 or more

19. 3.8 L 21. $4200 at 7%, $3150 at 6%, $3000 at 5% 23. 8 L

Page 368, Written Exercises

1. $\{3\}$ 3. $\{3\}$ 5. $\{1\}$ 7. $\{4\}$ 9. $\{1, -2\}$

11. $\emptyset$ 13. $\{2, -6\}$ 15. $\{2\}$ 17. $\{-2\}$

19. $\{-18, \frac{1}{2}\}$ 21. $\emptyset$ 23. $x = 2\pi r^2 + 2\pi rh$

25. $x = \dfrac{A - P}{Pr}$ 27. $\{\frac{3}{13}\}$ 29. $\emptyset$ 31. $\emptyset$

33. $k = 3 + x$ 35. $b = c$, $a = 0$, given; $ab = 0$, $ac = 0$, mult. prop of 0; $ab = ac$, trans. ax. of equality 37. $b \neq c$, $a = 0$, given; $ab = 0$, $ac = 0$, mult. prop. of 0; $ab = ac$, trans. ax. of equality

Pages 370–371, Problems

1. 15, 18 3. 18 5. 5, 6 7. 20, 36

9. $\frac{4}{16}$ 11. 7, 17 13. $\frac{2}{3}$, $\frac{3}{2}$ 15. 12, 36 17. 2, 3

19. 4 and 16

Pages 372–373, Problems

1. $1\frac{7}{8}$ d 3. $3\frac{3}{5}$ h 5. 1.36 h 7. 12 min, 24 min

9. $3\frac{1}{3}$ h 11. 2 h 13. $9\frac{3}{5}$ h 15. 6 h

Pages 374–375, Problems

1. 5 km/h 3. express, 100 km/h; freight, 80 km/h 5. 27 km 7. 1440 km 9. 80 km/h

11. 96 km 13. 200 km/min 15. faster car, 164.8 km/h; slower car, 155.2 km/h 17. 7.5 min

Page 378, Written Exercises
1. $\frac{12}{5}$ **3.** $\frac{140}{3}$ **5.** 24 **7.** 6 **9.** $22\frac{1}{2}$ **11.** $\frac{1}{30}$
13. $6e$ **15.** x is halved. **17.** Yes

Page 379, Problems
1. 750 kHz **3.** 6 m **5.** 2500 r/min
7. 1818.1 m

Pages 382–383, Written Exercises
1. $153\frac{1}{8}$ **3.** 8 **5.** $43\frac{3}{4}$ **7.** $6\frac{6}{19}$ **9.** $1\frac{1}{14}$
11. x is divided by 4. **13.** No; $x = ky^2$ and

$y^2 = \frac{1}{k}x$ **15.** x is multipled by 12. **17.** Q is
multiplied by $\frac{3}{2}$. **19.** Q is multiplied by 4.

Page 383, Problems
1. 98π cm^3 **3.** 195 J **5.** 16.384 t

Pages 385–386, Chapter Review
1. c **3.** d **5.** a **7.** d **9.** c **11.** c **13.** a

Chapter 11   Irrational Numbers and Radicals

Page 391, Written Exercises
1. 8 **3.** -9 **5.** 14 **7.** 5 **9.** $\frac{3}{8}$ **11.** $-\frac{6}{7}$
13. $\sqrt{4}$ **15.** $-\sqrt{144}$ **17.** $\sqrt{\frac{9}{25}}$ **19.** $4|a|$
21. $7a^2$ **23.** $\left|\frac{a}{b}\right|$ **25.** $3|ab|$ **27.** $|a + b|$
29. $\sqrt{9x^2 + 12xy + 4y^2} = \sqrt{(3x + 2y)^2} = |3x + 2y|$
for all real values of x and y.
31. Yes; domain: $\{x: x \geq 0\}$; range: $\{y: y \leq 0\}$
33. $\sqrt{(-a)^2} \neq -a$.

Pages 393–394, Written Exercises
1. $\frac{73}{100}$ **3.** $\frac{23}{99}$ **5.** $\frac{7207}{999}$ **7.** $\frac{25}{33}$
9. $\{\sqrt{5}, -\sqrt{5}\}$ **11.** $\{\sqrt{14}, -\sqrt{14}\}$
13. $\{-\sqrt{5}, \sqrt{5}\}$ **15.** $\frac{19}{30}$ **17.** $\frac{13}{24}$ **19.** $\frac{11}{10}$
21. $\frac{1}{1}$ **23.** $\frac{76}{45}$ **25.** $\frac{10}{9}$ **27.** $\frac{1}{10}$ **29.** $\frac{11}{15}$
31. $\frac{1}{3}$ **33.** $\frac{1}{1}$ **35.** $\frac{101,803}{9801}$ **37.** $x + 1$

Page 394, Diversion
4,444,355,556

Page 394, Programming in BASIC
```
10   PRINT "INPUT NUMERATOR N:";
20   INPUT N
30   PRINT "INPUT DENOMINATOR D (> N):";
40   INPUT D
50   PRINT N; "/"; D; "=.";
60   LET K = 0
70   LET Q = 10*N/D
80   LET M = INT (Q)
90   PRINT M;
100  LET N = 10*N − M*D
110  IF N = 0 THEN 150
120  LET K = K + 1
130  IF K > D THEN 150
140  GOTO 70
150  END
```

Page 397, Written Exercises
1. 4.12 **3.** 7.28 **5.** 8.10 **7.** 9.39 **9.** 6.49
11. 0.12 **13.** 0.56 **15.** 0.94 **17.** 1.1 **19.** 0.5
21. 2.63 **23.** Any two rational numbers x and y,
not zero, such that $x = -y$. **25.** Computed
results for the given values are $-5, 5, -9, 32, 0$.
These suggest that
$(\sqrt{a} + \sqrt{b})(\sqrt{a} - \sqrt{b}) = a - b$.

Page 398, Diversion
Answers will vary. Example: (4, 1)

Page 398, Programming in BASIC
```
10   PRINT "INPUT N";
20   INPUT N
30   PRINT "INPUT FIRST TEST FACTOR F";
40   INPUT F
50   LET Q = N/F
60   PRINT "TEST FACTOR ="; F
70   PRINT "QUOTIENT ="; Q
80   IF ABS(Q − F) < .00001 THEN 120
90   LET F = (F + Q)/2
100  PRINT "AVERAGE ="; F
110  GOTO 50
120  PRINT "SQUARE ROOT OF"; N;
        "IS APPROXIMATELY"; (F + Q)/2; "."
```

Page 401, Written Exercises
1. 6.71 **3.** 8.06 **5.** 7.75 **7.** 4.12 **9.** 5.20
11. No **13.** No **15.** No **17.** Yes **19.** Yes
21. No

Page 402, Problems
1. 10 m **3.** 7.48 m **5.** 15 km **7.** 24 m
9. 5.66 cm **11.** 8.94 mm by 17.89 mm
13. 1.73 cm **15.** 10.77 cm **17.** $d = k\sqrt{3}$

Page 404, Written Exercises
1. 5   3. $\sqrt{26}$ 5. $\sqrt{26}$ 7. $\sqrt{130}$ 9. $\sqrt{18}$
11. $5^2 = 3^2 + 4^2$ 13. $(\sqrt{50})^2 = (\sqrt{32})^2 + (\sqrt{18})^2$
15. $(\sqrt{180})^2 = (\sqrt{90})^2 + (\sqrt{90})^2$
17. $(\sqrt{18})^2 + (\sqrt{18})^2 = (\sqrt{36})^2$ 19. 5, -1
21. Origin coordinates are (0, 0);
$\sqrt{(x-0)^2 + (y-0)^2} = \sqrt{x^2 + y^2}$

Pages 406–407, Written Exercises
1. $4\sqrt{3}$ 3. $2\sqrt{6}$ 5. $4\sqrt{2}$ 7. $a\sqrt{7}$ 9. $2\sqrt{7}$
11. $c\sqrt{c}$ 13. $3e\sqrt{e}$ 15. $7\sqrt{g}$ 17. $4s\sqrt{2s}$
19. $3m\sqrt{6n}$ 21. $a + b$ 23. $2(x^2 + y^2)$
25. $4(x + 2y)^2$ 27. $2(r - s)^3\sqrt{6}$ 29. $a + b$
31. $|3 - 4x|$ 33. $r(r^2 + s^2)\sqrt{r^2 - s^2}$
35. Law of exponents for multiplication.
37. Substitution principle   39. $|a - 1|\sqrt{a + b}$

Pages 408–409, Written Exercises
1. $\dfrac{3\sqrt{2}}{2}$ 3. $\dfrac{8\sqrt{a}}{a}$ 5. $\dfrac{\sqrt{2c}}{c}$ 7. 3   9. $\dfrac{2\sqrt{x}}{x}$

11. $\dfrac{3\sqrt{5z}}{5z}$ 13. 15.88   15. 17.32   17. 4.11

19. $\dfrac{2\sqrt{3}}{a}$ 21. $\dfrac{2a\sqrt{b}}{b}$ 23. $\sqrt{6s}$ 25. $2\sqrt{xy}$

27. $\dfrac{\sqrt{xy}}{xy}$ 29. $\dfrac{7\sqrt{10}}{2x}$ 31. $\dfrac{\sqrt{m+n}}{m+n}$

33. $\dfrac{m\sqrt{m-n}}{(m-n)^2}$ 35. $\dfrac{\sqrt{r(r+s)}}{(r+s)^2}$

Pages 410–411, Written Exercises
1. $4\sqrt{3}$ 3. $-2\sqrt{2}$ 5. $-5\sqrt{2}$ 7. $2\sqrt{2a}$
9. $12\sqrt{6}$ 11. $6\sqrt{2} - 6$ 13. $2\sqrt{2} - 1$
15. $21 - 11\sqrt{5}$ 17. 2   19. $16\sqrt{7} + 44$
21. $4x - 8\sqrt{xy} + 4y$ 23. $5\sqrt{5} - 10$
25. $\dfrac{3\sqrt{2} + 3\sqrt{3} - 2 - \sqrt{6}}{7}$ 27. $\dfrac{x - 2\sqrt{xy} + y}{x - y}$

29. $\sqrt{15}$ 31. $4m^3\sqrt{mn}$ 33. $\dfrac{s\sqrt{55}}{10}$

35. $(56 - 16\sqrt{6})$ cm^2 37. $2\sqrt{3}$ 39. $-8\sqrt{2} + 14$
41. Checking by substitution:
$(-3 + \sqrt{5})^2 + 6(-3 + \sqrt{5}) + 4 = 9 - 6\sqrt{5} +$
$5 - 18 + 6\sqrt{5} + 4 = 0; (-3 - \sqrt{5})^2 +$
$6(-3 - \sqrt{5}) + 4 = 9 + 6\sqrt{5} + 5 - 18 -$
$6\sqrt{5} + 4 = 0$

Page 413, Extra for Experts
1. 3   3. 0   5. -1 7. $3\sqrt[3]{2}$ 9. $\dfrac{3\sqrt[3]{4}}{2}$

11. $\dfrac{\sqrt[3]{20}}{4}$

Pages 414–415, Chapter Review
1. c   3. b   5. c   7. d   9. b   11. d   13. c

Page 415, On the Calculator
1. 17   3. 12

Chapter 12   Quadratic Equations and Functions

Page 419, Written Exercises
1. $\{2\sqrt{2} - 1, -2\sqrt{2} - 1\}$
3. $\{3\sqrt{2} - 2, -3\sqrt{2} - 2\}$ 5. $\{2, -3\}$
7. $\{\sqrt{5} - 3, -\sqrt{5} - 3\}$ 9. $\{19, 1\}$ 11. $\{-\frac{1}{3}, 1\}$

13. $\{0, \frac{3}{2}\}$ 15. $\left\{\dfrac{-1 + \sqrt{61}}{10}, \dfrac{-1 - \sqrt{61}}{10}\right\}$

17. $\left\{\dfrac{3 + \sqrt{13}}{2}, \dfrac{3 - \sqrt{13}}{2}\right\}$

19. $\left\{\dfrac{-b + \sqrt{b^2 - 4c}}{2}, \dfrac{-b - \sqrt{b^2 - 4c}}{2}\right\}$

21. $\{k + \sqrt{k^2 - 9}, k - \sqrt{k^2 - 9}\}$

23. $\left\{\dfrac{a + \sqrt{a^2 + 4ab}}{2b}, \dfrac{a - \sqrt{a^2 + 4ab}}{2b}\right\}$

25. $\left\{\dfrac{-a + \sqrt{a^2 - 4ab^2}}{2b}, \dfrac{-a - \sqrt{a^2 - 4ab^2}}{2b}\right\}$

27. $\{-\sqrt{2} + \sqrt{b - a}, -\sqrt{2} - \sqrt{b - a}\}$

Page 419, Diversion

$$124)\overline{11260316} \quad \begin{array}{r} 90809 \\ \hline 1116 \\ \hline 1003 \\ 992 \\ \hline 1116 \\ 1116 \end{array}$$

Pages 422–423, Written Exercises
1. $\left\{\dfrac{-5 + \sqrt{13}}{6}, \dfrac{-5 - \sqrt{13}}{6}\right\}$ 3. $\left\{-\dfrac{1}{8}, 1\right\}$

5. $\left\{-\dfrac{\sqrt{10}}{2}, \dfrac{\sqrt{10}}{2}\right\}$ 7. $\left\{\dfrac{4 + \sqrt{26}}{5}, \dfrac{4 - \sqrt{26}}{5}\right\}$

9. $\left\{\dfrac{5}{2}\right\}$ 11. $\left\{\dfrac{1}{3}, 0\right\}$ 13. Two   15. Two

17. One   19. Two   21. Two

23. $\left\{\dfrac{-5 + \sqrt{7}}{6}, \dfrac{-5 - \sqrt{7}}{6}\right\}$

25. $\left\{\dfrac{1 + \sqrt{33}}{4}, \dfrac{1 - \sqrt{33}}{4}\right\}$

27. $\{-3 + \sqrt{5}, -3 - \sqrt{5}\}$ **29.** $\left\{\dfrac{1}{2}, 3\right\}$

31. $\left\{\dfrac{\sqrt{6} + \sqrt{2}}{2}, \dfrac{\sqrt{6} - \sqrt{2}}{2}\right\}$

33. $\left\{\dfrac{-\sqrt{3} + \sqrt{11}}{2}, \dfrac{-\sqrt{3} - \sqrt{11}}{2}\right\}$

35. $\left\{\dfrac{\sqrt{5} + \sqrt{13}}{2}, \dfrac{\sqrt{5} - \sqrt{13}}{2}\right\}$ **37.** $k = 4$

39. $\dfrac{-b + \sqrt{b^2 - 4ac}}{2a} + \dfrac{-b - \sqrt{b^2 - 4ac}}{2a} =$

$-\dfrac{2b}{2a} = -\dfrac{b}{a}$

Pages 423–424, Problems

1. $\frac{3}{2}$ or $\frac{2}{3}$ **3.** Width, 60 cm; height, 40 cm.
5. 8 km/h   **7.** 40 km/h   **9.** 12 shares

Page 424, Programming in BASIC

```
10   PRINT "INPUT A, B, C"
20   INPUT A, B, C
30   LET D = B↑2 − 4*A*C
40   IF D < 0 THEN 100
50   PRINT "THE SOLUTIONS OF"; A; "X↑2 +";
         B; "X +"; C; " = 0";
60   LET S = SQR(D)
70   PRINT "ARE"; (−B+S)/(2*A);
80   PRINT "AND"; (−B−S)/(2*A); "."
90   STOP
100  PRINT "B↑2 − 4*A*C < 0; NO REAL
         SOLUTIONS"
120  END
```

Note: A $\ne$ 0, by definition of a quadratic equation.

Pages 428–429, Written Exercises

1. 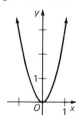   **3.**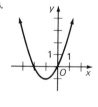

5.

7.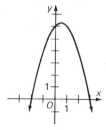

9. A point on a curve is a maximum point if its y-coordinate is greater than or equal to the y-coordinate of every other point on the curve.

11.   No; no; no, since each element of the domain does not have exactly one element of the range assigned to it.

13. Answers will vary. (Graph is a non-horizontal line.)
15. Answers will vary. Example: (1,000,000, 1,000,000,000,000) and (1,000,001, 1,000,002,000,001)
17. 50 m by 50 m

Page 431, Written Exercises

1. 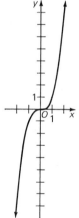   **3.**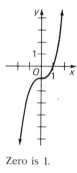

Zero is 0.   Zero is 1.

5. 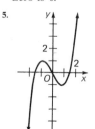   Zeros are 0, $\sqrt{2}$, $-\sqrt{2}$.

7.

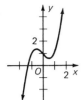

Zero is −1.3.

9.

Zero is 0.

Zeros are 0, 1, −1.

11.

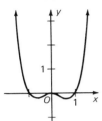

7.

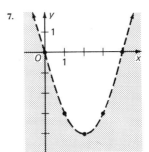

9.

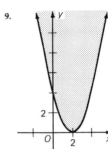

11.

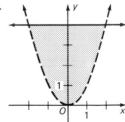

Page 432, Written Exercises

1.

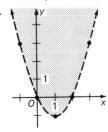

3.

13.

5.

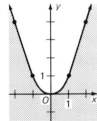

15.

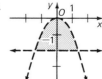

17.

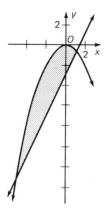

19.

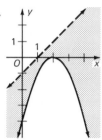

21.

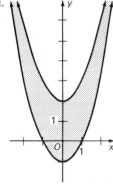

23.

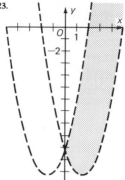

Page 433, Programming in BASIC

1. a.
```
10  FOR X = 0 TO 5
20  LET Y = X↑2
30  PRINT "("; X; "," Y; ")",
40  NEXT X
50  END
```

b. Use program for Exercise 1a, but change lines 10 and 20 to:
```
10  FOR X = 0 TO 2.5 STEP 0.5
20  LET Y = 2*X↑2
```

c. Use program for Exercise 1a, but change lines 10 and 20 to:
```
10  FOR X = 0 TO 10 STEP 2
20  LET Y = .5*X↑2
```

2. a. (0, 0) and (1, 1)

b. (0, 0) and (0.5, 0.5)

c. (0, 0) and (2, 2)

$\left(\dfrac{1}{a}, \dfrac{1}{a}\right)$, because $a\left(\dfrac{1}{a}\right)^2 = \dfrac{a}{a^2} = \dfrac{1}{a}$.

Page 436, Extra for Experts

1. 2   **3.** −2   **5.** Not real   **7.** 8   **9.** $\frac{1}{6}$ **11.** 9
13. $4y^4$ **15.** $4x^2$

Pages 436–437, Chapter Review

1. c   **3.** a   **5.** d   **7.** a

Pages 437–439, Cumulative Review

1. c   **3.** a   **5.** b   **7.** b   **9.** a   **11.** a   **13.** c
15. d   **17.** c   **19.** b   **21.** c

Chapter 13 Trigonometry and Vectors

Pages 443–444, Written Exercises
1. 680° **3.** −410° **5.** −280° **7.** I or II
9. 90°, 180° **11.** 90° **13.** 345° **15.** 810°
17. II or III **19.** 0° **21.** −50°
23. 600° **25.** 252°

Pages 447–449, Written Exercises

1. $\sin A = \dfrac{4}{5}$, $\cos A = \dfrac{3}{5}$, $\tan A = \dfrac{4}{3}$ **3.** $\sin A = \dfrac{3}{5}$,

$\cos A = -\dfrac{4}{5}$, $\tan A = -\dfrac{3}{4}$ **5.** $\sin A = \dfrac{12}{13}$,

$\cos A = \dfrac{5}{13}$, $\tan A = \dfrac{12}{5}$ **7.** $\sin A = -\dfrac{\sqrt{2}}{2}$,

$\cos A = \dfrac{\sqrt{2}}{2}$, $\tan A = -1$ **9.** $\sin A = \dfrac{\sqrt{2}}{2}$,

$\cos A = \dfrac{\sqrt{2}}{2}$, $\tan A = 1$ **11.** $\sin A = \dfrac{1}{2}$,

$\cos A = \dfrac{\sqrt{3}}{2}$, $\tan A = \dfrac{\sqrt{3}}{3}$ **13.** $\sin A = \dfrac{1}{2}$,

$\tan A = \dfrac{\sqrt{3}}{3}$ **15.** $\cos A = \dfrac{\sqrt{2}}{2}$, $\tan A = -1$

17. $\sin A = -\dfrac{\sqrt{2}}{2}$, $\cos A = -\dfrac{\sqrt{2}}{2}$

19. $\sin A = \dfrac{5}{13}$, $\tan A = -\dfrac{5}{12}$ **21.** $\sin A = -\dfrac{\sqrt{2}}{2}$,

$\cos A = -\dfrac{\sqrt{2}}{2}$, $\tan A = 1$ **23.** $\sin A = \dfrac{4}{5}$,

$\cos A = \dfrac{3}{5}$, $\tan A = \dfrac{4}{3}$ **25.** $\sin A = \dfrac{\sqrt{3}}{2}$,

$\cos A = \dfrac{1}{2}$, $\tan A = \sqrt{3}$ **27.** $\dfrac{4}{3}$ **29.** −2

Page 450, Written Exercises
1. 0.6018 **3.** 0.1584 **5.** 0.9816 **7.** 0.8693
9. 0.5736 **11.** 0.9659 **13.** 44° **15.** 28°
17. 18° **19.** 88° **21.** 83°

Pages 454–456, Written Exercises

1. a. 1 **b.** $\dfrac{\sqrt{2}}{2}$ **c.** $\dfrac{\sqrt{2}}{2}$ **d.** 1 **3. a.** 13 **b.** $\dfrac{12}{13}$

c. $\dfrac{5}{13}$ **d.** $\dfrac{12}{5}$ **5. a.** $\sqrt{2}$ **b.** $-\dfrac{\sqrt{2}}{2}$ **c.** $-\dfrac{\sqrt{2}}{2}$

d. 1 **7.** $\sin A = \dfrac{\sqrt{2}}{2}$, $\cos A = \dfrac{\sqrt{2}}{2}$, $\tan A = 1$

9. $\sin A = \dfrac{3\sqrt{34}}{34}$, $\cos A = -\dfrac{5\sqrt{34}}{34}$, $\tan A = -\dfrac{3}{5}$

11. $\sin A = \dfrac{\sqrt{2}}{2}$, $\cos A = -\dfrac{\sqrt{2}}{2}$, $\tan A = -1$
13. 6.4 **15.** 6.6 **17.** 48° **19.** 23°
21. $m° \angle B = 60°$, $AB = 4$, $AC \approx 3.5$
23. $m° \angle B = 25°$, $CB \approx 8.2$, $AC \approx 3.8$
25. $m° \angle B = 34°$, $m° \angle A \approx 56°$, $BA \approx 7.2$
27. $x \approx 7.1$

29. $x \approx 8.2$

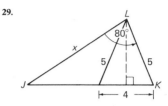

31. $x \approx 4.8$

Pages 456–458, Problems
1. 70.3 m **3.** 12° **5.** 19.5 m **7.** 2.91 m
9. 100 m **11.** 2.3 km; 828 km/h

Pages 461–462, Written Exercises
1. **3.**

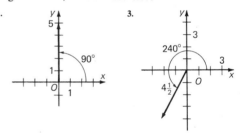

5. $2\sqrt{2}$ **7.** 2 **9.** $4\sqrt{2}$, 45° **11.** 3, 180°
13. 6.4, 129° **15.** 4.1, 76°

Pages 464–466, Written Exercises
1. a. x-component, 1; y-component, 2 **b.** 5
3. a. x-component, 1; y-component, 4 **b.** $2\sqrt{5}$
5. (Ex. 1) 37° **7.** (Ex. 3) 27° **9.** (6, 6)
11. a.

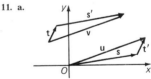

b. The vectors are equivalent.
13. $(-7, 3)$ **15.** $p = -5$, $q = 5$

17.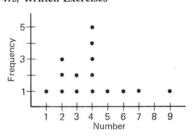

$\|u + v\| = 9.2$; direction, $349°$

Page 466, Diversion

Detective Watson divided the coins into three groups of three. He weighed two of the groups. If they weighed the same, the counterfeit coin was in the third group. If not, the counterfeit coin was in the lighter of the two groups. He then weighed two coins in the light group of three. If the coins weighed the same, the counterfeit was the third coin. If not, the counterfeit was the lighter of the two coins weighed.

Pages 467–468, Problems

1. Speed, $10.4\,\text{km/h}$; bearing, $107°$ **3.** Distance, $3.2\,\text{km}$; bearing, $198°$ **5.** $225\,\text{kg}$ **7.** $84.9\,\text{kg}$ **9.** $4.9\,\text{km}$ wide; $3.4\,\text{km/h}$

Page 466, On the Calculator

1. 15.1, $33°$ **3.** 15.1, $47°$

Pages 469–470, Chapter Review

1. c   **3.** c   **5.** c   **7.** b   **9.** b

Chapter 14   Statistics and Probability

Page 475, Written Exercises

1.

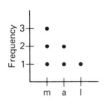

3.

No.	Freq.	Rel. Freq.	%
1	1	$\frac{1}{15}$	$6\frac{2}{3}$
2	3	$\frac{3}{15}$	20
3	2	$\frac{2}{15}$	$13\frac{1}{3}$
4	5	$\frac{5}{15}$	$33\frac{1}{3}$
5	1	$\frac{1}{15}$	$6\frac{2}{3}$
6	1	$\frac{1}{15}$	$6\frac{2}{3}$
7	1	$\frac{1}{15}$	$6\frac{2}{3}$
9	1	$\frac{1}{15}$	$6\frac{2}{3}$

5.

7.

Ltr.	Freq.	Rel. Freq.	%
m	3	$\frac{3}{6}$	50
a	2	$\frac{2}{6}$	$33\frac{1}{3}$
ℓ	1	$\frac{1}{6}$	$16\frac{2}{3}$

9. a.

Digit	Freq.	Rel. Freq.	%
0	1	$\frac{1}{16}$	$6\frac{1}{4}$
1	2	$\frac{2}{16}$	$12\frac{1}{2}$
2	2	$\frac{2}{16}$	$12\frac{1}{2}$
3	1	$\frac{1}{16}$	$6\frac{1}{4}$
4	2	$\frac{2}{16}$	$12\frac{1}{2}$
5	2	$\frac{2}{16}$	$12\frac{1}{2}$
6	1	$\frac{1}{16}$	$6\frac{1}{4}$
7	2	$\frac{2}{16}$	$12\frac{1}{2}$
8	2	$\frac{2}{16}$	$12\frac{1}{2}$
9	1	$\frac{1}{16}$	$6\frac{1}{4}$
Total:	16	$\frac{16}{16} = 1$	100

b.

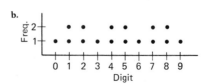

Page 476, Diversion

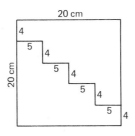

Pages 478–479, Written Exercises

1. a.

Int.	Freq.	Rel. Freq.	%
5–35	5	$\frac{5}{20}$	25
35–65	11	$\frac{11}{20}$	55
65–95	4	$\frac{4}{20}$	20
Total:	20	$\frac{20}{20} = 1$	100

b, c.

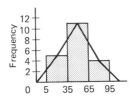

3. a.

Int.	Freq.	Rel. Freq.	%
4.5–34.5	6	$\frac{6}{20}$	30
34.5–64.5	5	$\frac{5}{20}$	25
64.5–94.5	9	$\frac{9}{20}$	45
Total:	20	$\frac{20}{20} = 1$	100

b, c.

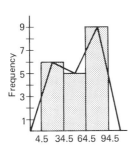

5. a.

Interval	Freq.	Rel. Freq.	%
0–1000	5	$\frac{5}{25}$	20
1000–2000	5	$\frac{5}{25}$	20
2000–3000	6	$\frac{6}{25}$	24
3000–4000	3	$\frac{3}{25}$	12
4000–5000	3	$\frac{3}{25}$	12
5000–6000	1	$\frac{1}{25}$	4
6000–7000	2	$\frac{2}{25}$	8
Total:	25	$\frac{25}{25} = 1$	100

b, c.

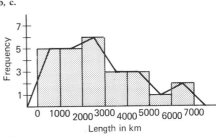

Pages 481–483, Written Exercises

1.

No.	Freq.	%	Cum. Freq.	Cum. %
1	1	10	1	10
2	2	20	3	30
3	1	10	4	40
4	2	20	6	60
5	1	10	7	70
6	3	30	10	100

3. 10%   5. 40%   7. 40%

9.

Interval	Freq.	%	Cum. Freq.	Cum. %
115–130	3	12	3	12
130–145	1	4	4	16
145–160	6	24	10	40
160–175	5	20	15	60
175–190	8	32	23	92
190–205	2	8	25	100

11.

Adjusted Gross Income	Freq.	%	Cum. Freq.	Cum. %
0–5000	24	29.6	24	29.6
5000–10,000	20	24.7	44	54.3
10,000–15,000	15	18.5	59	72.8
15,000–20,000	11	13.6	70	86.4
20,000–25,000	7	8.6	77	95
25,000–30,000	4	4.9	81	100

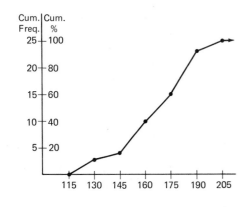

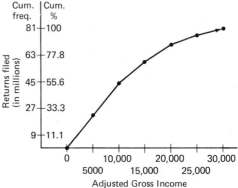

Page 483, Diversion

3 socks

Pages 485–486, Written Exercises

1. a. 8 b. 8 c. 7   3. a. 15 and 29 b. 27
c. 24.8   5. 61.7
7. $M = (r_1 + r_2 + r_3 + \cdots + r_n) \div n$
9. 400   11. 15   13. Mode. This is the most common shoe size bought.

15. Given $x < y$, $\dfrac{x + y}{2} = z$; $x + y = 2z$;

$y = 2z - x$; $x < 2z - x$; $2x < 2z$; $x < z$. Also,
$x = 2z - y$; $2z - y < y$; $2z < 2y$; $z < y$.
$\therefore x < z < y$.

Page 488, Written Exercises

1. 3.5   3. 2.5   5. $r_1 - m$

7. $\dfrac{(r_1 - m)^2 + (r_2 - m)^2 + \cdots + (r_n - m)^2}{n}$

Pages 491–492, Written Exercises

1. 1, 2, 3, 4, 5, 6; $\frac{1}{6}$ 3. $\frac{1}{3}$ 5. $\frac{1}{3}$ 7. 0   9. $\frac{3}{10}$
11. $\frac{3}{5}$ 13. 0   15. $\frac{1}{13}$ 17. $\frac{1}{26}$ 19. $\frac{1}{13}$ 21. $\frac{25}{26}$
23. No; P(both heads) $= \frac{1}{4}$; P(not both heads) $= \frac{3}{4}$
25. $\frac{1}{2}$ 27. $\frac{1}{2}$

Page 494, Written Exercises

1. $\frac{81}{100}$ 3. No   5. $\frac{1}{9}$ 7. $\frac{1}{3}$ 9. $\frac{1}{2}$
11. green $= 30$; blue $= 20$ 13. $P(E) = \frac{15}{19}$

Pages 496–497, Chapter Review

1. c   3. a   5. b   7. a

Pages 498–500 Comprehensive Review, Chapters 1–14

1. c   3. c   5. d   7. b   9. b   11. b
13. d   15. a   17. b   19. a   21. b
23. a   25. d   27. c   29. d

Extra Practice

Chapter 1, page 505
1. 745 **3.** 612 **5.** 8 **7.** 6 **9.** 47 **11.** 288
13. 529 **15.** 24

16.–19.

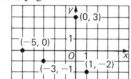

21. 0.52 **23.** $-0.86\overline{1}$ **25.** -1.75 **27.** ⊂
29. ⊂

31. finite

33. infinite

35. finite

37. $\{0, 2\}$

39.

41.

43. $7b = 70$
45. The sum of a number r and eight is eleven.
47. The difference of twenty and a number w is ten.

Chapter 2, page 506
1. Answers may vary. For example, $n = 5$.
3. Answers may vary. For example, $z = 0$.
5. Answers may vary. For example, $n = 2$.
7. 50 **9.** 72 **11.** $2t^2 + 4t + 9$ **13.** 27
15. $9k + 8$ **17.** -6 **19.** 8 **21.** 4 **23.** $\{-8\}$
25. $\{7\}$ **27.** 5 **29.** 12 **31.** -5 **33.** -4
35. -6 **37.** -11 **39.** $\{8, -8\}$ **41.** ∅

Chapter 3, page 507
1. 15 **3.** 16 **5.** -17 **7.** 34 **9.** $\frac{5}{8}$
11. $-4r + s$ **13.** $m + (-2)$ **15.** $10k^3 + (-13)k$
17. 192 **19.** -149 **21.** $-10x + 18y$
23. $s + 3s^4$ **25.** $-2p^4 + 3p^2$ **27.** $-44c + 24d$
29. -5 **31.** -5 **33.** $-xy$ **35.** $-3xz$ **37.** 6
39. $10 + 5t$ **41.** $9s^2 + (-7)t^2$

Chapter 4, page 508
1. $\{8\}$ **3.** $\{0\}$ **5.** $\{24\}$ **7.** -32 **9.** 1
11. -22 **13.** $\{-9\}$ **15.** $\{-24\}$ **17.** $\{9\}$
19. $\{57\}$ **21.** -50 **23.** $-\frac{2}{3}$ **25.** $-35a$
27. $\{-27\}$ **29.** $\{360\}$ **31.** $\{-6\}$ **33.** $\{6\}$
35. $\{-2\}$ **37.** $\frac{1}{5}p = 3.20$ **39.** $n + (n - 2) = 54$
41. $P = 16$ **43.** $n = 28$ **45.** $\{-6\}$ **47.** $\{5\}$
49. $\{8\}$ **51.** $y = \dfrac{A}{\pi r}$ **53.** $x = \dfrac{s^2}{2p}$

55. $z = \dfrac{a}{p} - a$

Chapter 5, page 509
1. Given **3.** Assoc. axiom of add. **5.** Identity axiom of add.
7. $x \geq 4$

9. $z \geq -6$

11. $m < 0$

13. $\{1\}, \{1, 4, 5, 10\}$ **15.** ∅, $\{0$ and the odd integers$\}$ **17.** $\{$the real numbers between -2 and $3\}$ **19.** $-1 \leq a \leq 3$ **21.** $-3 < z < 2$
23. $a \leq -4$ or $a > \frac{1}{2}$ **25.** $y < 3$ or $y > 3$
27. $\{0, 5\}$ **29.** $t < -1$ or $t > 1$ **31.** 6, 7
33. 58, 60, 62 **35.** In 1.25 h **37.** 121 adult, 287 children

Chapter 6, page 510
1.–4.

5. $D = \{-1, 2, 3\}$; $R = \{-2, 4, 6\}$; yes
7. $D = \{0, -1\}$; $R = \{0, 1, -1\}$; no **9.** $a = -3$, $b = -3$
11.

x	$f(x)$
-2	3
0	1
2	-1
4	-3

13.

x	$h(x)$
-2	16
0	4
2	0
4	4

15.

17.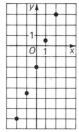

19.

21.   **23.**

25.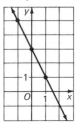

27. -15 **29.** 6 **31.** 7
33. $2x + y = 3$

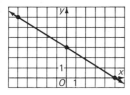

35. $y = 5x - 19$ **37.** $y = -\frac{3}{7}x - \frac{11}{7}$
39. $y = -1$ **41.** $y = -x + 3$ **43.** $y = 4x + 1$
45. $y = -\frac{6}{5}x$

47.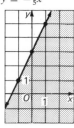

49.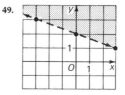

Chapter 7, page 512
1. $\{(1, 4)\}$

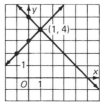

3. $\{(-3, 3)\}$

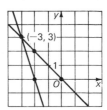

5. $\{(0, -6)\}$ **7.** $\{(-2, 0)\}$ **9.** $\{(-8, 6)\}$
11. $\{(7, 1)\}$ **13.** $\{(6, -5)\}$ **15.** $\{(-1, -8)\}$
17. $1.48 \, \text{kg/m}^3$ **19.** $2100 \, \text{km}$ **21.** 25

23.

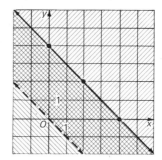

25. $(0, 6), (0, 4), (2, 1), (4, 0)$
27. $(0, 6), 12; (0, 4), 8; (2, 1), 4; (4, 0), 4$
29. $\{(-1, -2, 4)\}$

Chapter 8, page 513
1. $3x^4 - x^3 - x^2 + 6x$ **3.** $4st^2 - 2st$ **5.** $27x^5$
7. $-30c^5$ **9.** $-j^7k^6$ **11.** $2f^7 - f^6 - 5f^5$
13. $-12m^5n + 4m^3n^3 - 28mn^3$ **15.** $q^2 + q - 12$

17. $6x^2 + 5x - 4$ **19.** $9t^2 + 12ty + 4y^2$
21. $a^2b^2 - 2abc + c^2$ **23.** 33   **25.** $9c^2d$
27. $2(3a^2 + a - 2)$ **29.** $3ab(4a^2b + 7abc - c^2)$
31. $(y - b)(y + 8)$ **33.** $(6c + 5d)(6c - 5d)$
35. $2z(5z - 1)^2$ **37.** $(5x^2 + 1)(7x + 8)$
39. $(t + 6)(t + 2)$ **41.** $p(p - 9)(p - 2)$
43. $-s^2(s - 3)(s - 1)$ **45.** $(3x + 2)(x + 3)$
47. $(3n + 2)(4n - 5)$ **49.** $(3y - 1)(5y + 2)$
51. $\{0, 7\}$ **53.** $\{-4, 6\}$ **55.** $\{-\frac{5}{6}, \frac{3}{4}\}$
57. height, 3 cm; base, 8 cm   **59.** 7 and 12

Chapter 9, page 514

1. $\dfrac{6a^4}{b}$ **3.** $\dfrac{1}{5rs^3}$ **5.** $-\dfrac{5x^6y^8}{3}$ **7.** $\dfrac{1}{11(a + 2b)^2}$

9. $3x^2y + 4x - 7$ **11.** $-4 + \dfrac{1}{d} + \dfrac{3}{d^2}$

13. $\dfrac{1}{2b} - 2a - \dfrac{4b^2}{a}$ **15.** $2y - 1$ **17.** $3y - 4$

19. $n^2 - n + 1$ **21.** $3x + 5$ **23.** $\dfrac{4}{a}$ **25.** $\dfrac{z}{z - 2}$

27. $\dfrac{y - 5}{2y + 5}$ **29.** $\frac{1}{72}$ **31.** $\dfrac{r^9}{t^6}$ **33.** $\dfrac{k^5}{9j^2r^7}$

35. $-\dfrac{15}{2a}$ **37.** -1 **39.** $\dfrac{4c + d}{3}$ **41.** $\dfrac{t + 2}{t}$

43. $\frac{1}{3}$ **45.** $-\frac{1}{2}$ **47.** $\dfrac{5a^3 + 3b^3}{a^2b^2}$ **49.** $\dfrac{p^2 + 2}{4p^2}$

51. $\dfrac{4}{5g - 20h}$

Chapter 10, page 515

1. $\{\frac{28}{25}\}$ **3.** $\{x\colon x > -33\}$ **5.** $\varnothing$ **7.** 4.5
9. 110   **11.** 37.5%   **13.** $\{\frac{3}{2}\}$ **15.** $\{-\frac{7}{25}\}$
17. $\{-3\}$ **19.** 20 and 28   **21.** 1 or -1
23. 12 min   **25.** 84 km/h   **27.** 45   **29.** 32
31. 48

Chapter 11, page 516

1. 7   **3.** $\frac{1}{10}$ **5.** $\sqrt{196}$ **7.** $-\sqrt{121}$ **9.** $\frac{62}{99}$
11. $\frac{4}{495}$ **13.** $\{-7, 7\}$ **15.** $\{\sqrt{13}, -\sqrt{13}\}$
17. 3.08   **19.** 27.8   **21.** 15.75   **23.** -0.77
25. No   **27.** Yes   **29.** $\sqrt{41}$ **31.** $\sqrt{194}$
33. $\sqrt{170}$ **35.** $2n\sqrt{2}$ **37.** $5\sqrt{7}$ **39.** $2\sqrt{5t}$
41. $4cd\sqrt{5d}$ **43.** $\dfrac{8\sqrt{a}}{a}$ **45.** $\dfrac{\sqrt{13}}{2}$ **47.** $\dfrac{2\sqrt{30y}}{3y}$
49. $16\sqrt{3}$ **51.** $10r\sqrt{r}$ **53.** $32 - 21\sqrt{2}$

Chapter 12, page 518

1. $\{5, -7\}$ **3.** $\{-3 + \sqrt{2}, -3 - \sqrt{2}\}$
5. $\{4 + \sqrt{13}, 4 - \sqrt{13}\}$ **7.** $\{6 + \sqrt{43}, 6 - \sqrt{43}\}$
9. $\left\{\dfrac{-1 + \sqrt{41}}{4}, \dfrac{-1 - \sqrt{41}}{4}\right\}$ **11.** $\{0, \frac{3}{5}\}$
13. $\{-\frac{1}{5}\}$ **15.** $\left\{\dfrac{3 + \sqrt{57}}{8}, \dfrac{3 - \sqrt{57}}{8}\right\}$

17. 0

19. $0, -3$

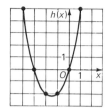

21. $-2, 6$

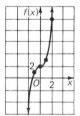

23. approx. -1.26

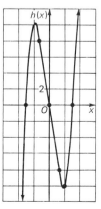

25. $0, 3, -3$

27. $-1, 1, 2$

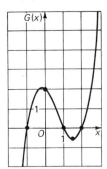

29.

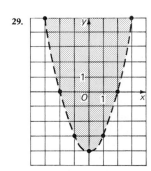

31.

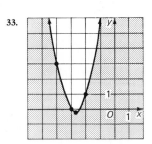

33.

Chapter 13, page 518

1.

3.

5. $\sin A = -\frac{15}{17}$, $\cos A = \frac{8}{17}$, $\tan A = -\frac{15}{8}$

7. $\sin A = \frac{\sqrt{2}}{2}$, $\cos A = -\frac{\sqrt{2}}{2}$, $\tan A = -1$

9. $\cos A = \frac{\sqrt{3}}{2}$, $\tan A = -\frac{\sqrt{3}}{3}$ **11.** $\cos A = \frac{3}{5}$,

$\tan A = \frac{4}{3}$ **13.** $\sin A = \frac{\sqrt{2}}{2}$, $\cos A = -\frac{\sqrt{2}}{2}$

15. 0.9781 **17.** 0.6691 **19.** $69°$ **21.** $8°$
23. $80°$ **25.** $m°\angle B = 40°$, $a \approx 3.6$, $c \approx 4.7$
27. $m°\angle A = 45°$, $b = 7$, $c \approx 9.9$
29. $m°\angle A \approx 31°$, $m°\angle B \approx 59°$, $c \approx 5.8$
31. $4\sqrt{2}$, $225°$ **33.** $\sqrt{34}$, $121°$ **35.** $7, 3, 13$
37. $-2, 1, \sqrt{41}$ **39.** $37\,\text{kg}, 19°$

Chapter 14, page 520

1.

Ltr.	Freq.	Rel. Freq.	%
i	3	$\frac{3}{8}$	37.5
n	1	$\frac{1}{8}$	12.5
t	2	$\frac{2}{8}$	25
a	1	$\frac{1}{8}$	12.5
e	1	$\frac{1}{8}$	12.5

3.

Ltr.	Freq.	Rel. Freq.	%
s	3	$\frac{3}{10}$	30
u	2	$\frac{2}{10}$	20
c	2	$\frac{2}{10}$	20
e	1	$\frac{1}{10}$	10
f	1	$\frac{1}{10}$	10
l	1	$\frac{1}{10}$	10

5.

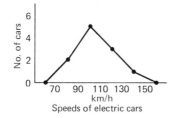

Speeds of electric cars

7.

Ltr.	Cum. Freq.	Cum. %
d	1	10
e	2	20
i	4	40
n	6	60
o	7	70
s	10	100

9. 6 and 12; 6.5; 7.375 **11.** 2; 3; $3\frac{1}{3}$
13. 7; 6.5; $\sqrt{6.5}$ **15.** 2; 0.5; $\sqrt{0.5}$
17. $\frac{1}{6}$ **19.** $\frac{2}{5}$ **21.** $\frac{2}{75}$ **23.** $\frac{7}{300}$